P9-AQM-580

# Operational Subjective Statistical Methods

WILEY SERIES IN PROBABILITY AND STATISTICS

Established by WALTER A. SHEWHART and SAMUEL S. WILKS



A complete list of the titles in this series appears at the end of this volume

# Operational Subjective Statistical Methods

## A Mathematical, Philosophical, and Historical Introduction

FRANK LAD
University of Canterbury
Christchurch, New Zealand

A Wiley-Interscience Publication
JOHN WILEY & SONS, INC.
New York · Chichester · Brisbane · Toronto · Singapore · Weinheim

This text is printed on acid-free paper.



***Library of Congress Cataloging in Publication Data:***

Lad, Frank.
Operational subjective statistical methods / Frank Lad.
p. cm.—(Wiley series in probability and statistics. Applied probability and statistics)
"A Wiley-Interscience publication."
Includes bibliographical references and index.
ISBN 0-471-14329-4 (cloth : alk. paper)
1. Mathematical statistics. I. Title. II. Series.
QA276.L26 1996
519.5—dc20 96-6197

Printed in the United States of America

10 9 8 7 6 5 4 3 2 1

To my schoolroom arithmetic teachers as I knew them.
They knew the value of money, and taught out of love.

Sister Mary Vincent
Sister Mary Jane
Miss Maher
Miss Dister
Sister Chaminade
Mrs. Klapec
Sister Ancilla
Brother Wuco
Brother Hildebrand
Brother Landin

# Contents

# Preface

*One wonders sometimes where Watt thought he was. In a culture park?*
*Watt*, SAMUEL BECKETT

Intellectually, we have been developing for some 100 years now our recognition of evolution as a feature of natural history, but we have yet to come to grips with this awareness emotionally. We still cling to categories of thinking in which nature is characterized as fixed, refusing to accept that everything is continually turning into something else.

The twentieth-century fascination with theories of objective probability and statistics has played on this sham, proposing that variation in nature is only superficial in that it is generated and governed by fixed but unobservable random generating structures, which can be identified by careful attention to an impersonal statistical method. Such an outlook has helped to create false hopes for science and, in some cases, to allow the development of abusive authority. To understand the true value of science, and to be honest with ourselves, we cannot escape facing the uncertainty that is a component of all our knowledge, scientific or otherwise.

There is no impersonal objective scientific procedure for making decisions about issues that concern us. Discerning and avowing our beliefs and value judgments are necessary if we are to deal forthrightly with one another in developing and sharing our knowledge, without relying on games of administrative power for the resolution of scientific disputes.

In the scientific statistical tradition passed on through this century, scientists who require statistical consulting in their empirical work often find themselves visiting a probabilist/statistician to ask what the unobservable probabilities relevant to their problems might be. In the operational subjective mode of empiricism presented in this book, the probabilist and statistician renounce any claim to knowledge of unobservable probabilities. All that we can be skilled at doing is asking you, the scientist or concerned decision maker, what *your* probabilities are, helping you to be consistent and efficient in the organizing of your uncertain knowledge about practical matters.

Though subjectivist attitudes were commonplace among the early emerging ideas of probability in Europe, the operational subjective theory of probability and statistics has developed slowly in a minority tradition throughout the past two centuries, even while mainstream objectivist theories of scientific statistics have flourished. On many issues, to understand properly the valid role that statistical methods can play in the development of science and commerce, it may be necessary to reverse completely some basic conceptions that you perhaps cherish. Among these are commonly presumed distinctions between religious and scientific activity on the basis of belief and a-belief. The younger and less knowledgeable and experienced you are in mainstream statistical theory, the easier perhaps you will find it to make sense of the subjective viewpoint. I believe that it will come naturally as a way of thinking. While I hope that this text will be found intriguing to university-level teachers of mathematical probability and statistics, and I have written many paragraphs in the book precisely for them, in the main I feel I am presenting mathematical details for a new generation of students who may feel comfortable with the outlook they presuppose.

This text is written for four distinct audiences. In whichever of them you find yourself, I hope you will forgive the time spent particularly with another one of the audiences. I expect you to judge for yourself what is relevant and what should be skipped over for someone with your background. First, I am writing for an honest undergraduate student with a decent mathematics background in linear algebra and calculus. Today, no doubt, the facility with linear algebra is more important. But I do recognize that you may well have learned this material only recently, so I use it rather gingerly, especially at first. Second, I am writing for an advanced undergraduate and even a graduate audience. Some of the more detailed presentations of algebraic results are meant expressly for you. Though you will surely find some of the examples and applications to be technically quite simple, I hope they will aid your conceptual development. For more advanced mathematical and statistical researchers and prospective teachers of this material, I hope to intrigue you to a fresh way of thinking about subject matter that you have already spent much time considering yourselves, from many possible points of view. Finally, I write to some extent for the generally educated adult who is concerned with questions regarding the role of personal beliefs and values in the conduct of science and who wonders just how statistical methods might proceed by insisting on their recognition rather than their suppression. You should feel free to skip through the book, focusing on the many passages that I hope will engage your interest, while ignoring extensive detail.

Conceptually this book is sophisticated, and perhaps difficult, the weaker your philosophical background. I don't apologize for this. There are many serious matters that are worth thinking deeply about, and I believe you will enjoy doing so. I should remark that most every book on probability and statistics, and even those on mathematical topics other than probability, involve deep philosophical attitudes and understandings, especially those that either pretend not to or make efforts to hinder their recognition.

This book is meant to provide a basis for teaching mathematical statistics at the university and graduate level. I have attempted to produce a text that contains formal developments of this material as it is understood in the subjectivist perspective. It is presented in such a way that it is accessible to university students who are developing their mathematical sophistication as well as technique while they are studying it. The mathematical formality of this text, while appearing belabored in places, has been intended to address an unfortunate attitude that runs through a broad swath of the scientific and statistical community. When discussing the perspective of subjective probability, almost invariably one is presented with the comment: "Oh, that sounds like philosophy that you are talking about. But if we really want to study *mathematical* probability, that would be something else entirely." Nothing could be further from the truth. The subjectivist understanding of our subject motivates a mathematical characterization of probability that is surely different from the set and measure-theoretical approach. But it is eminently mathematical and formalizable, nonetheless.

Before beginning the mathematical development in a second-year course, I always begin with two lectures that outline the content of Chapter 1 and overview the subjectivist perspective, providing simple examples of the central difference between the approach of a subjectivist who regards a sequence of events exchangeably, and the objectivist characterization of a sequence of independent events. As a practical and conceptual motivation for technical developments that will defy much of what the university student has already learned about probability from high school and first year introductions, I find this difference most persuasive.

The second-year class would then study Chapter 2 through the fundamental theorem of prevision in Section 2.10, followed in Chapter 3 by Sections 3.1, 3.2, 3.4, 3.7, 3.8, and 3.11. At this level of study, developments should focus on conceptual understanding of the basic principles and on the formalities of expressing their implications mathematically. The specific study of exchangeability stresses the examples that begin Section 3.8, along with its urn problems, and proceeds through the statement of its formal definition in Section 3.8.2. Moving directly then to the application of this judgment in the examples and problems of Section 3.11.2.1 concludes the subject at this level. From here, attention shifts to distribution theory in Chapter 5. At the second-year level, however, this chapter is not meant to be studied page by page. Rather, I lecture on the subject of distributions and their moments, referring to specific places in the text where the relevant material appears. In the study of parametric distributions which the student is typically seeing formally for the first time, the usual features of shapes and moments of the various parametric family members are studied by skipping about in Chapters 5 and 7 where they are presented. (The format of Chapter 5 is designed for the next level of study, a third- or fourth-year university student, who would by then have already been introduced formally to many of its specific contents.) Finally, results such as variances for linear combinations, Jensen's inequality, and the Bienaymé-Chebyshev inequality in Sections 4.2 and 4.3 are covered when they are appropriate to the study of distributions at the second-year level. This amount of material would complete a second-year course in mathematical probability over a series of 50 fifty-minute lectures with 12 one-hour tutorial sessions (including computing problems using MATLAB, with no previous background in MATLAB being presumed).

As a third-year course, with the previously mentioned studies now presumed, the students can be directed straightforwardly through Chapter 5 and Chapter 7 through Section 7.4.5.4, developing a fairly thorough understanding of the structures of exchangeable multivariate inferences in many standard contexts. Although our students would have already studied structures of linear regressions with a geometrical orientation in another course, I do not proceed with Chapter 8 at the undergraduate level. (This chapter, in tandem with Chapter 9, is written in a style more appropriate for a graduate seminar, with students who are faily comfortable with contemporary statistical practice but are willing to rethink some of its tenets.) Rather, for practical reasons I conclude the second undergraduate course with a study of multivariate statistical analysis as currently practiced, to aid students' development of widely used statistical programming skills and their understanding of current styles of statistical reporting. To the extent that it is possible, we focus on the relevance of these statistics to the subjectivist outlook they have developed and on ways that the subjectivist perspective aids in their useful interpretation.

Finally, this book can be used effectively at the honours level, which is equivalent to a first-year graduate seminar, for students who have been introduced to the material at the level just described. This course can begin with a terse review of the fundamental theorem of prevision, aiming toward its extensions involving inferential probabilities that are formalized in Section 3.3. Then we linger on the algebraic and geometrical features of exchangeable inference structures that appear in Sections 3.9 and 3.10, and their extension to partially exchangeable assertions in Section 3.12. Chapter 6 can then be studied completely, moving at a good pace, and perhaps skipping some side topics. The honours course concludes with Chapter 8, including further computational applications.

While I have described a way to use this book in three distinct sequential courses, it can be used in many other ways too, particularly at the graduate level in applied fields, for students who have already been introduced to mathematical probability and statistics at the university degree level of technicality, but from the more common formalist perspective. Graduate-level students of statistics and researchers who are already familiar with much of the technical mathematical detail can study the text as a provocative development of a new perspective.

To allow the sharing of experience with the teaching of subjective probability by using this text, I shall introduce an electronically accessible support service that can aid in the teaching endeavor. This will include a file of definitions and theorems that can be reproduced on overhead transparencies for classroom use; a file of solutions to all problems in the text; a forum for the sharing of further examples and problems appropriate for examinations; a bank for sharing computer programs; and perhaps a challenge board in which controversial ideas and problems developed in the text can be discussed. At present, some of this material can be obtained from me by email to help your presentation if you are teaching a course using this text. My address is F.Lad@math.canterbury.ac.nz.

I began writing this text in October 1985 when I was a visiting scholar in the Department of Mathematics and Statistics and the Department of Economics at the State University of New York at Albany, an opportunity for which I am very grateful. My studies and my planning for it began much earlier as a graduate student in economics and statistics at the University of Michigan, 1970–1974, progressing through my years of teaching and research at the University of Utah, 1974–1984. I am most grateful for the wonderful support for this work provided by my colleagues at the University of Canterbury since 1988, especially from John Deely who encouraged me to say what I have to say. I have had the opportunity to teach from this text to students from the second-year through the honours level. Thanks to the many students who have made spirited efforts at studying earlier editions of this work, especially to Andrea Piesse, Grace Johnson, Rachel Hamlin, Susan Lindsay, and James O'Malley for their contributions.

I would like to recognize especially the welcome support, comments, and encouragement I have received from Jim Dickey, Dennis Lindley, and Gail Blattenberger throughout the development of this book. Over the years I have also received helpful and extensive comments on content, organization, and style from several colleagues, especially John Lad, Vivian Shayne, David Lane, Marji Lines, Pierre Crépel, Michael Goldstein, P. A. V. B. Swamy, Peter Walley, Jim Young, John Fountain, Duncan Foley, and Romano Scozzafava. None of them should be presumed to agree with the entire text of the current edition. The University of Canterbury Department of Mathematics and Statistics has provided institutional support throughout my employment over the past eight years. Thanks also to the E. N. Huyck Nature Preserve and Biological Field Station in Rensselaerville, New York; the University of Minnesota School of Statistics; and friends at the Shell Inn, Westerlo, New York.

Figures in the text were produced using Aldus Freehand, with the collaboration of Paul Connell. Steve Buerle provided help with text format translations.

Thanks to the editing and production staff at John Wiley in New York for their careful work, and to Kate Roach for her interested encouragement of my efforts.

Most of all, thanks to my mate, Belle, for her loving support over all these years.

FRANK LAD

*Christchurch, New Zealand*

CHAPTER 1

# Philosophical and Historical Introduction

*And Watt's need of semantic succour was at times so great that he would set to trying names on things, and on himself, almost as a woman hats.*
*Watt*, SAMUEL BECKETT

## 1.1 AN INTRODUCTION TO THIS BOOK

In the twentieth century, philosophy has come to be regarded once again as an activity rather than a subject matter. The task of philosophy is to uncover the meaning we attach to controversial words by examining the contexts in which we use them and the use we make of them. The analysis may identify contradictions in our usage, and even syntactic inconsistencies. In these situations, it may point to avenues of more appropriate usage.

*Probability*: It would be difficult to suggest another word that has exhibited so regularly since the European enlightenment such a controversial importance in the domain of science. Indeed, the word itself has been used multifariously, both synonymously with and distinctly from chance, frequency, likelihood, proportion, possibility, accident, risk, hazard, belief, knowledge, uncertainty, and ignorance. Some large part of the controversy has derived from the inattention paid by practitioners of science to the precise meaning associated with their uses of the word. This inattention is understandable to some extent. Scientists' attentions are typically focused not on the precise meaning of words but on substantive questions of empirical evidential concern in whatever subject area, whether physics, chemistry, geology, biology, demography, anthropology, economics, sociology, or astronomy. But scientists need beware of ignoring issues of "meaning" completely. They would run the risk that the language issuing from their lips and from their word processors can be identified as incapable of conveying scientific meaning. For philosophers, the consistent usage of words such as probability is itself a focus of study. Controversy among them derives from the fact that reasonable people can and do disagree in their analyses and conclusions.

Among applied scientists, there has emerged in this century a widespread attitude that the philosophical issues surrounding the meaning of probability can be relegated to the realm of sometimes enjoyable but largely irrelevant discussion. For it has appeared that the various points of view on the correct usage of the word "probability" support closely related modes of statistical analysis for specific empirical scientific questions. If the identification of the meaning of probability makes no difference for the conduct of science, you may well use the word however you please as long as you are consistent. Perhaps only your computational activities need be understandable to your scientific colleagues. Then they might benefit from your empirical research in whatever translated linguistic forms they prefer.

To be sure, there has been a remarkable apparent unification in the statistical methods typically used to evaluate empirical evidence as it bears upon a variety of scientific questions. Whatever the topic of investigation, it has become common to conceive of the workings of nature as a random process generated by some unknown probability distribution. Scientific questions are ubiquitously posed in the form of "whether the unknown probability distribution that generates these data is the distribution $A$ or the distribution $B$." In more sophisticated cases it is wondered "whether the generating distribution is a member of the family of distributions **A** or of family **B**." There is widespread scientific familiarity with standard statistical methods that purport to answer such a question, albeit tentatively. Once having decided to settle on the hypothesis that the generating distribution is a member of family **A**, the determination of a particular member of the family to represent scientific conclusions from the data is achieved via procedures for estimating the values of the parameters that index the members of the specified family.

Though there has been some acrid discussion of what might be the *meaning* of this reference to "probability distributions" that are presumed to be "generating" these data, the discussion has been of little practical consequence in applied science.

Many consider this to be a happy state of affairs. Universities throughout the world generally require science and commerce students to complete at least one course in statistical methods at both the undergraduate and graduate levels, whatever be their major fields of study. This requirement has largely supplanted the requirement of formal coursework in deductive logic, which was virtually universal even only 25 years ago. Students who would once have puzzled over BARBARA and CELARENT, today puzzle over "efficient estimators of probabilities," "confidence intervals," and the "statistical significance" of $\chi^2$ and $t$-statistics.

The syllabus for basic statistics courses typically includes standard topics: summary statistics and their graphical display; the rules of probability; distribution theory; procedures for hypothesis testing and the estimation of probabilities; linear models of regression, the analysis of variance, and factor analysis; and methods of survey sampling design. Given the spectrum of mathematical capabilities among university students, the level of presentation may range from intuitive verbal to simple computational, to algebraic and geometrical, to abstract. Undergraduate-level competence is presumed to mean the capability of understanding the meaning of empirical research, reported using the terminology of statistical theory. Graduate-level competence is presumed to mean the capability of designing and performing the computations for which the statistical theory calls in the evaluation of empirical scientific questions. Graduates whose major field of study is mathematical statistics are presumed to achieve competency in theoretical formulations as well as these applied activities, though they may not be expert in the detailed subject matter of any particular applied science.

Moreover, the statistical conceptions described above lie at the base of public understanding of virtually every practical, important question regarding human activity today, whether East or West, North or South. Consider the questionable efficacy of various therapies in treating a medical condition; the relative productivity of different agricultural procedures; demographic movements of populations throughout the globe, whether of people, of insects, or of microbes; and the efficiency and safety of engineered products, whether transportation methods, architectural structures, communication systems, or energy sources. The analysis of each of these issues has succumbed to description and manipulation in commonly used statistical terms. Scientists who are especially facile in representing their ideas within this statistical framework are often among the leading theoreticians in their respective fields. The development and dissemination of standardized computer software such as SAS, SPSS, BMDP, S, GAUSS, LISREL, ESP, SPEAKEZ, SHAZAM, and whatever, have been a great boon to the unification of computational technique.

But there is something odd about this apparent unification of statistical practice. Philosophically speaking, the methods that have won the day have emerged from efforts of theoretical and applied statisticians whose own understandings of what they were about were as different as chalk and cheese. The pantheon of generally recognized modern statistical immortals whose names grace the codification of now commonplace procedures and concepts includes Francis Galton, Karl Pearson, A. L. Bowley, Emile Borel, Paul Levy, Ronald Fisher, A. N. Kolmogorov, Harald Cramer, P. C. Mahalanobis, Jerzy Neyman, Egon Pearson, Harold Hotelling, and Abraham Wald, to mention only a few. Yet even among these, conceptions of theoretical statistics were so different that there is no common syntax of language that can represent each of their achievements as they each understood them. Public communication that actually passed among some of them was at times vituperative. For an example, see the exchange between Bowley and Fisher described in O'Brien and Presley (1981). Perhaps inevitably, the body of contemporary statistical theory and practice, which derives from traditions honoring all of them, exhibits philosophically the character of a potpourri.

There is yet another fly in the wax. The contemporary synthesis of computational methodology is on the rebound from four decades of an invigorating challenge to its evident melange. In 1954, L. J. Savage published a mathematical investigation, entitled *The Foundations of Statistics*, ushering in some decades of soul searching among statisticians. Though he shared some ideas already proposed in the then not-so-ancient works of Frank Ramsey (1926), Bruno de Finetti (1937), and Harold Jeffreys (1939), as well as his contemporary authors I. J. Good (1950) and Dennis Lindley (1953), it has been Savage's spirited arguments that have led the way to the current acceptability of certain "Bayesian" clarifications, modifications, and extensions of statistical practice. His provocative role in these developments has been recognized in the remarks of several distinguished statisticians introducing *The Writings of Leonard Jimmie Savage: A Memorial Selection* (Savage, 1981). Today, the success of the personalist-Bayesian approach to probability and statistics may be gauged by the fact that "Bayesian estimators" of probabilities and parametric distributions subject to various loss structures, and "Bayesian tests" of statistical hypotheses have taken their place in the panorama of recognized computational statistical procedures. Computational software routines based upon Bayesian ideas now circulate widely, some already in versions of several generations, for example BRAP, SEARCH, RATS, and BATS. Most of the larger eclectic packages also include some Bayesian variations on their themes.

The statistician's toolbag today is more comprehensive than ever. Therein can be found Bayesian computational procedures alongside of those based upon other organizing principles such as maximum likelihood, maximum entropy, best invariant and minimax estimation, and the most powerful specifications of tests. Theoretical statistical research over the past 40 years has detailed the extent to which the various procedures agree and disagree. When they do disagree, the merits and demerits of each have been identified in terms of computational cost and accuracy. From the very inception of the modern Bayesian dissent, this development was expected. In the first edition of his *Foundations*, Savage promised (1954, p. 4) that his personalist theory of probability would support statistical procedures which, on the whole, are consistent with those followed by promoters of the then- reigning objectivist theories of probability. Although he reneged on this promise in his preface to a 1972 reprint edition, his sense of the complementarity of various foundational (philosophical) approaches to probability and statistics was still only waning in his final posthumous publication (1977), "The Shifting Foundations of Statistics." Another eminent Bayesian proponent, I. J. Good (1983), has suggested that future developments in statistics would proceed along lines of a "Bayes/non-Bayes synthesis."

Savage's sometimes Panglossian attitude could be paraphrased: Statisticians must be doing *something* right. Just look at their practical successes in science, business, and industry. As statistical theorists, let us only be careful in analyzing what it is they *are* doing right and gently nudge them away from their errors. This sentiment was echoed by Dennis Lindley (1965) in the introduction to his Bayesian text. In turn, Lindley was quoted to this effect by Arnold Zellner (1971) in one of the first modern expansive applications of Bayesian concepts to an empirical science. (Mosteller and Wallace, 1964, had provided another.) This has also been the basic viewpoint of Ed Leamer (1978) in his often brilliant investigation of the activities of econometricians and his expansion of their empirical sights. Lindley expressed rather different feelings some two decades later in his 1988 Wald lectures to the Institute of Mathematical Statistics (Lindley, 1990).

Embarrassingly aloof from the productive give and take among the advocates of the various schemas of probability estimation are the extensive writings of Bruno de Finetti. Though he has now become somewhat of a mythical hero to the Bayesian movement in statistics, and though he did collaborate directly with Savage in extensive discussion and joint research, the thrust of de Finetti's work has largely remained outside the arena of constructive engagement among the several statistical factions. Much of the emerging "Bayes/non-Bayes synthesis" he would classify as "verbal diarrhea" in the taxonomy of his retirement address (de Finetti, 1976, Section IX). If "probability does not exist" in an absolute sense, as de Finetti proclaimed in the introduction to his *Theory of Probability* (1974, 1975), what is one to make of *any* estimate of a probability, whether it be a Bayesian or non-Bayesian estimate? To the current paradigmatic statistical question of "whether this scientific data been generated by a probability distribution that is a member of family **A** or of family **B**," he would unabashedly answer "No!".

De Finetti lived wholly in the twentieth century (1906–1985), and he wrote extensively and regularly from 1926 through 1976. The theory of prevision, or probability and statistics, dominated his concerns, but he also wrote serious works in genetics, economics, demography, educational psychology, philosophy, and mathematics. By now, much of his important work has been translated into English from the original Italian or French. An exception is his work in political economy, particularly his critical evaluation and extension of Pareto's ideas. These are available in English only in the cryptic form of summary notes (de Finetti, 1974b). De Finetti's ideas have attracted by now a small following of statisticians who take them seriously. Some of their developments of his ideas are explicated in the present book. But by and large his writings have been relegated by the applied statistical profession as irrelevant philosophy. The problem with de Finetti, charge his critics, is that neither he nor his followers have proposed practical and exciting new computable probability estimators or hypothesis testing procedures.

The concluding chapter of de Finetti's synthetic treatise, *Theory of Probability*, addresses specifically the topic of mathematical statistics. For reasons that he explains, the commentary is rather awkward. To a seasoned statistician who looks directly into his final chapter, the discussion appears needlessly verbose, yet quite limited in content. His ideas expressed on the Bayesian approach to "estimation" and "hypothesis testing" are set in an idiosyncratic context relative to common practice. And they seem even childish compared to the detailed discussions presented by other Bayesian statisticians.

The incongruity of this "problem with de Finetti" is that his work actually does develop the basis (and much extraordinary mathematical detail) for a completely different theory and practice of statistical method from that which engages the practitioners of the current reigning synthesis. The book you are now reading amounts to a text-book style presentation of his "operational subjective" method of statistical activity. Of course, the details and even the overall imagination behind the method are not due solely to de Finetti. Several theoretical statisticians and philosophers have participated, and their contributions shall be noted. But most other researchers have also found some "semantic succor" in modes of analysis based on different philosophical foundations as well. Sometimes I shall be presenting their results in a different context from that in which they were originally developed. De Finetti was the champion of the operational subjective statistical method as the only procedure that really makes sense, no matter whether you believe in the existence of unobservable probabilities or not. I share this point of view, and I offer this text as an introduction to the practical consequences of this way of thinking.

In the midst of his circuitous journey through the *Theory of Probability*, de Finetti wistfully opined that his exposition of the method could be made much more clearly in an abbreviated and nonargumentative form—if only foolish misconceptions were not so rampant, and if only there were not some very delicate matters which need to be pursued in great analytic detail in order to achieve full understanding. His clear awareness of the many implications of taking a stance on any particular issue imbued his own writing style with the distracting tendency to leave the track of his argument, often even numerous times in the same sentence! (It has been said that de Finetti loved to open parentheses for side remarks in the middle of a thought, but that he was not always careful to close them.) In several ways, I have attempted to respond to his suggestions for what might be useful pedagogically if a more direct textbook were to be written. For the most part, I believe he would have smiled on this text.

Because an attempt at philosophical clarity and consistency is central to the presentation of this book, the next section of this chapter outlines briefly some basic distinctions among the various points of view on probability. Their associations with related distinctions in the philosophy of science and the foundations of mathematics are also indicated. The final introductory Section 1.3 outlines the historical context in which subjectivist ideas developed. My aim in this chapter is not to make a balanced presentation of all points of view on controversial issues, but to present the issues as they appear to a subjectivist viewpoint, to provide a basis for the technical detail that is developed in the remainder of the book. I do hope that you find the viewpoint appealing and appropriate to your own scientific and commercial investigations, and that you will work to incorporate the statistical procedure into your empirical activities.

Once we are settled on distinctions that need be made in order to understand the operational subjective viewpoint, we can get down to technical development. Beginning with Chapter 2, we shall suppose the operational subjective point of view and pursue its implications *without arguing* for its adoption or discussing alternatives at every turn. This pedagogical approach has been suggested by my own studies of Morris DeGroot's (1970) *Optimal Statistical Decisions*, the most influential textbook in the spread of Bayesian statistical methods over the past two decades. Methodical, detailed, and pleasant, DeGroot worked through the implications of one formulation so that applications can be made by whoever finds them relevant. Arguments are left to other times and other places. As a result, statisticians with a variety of foundational attitudes have appreciated the techniques in a similar way that Bayesians have profited from the important non-Bayesian statistical texts published since the Second World War.

Specific applications of operational subjective statistical methods that appear in my examples and problems are drawn from several scientific subjects. Though applications in economics may predominate, the examples interspersed from a variety of scientific fields are meant to show that the method is applicable to recorded evidence on *any* substantive empirical question. The scientific forum is one in which experiential evidence can be related to theoretical arguments. An important feature of the operational subjective statistical methodology is that it presents a formal language in which the activities of scientific investigation and disputation can be engaged and in which disputation can be recognized publicly for what it is.

As an economist, I have participated in the scientific forum, particularly with respect to the theories of rational expectations, efficient markets, and labor productivity (Lad, 1983a, 1985; Blattenberger and Lad, 1986, 1988). In these articles, careful attention is given to the application of operational subjective statistical concepts, both in theory formulation and in the evaluation of empirical evidence. Even if these statistical applications involved simplifications that can now be removed due to recent computational developments, I believe they are still worth reading as examples of applied work in this statistical tradition. The article of Lad and Brabyn (1993) provides another example in zoology.

To conclude this introduction, let me recommend for your leisure reading the amusing and poignant address by Frank Ramsey to a Cambridge discussion society in 1925 on the subject of philosophical activity. It appears as an epilogue in some edited papers of Ramsey (1931).

## 1.2 PHILOSOPHICAL DISTINCTIONS

What shall we mean by the word "probability"? Let us first distinguish three general classes of response to this question: the objectivist formulation, the subjectivist formulation, and the formalist formulation. We shall then return to each of these to discuss certain aspects in more detail. On our first brief pass through each formulation in the next few paragraphs, we shall sometimes use terms informally without explicit definition. Precise technical definitions of terms such as "event," for example, depend on the formulation of the response in which they are used. These will be discussed in detail only to the extent that will be found appropriate.

### 1.2.1 Three Formulations of Probability Theory

The *objectivist formulation* specifies probability as a real property of a special type of physical situations, which are called random events. Random events are presumed to be repeatable, at least conceivably, and to exhibit a stable frequency of occurrence in large numbers of independent repetitions. The objective probability of a random event is the supposed "propensity" in nature for a specific event of this type to occur. The propensity is representable by a number in the same way that your height or your weight is representable by a number. Just as I may or may not know your height, yet it still has a numerical value, so also the value of the objective probability of a random event may be known (to you, to me, to someone else) or unknown. But whether known or unknown, the numerical value of the probability is presumed to be some specific number. In the proper syntax of the objectivist formulation, you and I may both well ask, "What is *the* probability of a specified random event?" For example, "What is the probability that the rate of inflation in the Consumer Price Index next quarter will exceed the rate in the current quarter?" It is proposed that there is one and only one correct answer to such questions. We are sanctioned to look outside of ourselves toward the objective conditions of the random event to discover this answer. As with our knowledge of any physical quantity such as your height, our knowledge of the value of a probability can only be approximate to a greater or lesser exent. Admittedly by the objectivist, the probability of an event is expressly not observable itself. We observe only "rain" or "no rain"; we never observe the probability of rain. The project of objectivist statistical theory is to characterize good methods for estimating the probability of an event's occurrence on the basis of an observed history of occurrences and nonoccurrences of the same (repeated) event.

The *subjectivist formulation* specifies probability as a number (or perhaps less precisely, as an interval) that represents your assessment of your own personal uncertain knowledge about *any* event that interests you. There is no condition that events be repeatable; in fact, it is expressly recognized that no events are repeatable! Events are always distinct from one another in important aspects. An event is merely the observable determination of whether something happens or not (has happened, will happen or not). Operational procedures for assessing your uncertain knowledge of an event will be described in detail in Chapter 2. Although subjectivists generally eschew use of the word "random," in subjectivist terms an event is sometimes said to be random for someone who does not know for certain its determination. Thus, randomness is not considered to be a property of events, but of your (my, someone else's) knowledge of events. An event may be random for you, but known for certain by me. Moreover, there are gradations of degree of uncertainty. For you may have knowledge that makes you quite sure (though still uncertain) about an event, or that leaves you quite unsure about it. Finally, given our different states of knowledge, you may be quite sure that some event has occurred, even while I am quite sure that it has not occurred. We may blatantly disagree, even though we are each uncertain to some extent. About other events we may well agree in our uncertain knowledge. In the proper syntax of the subjectivist formulation, you might well ask me and I might well ask you, "What is *your* probability for a specified event?" It is proposed that there is a distinct (and generally different) correct answer to this question for each person who responds to it. We are each sanctioned to look within ourselves to find our own answer. Your answer can be evaluated as correct or incorrect only in terms of whether or not you answer honestly. Science has nothing to do with supposed unobservable quantities, whether "true heights" or "true probabilities." Probabilities *can* be observed directly, but only as individual people assess them and publicly (or privately, or even confidentially) assert them. The project of statistical theory is to characterize how a person's asserted uncertain knowledge about specific unknown observable situations suggests that coherent inference should be made about some of them from the observation of others. Probability theory is the inferential logic of uncertain knowledge.

The *formalist formulation* specifies probability as a symbol which plays a specified role in a system of mathematical axioms. It has no meaning in itself. It might be characterized one way in one axiom system (such as a function over a space of sets) and another way in another axiom system (such as an operator on a space of functions). In the proper syntax of the formalist formulation, we might well ask, "What is the numerical value of $P(A \cup B)$?", or "What is the numerical value of $P(A \wedge B)$?", or "What is the numerical value of $P(A + B)$?", depending on the axiomatic specification of the use of the symbol $P$. Whatever the correct syntax for posing the question within a system, there can be only one correct answer if the axiom system is to be considered consistent. If there is no answer to a well-formed formula, the axiom system is said to be incomplete. At any rate, we are sanctioned to look neither to any actual situation outside of ourselves nor to uncertain knowledge within ourselves for the answer to any of these probability problems. Rather, we are instructed to look to the axioms for the answer. To be sure, an axiomatic probability structure may be applied by way of analogy to any variety of scientific or even artistic or theological questions. But its applicability or inapplicability in any instance can have no critical relevance to the axiom system *per se*. Symbols with well-defined usage within the axiom system are what they are—no more and no less than marks on paper with specified syntactic rules for their use in combination with each other.

Of course, we may have scientific motivation to study an axiom system that is applicable metaphorically to some troublesome empirical questions. But alternatively, mathematicians sometimes discover a set of rules which allow for intriguing and consistent manipulations but for which there is yet no known realm of applicability. Most people who take the formalist view of mathematics regard the activities of both applied and theoretical mathematics to be fruitful. But these are regarded to be completely different types of activity. As with probability, the structure of mathematical statistics is also considered to be a formalism. Statistical practice is considered to be the art of learning about the world through observations. The art uses the formalism only as an ideal model, to be mimicked when it appears applicable. The practice of statistics may suggest new problems for formal theoretical analysis when no guide can be found for some desirable artistry.

It should be evident that each response to the question of the meaning of probability is intimately related to a resolution of larger questions regarding the foundations of science and of mathematics. The questions concern the existential status of theoretical scientific constructs and their empirical measurement, the role of personal belief in the conduct of science, and the relationship of mathematics to human activity in the world. Let us discuss these issues to the extent that the situating context for the operational subjective theory of probability will become apparent.

### 1.2.2 Larger Issues in the Philosophy of Science and the Foundations of Mathematics

The resolution to these questions that we will adopt in this text has developed out of the interaction among several distinct modern traditions in philosophy, notably positivism and conventionalism in France, logical empiricism in Austria, analytic philosophy in England, and pragmatism in Italy and the United States. Though geographical locations have been associated with these traditions due to certain eminent adherents, the enforced mobility of so many scholars throughout the first half of this century has ensured the fertilization of the various lines of thought almost everywhere. In fact the interaction was actively sought out, as evidenced, for example, by the remarkable International Congress on Scientific Philosophy which convened in Paris in 1935. Though that Congress did establish a committee to prepare an International Encyclopedia on the Unity of Science (see Carnap, 1939) and a committee to direct a unification of logical symbolism, the papers presented to the Congress show that a unified set of conclusions had not yet been reached. (See the *Actes du Congrès International de Philosophie Scientifique*, 1935.) What was generally agreed was that the old order of understanding science, constituted by claims about metaphysical entities, was to be rejected. Precisely what position could constitute a rallying point for a new understanding was still questionable.

The resolution described in the present text as the basis for the operational subjective theory of probability does not follow completely the prescriptions of any one of these modern philosophisms. In fact the historical advocates of each one of the positions mentioned in 1930, say, would have specifically renounced some aspect of the position that we now describe in a discussion of the following three basic statements:

1. Scientific knowledge amounts to meaningful claims about prescribed measurements.
2. Personal judgments and beliefs are central to the conduct of scientific activity.
3. Mathematics is a language.

These simple-sounding statements constitute basic presumptions underlying the operational subjective theory of probability and statistics. Perhaps seemingly innocuous, they are quite controversial. Let us discuss them in turn.

## 1. Scientific Knowledge Amounts to Meaningful Claims about Prescribed Measurements

The objectivist formulation of probability presumes the existence of a "real world," which is beyond the pale of our ability to experience it and to measure it directly. Our ability to know about this "real world" is limited by the fact that whenever we try to measure some aspect of it, our measurement is inevitably compounded by some unidentifiable component of error. Though we may like to know the actual value of a quantity, $Q$, our measuring instrument yields a measurement, $M$, which equals $Q + E$, where $E$ is a measurement error of unknown magnitude. Taken by itself, the measurement $M$ can be considered as an estimate of the quantity $Q$, which is unmeasurable directly. It is supposed that quantities of probability are in this way no different from any other quantity investigated by science. A probability is unobservable and unmeasurable in itself. It can only be estimated on the basis of measurements of random variables that can be made. The probabilities in the objectivist's loudly proclaimed "real world" turn out to be imagined items, imagined for convenience to account for variations that we observe in nature. But the probabilistic reasoning that supposedly allows for our "almost certain" knowledge of their values is circular.

The subjectivist formulation of probability and statistics is based upon a radical break with this conception of empirical science as yielding approximate results about a real world beyond our experience. We presume that scientific investigations constitute our studies about the world of human experience, not some supposed world outside of our experience. Empirical measurements are merely the recording of human activity in the world, conducted according to operationally described procedures. The measurement procedure is itself part of the activity being recorded. Scientific arguments about the result of any specific operationally defined activity can be resolved empirically by performing the prescribed procedure. Arguments about the value of quantities that are in principle unobservable cannot be resolved. Though there may well be some reality inaccessible to human experience, there is, by definition, no way for humans to know anything about it. Supposed scientific claims that might be made about such reality or how it relates to the world of our experience can only be idle.

Any proposed quantity capable of scientific investigation must be accompanied by an operational definition of how it is to be measured. The subjective formulation of probability prescribes that your probability for an event is measured as the price at which you value a specific economic transaction. It is because the specific formulation of probability proposed in this textbook provides such an operational definition that we designate the theory as the *operational* subjective theory of probability. If there was a fundamental philosophical commitment in de Finetti's works, it was not so much to a subjective point of view per se but to the methodological stance that the meanings of scientific assertions must be identified in terms of the verifiable results of engaging in well-specified actions. Only thus can any assertion be evaluated by recourse to observation. It was de Finetti's conclusion from his study of the various proposed notions of probability that the subjectivist formulation is the only one that could be proposed meaningfully in terms of an operational definition. The only concrete fact beyond dispute is that you (me, someone else) find yourself in a state of uncertainty about something. Your probability is a measurement of your state of uncertainty, performed in a specified way.

An insistence upon operational definitions of scientific concepts is generally associated with the readable arguments of the physicist Percy Bridgman (1927, 1936). But the idea has deeper roots in the French positivism of the nineteenth century and in the turn-of-the-century ideas of Ernst Mach in Vienna. Anatol Rapaport (1954) provides enjoyable bedtime reading on the subject. The discussions at the meetings of the American Association for the Advancement of Science, edited by Philipp Frank (1954), include the ideas of an important critic, Carl Hempel. See also the collection of Hempel (1965).

## 2. Personal Judgments and Beliefs Are Central to the Conduct of Scientific Activity

The objective formulation of probability is generally associated with the characterization of science as an impersonal, value-free, and skeptic process whereby participants impassively and systematically investigate the empirical truth of falsifiable propositions that they regard as critical to the furthering of our knowledge. The extensive works of Karl Popper (1959, 1963, 1983) which promote this view are well known among scientists who are interested in the philosophy of science. This skeptic tradition in scientific methodology, which dates from Descartes (1637), admittedly has been instrumental in freeing scientific investigation from the dogmatic clutches of the Church in the age of decadent scholasticism. Its favor among applied scientists even today is a tribute to that glorious role. But it has served its day, and the day is now gone.

The subjectivist formulation of probability is associated with the opposite view. Scientific activity necessarily involves the personal judgment and belief of its participants, who commit themselves to some manner of understanding the world as opposed to another—even at the risk of being considered to be wrong by some or most of their respected colleagues who choose to believe otherwise. The subjective formulation of probability proposes that probability is the measurement of a scientist's degree of belief. The fiducial activity of science is distinguished from that of metaphysics or a dogmatic religion in that the objects of the scientist's belief assertions are expressly observable quantities, as discussed above. We have observed what we have observed. This is regarded to be a matter of fact. The art of scientific observation amounts to the skill of distinguishing and describing what it is that we are observing. What we shall observe next is a matter of anyone's conjecture. "That remains to be seen." What we should learn from our past experience about what we are to experience next is a matter of belief, of assertion. As would any group of people in uncertain situations, scientists sometimes agree on such matters, and sometimes they do not.

The logic of probability merely formalizes the coherent implications of specified beliefs about events when the beliefs are expressed in a graded scale from 0 (complete denial) to 1 (complete affirmation). This logic is relevant to *anyone's* uncertain knowledge about any unknown observable quantities. The "scientific method" empowers the scientist with precisely the same inferential principles that it empowers the person on the street in the conduct of daily affairs, no more and no less.

The philosophical writings of Michael Polanyi (1958, 1966, 1974) spiritedly advanced the view that science is a fiducial activity, in marked disputation with the ideas of Popper. Similar ideas on faith and commitment in science are shared by Thomas Kuhn (1962), whose works are perhaps more widely read. Nonetheless Polanyi was an adamant critic of operationalism in science, which we advanced in Statement 1. And Polanyi's ideas on probability (1958, Chapter 2) do not correspond to the operational subjective viewpoint.

### 3. Mathematics Is a Language

Western intellectual tradition had long been saddled by the platonic doctrine that the truths of mathematics are the secrets of a world of pure ideas, in which reside such objects as numbers, angles, and parallel lines. Such entities are evidently inaccessible directly to human experience. Our knowledge of them was considered to be only inferable from our experience in the imperfect material world, in which they exhibit themselves by weak reflection. For example, the imperfect "lines" that we can see have some width, whereas real "ideal" lines do not. In this spirit, for example, mathematical physics has long held a place of special honor, due to its capacity for embodying "pure truths." In comparison, experimental physics has been considered a lower-order activity, being relegated to the suspicious shadows of the mundane world where false perceptions lurk. Apropos to our subject, the objectivist formulation of probability, which has found applicability in all sciences, lies within this platonic tradition. Objectivist doctrine holds that the truths of probability, expressible in mathematical form, can only be inferred from observed values of random variables which are contaminated by the particular idiosyncracies of historical happenstance. The discovery of the laws of probability in random behavior is deemed in this view as one of the great twentieth-century scientific achievements.

However, the scandalous creations of non-Euclidean geometries in the nineteenth century by Nicholas Lobachevsky and Johann Bolyai, and the determination in our own century by Kurt Gödel that the theorems of mathematics cannot be completely reduced to necessary truths of pure logic, have by now largely undermined the platonic doctrine of mathematics. [Not to overstate the present historical situation, the platonic outlook underlies the work of many practicing mathematicians who do not care to think about philosophical foundations of their efforts. And platonic ideas do continue to dominate the institutions of French mathematics through the purely abstract systematization known as Nicholas Bourbaki (1950).] Rather than think of mathematics as a body of self-existing truths to be discovered, the twentieth-century tradition we shall follow is that mathematics is a language we construct for expressing certain kinds of ideas. Within this linguistic interpretation of mathematical activity, there are three generally recognized categories of conversation topic.

First, mathematics is the language in which you can state what you want to assert about the observations of human experience as precisely as you are able. "This New Zealand Grey Duck laid 11 eggs in 11 days. These eggs all hatched live ducklings over a period of 6 hours, on the 29th day after the first egg was laid. Six days after hatching, nine of the chicks are still living." As we shall learn in this text, your assertion may range from an explicit declaration of a precise observation, as just exemplified, to an uncertain surmise of the numerical value of an observation that is still to be made or that has been forgotten. We define such a surmise as your "prevision." In addition, the language of mathematics allows you to evaluate the quality of your assertions about this experience in terms of the relation between the numerical representation of the experience and the numerical representation of your assertion about it. The quality of your assertion is measured in terms of its efficacy for achieving ends that you expressly desire.

Then, again, the language of mathematics can also be used to create images that are not meant to refer to our experience of nature, but whose creation amount to part of the human contribution to nature. Such mathematical imagery is a creation of art, as is poetry, music, and painting. Think of the beautiful imagery constructed by fractal geometry, topological figures, or space-filling lines. These mathematical images can be evaluated in terms of beauty, as can the song of a bird lilting through the forest early on a summer morning.

Finally, the language of mathematics can also be used to express properties of itself as a language. Mathematics is called a formal language, as opposed to a natural language. This is to say that rules for the correct usage of mathematics can be expressed within the language itself. Any expression must obey these rules of usage if it is to be recognized as conveying any meaning within the language. A natural language, on the other hand, is one which is constructed by its very use. It has no immutable rules of correct usage, in that its usage is always changing. A natural language can be considered to be alive, as are the people who use it. Approximate "rules" for its usage, grammar, can be codified by paying attention to how it is used in practice. But this code of linguistic practice can be broken, and the code itself changes over time as the social structures supporting the language change. Moreover, on the contrary, artistic creation is sometimes achieved in a natural language through self-imposed adherence to extremely restricted forms of usage, which are in no way necessary for comprehensibility within the language—to wit, the poetic forms of the sonnet and ghazal.

In this century, distinctions among these three uses of mathematics have been the source of long debate about what is true in mathematics and what is not. But it has been with the insight of this century that we have identified mathematics as a language at all. Actually, several alternative viewpoints on the foundations of mathematics emerged in the froth and in the wake of nineteenth-century developments. We should remark in greater detail on an important competitor to the linguistic tradition—the formalist viewpoint—and on two distinct but supportive traditions—the intuitionist and constructivist viewpoints. These have direct relevance for our tripartite taxonomy of formulations of probability.

### 1.2.3 Alternative Views on the Foundations of Mathematics

The formalist viewpoint on mathematics, championed in the twentieth century by David Hilbert, proposes that any system of well-formed formulas is fit to constitute a subject for mathematics. The contents of such a system are the valid relations among the marks on paper in which they are expressed. Mathematical statements have no meaning. Mathematics refers to nothing other than itself. It can be considered as a game. The project of research in the foundations of mathematics is conceived as a search for unproved axioms which, taken as given, can be used to prove completely and consistently the results of such arbitrary systems. Historical commentary appears in Fang (1970). In our tripartite division of formulations of probability theory, the formalist formulation falls within this understanding of mathematics. Kolmogorov's (1933) axiomatization of probability is widely understood to provide the rules of the game, and the mathematical theory of probability and statistics is taken to be the derivation of formal results that follow from these axioms. The results have no meaning in themselves. The relation of this viewpoint and these "results" to objective theories of probability is that objectivists typically assert or presume that their unobservable probabilistic "propensities" enjoy the properties of probability specified in Kolmogorov's axioms.

Hilbert's formalist program of mathematical analysis was challenged by the proponents of intuitionist mathematics early in this century, notably by L. E. J. Brouwer and Arend Heyting. They argued that mathematics is a representation of interpretable human activities, which they termed "intuitions." It made no sense to them to study the implications of arbitrary axioms that do not represent any specifiable activity. In fact, the intuitionists exposed several ways in which the blind acceptance of axioms that have no interpretation could yield results inconsistent with themselves. The intuitionist insistence on interpretability made important restrictions on the systems of axioms its proponents found amenable to mathematical analysis. For example, since human activities are necessarily finite in extent, the axioms of intuitionist mathematics express only requirements that impinge on a finite number of operations. Thus, the results they accept are much less confining than those of formalist mathematics. For the formalists have no qualms about stating as axioms requirements that could never be brought to bear on a finite number of operations. One case in point relevant to the study of probability concerns whether the finite additivity requirement of a probability function should be extended to a requirement that holds for an unlimited number of events as well. The operational subjective theory of probability does *not* support this extension as representing any universal logical *necessity*, though it can be an acceptable assertion in any specific instance. Many formalists and objectivists regard countable additivity as a valuable axiom. But the issue is hardly relevant to applied scientific applications, and we shall not address its detail in this text. It is discussed in an enjoyable article by Scozzafava (1984).

On some counts the operational subjective theory of probability differs from intuitionist foundations of mathematics. Intuitionists argued, for example, that there is no intuition supporting "the principle of the excluded middle." So they renounced the practice of mathematical proof by contradiction. To the contrary, as we shall see, the operational subjective theory supports an operational meaning to proof by contradiction through its economic characterization of the "coherency" of uncertain knowledge.

Interestingly, intuitionist mathematicians and subjectivist probabilists have been similarly criticized by empiricists who challenge "exhibit for me these intuitions in the mind," or "exhibit for me these people whose uncertain knowledge satisfies the properties of coherent subjective probability." See, for example, Kyburg (1983) and related arguments in Kahneman, Slovic, and Tversky (1982). Whereas some intuitionists became lost in a philosophical search to substantiate intuitions as some sort of Kantian categories of the mind, the operational subjective theory is not proposed as a descriptive characterization of all human thought. Rather it is proposed as an enviable characterization of uncertain knowledge, to be used in private practical decisions whenever the gravity of the situation and the valuation of the decision maker merit it, and to be systematically required in scientific discourse.

Finally, we should mention the development of the intuitionist tradition in the form of the constructivist philosophy of mathematics. The clarion manifesto of Bishop (1967), introducing his *Foundations of Constructive Analysis*, expressed frustration with idealist tendencies in mathematics that take pleasure in the proof that specifiable mathematical objects "exist" without specifying constructive procedures whereby embodiments of the existing objects can be computed. (If such and such conditions hold, then there exist functions that have such and such properties.) The constructivist position requires that mathematical proof be couched in terms that allow computable representation of the relations under discussion. As it turns out, operational subjective statistical methods are representable by such a computational program.

Although de Finetti never wrote directly on the formulation of these arguments in the foundations of mathematics within the Anglo-Saxon literature, he always insisted in his work that the mathematical formulation of statistical theory submit itself only to the exigencies of what is required by a practical application, never to artificial requirements of what a unique mathematical conclusion should look like. This is essentially a constructivist and intuitionist attitude. It is not to deny the beauty of art for art's sake, or mathematics for mathematics' sake. It is only to delineate clearly the role of mathematics in representing scientific knowledge, and the meaningful requirement that it be coherent. De Finetti refers to other Italian mathematicians early in this century who agreed in this outlook, notably Giovanni Vailati and Mario Calderoni, and others in the Italian pragmatist tradition.

Further discussion of the history of modern development of the various points of view about the foundations of mathematics would distract us from the limited issues that concern us here. A more extensive summary can be found in the masterful sourcebook of translations of classic articles in the volume edited by van Heijenoort (1967) and in the collection of Benacerraf and Putnam (1983). The latter includes a lively portion of Heyting's *Intuitionism: An Introduction* (1956), which is written in the lively, classic, philosophical form of a dialogue. The relevance of these issues to statistical theory is discussed explicitly in a beautiful article by Hill (1994a).

### 1.2.4 Still Two More Traditions in Probability Theory: The Objective Logical View and the Marxist Materialist View

To conclude this presentation of philosophical distinctions, we should remark on two important traditions that have escaped explicit notice until now. The first involves a notable variation on two themes of probability as a many-valued inferential logic, mixing aspects of two of the formulations of probability which we have discussed. It is referred to as the logical view of probability. The second is a major politico-philosophical tradition of our century which has gone awry in identifying the role of probability in science and the procedure for its mathematical construction. I refer to the Marxist materialist theory of probability as it had been proposed in Soviet writing.

In discussing the objectivist formulation of probability, we have referred to its identification of the supposed objective entity "probability" as the "propensity" in nature for random events to occur. Within the objectivist tradition there is another viewpoint which identifies the objective probability as a unique logical relation among sentences—the relation specified by the degree of "support" that the truth of one sentence lends to the truth of another. The development of this viewpoint had two distinct antecedents.

The notion that probability theory is a form of many-valued logic received impetus at the turn of this century due to considerations of Russell's paradox. This suggested that in the mathematical logic of set theory, the use of two-valued truth functions could yield annoying contradictions. Research was pursued in many directions attempting to resolve such difficulties. The works of Jan Lukasievicz (1929) and Hans Reichenbach (1934) focused on the extent to which the paradoxes rest on the supposition of merely two truth values. Their developments in logic were based on the supposition that the observable truth value of a proposition may be determined not only as true or false (1 or 0), but as "something else," such as "perhaps," as well. A second development in the logical tradition still requires that the truth value for a proposition be either true or false (1 or 0). But it proposes that the degree of conditional probability $P(A|B)$ represents the objective degree of logical support that the truth of sentence $B$ lends as evidence for the truth of sentence $A$. This idea has been central to the work of Keynes (1921), Jeffreys (1939), R. T. Cox (1961), Hacking (1965), and Kyburg (1974). Bruno de Finetti wrote beautiful pointed papers on both of these logical objectivist developments related to probability (1936, 1938a). He showed deep sympathies with the ideas of logical objectivists, but he argued that this degree of logical support relies on a personal judgment to be asserted by each individual, and it is not a unique objective relation between sentences, perceivable to everyone. In this he concurred with Ramsey's (1926) critique of Keynes. Moreover he found the work in many-valued logic to be intriguing but misdirected in the probabilistic context. Many-valued logics have a role in the definition of quantities rather than in the specifications of our opinions about quantities, a distinction we study in Chapter 2.

Our concluding philosophical remark concerns the objectivist interpretation of probability asserted by Soviet writers and their explicit arguments against the subjectivist formulation. Given the historical context of the Russian people's revolution against their Czarist overlords, it is understandable that their academicians would have been wary of any tendency toward what they termed "subjective idealism." This philosophical doctrine was used explicitly in probability theory by reactionary mathematicians who were supporters of the czar. See, for example, Maistrov's (1974) discussion of the Nekrasov affair, and the progressive challenges organized by A. N. Markov. The often-printed and well-known textbook of B. V. Gnedenko (1964) erroneously tried to identify the subjective formulation of probability with this doctrine:

> ...If mathematical probability is the quantitative measure of the degree of uncertainty of the observer, then the theory of probability is something not unlike a branch of psychology. The final outcome of consistently using such a purely subjectivistic interpretation of probability is inevitably subjective idealism. Indeed, if we assume that the evaluation of probability only concerns the state of the observer, then all conclusions based on probabilistic judgments are deprived of the objective meaning that they have independently of the observer. Meanwhile, science has established many positive results on the basis of probabilistic judgments.... For example, in physics all "macroscropic" properties of gasses are derived from assumptions concerning the probabilistic nature of the behavior of the individual molecules. If we are to ascribe an objective meaning independent of the observer to these deductions, then the initial probabilistic hypotheses concerning the course of the "macroscopic" molecules must be something more than just a statement of the psychological state into which we are thrown when we think about the motion of molecules.
>
> To those who take the point of view that the external world has a reality independent of ourselves and is in principle cognizable, and who can take into account the fact that probabilistic judgments can be used successfully to obtain knowledge about the world, it should be perfectly clear that a purely subjective definition of mathematical probability is quite untenable.... (Gnedenko, 1964, pp. 25–26)

In contrast to the subjective interpretation of probability, Gnedenko presents an objectivist view of probability as a scientific theory of mass phenomena and of repeatable experiments whose outcomes exhibit a stable frequency of occurrence. He supposes throughout that such a formulation is in keeping with the Marxist materialist view of science. But in developing such a theory, the fundamental characteristic of reality, supposed to be "probability," is identified as being in principle unobservable, quite an odd category for a Marxist materialist. And the assertion of repeatable experiments is directly opposed to a central Marxist observation that nature is constantly in flux. Nothing is repeatable. In fact, the Soviet objectivist doctrine of probability amounts to an inversion of subject and object of the type that Marx criticized in Hegel. Human ideas about the world (that is, our personal probability assertions) are projected outside of ourselves in Gnedenko's theory and made to appear as objects (objective features of nature). In Marxist terminology, human uncertainty is "reified" as an object, "probability."

Perhaps surprisingly, it is the operational subjective theory of probability that properly associates what is real with the experience of history and recognizes probability as a measure of our own uncertain ideas about it. A detailed discussion of this complicated matter appears in my essay "The Foundations of Probability: A Marxist Discussion" (Lad, 1983b). Although Gnedenko's ideas had often been repeated by other Soviet writers (see the papers edited by Tavanec, 1970, for example), the translated publications by Nalimov (1981a, b, 1982) suggest that subjectivist ideas of probability as uncertainty may have long enjoyed more support among the Soviet thinkers than Gnedenko's diatribes had let on. The moral of this discussion is that we should beware of misleading connotations associated with the names "subjective" and "objective" which are associated with formulations of probability.

To summarize this section, the operational subjective statistical viewpoint is set in the context of an operational and positivist approach to science, an analytic approach to philosophy, and a constructivist, finitist, and intuitionist approach to mathematics. This is a twentieth-century resolution that has emerged from a sea of developments throughout the past 300 years. The concluding section traces important lines of development of the subjective theory of probability over this period. It is not meant to outline the general history of probabilistic reasoning, as many facets of the development of contending points of view are ignored. These have been extensively described elsewhere by others.

## 1.3 HISTORICAL DEVELOPMENT OF OPERATIONAL SUBJECTIVE STATISTICAL METHODS

In this presentation we shall not linger on any detail, but make sufficient references to works you can examine as extensively as your pleasure and your leisure permit. Further detailed historical remarks are interspersed throughout the text.

During the European enlightenment of the seventeenth century, probability and statistical methods emerged as related subjects with broad scope of application. With little intellectual forewarning, a variety of newly relevant questions began to surface as topics amenable to mathematical analysis and resolution—questions regarding optimal play in games of chance, the just division of a purse after only partial completion of play, the appropriate value for life insurance contracts and for interest rates charged in the funding of risky economic ventures, the unavoidable errors in measurement of newly discovered scientific phenomena, and the influence of public laws on the health of citizens. Recorded statistical results and various problems for consideration circulated among the new literati, including Galileo, Pascal, Fermat, Huygens, several Bernoullis, Graunt, Petty, Halley, and De Moivre. Isaac Todhunter (1865) listed chronologically the problems and results that passed among them and among investigators of the next century, including Bayes and Price, Condorcet, D'Alembert, Lagrange, and Laplace. Frustrated with the mere listing of technical results by Todhunter, Karl Pearson produced an ambitious series of lectures at the University College of London during the academic sessions of 1921–1933 on The history of statistics in the seventeenth and eighteenth centuries against the changing background of intellectual, scientific, and religious thought. Scholarly, opinionated, sometimes speculative, and thoroughly enjoyable reading, these lectures were not printed until 1978 after their editing by his son, Egon.

Ian Hacking's historical study (1975) of the early philosophical developments notes that throughout this period probability emerged as a dual concept, both as a name for the apparently stable frequency of occurrence of similar events and for the appropriate degree of belief in the outcome of a chance event. It was in the nineteenth century that concern focused on the incompatibility of these two concepts, and arguments for the fundamental correctness of one or the other became prevalent. The practitioners of the seventeenth and eighteenth centuries had proceeded, as have the latter-day practitioners of our century, to solve practical problems using whatever methods seemed appropriate to each, without agonizing over the meaning of words. Argumentation was confined to considerations of the technical correctness of solutions to the problems as posed. But even the syntax of technical formulations varied. Many authors considered probability as a fundamental concept. Others, notably Christian Huygens (1657), considered "the expectation of a quantity" rather than the probability of an event as the basic concept. The mathematical formulation of the operational subjective theory described in this book is related formally to this line of development. Leibniz (1704, 1714) foresaw the possibility of applying Huygens' technical formulation to characterizing the logic of partial knowledge. See Couturat (1901).

To be sure, the notion of probability arose at times and places other than in "enlightened" Europe. Edmund Byrne (1968) details the use of the concept in the scholastic philosophy of Aquinas, based on foundations laid by Aristotle and Plato. Eventually, the casuistic doctrine of probabilism became bothersome to the canon lawyers in the sixteenth and seventeenth centuries. See von Harnack (1885, Vols. V, VI, and VII) and Hacking (1975, Chap. 3). But both these developments were set in the context of premodern notions of the fundamental fixity of "being" and that knowledge of primordial eternal truth is capable of embodiment only in certainty.

Early in the development of Chinese culture there had emerged a confluence of ideas regarding change, uncertainty, action, and determinism, exhibited in the ancient book known as *I Ching*. The basic idea is that since all nature is a unity, if you are confused about how to behave in some perplexing instance, you can gain insight by observing a casting of lots. For the casting of lots is a participation in the happening of the moment and is thus necessarily informative about the status of the whole. Participation in *I Ching* by the throwing of coins or the separating of straws has been a central cultural event throughout the varied political economic history of the Chinese people over at least 3000 years. Though the ideas and the practice developed within the Chinese scientific tradition, Joseph Needham (1956, Vol. 2) opined that the use of *I Ching* in modern China has had a stifling influence on inquisitive and creative scientific thinking. It surely would be contrary to the common current Western understanding of the throwing of coins as the generation of a probabilistic chance event. The basis for the practice lies in a different understanding of probability not as randomness but as personal uncertainty. Jung's forward to the Wilhelm/Baynes English edition (Jung, 1949) makes stimulating reading. He discusses *I Ching* within the concept of historic activity that he termed "synchronicity." Interestingly, Elaine Pagels' (1979) study of the gnostic tradition among early Christians identified an understanding that is similar to that of the Chinese, for they used the casting of lots to determine who would perform various liturgical functions as an exhibition not of chance, but of divine will, which thus enters history.

Indian civilization, too, has grappled with the concept of probability. The ancient epic *Mahabharata* is set in the context of a gambling challenge. Hacking describes insightful details of one minor story of Nala and Kali, interpreting it to exhibit an awareness of the relationship between strategy in gambling and estimation methods based upon survey sampling. But aside from the papers of Mahalanobis (1957) and Haldane (1957) discussing the Jaina theory of uncertain predication (note this is *not* a typographical error for the word "prediction"), technical formulations of probabilistic mathematics from ancient India are not well known. The multiple-form syntax of predication seems directly related to the idea of many-valued logic studied by Lukasievicz and Reichenbach, which we mentioned in the previous section. The fact that uncertainty of predicate was mentioned by Jaina philosophers may be relevant to our understanding of the modern development of probabilistic concepts in Europe. For Jaina and Buddhist ideas emerged among the new *merchant* families of India during the sixth-century B.C., as opposed to the settled agricultural families who remained largely Hindu. During the Mauryan (Buddhist) all-India empire of the fourth and third centuries there was a minimal institution of private property in land, usury flourished, and merchant wealth accumulated. Trading prices had to be regulated to keep the merchants from garnering too much profit. (See Thapar, 1961, and 1966, chap. 4.)

The importance of mercantilism in the development of probabilistic concepts was also evident in Europe, to which we now return. Concomitant with the emergence of probability as a radically new concept was the economic transformation of Europe from its medieval feudal organization to the market-based social organization of the mercantile era. Though reacting negatively to what he considers a crudely causal statement of this fact by Maistrov (1974), Hacking (1975) concurs (with notable reservations, p. 4) that "an undogmatic version of this doctrine must be right." On the one hand, the experience of change itself in social organization was disorienting to the life prospects of individuals and of families. The development of markets in which an increasing number of consumption items became commodities, as opposed to a socially organized feature of feudal distribution, was a liberating factor for human imagination. Not without struggle, realms of possibility were generated surrounding many personal and social events that had been severely regulated in feudal Europe, particularly surrounding marriage and occupation. During this time of transition, the once "inevitable implications" of one's actions became uncertain and controversial. Typically different were the presumptions of parents and their children. The social security of the village unit began to disintegrate in the process.

On the other hand, the emergence of probabilistic concepts was accompanied not merely by social change. The decay and collapse of the Roman Empire had ushered in social change too. Yet probabilistic ideas did not flourish. Hacking notes with pleasure, in mildly contesting an economic/social theory of the emergence of probability, that a third-century Roman state nonetheless did sell annuities to finance its administration. To be sure, the selling of annuities must have been a wily move. In the middle of the third century the Roman empire was already under pressure from regular strikes by Franks, Goths, and Alemanni raiding parties. The selling of annuities could well have been a retrograde support for the security and consolidation of a crumbling social structure rather than an impetus to new possibilities of liberation. One must wonder whether those annuities were ever paid off by the state before it collapsed. The European territory of the empire had to be consolidated by the emperor Aurelian during the years A.D. 270–275. By A.D. 410, Rome itself was sacked by the Goths. It may not have mattered much how reasonably the value of the annuities had been assessed. But the socioeconomic transformation of medieval Europe was more than just an era of social change. The development of a market-based economy offered the possibility of previously unimaginable experiences to people at virtually every level of feudal society.

The experience of the mercantile cities was also telling. It was only in 1498 that Vasco da Gama first led a European fleet around the Cape of Good Hope through the notoriously difficult seas merging the Atlantic and the Indian Oceans. It took a century for the economic implications of this event to become fully recognized and exploitable. By the early seventeenth century, the merchants of many European nations were regularly outfitting ships to trade in India, Indonesia, and, to a lesser extent, even China. The many Middle Eastern merchants of the ancient overland trade routes could be circumvented. There were fortunes to be made in doing so—if only the ships would return with their bounty rather than be dashed by the seas or otherwise succumb during the treacherous journey. Risk became recognized as a commodity with a salable value. The French jewel trader Jean Baptiste Tavernier (1676) discussed ways to circumvent interest payments for insured risk at various bottlenecks in trade negotiations. Of course, the institutional acceptability of loans for profit was itself an innovation. For interest payments had been morally censured as "usury" in medieval Europe and are deemed so even today by some in orthodox Islamic countries. But we are, ourselves, now drifting toward the vortex of discussing the historical mechanism of the socioeconomic transformation of Europe. The relevant point here is that sometimes the ship would come in, and the community which depended on it would flourish. And sometimes it would never come, though surviving stragglers from the expedition might return one by one years later. How great a "return" would you demand on your investment to outfit such a fleet at that time? To how many sources could the merchant adventurers turn for financing? What alternative lucrative outlets did the financier have? Practical matters.

By the eighteenth century, the new methods of economics and engineering had become subjects of philosophical consideration. The first half of the century was the period in which the so-called British empiricist philosophers, Locke, Berkeley, and Hume, led the attempts to sort out an understanding of the new scientific knowledge. For although the new knowledge evidently had substantive practical content, it also involved some degree of uncertainty about its details on the part of its proponents. The important break with the scholastic notion that distinguished "certain knowledge" from "mere opinion" was achieved by the radical epistemological doctrines of Hume. Our scientific knowledge of an impending empirical experience has no logical (necessary) foundation. Rather, its basis is animal instinct, convention, habit, and belief. Hume's "inductive skepticism" refers to his doubting that there can be any purely logical ground for inferring universal statements from particular observations. The mode of scientific knowledge, uncertainty, was identified as categorically different from the previously acclaimed certainty of the necessary truths of metaphysics.

Given the perceived ultimate importance of the implications of this trend of thought in a religious age, it is not surprising that the issue was taken up among the scientifically trained clergy in England, particularly among the nonconformist ministers and their congregations, free from explicit political alliances. Thomas Bayes and his friend Richard Price, who both puzzled over problems of probability from the perspective of reasonable uncertain knowledge, were among the discussants. Gillies (1987) outlines the known personal relations between Price and Hume, who met first in 1761, shortly before Bayes' death. But he surmises that Bayes also knew of Hume's (1748) *Enquiry Concerning Human Understanding*. Gillies argues that Price, at least, conceived of Bayes' "Essay toward Solving a Problem in the Doctrine of Chances" as an answer to Hume's celebrated pondering: "Now where is that process of reasoning which, from one instance, draws a conclusion so different from that which it infers from a hundred instances that are nowise different from that single one?"

Bayes' essay (1764) was published posthumously, having been submitted for publication to the Royal Society of London by Price, who wrote a preface and an appendix. The highlight of the essay is a theorem identifying how mixed evidence from a sequence of experiments can be used to inform one's probability for the outcome of the next experiment, detailing precisely how the evidence from 100 observations can be recognized as different from the evidence in a single one. Price drew attention in his preface to a postulate that Bayes had used in an earlier version of the essay but had eliminated since he had himself found it disputable. Nonetheless, when interest in Bayes' work rekindled in Britain some one hundred years later, the rejected argument drew more critical attention than did the inferential logic that was the subject of the paper. At issue was the postulate, later termed "the principle of indifference," stating that the absence of knowledge about the truth of several possibilities should be represented by a uniform probability distribution over those possibilities. The operational subjective theory of probability and statistics is *not* based upon the principle of indifference. Stigler (1982, 1983, 1986) presents several enjoyable discussions of Bayes' analysis and the historical context of the essay. At the moment, let us note only that the full statement of what has come to be known as Bayes' theorem first appeared in the writing of Laplace. [See Deming's introduction to Bayes' essay in the edition listed in the references, and the pointed remarks of Fréchet (1947, p. 372).]

For while the English brooded over foundational questions, the eighteenth-century French mathematicians Condorcet, d'Alembert, Lagrange, and Laplace pressed the uncertain belief point of view through to the technical resolution it could provide for several detailed practical problems in engineering and economic planning as well as in chemistry, physics, and astronomy. Lagrange and Laplace survived the French revolution, working into the next century. Karl Pearson's (1978) *History of Statistics* provides such sophisticated and enjoyable commentary that I will not spoil it here. The masterful historical analysis by Daston (1988) is even more mature and thorough on philosophical detail. The development of French ideas was very complex, as were the social and economic issues of their day. Pearson's evaluations are amusingly English, and they do not always please the French, sometimes with good reason. See, for example, Pierre Crépel's (1988) interesting reevaluation of contributions by Condorcet to the mathematics of finance. But Pearson, in amusing agreement with Emile Borel (1924) about the same time, avows that there really is something incomprehensible in each culture (French and English) to natives of the other. At any rate, you should bring the histories by Pearson and by Daston home from the library sometime for bedtime reading. Since it is so apropos to our theme, let us only allude to an idea of this era of French genius with the following beautiful quotation from d'Alembert's *Elements of Philosophy* as translated by Pearson (p. 523). d'Alembert is describing the importance of uncertain knowledge as a valuable gradation of certainty:

> It is only by accustoming ourselves to recognize truth in all its purity that we can afterwards distinguish what approaches it more or less. The sole thing that we must fear is that the habitude too great and too long continued of seeking absolute and rigorous truth blunts the feeling for that which is not; ordinary eyes habitually accustomed to a bright light fail to discriminate the graduations of feeble light, and see only heavy

darkness, where others distinguish still some illumination. The mind which recognises the truth only when it is immediately in contact with it, is distinctly inferior to the mind which knows not only how to recognize it at close quarters, but as well how to foresee it and note it in the distance by its fugitive characters. It is this which distinguishes in the first place the mathematical spirit from the spirit of the mere mathematician, whose talent is restricted to a narrow and limited sphere.

By the nineteenth century, a recognition of the acceptability and the value of uncertain knowledge, as opposed to certainty, was well established.

In the year 1827 Laplace died and the English mathematician Augustus De Morgan completed his formal education as fourth wrangler in mathematics at Cambridge. A peculiarly delightful man, De Morgan was said to be unsurpassed as a teacher throughout his professorial career at the nondenominational University of London (today University College) from its inception in 1828 to his second resignation on issues of principle in 1866. He had resigned his chair first in 1831, spending a few years "in the wilderness" as an actuarial consultant until he was unanimously reelected to the post upon the accidental death of his successor. Peter Heath's brief introduction to the edition of De Morgan's six papers *On the Syllogism* (De Morgan, 1966) gives details and is well worth reading.

De Morgan was an adamant subjectivist in interpreting probability as a measure of an individual's uncertain knowledge about an event. He had the great insight to include probability theory as the second part of his *Formal Logic* (1847), writing:

When I am told that logic considers the validity of the inference, independently of the truth or falsehood of the matter, or supplies the conditions under which the hypothetical truth of the matter of the premises gives a hypothetical truth to the matter of the conclusion, I see a real definition, which propounds for consideration the forms and laws of inferential thought. But when it is further added that the only hypothetical truth shall be absolute truth, certain knowledge, I begin to see arbitrary distinction, wanting the reality of that which proceeded. Without pretending that logic can take cognizance of the probability of any given matter, I cannot understand why the study of the effect which partial belief of the premises produces with respect to the conclusion, should be separated from that of the consequences of supposing the former to be absolutely true. (1847, p. v)

Probability was identified by De Morgan as a part of formal logic, the many-valued logic of uncertain inference. In the field of mathematical logic, his work was concerned with the formalization of the theory of quantification.

The frontispiece to De Morgan's *Essay on Probabilities and their Application to Life Contingencies* (1838) is a pen-and-ink sketch of one romantic symbol of the era—a woman, seated on a bale of cotton at the dock, pensively glancing out to sea, as a fleet of ships, no doubt carrying "her man," sails out. Another work worthy of examination is De Morgan's (1872) *Budget of Paradoxes* which presents his witty commentaries on a wide variety of zany contributions to science, mathematics, logic, and social issues by various writers of the day. The

*Budget* is a stellar exhibition of educated wit in a literary style more akin to Jonathan Swift and Alexander Pope than to the academic commentaries on such matters in our own day. De Morgan was a major figure in the interstice between classical and modern scholarship. His popular lectures stimulated many well-known students, including Isaac Todhunter and W. Stanley Jevons, an important economist and logician who produced his own interesting tome on *Principles of Science* (1874) in a subjectivist framework. De Morgan also encouraged George Boole (see the *Boole–De Morgan Correspondence*, 1983), whose inclinations in *Laws of Thought* (1854a) were more objectivistic in spirit, based on the logical idea that probability should equal a frequency. Boole resisted particularly the principle of indifference, which he considered to be part of subjectivism. Nonetheless, his contributions (Boole, 1854b) to the algebra of probabilistic logic foreshadow the most important results of the operational subjective theory of probability, which we shall study under the guise of the fundamental theorem of prevision in Chapters 2 and 3.

De Morgan's work even precipitated a brilliant reaction in the treatise of John Venn, who could not agree with him at all. Venn raised a challenge that would come to dominate much of twentieth-century objectivist thought on probability. His prefatory comments to *The Logic of Chance* (1866) are worth quoting:

> ... Professor de Morgan's (sic) *Formal Logic*. He has there given an investigation into the foundations of Probability as conceived by him, and nothing can be more complete and precise than his statement of principles, and his deductions from them. If I could at all agree with these principles there would have been no necessity for the following essay, as I could not hope to add anything to their foundation, and should be far indeed from rivalling his lucid statement of them. But in his scheme Probability is regarded very much from the Conceptualist point of view; as stated in the preface, he considers that Probability is concerned with formal inferences in which the premises are entertained with a conviction short of absolute certainty. With this view I cannot agree.
>
> ... With what may be called the Material view of Logic as opposed to the Formal or Conceptualist,—I am in entire accordance. Of the province of Logic, regarded from this point of view, and under its widest aspect, Probability may, in my opinion, be considered to be a portion. The principle objects of this Essay are to ascertain how great a portion it comprises, where we are to draw the boundary between it and the contiguous branches of the general science of evidence, what are the ultimate foundations upon which its rules rest, what the nature of the evidence they are capable of affording, and to what class of subjects they may most fitly be applied. That the science of Probability, on this view of it, contains something more important than the results of a system of mathematical assumptions, is obvious. Venn (1866, pp. ix–xi)

Venn's conclusion was to develop the theory of probability as an empirical doctrine concerning rules governing the frequency with which irregular events occur. Mathematically he was interested in the limit properties of the proportion of 1's occurring in those infinite sequences of 0's and 1's that cannot be generated according to a specifiable determined rule. In his view, probability is a notion that is applicable not to individual events but only to sequences of repetitions of random events. Such sequences of events were later referred to as a "collective" by Richard von Mises (1928) and Jean Ville (1939), who developed related arguments. Venn's critical chapters in which he rejects the view of probability as the logic of uncertain beliefs are very much worth reading today. In his view, subjective probability assessment would founder on the influence that emotion and desires hold over human judgments. The objections which he identifies beautifully still rank among the serious difficulties with applying the operational subjective methods. We hope that they can be surmounted by practice.

Influential as De Morgan was in mid-nineteenth-century English mathematics, the sway of international opinion was moving away from subjectivist ideas. In France, the writings of Siméon-Denis Poisson, who largely followed in the tradition of Condorcet and Laplace, recognized explicitly the distinction between the objective and subjective senses of probability. But his own application of subjective probability to judicial judgments drew an onslaught of criticism from colleagues who ridiculed its validity, scorning the supposed logic as "the real opprobrium of mathematics" in the words of John Stuart Mill (1843). Although Mill eventually saw through the distracting difficulties, siding with the subjectivists in later editions of his *Logic*, it was Poisson's earlier results concerning "the laws of large numbers," and even his coining of the phrase, that attracted the greater further interest. By and large, the fascination of the second half-century was with the "stability of frequencies" in nature which were felt to provide the empirical basis for probability as an inherent tendency or propensity in nature for particular events to occur.

A recent systematic program of philosophical, historical, and sociological research has provided wonderful documentation and analysis of the submergence of subjectivist ideas in favor of the objective characterization of probability and statistics from 1830 through 1950. The very titles of Hacking's (1990) *The Taming of Chance* and *The Empire of Chance* by Gigerenzer et al. (1989) suggest the stark contrast in conception of probability to what had been studied classically under the guise of *Ars Conjectandi*. Eminently readable, and too rich to suffer my paraphrase, these works place in broad social context the attractive statistical and mathematical reasonings developed by Antoine-Augustine Cournot, Adolph Quetelet, Francis Galton, James Clerk Maxwell, Ludwig Boltzmann, Charles Peirce, and Richard von Mises. Indeed, the depth of the placement of formal mathematical ideas in these collections within the practical, social, literary, and physical scientific concerns of the time is a great tribute to the efforts and the value of serious research into the history of ideas. Recognition of this value had supported a very productive international research program, organized in Germany during 1981–1984, ostensibly to discern the evidence for describing the change in attitude toward probability as *The Probabilistic Revolution*. The intriguing results of the sustained interdisciplinary collaboration among some 30 scholars are collected in still two more volumes under this title, edited by Krüger et al. (1987).

The overriding theme of these directed studies, with only minor variation, is that concomitant developments in probability theory and myriad applied scientific fields led to the same conclusion: uncertainty of an individual's knowledge is insufficient to characterize what we have supposedly learned about nature. For nature is riddled with an objective indeterminacy within in it, governed by mathematically describable laws of chance. An emerging interest in systematic scientific social theory featured as strongly in objectivist developments as did advances in the physical theories of mass phenomena such as the gas laws and thermodynamics. Adolphe Quetelet's (1835, 1869) influential essays on "social physics" both stimulated and benefited from statistical theory along the lines described in Cournot's (1843) objectivist theory of probability. Quetelet argued that social and moral laws could be determined within a social system in which randomness prevails, mainly on account of the laws of large numbers which were being refined. It was his support for statistical moral laws that attracted utilitarian social reformers such as Florence Nightingale to his work (see Diamond and Stone, 1981; Eyler, 1979). Moreover, Quetelet's recognition of statistical regularities of mass phenomena stimulated the imaginations of Maxwell and Boltzmann (Gigerenzer et al., 1989).

Influential nineteenth-century thinkers who contested a purely statistical methodology for social science included Auguste Comte and Frédéric Le Play, but the content of their criticisms was quite different. The latter's imagination and intriguing outlook is described well by Hacking. Its importance for subjectivist statistical thinking is that Le Play focused on the detailed study of distinct individuals, reasoning that while the representation of social behavior as randomly generated variations in individual behavior from *l'homme moyen* may prove useful for social administration, such a conception could never form a basis for the scientific analysis of social life. Le Play was not averse to collecting statistics. His many monographs representing contemporary experience of French life were organized around the household budget as a framework for describing the social life experience of distinct families. Nonetheless he insisted that any component statistic is not to be understood as a mere deviation from the true mean. The numerical summaries of individuals' lives can indeed, being numbers, be added or otherwise manipulated. But that is not necessarily an informative thing to do. Le Play insisted that a scientist's judgment, impartiality, and knowledge of subject matter are essential to any statistical technique that will advance our knowledge of society. In this regard, his views were quite in keeping with the attitude we develop in the present text. It is an analyst's (or a scientific group's shared) judgment to regard a sequence of numerical summaries exchangeably, for example, that provides the logical basis for regarding a sum of households' measurements to be informative about the life experience of another household that has not been individually studied. The translation of Le Play (1982) and introductory essay by Catherine Silver are well worth your reading.

The criticisms by August Comte, on the other hand, allowed no scope to either frequentist or subjectivist motivations for statistical activity. Hacking focuses on Comte's contempt for the notion that nondeterministic probabilistic laws might govern society, exhibiting themselves by statistical regularities. But neither did Comte imagine any useful scope for the use of probability as a representation of uncertainty. In his influential *Cours de Philosophie Positive* he ridiculed the ideas of Bernoulli and Laplace as an "absurd doctrine."

> Would it be possible, indeed, to imagine a conception more radically irrational that that which consists in giving to the whole of social science, as philosophical basis... an alleged mathematical theory in which... we try to subject to calculation the necessarily sophistic notion of numerical probability, which leads directly to giving our own real ignorance as the natural measure of the degree of likelihood of our various opinions? (quoted by Cohen in Krüger et al., 1987)

Comte was stuck into a recognizably absurd program of his own, pontificating about a method that would lead to certainty about the course of history.

But objectors notwithstanding, the historical domination of the objectivist statistical viewpoint during the century from 1850 to 1950 cannot be contested. Even Le Play's organizing scheme of household budget statistics was appropriated by Ernst Engels, the director of the Prussian statistical office, for use as an objective measure of the prosperity of a population. Today this practice has become commonplace in virtually every country of the world and is now codified in terms of the United Nations System of National Accounts. In recent years the basis for these "objective measures" of prosperity and their use has been appropriately questioned (Waring, 1988, 1989; Daly and Cobb, 1989).

The historical account of Maistrov (1974) describes several developments in Russia during this era that are missed in these volumes. Especially of interest are the early nineteenth-century ideas of V. Ya. Bunyakovskii and M. V. Ostrogradskii, the latter most clearly a subjectivist. While their subjectivist tendencies are chided by Maistrov, their applications of probability to real problems are lauded. His criticism of subjectivism is in keeping with the misplaced modern Soviet concerns that attempted to identify unobservable probabilities as a real feature of matter, which we have already mentioned. Nonetheless, Maistrov's exposition of the materialist ideas of Chebyshev near the end of the century, and of Markov and Lyapunov through the revolutionary era, is most informative.

At least as a serious undercurrent, the subjectivist perspective did continue to develop among certain logicians in Germany and among applied statisticians in Belgium and France, notably within the military. Porter (1986) notes the work of German logicians von Kries (1886) and Sigwart (1895), who, along with the statistician Czuber (1910, 1914), were studied by Keynes. The use of textbooks in the tradition of Laplace, like Czuber's, may finally have been in decline, as argued by Kamlah (1987). But at least in some places the subjectivist tradition flourished throughout the century. It was not confined to philosophical dreamers either. The Belgian Captain J. B. J. Liagre, an "old boy" of l'Ecole Militaire Belgique, and permanent secretary of the Belgian Royal Academy of Science, systematically recognized probability as a representation of uncertain knowledge, even in applications of the type that lead objectivists to the unnecessary supposition of physical probabilities. Liagre's now little-known probability text (1852, revised in 1876) would stand well as a present-day text on

Bayesian statistics, using Bayes' theorem regularly in applications to geodesy and ballistics.

In France, the subjectivist tradition was championed in the military academies, too, by the famous General J. E. Estienne, chef d'escadron d'artillerie. In a series of four scintilating articles in *Révue d'Artillerie* (1903, 1904), he presented a formulation of probability and expectation as a linear operator using a style of argument that today reads as a precursing of ideas and lines of development that were formalized completely in the work of de Finetti. Outstanding among his insights were the explicit use of an individual's assessed betting price (*la cote*, or "bid") as the *definition* of a subjective probablity; the recognition of *le bon sens* to avoid contradictions in the prices so offered as providing the mathematical basis for the rules of probability, both the linearity of expectation and the product rule for conditional probabilities; and the regular use he makes of Bayes' theorem in solving decision problems such as the acceptance or rejection of a lot of 20,000 fuses on the basis of testing a sample of them. Another avid subjectivist, Estienne beautifully decried the statistical estimation methodology that had developed from the tradition of Quetelet and which continues through the parameter estimators of today:

> ...and when someone pretends to define by their experimental mean value certain constants, of which one considers rightly or wrongly the apparent variations as results of accidental errors of nature, we regard that one commits an error more or less important in practice, depending on whether the apparent variations are more or less extensive, but an error always real from the point of view of logic.
>
> Estienne (1904, p. 18)

Estienne regarded the theory of errors as nothing more than a branch of the art of conjecture, recognizing the role of judgment and prior opinion in dealing with any particular problem. An historically amusing sidelight occurs in the second part of his four-part essay, in which he was attracted to study and to achieve a result regarding the strategy of "bold play" in the game of roulette : "If one proposes to win a sum $a$ from the initial holding of capital $s$, one cannot do better than to wager the entire capital on a single throw." This very same result subsequently intrigued Dubins and Savage (1965), who studied it in more general contexts.

Ten years later, General Estienne was preoccupied with a different form of savagery, and was contributing no further to subjective probability in academic articles. But the subjectivist statistical tradition continued decisively at l'Ecole Militaire de Belgique during some 30 years of lecturing by Colonel Pierre Van Deuren, who was also a member of the Royal Association of Belgian Actuaries. His *Lecons sur le Calcul des Probabilités* was published only in 1934, though Van Deuren noted that the text had been long evolving as his lecture notes in the same classical (subjectivist) style. To a present-day subjectivist, it is wonderful to read his frank and explicit recognitions that the same event can be well known to someone at the same time it is uncertain for someone else; that the statistics deemed to be informative about the occurrence of an event may be judged differently by one person and by another, and from one moment to the next; and thus, since they depend on these statistics, people's probabilities can be different too.

Van Deuren's text was perhaps the last subjectivist attempt at systematization in the pre-Kolmogorov era, and it is well worth examining on that very count. Events are not characterized as sets (nor as numbers) but rather awkwardly as "facts that are susceptible to happen." While some of his distinctions and argumentation appear unduly verbose to a present-day mathematician, Van Deuren showed great clarity of perception, making important distinctions that were subsequently lost on most objectivist probabilists and statisticians during the next 30 years. Prime among these is the very important distinction between the independence of events and the conditional independence of events, a distinction which is now known to be central to the concept of judging events exchangeably. Although the concept of exchangeability had already been aired in the avant-garde studies of Jules Haag (1924) and de Finetti (1931), Van Deuren does not seem to have been aware of it, per se. Nonetheless, he did use the concept of conditional independence correctly in a very precise manner in deriving the posterior mixing distribution for a frequency parameter and in deriving the predictive distribution for another event that is conditionally independent of some given events. Equally astonishingly *au current* today was his willingness to leave probabilistic solutions to well-posed problems as intervals when the symmetries that someone regards relevant to a problem do not require a unique numerical answer. Van Deuren recognized that conditioning on specific values of other variables can sometimes only reduce the scope of the indeterminacy of a probability rather than requiring a precise value. In this regard he refuted Joseph Bertrand and Henri Poincaré, who had both criticized probability theory on this count.

As this discussion has now settled into our own century, we should remark especially on the stellar scholarship displayed in John Maynard Keynes' *Treatise on Probability*. Though published in 1921, he had written it largely during the year 1906. It is worthy of examination by every scholar. Keynes developed the viewpoint that the probability function $P(\mathrm{A}|B)$ represents the logical relation that the truth of proposition $B$ evidences for the truth of $A$. For too many years, I had not read Keynes myself because I had first read the brilliant, challenging, and humorous commentaries on Keynes' work by Frank Ramsey (1926) and Emile Borel (1924), both of which were reprinted in the important collection of papers, *Studies in Subjective Probability*, edited and translated by Kyburg and Smokler (1964). Disagreement with his viewpoint notwithstanding, Keynes' *Treatise* is a truly masterful development of an argument, all the while keeping in view and commenting on the works of nearly every nineteenth-century scholar writing on probability in English, French, German, or Italian. Outstanding among Keynes' positive contributions were his critique of the principle of indifference and his recognition of the role that interval probabilities can play in a logical theory of probability.

But concern with the meaning of probability was not to motivate the most influential developments of the first half of this century. Early on there was vocalized frustration with the long-running argumentation over methodology, appropriate application, and the meaning of probability. Moreover, unexplained paradoxes emerged in which several different numerical answers could be derived for seemingly well-posed problems in probability (such as Bertrand's paradox, nicely presented by Gnedenko, 1964), which led Bertrand to suggest that probability theory should not extend to problems involving infinite domains. Finally, there was professional distaste for varieties of raving gibberish that often passed for serious applications of probabilistic reasoning in public discussion. In this setting, the focus of mathematicians became centered on specifying a consistent formal axiom system that would allow at least a modicum of agreement on what constituted the mathematical substance of probability theory. In fact, David Hilbert's list of 23 problems meriting resolution, ceremoniously presented to the Second International Congress of Mathematicians in Paris, 1900, included as number 6 the project of treating in an axiomatic manner those sciences in which mathematics plays an important role, probability and mechanics ranking first among them. Ivo Schneider (1987) discusses the historical developments that led to Hilbert's regarding probability as a "natural science," providing the context in which the sixth problem was posed in this way.

The interest at the turn of the century in a systematic axiomatization of probability was part of a larger intellectual thrust among logicians, intent on the project of generating a complete and consistent axiomatization of mathematics more generally. The desire for completeness and consistency was in a way a reaction against the dangers recognized in subjectivism and personal opinion. In particular, Gottlob Frege, one of the leading logicians of Europe and a central figure in the axiomatic program, had bitterly contested the tendency toward what he called "psychologism" in the field of mathematical logic. His introduction to *The Basic Laws of Arithmetic* (1893) provides instructive reading, though Resnik's commentary (1980) qualifies some of Frege's sting. In Chapter 2 we shall recognize Frege's positive role in identifying the personal assertion of truth as a category for logical consideration. But the upshot of his views against psychologism was largely to deter philosophical logical inquiry from the subjectivist outlook for several decades in favor of formalist axiomatics. Isaac Levi (1980) refers somewhat facetiously to his influence as "the curse of Frege."

Thus, in the context of the widespread interest in mathematical systematization, it is not so surprising that A. N. Kolmogorov's (1933) formalist axiomatic characterization of *The Foundations of Probability* as a special type of measure space was to become so influential. For it seemed to provide a consistent framework, at least, in which can be identified those propositions regarding probability that follow logically from the presumed truth of others (axioms), taken as given, without regard to any meaning associated with the objects and syntax in the theory. While Kolmogorov's axioms were not complete in the technical sense of providing a unique resolution of every well-posed problem in probability, they could resolve paradoxes such as Bertrand's by pointing to this incompleteness. His seemingly contradictory solutions to the same problem result from an inadequate specification of the procedure for choosing a chord from a circle "at random." The choice procedures that yield Bertrand's several solutions are recognizably not equivalent.

In retrospect, however, and even at the time it was presented, Kolmogorov's axiomatization was not at all stunning. The mathematical formulation of measure theory (additive set functions), the context in which the axioms were stated, had been developing for at least 20 years, largely through the stimulation of the French analysts Emile Borel and Maurice Fréchet, whose work he mentions in the text. Kolmogorov believed that the central distinguishing feature of probability relative to measure theory was the formalization and use of the concept of stochastic independence of events. For even when they did not presume complete independence, other investigations of the time introduced analogous, if weaker, conditions to obtain substantive results. But Kolmogorov deliberately sidestepped the problem of "making precise the premises which would make it possible to regard any given real events as independent," focusing rather on the formal mechanics implied by his *definitions* of independence and of conditional probability.

What is startling is the extent to which Kolmogorov's axioms have become so widely recognized as *the* axioms of mathematical probability, and to which the technical formulation of problems in probability and statistics has become based upon them so universally. When recourse to the meaning and application of probabilistic concepts has been required, the most widely used texts have long resorted to a motivation in line with Kolmogorov's half-hearted claims about the frequency of occurence of an event in a long sequence of independent repetitions of the complex of conditions in which it might or might not occur. Arguments regarding the foundations and meaning of probability did continue regularly at international conferences on the philosophy and the mathematical formulation of probability, in which de Finetti participated throughout the 1930s and thereafter. But concern with foundational debates had become limited to the few, as interest focused on numerical methods for estimating probabilities in practical situations, now that abstract axiomatization had somehow tamed them.

A surprising feature of the rapid acceptance of Kolmogorov's formalism as a basis for applied work is that de Finetti was espousing a serious alternative which seems not to have been sufficiently discussed nor understood, even as a formalism. Born in 1906, de Finetti was already publishing works on probability by the mid-1920s, and had come to grips with the important concept of exchangeability and the role it plays in the subjectivist understanding of inference by the early 1930s. His ideas first came to international attention in a series of articles in which he argued with the eminent Maurice Fréchet regarding the analytical status of events assessed with probability zero, and the unneccessary limitations imposed on probability theory by the presumption of countable additivity of a probability measure when applied to infinite spaces of events. Intrigued by their exchange, Fréchet invited de Finetti to present a series of six lectures at the Institute Henri Poincaré in Paris in 1935. These lectures were published (de Finetti, 1937) in the now famous exposition entitled "Prevision: Its Logical Laws, Its Subjective Sources." The lectures provided not only the mathematical representation of the judgment to regard a sequence of events exchangeably, for which they are best known, but also a different axiomatic characterization of probability as a measure of personal belief based on the operational concept of coherency of opinions. Furthermore, they developed an operationally meaningful concept of conditional probability, and derived as a theorem the condition for its coherent assertion.

Of course, the Second World War intervened by the end of that decade, involving consequences far more serious than a disruption in academic exchanges among the major discussants of issues in probability, residing as they did in the U.S.S.R., Italy, France, and England. Nonetheless, the conclusion to the purely formalist discussions has been rather disappointing. In a brilliant article finally published in 1949, and appearing in English only in 1972, de Finetti addressed explicitly and exclusively the formalistic axiomatization of Kolmogorov. He critically analyzed the entire development, wondering aloud whether various aspects of the structure are necessary, or merely convenient, or even desirable. The investigation concerned not only the probability function and its properties but the algebra of events which constitutes its domain. Merely on formalistic grounds, but surely with an eye to the purpose and applicability of the axiomatization, it is found that Kolmogorov's axioms are overwrought with restrictions that unduly limit the applicability and disguise the scope of probability mathematics. The paper is easily available, but it has had little influence on the development of probability theory among those bent on formalism, nor on mathematical statistics among those bent on techniques for estimating probabilities. (This paper is written for mathematically sophisticated readers who are already familiar with Kolmogorov's axiomatization and its implications. The most important ideas in it are expressed more informally in the introductory chapter of de Finetti's *Theory of Probability*.)

Syntactically, a central difference between the two axiomatizations is that Kolmogorov identifies probability as a measure on a space of sets that provides a reference for a companion theory of random variables, their distributions and expectations, whereas de Finetti provides a unified theory of probability and expectation, termed prevision, which is characterized directly as a linear operator on functions of quantities (measurements). Of course, the semantic differences are much deeper than this.

Ironically, a central figure in shifting the focus of attention to applications was Emile Borel, a leading mathematician whose work in analysis had broken new ground in the abstract understanding of measure theory and the continuum. He did not see the axiomatization of probability as being of such central importance. While he had no time for charlatans or shoddy analysis, Borel was much more concerned with the development of fruitful applications of probability in science than he was with formalities of axiomatization per se. From 1924 to 1938, Borel directed the publication of 18 works, collected in four large volumes as *Traité du Calcul des Probabilités et de Ses Applications*, written by several esteemed collaborators, including himself. Recognized as a milestone in the history of science in its own day, the treatise developed the mathematics of probability, random variables, least squares, and general statistical procedures in particular application to classical mechanics, astronomy, quantum mechanics, life insurance, finance, health insurance, fertility, demography, ballistics, and games of chance.

In the final volume, Borel (1939) reflected in his inimitable style on the practical value and philosophy of probability. The most important ideas are reviewed in English by Eberhard Knobloch (1987). Although Borel and de Finetti surely did disagree on specific important issues, their basic agreement that probabilities represent human judgments, and that individuals' probability assertions should be identifiable from their willingness to act upon them is evident on Borel's part as early as his critical commentary on the work of Keynes (Borel, 1924). An important conclusion to Borel's wide-ranging considerations was that "the probability of an isolated event is the foundation of the calculus of probabilities. The notion is natural to each of us, just as the notions of hot and cold; but experience and reflection permits us to make it precise and to obtain close enough evaluations of probabilities in the same way that we can evaluate well enough the temperature of the air around us without using a thermometer." Borel referred to de Finetti's lectures (1937) for the specification of conditions under which the personal judgments so expressed are acceptable.

Nevertheless, in the volumes of Borel's edited treatise, probabilities are discussed in the syntax of sets and measures, and the applications are presented as problems of probability estimation. By mid-century, the influence of the subjectivist tradition was at low ebb. It had virtually died in France, due to the rising fascination of mathematicians with abstract formalism and the popular objectivist arguments of Maurice Fréchet, which we shall study in Chapter 6. Subjectivism never really died in England, even though the rising objectivist spirit surely began to hold sway. Harold Jeffreys' (1939) *Theory of Probability* developed extensively the statistical implications of the logical ideas of W. E. Johnson and Keynes, and has had a wide readership throughout the century in two subsequent editions and a reprint. Moreover, the utilitarian subjective tradition enjoyed brilliant exposition in the book of I. J. Good (1950), *Probability and the Weighing of Evidence*, and eventually the influential volumes of Lindley (1965) entitled *Probability and Statistics from a Bayesian Viewpoint*. But since 1940 the framework of mainstream statistical theory and practice has been aligned not with this British tradition but with the objectivist tradition largely associated with the work of Jerzy Neyman, based in Berkeley, and Egon Pearson in London. The foundational outlook in the influential work of Ronald Fisher is more difficult to characterize simply. See, for example, the essays of Savage (1976) and Lane (1980).

Indeed, the logic of "confidence interval estimates of probabilities" and efficient procedures of hypothesis testing, which frame the contemporary understanding of statistics presented in mass education, stem from this era. Although controversial in Europe when first proposed, these procedures of the "British-American school" quickly won the day, unifying the mainstream outlook of probability theory and practice. In a word, probability is presumed to exist in the world, generating historical happenstance according to the laws of chance, ordered by the laws of large numbers, obeying the axiomatic properties of Kolmogorov, and estimable for use in science and commerce via efficient and consistent computable procedures.

Already by 1954, Savage had to appeal for renewed attention to foundational questions by admitting to the generally accepted "success of the British-American school" and promising to provide a better foundation for these very practices and this formalism. During the subsequent 40 years, virtually every technical presentation of probability theory other than de Finetti's, even the Bayesian ones, has accepted the burden of showing that the ideas expressed are consistent with Kolmogorov's axiomatization. DeGroot (1970, Chap. 6) provides one such beautiful exposition. Whatever one's view about the meaning and application of probability, one is not expected to bat an eye in reading that probability is not understandable to anyone without a serious understanding of measure theory, or that probability is but a part of measure theory. For an example, see Fortet's preface to the text of Neveu (1970).

The remaining chapters of this book develop the theory and practice of operational subjective probability assessment, largely as de Finetti conceived it. Some minor new developments and some simplifications to the completely finite context differ from his treatment. Any real problem of probability, involving observable quantities that you know something about, yet are uncertain about and would like to learn more about, can be addressed and resolved using procedures that we shall study. Yet we have no reason to study any aspect of measure theory. Moreover, the applicability of the measure-theoretic conception of probability to any real problem is nil. It is regularly required that you conveniently forget or ignore all the fine distinctions that measure theory requires in order to apply the theory of probability in any real instance.

The subjectivist rejection of the measure-theoretic characterization of probability has not been flippant. De Finetti's extensive publications and lectures at international meetings constitute a monument to the attempt at dialogue on the issues. Repeated experiences of either being ignored or of being pompously criticized on the flimsiest of grounds were bound to take a toll in some way, resulting in a sorrowful protestation in the *Theory of Probability* that "a dialogue among the deaf is not a discussion." By the time this two-volume work was completed, de Finetti was approaching the teasingly jocular retirement age, but still intent on spreading his message that had largely slipped through the century unattended. A reader would do well to imagine the twinkle in his eye that was surely present when he wrote, for example, of the "Procrustean bed" of probability spaces of elementary outcomes.

In the chapters that follow, I develop the theory and application of the operational subjective view of probability in a sophisticated yet practical fashion that is meant to be accessible to anyone who has completed at least the first year of university-level mathematics, and in some places more specifically to a graduate student. You may need to rely on your teacher to help you read at this level if you have only recently been introduced to linear algebra and calculus. But all in all, I expect you to find little difficulty with the relevant mathematical technique, which is simple. If you find yourself struggling, I believe it will be on account of your coming to grips with the reality that you are uncertain about so many real empirical questions that bother you. Perhaps you might benefit from de Finetti's mirth in realizing there is no alternative:

> There is no way, however, in which the individual can avoid the burden of responsibility for his own evaluations. The key cannot be found that will unlock the enchanted garden wherein, among the fairy-rings and the shrubs of magic wands, beneath the trees laden with monads and noumena, blossom forth the flowers of *Probabilitas realis*. With these fabulous blooms safely in our button-holes we would be spared the necessity of forming opinions, and the heavy loads we bear upon our necks would be rendered superfluous once and for all.
> de Finetti (1974, Vol. 2, p. 42)

With this quotation we conclude this introductory history of the emergence of the operational subjective statistical theory. It has meant to display the fact that the tradition of subjectivist statistical thinking is a long and continuous one since the European renaissance. Aspects of more recent history that are relevant to our studies shall merit comment in historical remarks interspersed throughout the text. Historical surveys of developments on many topics in the last half-century are available in the popular journal of recent origin, *Decision Sciences*, and in writings of von Plato (1994) and Regazzini (1987). The recently collated *Encyclopedia of the Statistical Sciences*, edited by Johnson and Kotz (1982–1989), contains a wealth of modern historical information as well. Furthermore, the comprehensive bibliographical search facilities in the online electronic compilation of the *Current Index to Statistics* allows for easy tracing of topics over the past 25 years of statistical publications.

CHAPTER 2

# Quantities, Prevision, and Coherency

*Finding himself now alone, with nothing in particular to do, Watt put his forefinger in his nose, first in one nostril, and then in the other. But there were no crusts in Watt's nose tonight.*

*Watt*, SAMUEL BECKETT

## 2.1 QUANTITIES AND THEIR REALMS

What's happening? This is the all-inclusive scientific question. At every level of nature as we experience it, we ask—at the atomic level, at the molecular level, at the inorganic level, at the organic level, at the cellular level, at the individual level, at the societal level of plants and animals, at the level of ecological systems, at the global level, at the cosmic level.

We do our best to answer. We take a look, ... and listen, smell, touch, and taste. We think about it, and talk it over.

What's happening in the past, in the present, in the future?

The *initial records* of our investigations of most any topic are kept as field notebooks. The naturalist, the explorer, the experimental scientist, the journalist, anthropologist, sociologist, economist, the diarist of dreams and daydreams, the historian, each has kept one.

A field notebook: Unfettered by any restriction on the form or the content of the remarks, it is a free-wheeling commentary and observational identification of some aspect of what's happening. To the reader, it is not only a recording by someone of what has happened in particular, but it may provide an intimation as well of what other things may have been or may be happening. More specifically, it could provide an intimation of what may happen to you if you would do such and such right now! An intimation, not a declaration.

As more of us extend our experience deeply into any particular level of nature, we come to agree upon operational definitions for measurements of our activities therein. By sharing our knowledge of these measurements, we can communicate simply with each other in terms of concise summaries of our experience. Of course much detail and many nuances may be lost in the translation of a field notebook onto a data codebook. At the very least, we must recognize that reverse translation is impossible. History merely reduces "onto" our measurements of it. The fullness of what's happening seems hardly to be limited by our ability to measure it.

Nonetheless, a great deal of summary information about what's happening can be stored in numerical values of operationally defined measurements. Likewise, our personal scientific knowledge about what's happening can be summarized by our knowledge of the values of some measurements.

***Definition 2.1.*** An *operationally defined measurement* is a specified procedure of action which, when followed, yields a number. The number yielded by the performance of an operationally defined measurement is called a *quantity*. □

***Example 1.*** "Ride this bicycle around this track as quickly as you can, and observe the number (of tenths of seconds) recorded as the time elapsed on this digital timing device." This specific example has been the source of objections by some critics of "operationalism" in science. How can we specify operationally how to ride the bicycle as quickly as you can? After you have completed your attempt, someone might say, "You could have ridden more quickly if you had started more slowly, saving more of your energy for the backstretch when you were riding into the wind!" Moreover, some able-bodied people, perhaps you, cannot even ride a bicycle. Evidently we are incapable of specifying an operational procedure for riding a bicycle as quickly as you can. Otherwise these people would presumably be able to ride a bicycle merely by following the procedural instructions. This criticism carries too far our intention in insisting on an operational methodology. In riding quickly, you should just do the best you can. If you cannot ride a bicycle, we would have to say that you are incapable of making such a measurement unless and until you learn. As of now, your performance of the specified procedure is just not happening. □

***Example 2.*** "Drive a boat along the coast at a cruising speed, keeping about 1 kilometer off the shore. Keep a watch on the stretches of water 400 meters on either side of the boat. If you sight dolphins within this region, approach and move through them slowly, observing their movements and behavior as they engage you for a minute or two. Count the number of dolphins who are swimming in this group, and then move away at an increasing speed to break from them. Reduce to cruising speed and continue along the coastline." This is an operational definition of a quantity, the number of animals counted in the encounter. Riding away from one such encounter, an experienced "spotter" questions the experienced boat driver, "Were there five or six in that group?" When the question cannot be resolved, the quantity is recorded as "5 or 6." Fair enough. The operational definition could not be specific enough to explain to you just precisely how to perform a definitive counting. Nonetheless, the recorded quantity amounts to an informative summary of the field encounter. □

Each quantity is represented by a name which refers to the historical description of the procedure that generates it. "The quantity that resulted from my performing the operationally defined measurement specified in Example 1, on 23 June 1986" equals 326. The nominative clause of the previous sentence, enclosed by quotation marks, is the name of a specific quantity. This quantity happens to equal 326. We represent the names of quantities generically by capital letters such as $X$, $Y$, $Z$, $A$, $B$, or $E$.

***Example 3.*** We might denote by $X$ the number of pounds of milk output recorded as the average 305-day yield for two-year-old, registered Ayrshire milking cows on official U.S. Department of Agriculture test during 1982. (As of this date, milk yields were officially recorded in units of pounds.) We might denote by $E$ the number 1 if the president of the United States of America on 1 July 2002 is female; otherwise $E$ equals 0. What is the numerical value of $X$? You may not know it, but the numerical value of $X$ is 12,696. You can look it up in the U.S.D.A. official record. In fact I am actually looking at this number right beside me on a computer printout, even as I type. What is the numerical value of $E$? □

In every case of *actual measurement*, the quantity recorded as the result of an operational procedure must be one of only a finite collection of possibilities. According to the operationally defined procedure, there is a smallest recordable value for the measurement, a largest recordable value, and an integer number of distinct recordable values in between them. The creation of a new measuring instrument which identifies finer recognizable distinctions (or even just different distinctions) constitutes the definition of a new procedure, of another quantity. True, we can describe measurement procedures for which we are unable to specify either a maximum or a minimum possible value for the resultant quantities. "Burn this lightbulb under conditions of steady current, and record the number of the subsequent minute in which it burns out. Call this number the quantity $M$." Any positive integer is apparently a possible value for the quantity $M$, and the collection of all positive integers is not finite. The problem with considering this an *operational* definition of a quantity is that if the lightbulb burns long enough, neither you nor anyone else would be able to observe its burning out nor complete the procedure that is supposed to yield a number. After any $T$ minutes, the most we can observe is that the quantity is not less than $T$. In practice, it must be specified (often merely understood) that the measurement procedure should stop at some point, and the recorded result denote either the awaited burnout time or that the awaited happening had not yet occurred when the experiment was terminated.

Although we sometimes consider quantities for which no precise limit on the number of elements in the realm of possible measurements can be stated, we always expressly recognize that the set of all possible recordable values is discrete. With the contemporary ubiquity of digital watches and scales, the discreteness of actual measurements may seem readily apparent. But in the day of sweep-hand indicators for wristwatches and scales, it was more customary for people to think that, at least in principle, we could make continuously graded measurements, while in practice we merely record a finite number of digits up to the point that finer distinctions are inconsequential for whatever purpose we are making the measurement. Our viewpoint is now different. Consider photographs taken with an electron microscope. These show the finest distinctions that we are capable of making under very rarified procedures. They exhibit discrete distinctions. In fact, the photographs are typically constructed using digital recording methods. At macroscopic levels our distinctions are also evidently discrete and typically much more crude in their scale of distinctions.

Thus, the operational definition of a measurement procedure specifies a discrete set of numbers which are possible results of the procedure. This set is called the *realm* of the quantity. In the context described in Example 3, the realm of the quantity $E$ is the set $\{0, 1\}$. What about the realm of $X$? The operational definition of $X$ is a number resulting from testing procedures that are detailed in the U.S. Dairy Herd Improvement program manuals. By now it has been recorded, and is listed in the U.S.D.A. record as "the average 305-day milk yield for Ayrshire cows on official test during 1982." Now that I have told you the value of $X$, you might say that the realm of $X$ is $\{12{,}696\}$. Had I not told you the value of $X$, you might have reasoned that the realm of $X$ is $\{0, 1, \ldots, 610{,}000\}$. For 610,000 pounds equals roughly the total premilking weight of a two-year-old Ayrshire cow, multiplied by 305. Clearly, no milking cow can yield more than its own weight in milk each day. Typically we consider the realm of a quantity to be the set of possible values arising from the performance of its operational definition, irrespective of whether anyone knows its recorded value or not.

This example is poignant, for it highlights the fact that in some sense the concept of the realm of a quantity is relative to one's knowledge about the operational definition that defines the quantity. This qualification has relevance when we speak "roughly" about quantities, as we often do in scientific discussion, referring to quantities in terms of shorthand names that are not fully specific of the measurement process. (In the extreme, the shorthand names are merely the symbols such as $X$, $E$, or $M$.) Throughout this text we presume that scientists who investigate the values of specific quantities know at least enough about the operational definition of each measurement process to know what is the realm of possible values that can result from it. The specification of the realm of a quantity is considered to be included in the longhand statement of its operational definition.

***Definition 2.2.*** The set of all numbers that can possibly result from performing the operational measurement procedure that defines a quantity $X$ is called the *realm* of the quantity, denoted by $\mathcal{R}(X) = \{x_1, x_2, \ldots, x_K\}$. The number of elements in $\mathcal{R}(X)$ is called the *size of the realm* of $X$. □

***Example 4.*** A fine-quality, digital, bathroom scale has been constructed to yield distinct readings in tenths of kilograms between 0.0 and 199.9. Let $X$ be the number that appears when you stand centered on this scale. The realm of $X$ is $\mathscr{R}(X) = \{0.0, 0.1, 0.2, \ldots, 199.8, 199.9\}$. The size of this realm is 2000. □

The simplest type of measurement can yield only two distinct values: 1 or 0. These numerals may identify any dichotomous observation activity such as on or off, north or south, inside or outside, yes or no. The quantity defined by such a measurement is also sometimes called an event.

***Definition 2.3.*** If $E$ is a quantity for which $\mathscr{R}(E) = \{0, 1\}$, then $E$ is also called an *event*. □

Notice that an event $E$ is also still a quantity. All events are quantities, but not all quantities are events. We often use the symbol $X$ to denote a generic quantity, whereas $E$ is used to done to a generic event.

Every event is distinct from every other event. Otherwise they would be the same event. The child of Sushila and Krishna is a girl. The first child of Cynthia and David is a boy. The child of Mirabell and Tom is a boy. We will *not* consider these to be "repeated trials of the same event with distinct outcomes," which is the common language usage in most objectivist theories of probability. These are distinct events, in no way regarded as repetitions of the same event. "No, I cannot find the sponge." "Are there more than seven dead flies on the windowsill?"

Measurement procedures that can yield one of three or more distinct recordable values are only *slightly* more complicated than events. "Milk this cow dry in the morning and evening today and record the total kilograms of milk she yields, in units of tenths (100 grams). Use a scale that can distinguish weights between 0.0 and 99.9 kilograms. Denote the resulting quantity by $M$." Evidently, the realm of $M$ is $\mathscr{R}(M) = \{0.0, 0.1, \ldots, 99.8, 99.9\}$. It turns out that *any such quantity can be expressed as a linear combinations of events.* Let us formulate this statement as a theorem, using in doing so an important notation for events, which will be used throughout this book.

USEFUL EVENT NOTATION. The notation of parentheses around an arithmetical statement, such as $(M = 43.6)$, or generically $(X = x_i)$, is used to denote an *event* defined to equal 1 if the expression is correct and to equal 0 if it is not correct.

***Example 5.*** In the context of the cow milking, $(M = 43.6)$ denotes the event that equals 1 if the milk yield is 43.6 kilograms, and equals 0 if not. Similarly, $(M < 40.0)$ denotes the event that equals 1 if the milk yield is less than 40.0 kg, and equals 0 if not. Since events are numbers, such parenthetical statements can be manipulated arithmetically. The product $(M \geqslant 40.0)(M < 50.0)$ is an event, too, since each of its multiplicands can equal only 0 or 1; thus, the product can equal only 0 or 1. This event can be written equivalently as $(40.0 \leqslant M < 50.0)$. The sum of these same two events, $(M \geqslant 40.0) + (M < 50.0)$, would be a quantity but not an event, for its realm of possibility is $\{1, 2\}$. Do you understand the *syntax* of parentheses around arithmetical statements? □

An additional feature of this notation is that functions that are defined in different ways over distinct subdomains can be written quite efficiently. For example, the function

$$F(x) = \begin{cases} 0 & \text{for } x \leqslant 0, \\ x^2 & \text{for } 0 < x < 1, \\ 1 & \text{for } x \geqslant 1, \end{cases}$$

can be conveniently written as $F(x) = x^2(0 \leqslant x \leqslant 1) + (x > 1)$. We use this parenthetic event notation whenever it is convenient as well as conducive to understanding. The following theorem is an example.

**Theorem 2.1.** Any quantity $X$ with realm $\mathscr{R}(X) = \{x_1, x_2, \ldots, x_K\}$ can be written as a linear combination of events:

$$X = x_1(X = x_1) + x_2(X = x_2) + \cdots + x_K(X = x_K) = \sum_{i=1}^{K} x_i(X = x_i).$$

*Proof.* Since $\mathscr{R}(X)$ contains all the possible values of the quantity $X$, one and only one of the events $(X = x_i)$ equals 1. All the others equal 0. Since this unique event is multiplied in the expression above by $x_i$, the very number that $X$ happens to equal, the entire sum necessarily equals the value of $X$. □

Any numerical-valued function, $G$, whose domain includes the realm of a quantity $X$, can be used to define another distinct quantity, $Y \equiv G(X)$. The operational definition of $Y$ begins with the operational definition of $X$ itself. To this is appended the additional injunction that once the value of $X$ is determined, the quantity $Y$ should then be determined according to the relevant application of the rule prescribed by $G$. That is, $G\colon \mathscr{R}(X) \to \{G(x_1), G(x_2), \ldots, G(x_K)\}$. Obviously, the realm of such a quantity $Y \equiv G(X)$ is the image of $\mathscr{R}(X)$ under $G$. That is, $\mathscr{R}(Y) = \{G(x) | x \in \mathscr{R}(X)\}$. For example, just as the quantity of milk yielded today by this cow is a quantity, $M$, so the logarithm of $M$ is also a quantity. Similarly, $M^2$ and $M^3$ are quantities, too.

In the development of operational subjective statistical methods, we often find ourselves considering quantities that are defined by functions of other quantities. Let us begin our familiarity with such considerations here by stating the following corollary to Theorem 2.1, which you can prove yourself.

**Corollary 2.1.1.** Let $X$ be a quantity with realm $\mathscr{R}(X) = \{x_1, \ldots, x_K\}$, and let $G$ be any numerical-valued function whose domain includes $\mathscr{R}(X)$. Then the quantity

$$Y \equiv G(X) = G(x_1)(X = x_1) + G(x_2)(X = x_2) + \cdots + G(x_K)(X = x_K)$$
$$= \sum_{i=1}^{K} G(x_i)(X = x_i).$$

***Example 6.*** If $\mathscr{R}(X) = \{1, 2, 3, 4, 5\}$, then $Y \equiv X^2 = \sum_{i=1}^{5} i^2(X = i)$, and $\mathscr{R}(Y) = \{1, 4, 9, 16, 25\}$. Similarly, $Z \equiv \log(X) = \sum_{i=1}^{5} \log(i)(X = i)$, and $\mathscr{R}(Z) = \{\log(1), \log(2), \log(3), \log(4), \log(5)\}$. □

For completeness, we should note that some quantities are even simpler than events. These are the results of "operationally defined measurements" that can yield only one number. "Record the number 54," for example, is a procedure that can yield only the number 54. Similarly, for any event $E$, the quantity defined as $F \equiv E(1 - E)$ can equal only 0. For $E$ equals either 0 or 1, and in either case $F$ equals 0. Thus, the quantity $F$ is a constant.

***Definition 2.4.*** If $\mathscr{R}(X)$ has only one member, then $X$ is a *constant*. □

The example of a constant quantity, $F \equiv E(1 - E) = 0$, implies, via simple algebra, that $E^2 = E$. Algebraically, we say that an event $E$ is *idempotent*. This is an important and useful property of events that is worth stating explicitly as a theorem.

**Theorem 2.2.** Every event $E$ is idempotent. That is, $E^2 = E$. □

Generally, the numerical value of any operationally defined measurement is called a quantity, whether it is an event, a constant, or whatever.

As our many examples suggest, in most serious investigations we are interested in the values of several quantities. It is worthwhile from the very start to have a notation with which we can refer to several quantities at the same time. The language of vectors of quantities is most convenient. Throughout this text, we typically define vectors as column vectors, their transposes thus representing row vectors. Explicit notice will be given at the few places where an exception to this convention is made.

***Definition 2.5.*** *A vector of quantities*, $\mathbf{X}_N = (X_1, \ldots, X_N)^{\mathrm{T}}$, is a vector, each of whose components is a quantity. Boldface type distinguishes a vector from one of its components, e.g., the vector $\mathbf{X}_N$ from the single quantity, $X_N$. □

***Definition 2.6.*** The *realm of a vector of quantities*, denoted by $\mathscr{R}(\mathbf{X}_N)$, is the set of all vectors that can possibly result from the performance of all their operational definitions. The number of vector elements in $\mathscr{R}(\mathbf{X}_N)$ is called the *size of the realm*. □

We postpone until later in this chapter an extensive discussion of a very important topic, the logical relations among several quantities. But in order to ensure your familiarity with vector notation and to extend our use of it, let us introduce the following new definitions and notation. Your understanding of these definitions will develop as we work our way through this chapter. They will be used just now in stating a very important multivariate extension of Theorem 2.1, which will conclude the technical content of this section. Some final matters of discussion appropriate to this section will be addressed after you have tried some problems.

***Definition 2.7.*** Consider a vector of quantities $\mathbf{X}_N$ whose realm is denoted by $\mathscr{R}(\mathbf{X}_N) = \{\mathbf{x}_{N_1}, \mathbf{x}_{N_2}, \ldots, \mathbf{x}_{N_K}\}$. The *realm matrix* for $\mathbf{X}_N$, denoted by $\mathbf{R}(\mathbf{X}_N)$, is the $N \times K$ matrix whose columns are the vector elements listed in $\mathscr{R}(\mathbf{X}_N)$, viz.,

$$\mathbf{R}(\mathbf{X}_N) \equiv [\mathbf{x}_{N_1} \quad \mathbf{x}_{N_2} \quad \cdots \quad \mathbf{x}_{N_K}].$$

The related *vector of events constituting the partition generated by* $\mathbf{X}_N$, denoted by $\mathbf{Q}(\mathbf{X}_N)$, is the $K \times 1$ vector of events

$$\mathbf{Q}(\mathbf{X}_N) \equiv [(\mathbf{X}_N = \mathbf{x}_{N_1}), (\mathbf{X}_N = \mathbf{x}_{N_2}), \ldots, (\mathbf{X}_N = \mathbf{x}_{N_K})]^{\mathrm{T}} \qquad \square$$

Again, we defer detailed discussion of partitions to Section 2.8. For now, you ought merely recognize that since $\mathscr{R}(\mathbf{X}_N)$ exhausts the possible values of $\mathbf{X}_N$, one of the components of $\mathbf{Q}(\mathbf{X}_N)$ must equal 1 while all the others equal 0. In general applications, we just don't know which of the components it is that equals 1. Nonetheless, the structures of the vectors and matrices we have defined, denoted by $\mathbf{X}_N$, $\mathbf{R}(\mathbf{X}_N)$, and $\mathbf{Q}(\mathbf{X}_N)$, ensure the following result.

**Theorem 2.3.** Any quantity vector $\mathbf{X}_N$ is representable as $\mathbf{X}_N = \mathbf{R}(\mathbf{X}_N)\mathbf{Q}(\mathbf{X}_N)$.

For whichever event $(\mathbf{X}_N = \mathbf{x}_{N_i})$ it is that equals 1, it is being multiplied on the right-hand side of this equality by the appropriate column $i$ of $\mathbf{R}(\mathbf{X}_N)$ that equals the quantity vector $\mathbf{X}_N$. All other columns of $\mathbf{R}(\mathbf{X}_N)$ that do not equal $\mathbf{X}_N$ are being multiplied by events that happen to equal 0. Thus, the representation holds no matter which event component $(\mathbf{X}_N = \mathbf{x}_{N_i})$ of $\mathbf{Q}(\mathbf{X}_N)$ it is that equals 1.

***Example 7.*** Are home mortgage interest rates high or low? Are they rising or not? Let $X$ denote the interest rate for first home mortgages offered by a certain bank on 1 July 1998, and let $Y$ denote the interest rate offered by the same bank on 1 August 1998. Define two further quantities, $A$ and $B$, by

$$A \equiv -(X < .07) + (X > .08) \qquad \text{and} \qquad B \equiv (Y > X).$$

In words, $A$ is a quantity that equals $-1$ if the interest rate $X$ is less than 7%, equals 0 if $X$ is between 7% and 8% inclusive, and equals $+1$ if $X$ exceeds 8%; $B$

is an event that equals 1 if the interest rate $Y$ exceeds $X$, and equals 0 otherwise. In principle, there is nothing to prevent the offered interest rate $X$ from being any particular number, nor is there anything to prevent the offered interest rate $Y$ from decreasing or increasing relative to $X$. Thus, the realm matrix, $\mathbf{R}$, and the partition vector, $\mathbf{Q}$, associated with the vector of quantities $(A, B)^{\mathrm{T}}$ are

$$\mathbf{R}\begin{pmatrix} A \\ B \end{pmatrix} = \begin{pmatrix} -1 & 0 & 1 & -1 & 0 & 1 \\ 0 & 0 & 0 & 1 & 1 & 1 \end{pmatrix}$$

and

$$\mathbf{Q}\begin{pmatrix} A \\ B \end{pmatrix} = \begin{pmatrix} ((A,B) = (-1,0)) \\ ((A,B) = (\ \ 0,0)) \\ ((A,B) = (\ \ 1,0)) \\ ((A,B) = (-1,1)) \\ ((A,B) = (\ \ 0,1)) \\ ((A,B) = (\ \ 1,1)) \end{pmatrix}.$$

You should notice that $\mathbf{R}$ is a matrix whose components are numbers, whereas $\mathbf{Q}$ is a vector whose components are events. For example, the third-row component of the column vector $\mathbf{Q}$ is the event $((A, B) = (1, 0))$. In the notation we have defined, the outer set of parentheses around the arithmetic statement $(A, B) = (1, 0)$ identifies this component as the event that equals 1 if $(A, B) = (1, 0)$, and equals 0 otherwise. That is, this is the event that equals 1 if the interest rate $X$ turns out to exceed 8% *and* the interest $Y$ does not exceed $X$. Otherwise this event equals 0. Since the six columns of $\mathbf{R}$ exhaust all the possible values for the vector $(A, B)^{\mathrm{T}}$, one and only one of the components of $\mathbf{Q}$ must equal 1, while the remainder equal 0. Thus, the identity specified by Theorem 2.3 must hold, no matter what happens—no matter what are the values of $X$ and $Y$ and of $A$ and $B$.

□

In due time, you will be able to simplify much of the detail of this notation for ease of your expression. But for now, it is well worth your while to pay attention to fine notational distinctions. The following set of problems should help you clarify your understanding.

## PROBLEMS

1. What are the differences among our notations $x_N, X_N, \mathbf{x}_N$, and $\mathbf{X}_N$; between $\mathscr{R}(\mathbf{X}_N)$ and $\mathbf{R}(\mathbf{X}_N)$; between $\mathbf{R}(\mathbf{X}_N)$ and $\mathbf{Q}(\mathbf{X}_N)$; among $\mathscr{R}(X_N)$, $\mathbf{R}(X_N)$, $\mathscr{R}(\mathbf{X}_N)$, and $\mathbf{R}(\mathbf{X}_N)$; between $\mathbf{X}_N$ and $\mathbf{E}_N$?

**2.** Returning to the setup of Example 4, let $X_1$ denote your weight measurement and let $X_2$ denote my weight measurement. Identify precisely the realm of the vector of our two weights, $\mathscr{R}(\mathbf{X}_2)$, the realm matrix, $\mathbf{R}(\mathbf{X}_2)$, and the vector of events constituting the partition generated by $\mathbf{X}_2$, $\mathbf{Q}(\mathbf{X}_2)$.

**3.** Suppose that two quantities have the identical realm, $\mathscr{R}(X_1) = \mathscr{R}(X_2) = \{1, 2, 3\}$ and that any value of $X_2$ is possible with any value of $X_1$. Define the quantity vector $\mathbf{X}_2 = (X_1, X_2)^{\mathrm{T}}$. Write out the realm matrix $\mathbf{R}(\mathbf{X}_2)$. Now define two more quantities as $X_3 \equiv (X_1 + X_2)/2$ and $X_4 \equiv \sum_{i=1}^{2} (X_i - X_3)^2/2$. Append two appropriate rows to $\mathbf{R}(\mathbf{X}_2)$, yielding $\mathbf{R}(\mathbf{X}_4)$.

**4.** Suppose that $\mathbf{E}_3$ is a vector of three events. Write out the realm matrix $\mathbf{R}(\mathbf{E}_3)$ and the partition vector $\mathbf{Q}(\mathbf{E}_3)$, noting how Theorem 2.3 applies to this case. What is $\mathbf{R}(\mathbf{Q}(\mathbf{E}_3))$?

**5.** Notice for yourself that if $N = 1$ the statement in Theorem 2.3 reduces to Theorem 2.1, $X = \mathbf{R}(X)\mathbf{Q}(X)$, using the notation of vector multiplication in place of the summation of products.

**6.** The following computer program in MATLAB syntax will generate the realm matrix $\mathbf{R}(\mathbf{X}_N)$ for $N$ quantities, each with realm $\{0, 1, \ldots, K\}$, where each quantity can be any of its possibilities no matter what the values of the other quantities are.

```
% realm.m A MATLAB program to generate realms of dimension N x (K + 1)^N.
RXN = [ ]; R = [ ]; RRXN = [ ];
for i = 1:N, z = (K + 1) ^ (i - 1); for j = 0:K, R = [R j*ones(1, z)];
                                          RRXN = [RRXN RXN];
                                  end,
        RXN = [RRXN; R], R = [ ]; RRXN = [ ];
   end
```

Enter this into a MATLAB procedure file called "realm.m." Then within MATLAB set $N = 4$, $K = 1$, and enter the command realm. Study the MATLAB output, seeing that it has sequentially generated $\mathbf{R}(\mathbf{E}_1)$, $\mathbf{R}(\mathbf{E}_2)$, $\mathbf{R}(\mathbf{E}_3)$, and $\mathbf{R}(\mathbf{E}_4)$. Using the help facility, study the syntax of any part of the code that you do not understand. Now clear the matrix $RXN$, reset $K = 3$, and enter the command realm again. The program will sequentially generate the realm matrices $\mathbf{R}(\mathbf{X}_1)$, $\mathbf{R}(\mathbf{X}_2)$, $\mathbf{R}(\mathbf{X}_3)$, and $\mathbf{R}(\mathbf{X}_4)$ with each component allowed to be 1, 2, or 3. But the realm matrix has grown so large that it cannot be seen on a single screen. So enter the command RXN′, asking for the transpose of $RXN$. You will see the realm $\mathbf{R}(\mathbf{X}_4)^{\mathrm{T}}$ more easily. Can you imagine how this realm program could be used to begin your generation of $\mathbf{R}(\mathbf{X}_4)$ in problem 3? (Set $N = 2$ and $K = 2$, run realm, and add the matrix "ones(2, 9)" to the resulting RXN.) How could you use MATLAB to generate the next two rows of $\mathbf{R}(\mathbf{X}_4)$?

Although we insist on operational definitions for measurements, it is necessary to recognize that our capacity for operational specification is limited. Moreover, we usually stop trying well before we reach our capacity. Consider the following operational definition of an event $F$: "Count the number of dead flies on the windowsill and define the quantity $F$ to equal 1 if there are more than seven dead flies, and define $F$ to equal 0 if not." Now suppose you examine the windowsill and you find on it six complete dead flies with body, legs, and wings intact, one dead fly missing a left wing, and two single right fly wings along with plenty of fly particle dust. Do you record 1 or 0 for the event in question? In the context of what you want to know about what's happening, reasonable arguments could be made for having specified one value or the other in your operational definition. But not having decided beforehand, you, as the measurement technician, are in a quandary as to what number is yielded by performing the "specified operational procedure" of counting the dead flies on the windowsill. There is an ever-present incompleteness to the operational specifications defining any measurement procedure. In defining, performing, and reporting measurements, we can only do the best we can, and report any perceived difficulties as qualifications to the reported list of measured quantities. The activities of experimentation, fieldwork, enumeration, and statistical accounting always involve numerous on-the-spot decisions about such matters.

The U.S. Bureau of Standards Publication, *Precision Measurement and Calibration: Statistical Concepts and Procedures*, provides excellent detailed examples of specific operationally defined measurement procedures. In particular, the example of preparing an ice-water bath of 0°C "slush" for experimental purposes is virtually paradigmatic. The detail of this operational definition is required by our common experimental observation of a large jump in the energy-temperature relation during the transformation of a material from a solid to liquid state. If an experimenter does not take the operationally specified care in preparing the slush, the numerical outcomes resulting from the associated experiments will not summarize an experience in a way useful to someone who reads them, but will rather misconstrue what has happened.

Although the subject of operational definitions of quantities may seem inconsequential and perhaps boring, you should be aware that the views stated in this discussion constitute a major break with widespread contemporary attitudes toward statistical measurements. It is commonly supposed that "true values" of quantities are in principle not observable. Actual measurement procedures are considered to yield a number that equals the true quantity value plus an unobservable component of "measurement error." How tall are you? In answering this question, it is supposed that you do have some "true height," but that your measured height equals this true height plus some unobservable measurement error. In symbols, this outlook is typically denoted by writing $X = \mu + \varepsilon$, where $X$ represents the actual measurement, $\mu$ represents the true height, and $\varepsilon$ represents the unobservable measurement error. If you look at the National Bureau of Standards' publication *Precision Measurement*, you will find a discussion of measurement that presupposes this viewpoint. In fact, virtually all of contemporary statistical analysis presumes this too, using precisely this notation.

We are expressly rejecting this viewpoint as operationally meaningless. Although there may well be many definable quantities that are not actually observed, reference to in principle unobservable quantities is meaningless. It is akin to talking about "the little man who wasn't there," the subject of a 1930s jingle: "Last night I saw upon the stair a little man who wasn't there. He wasn't there again today.—Oh, how I wish he'd go away."

As an economic example, it is widely supposed that there exists a true rate of inflation but that it is unmeasurable. The rates of increase in various price indices are supposed to provide various measurements of this rate, each of which is augmented by its own unobservable component of measurement error. Another example with even more emotive connotations is the common conception that there is some true amount of "economic product" generated by a society, and that GNP and GDP statistics constitute measurements of this true amount, each contaminated by measurement error in a different way. Discussions often focus on the statistical properties of the probability distribution of unobservable errors associated with various indices or measurements.

Quite the contrary, we presume herein that each index or measurement is what it is, the outcome of a specified procedure of activity that yields a number, no more and no less. We completely reject the possibility of any meaningful definition of "a true rate of inflation," "a true amount of economic product," or, more generally, of "unobservable measurement error"! Someone's willingness to use GNP statistics as a summary of the production and economic welfare of a society, for example, must be based on a personal commitment that the operational activities involved in generating these numbers is a reasonable summary of what's happening in these regards. Such a commitment is surely not required by any rule of logic or empiricism. Marilyn Waring's (1988, 1989) brilliant historical account of the so-called United Nations Standardized System of National Accounts presents a wide array of reasons why *not* to make this commitment. Daly and Cobb (1989) agree. Similar statistical concerns had been raised by Magdoff (1939).

An important survey article by Robert Eisner (1988) documents the extent to which these ideas are taken seriously by economic statisticians and exemplifies the attitude toward operationally defined measurements that we are presuming in this text. The volume edited by Churchman and Ratoosh (1959) also contains ideas on measurement that are sympathetic to the approach we are following, whereas Kyburg (1984) is opposed.

In response to changes in world economic structures, the U.S. Departments of Labor and of Commerce are currently in the process of changing the operational procedures that define the CPI, GNP, and GDP.

## 2.2 ARITHMETICAL REPRESENTATION OF LOGICAL OPERATIONS

In every field of science and commerce we are concerned with logically related measurements. Are home mortgage interest rates currently above 8%, *and* are they rising? Does placing a salamander in a terrarium in which the pH measurement of the soil is 3.4 *imply* that it will die within three days? Will your property be damaged by fire *or* by flood *or* by earthquake during the next year? It is important to examine how logical relations among measurements can be identified.

We have already noticed that new quantities can be defined in terms of functional images of other quantities. If $X$ is a quantity and $G$ is a function whose domain includes the realm of $X$, then $Y \equiv G(X)$ is also a quantity, with realm $\mathcal{R}(Y) = \{G(x) | x \in \mathcal{R}(X)\}$. In the present section we focus specifically on *functional* relations among events. We conclude that all the results of finite propositional logic can be represented as results of finite arithmetic. Further in this chapter we expand our considerations to include logical relations among quantities, superseding logical relations that can be defined functionally.

To begin, notice that any operationally verifiable proposition can be used to define an event merely by identifying the event as the "truth value" of the proposition: If $q$ is a verifiable proposition whose truth is unknown, define the quantity $Q \equiv T(q)$, where $T$ is the truth function with range $\{0, 1\}$. This would define $Q \equiv T(q) = 1$ if the proposition $q$ is observed to be true, and $Q \equiv T(q) = 0$ if $q$ is observed to be false. Thus, $Q$ is an event. Reversing this process, any quantity $Q$ can be used to formulate a verifiable proposition: If $Q$ is an event, define $q$ as the proposition "$Q$ equals 1." With this understanding, we can regard events as the unknown truth values of verifiable propositions.

Since we have defined events as numbers, we can operate upon them with arithmetical operators. A pleasing benefit of this construction is that the operations of propositional logic can be represented arithmetically. For every sentence of propositional logic can be expressed in terms of the two logical operators that are denoted by $\sim$ (meaning "not") and $\wedge$ (meaning "and"). Thus, the arithmetical representation of the results of propositional logic is achieved merely by identifying the arithmetical representation of these two operations. We do this by building on the equivalence of quantities with the truth values of propositions, as described previously. Since for any proposition $a$, the truth value of its negation is $T(\sim a) = 1 - T(a) = 1 - A$, it makes sense to define the logically related quantity $\widetilde{A}$ arithmetically by $\widetilde{A} \equiv 1 - A$. Similarly, since the truth value of the conjunction of two propositions is $T(a \wedge b) = T(a)T(b) = AB$, we define the quantity $A \wedge B$ by the product function $A \wedge B \equiv AB$. Using these definitions, we can define any logical function by arithmetical operations. For examples, the logical operations of alternation and implication are represented arithmetically by

$$C = A \vee B \equiv \sim(\widetilde{A} \wedge \widetilde{B}) = 1 - (1 - A)(1 - B) = A + B - AB \qquad \text{(inclusive or)},$$

and

$$D = (A \Rightarrow B) \equiv \sim(A \wedge \widetilde{B}) = 1 - A(1 - B) = 1 - A + AB \qquad \text{(implication)}.$$

Finally, *using the important fact that events are idempotent*, that is, $A^2 = A$, no matter whether $A = 0$ or $A = 1$, the results of propositional logic can be seen to be results of simple integer arithmetic.

Logically, a proposition may have the status of a tautology, an impossibility (a contradiction), or a contingency. Arithmetically, logical tautologies are represented by the function of events that is constant and equal to 1, no matter what values the events happen to be. For example, the transitivity of implication is tautological, since

$$\begin{aligned}([(A \Rightarrow B) \wedge (B \Rightarrow C)] \Rightarrow (A \Rightarrow C)) &= 1 - (1 - A + AB)(1 - B + BC) \\ &\quad + (1 - A + AB)(1 - B + BC)(1 - A + AC) \\ &= 1.\end{aligned}$$

The first equality is achieved by replacing the implication and conjunction symbols by their equivalent arithmetical operations. The conclusion is then reached via multiplicative expansions and use of the idempotency property.

Logical impossibilities, or contradictions, are represented by the function that is constant and equal to 0. For example, the principle of noncontradiction results from the arithmetical derivation that

$$A \wedge \tilde{A} = A\tilde{A} = A(1 - A) = A - A^2 = 0.$$

No matter whether $A = 0$ or $A = 1$, $A\tilde{A} = 0$.

Finally, logically contingent propositions determine events that are representable by an arithmetical function that may equal 0 or 1, depending on the numerical values of its arguments. For example, the equivalence of two events depends arithmetically on their two values, since

$$A \Leftrightarrow B \equiv (A \Rightarrow B) \wedge (B \Rightarrow A) = (1 + AB - A)(1 + BA - B) = 1 + 2AB - A - B.$$

The equivalence of $A$ and $B$ depends on the values of $A$ and $B$. If both equal 0 or if both equal 1, $A \Leftrightarrow B = 1$; but if either equals 1 while the other equals 0, then $A \Leftrightarrow B = 0$. (Evaluate the far right-hand equality in each of these instances!)

On the basis of these results, we have a motivation for the following definition.

***Definition 2.8.*** Any function $f: \{0, 1\}^N \rightarrow \{0, 1\}$ is called a *logical function.* □

Any function of this type can be used to define a new verifiable proposition (an event) whose truth or falsehood (numerical value) depends functionally on the constellation of truth and falsehood among $N$ other verifiable propositions: $E_{N+1} \equiv f(\mathbf{E}_N)$. If the logical function is the "into" function with range $\{1\}$, it defines the necessary proposition, or tautology, a proposition that is *true* for *every* constellation of truth and falsehood among $N$ other verifiable propositions. Similarly, the into logical function with range $\{0\}$ defines the logically impossible proposition, or contradiction, a proposition which is *false* for *every* constellation of truth and falsehood among $N$ other verifiable propositions. Otherwise, if a logical function is an "onto" function, it defines a contingent proposition whose truth depends on the constellation of truth and falsehood among $N$ other verifiable propositions.

Every logical function has an arithmetical representation. Table 2.1 exemplifies all the logical functions that can be defined on two events, showing their arithmetical and logical forms.

Of course, the logical forms for the functions displayed in Table 2.1 are not unique. For example, the function $f_2(A, B)$ could also be specified by the logical form $\sim(A \vee B)$. The forms displayed are merely meant to give you an idea of how the usual theory of arithmetic on variables taking only the values of 0 and 1 can represent all of propositional logic. The theory of prevision that we develop actually generalizes all of arithmetic involving quantities that can take on many numerical values, but in a unified way that does not treat the logic of events, of 0 and 1, differently than the arithmetical logic of any other quantities. For now, let us solidify your understanding with some problems.

**Table 2.1. Sixteen Logical Functions of Two Events, $f$: $\{0, 1\}^2 \rightarrow \{0, 1\}$, Their Logical Names, and Their Arithmetical Representations**

| Event | Realm Element | | | | | |
|---|---|---|---|---|---|---|
| $A$ | 0 0 1 1 | Each column pair is an element | | |
| $B$ | 0 1 0 1 | of the function domain, $\{0, 1\}^2$ | | |
| **Function** | **Function Value** | **Arithmetical Form** | **Logical Form** | **Function Name** |
| $f_1(A, B)$ | 0 0 0 0 | $0$ | $(A \wedge \tilde{A}) \vee (B \wedge \tilde{B})$ | Contradiction |
| $f_2(A, B)$ | 1 0 0 0 | $1 - A - B + AB$ | $\tilde{A} \wedge \tilde{B}$ | |
| $f_3(A, B)$ | 0 1 0 0 | $B - AB$ | $\tilde{A} \wedge B$ | |
| $f_4(A, B)$ | 0 0 1 0 | $A - AB$ | $A \wedge \tilde{B}$ | |
| $f_5(A, B)$ | 0 0 0 1 | $AB$ | $A \wedge B$ | Conjunction |
| $f_6(A, B)$ | 1 1 0 0 | $1 - A$ | $\tilde{A}$ | Negation |
| $f_7(A, B)$ | 1 0 1 0 | $1 - B$ | $\tilde{B}$ | |
| $f_8(A, B)$ | 1 0 0 1 | $1 + 2AB - A - B$ | $A \Leftrightarrow B$ | Equivalence |
| $f_9(A, B)$ | 0 1 1 0 | $A + B - 2AB$ | $\sim(A \Leftrightarrow B)$ | |
| $f_{10}(A, B)$ | 0 1 0 1 | $B$ | $B$ | |
| $f_{11}(A, B)$ | 0 0 1 1 | $A$ | $A$ | Identity |
| $f_{12}(A, B)$ | 1 1 1 0 | $1 - AB$ | $\sim(A \wedge B)$ | |
| $f_{13}(A, B)$ | 1 1 0 1 | $1 - A + AB$ | $A \Rightarrow B$ | Implication |
| $f_{14}(A, B)$ | 1 0 1 1 | $1 - B + AB$ | $B \Rightarrow A$ | |
| $f_{15}(A, B)$ | 0 1 1 1 | $A + B - AB$ | $A \vee B$ | Alternation |
| $f_{16}(A, B)$ | 1 1 1 1 | $1$ | $(A \vee \tilde{A}) \wedge (B \vee \tilde{B})$ | Tautology |

## PROBLEMS

1. The logical operator "exclusive or" is defined by $C \equiv \sim(A = B)$, which is equivalent to $[(A \vee B) \wedge \sim(A \wedge B)]$. Using algebraic methods, express the event $C$ as an arithmetical function of $A$ and $B$.

2. Determine a simple arithmetical expression for the logical proposition $\{[(A \Rightarrow B) \wedge (B \Rightarrow C)] \Rightarrow C\}$. Compare your answer with your solution to problem 3.

3. A common problem in actuarial probabilities involves determining the price of policies that cover multiple risks. For example, you might want to insure your household goods against damage by fire or by flood or by earthquake. We have learned in this section that $A \vee B = A + B - AB$. If $C$ is a third event, determine the arithmetical expression for the event defined logically by $A \vee B \vee C$. Compare your answer with your answer to the previous problem. What would you suspect is the arithmetical form for the event that at least one of four events equals 1: $E_1 \vee E_2 \vee E_3 \vee E_4$? Try to prove the general formula as your next problem.

4. Beginning in the same way you began problem 3, use a simple multiplicative expansion to prove more generally that

$$E_1 \vee E_2 \vee \cdots \vee E_N = \sum_{i=1}^{N} E_i - \sum_{j>i=1}^{N} E_i E_j + \sum_{k>j>i=1}^{N} E_i E_j E_k - \cdots + (-1)^{N-1} E_1 E_2 E_3 \cdots E_N.$$

(Notice the *alternating signs* of the summands in this expression.) *Hint*: If you know the formula for the product of sums of two quantities, $\prod_{i=1}^{N} (a_i + b_i)$, the result should be simple after you write the first line. If you don't know this formula, you may prove this result by induction, presuming it is true for $N$ and then using the result that $E_1 \vee E_2 \vee \cdots \vee E_{N+1} = (E_1 \vee E_2 \vee \cdots \vee E_N) \vee E_{N+1}$ and some algebra, proving that it must be true for $N+1$.

5. Use arithmetical operations on events to prove that the following six famous (see below) propositions are tautologies:

   a. $A \Rightarrow (B \Rightarrow A)$
   b. $\{A \Rightarrow (A \Rightarrow B)\} \Rightarrow (A \Rightarrow B)$
   c. $\{A \Rightarrow (B \Rightarrow C)\} \Rightarrow \{B \Rightarrow (A \Rightarrow C)\}$
   d. $(B \Rightarrow C) \Rightarrow \{(A \Rightarrow B) \Rightarrow (A \Rightarrow C)\}$
   e. $A \Rightarrow (\tilde{A} \Rightarrow B)$
   f. $(A \Rightarrow B) \Rightarrow \{(\tilde{A} \Rightarrow B) \Rightarrow B\}$

   You may quit once you get the hang of it.

These propositions were the six axioms proposed by David Hilbert to generate what he called "a logic of judgments." The first four were called axioms of implication, whereas the last two were called axioms of negation. See the sourcebook of van Heijenoort (1967).

**6.** Many computing languagues such as MATLAB contain the logical functions, defined arithmetically, accessible for use. Use the command "help ops" to learn about them. For example, in MATLAB, the symbol "&" represents the logical "and," the symbol "|" represents the logical (inclusive) "or," and the relational operator "< =" represents "implies." Following is a programming suggestion to help you learn about it. You will begin by using the realm.m program you created in problem 6 of Section 2.1. After identifying the events $A$ and $B$ with the first two rows of the realm matrix, $RXN$, you will define the variables CL, DL, and EL using the logical operators, and the variables CA, DA, and EA using the usual arithmetic operators. You should see that they are identical. Also worth seeing here is that the definition of implication via the relational operator $< =$ makes precise use of the same notation involving parentheses that we adopt in this text. Defining EL by $(A < = B)$ identifies any component of the vector EL as 1 if the corresponding element of $A$ is less than or equal to the corresponding element of $B$, and as 0 if not. Get on to MATLAB and try the following. Use the help facility to learn about the syntax of any operation you do not understand.

```
N = 2; K = 1; realm
A = RXN(1,:), B = RXN(2,:)
CL = A&B, CA = A.* B
DL = A|B, DA = A + B - A.*B
EL = (A < = B), EA = ones(1,4) - A + A.*B
```

***Historical Notes***

George Boole was one of the first to devise an algebraic representation of logical operations in his treatise *The Mathematical Analysis of Logic* (1847). This work was an important milestone in modern study of the relations between logic and mathematics. Boole constructed an abstract algebraic system that, subject to appropriate restrictions of "interpretability," could represent the known laws of deductive logic. Hailperin (1986) provides a brilliant historical and technical exposition of the work. Whereas Boole understood his algebraic operators as acting upon symbols that represent statements of class membership, we understand the algebra simply as the arithmetic of 0 and 1 applied to events, which are recognized in our syntax as numbers. In our development here, the meaningful truths of propositional logic are merely truths of finite integral arithmetic when it is restricted to the domain of $\{0, 1\}$. So the logic expands quite simply and easily to the general arithmetic of quantities, with nothing of fundamental interest gained by the restriction to $\{0, 1\}$. This is precisely the opposite of the view proposed by Gottlob Frege (1878, 1884, 1893, 1903) at the end of the nineteenth century—that the mathematical structure of arithmetic rests upon a foundation provided by a theory of logic, generated from a finite number of specifiable axioms. For a variety of reasons, Frege's outlook (if not the detail of his theory) won the day, motivating the important developments of mathematical logic generated in the era of Bertrand Russell and Kurt Gödel. The notations of propositional logic and of set theory, with which students generally are more familiar today, were developed expressly for the purpose of developing this outlook. Nonetheless, the algebraic and arithmetic view of logic did survive the nineteenth century, at least in the work of the French mathematician and logician, Louis Couturat (1905). Its current reemergence is at least partly due to the direct applicability of arithmetic syntax to computer programming languages.

## 2.3 MODES OF KNOWING—A MEDITATION

To this point, we have been considering operational definitions of quantities that summarize research investigations into what's happening. As distinct from what's happening, we may also consider your or my (or some other person's) knowledge of what's happening. Statistical theory provides an analysis of the mathematical structure of an operationally defined *measurement of your scientific knowledge* of what's happening, albeit uncertain knowledge. We call this measurement of your knowledge your *prevision*. When we define the operational procedure for generating this measurement of your knowledge, beginning in Section 2.4, it is you who must decide whether the value of its use outweighs the (minimal?) confusion it provides via its inevitably incomplete operational specification. As in the use of any quantity, we need request such a commitment from any scientist who would use the operational subjective statistical methods developed in this book. While you study this text, you will develop the understanding and experience needed to determine whether you want to make this commitment. I trust that you do.

Admittedly, your knowledge of what's happening may well be more extensive than knowledge merely of the numerical values of specific quantities. We only argue that your knowledge can be *summarized* by knowledge of the values of quantities. Any summary involves some loss of detail, just as quantities are merely empirical summaries of a richer experience. In this section, let us digress briefly to reflect on the position that "scientific knowledge" holds in a complex of many recognizable ways of knowing.

The Russian statistician V. V. Nalimov (1981a) has proposed a beautiful image of a very different mode of human knowing that completely defies summary in terms of knowledge about the values of quantities. He describes this mode of knowing as a personal dwelling in "the semantic vacuum." Perhaps best described as a meditative or dream state, it involves learning by following the somewhat self-contradictory prescription of "do nothing." It is a recognizable form of knowledge of which virtually nothing operationally defined can be said. This mode is one extreme of a partially ordered array of forms which includes the mode of "fully semantic knowledge" at another extreme. Still other modes of human knowledge are involved in the learning of skills. To some extent, the knowledge of how to hit a baseball or to play the violin can be conveyed through operationally defined summaries. But in another way, such knowledge is transmitted by the conveying of a feeling that cannot be described, though you know it when you've got it, as we say. Statistical theory is limited in its relevance, pertaining to scientific knowledge that is representable in the fully semantic mode. At the very least, we should recognize that fully semantic knowledge is only one form of human knowledge, of which all forms constitute something like a semantic spectrum.

The generation of knowledge in the fully semantic mode and in the semantic vacuum proceed along contradictory paths. We shall learn how measurements of fully semantic knowledge are calibrated according to one's personal utility valuation of experience, that is, the distinguishing among experiences according to one's assessment of their desirability. In the semantic vacuum, however, the assertion of one's personal utility valuation of experience actually *precludes* the achievement of knowledge. Lao Tsu expressed this idea clearly some 2500 years ago in a statement that has resounded from several people over many times and cultures. In reference to personal knowledge of "the name that cannot be named" as "the mother of all things," he wrote

> Always without desire we must be found,
> if its true mystery we would sound.
> But if desire within us be,
> Its outer fringe is all that we would see!
>
> *Tao Te Ching*, James Legge (tr, 1891)

At the opposite extreme lies fully semantic scientific knowledge, measurable by a process of defining distinctions that are graduated according to one's personal value judgments. For someone who would never make such a value distinction of better and worse, no distinction between gradations of knowledge is possible, and science is impossible. This is not to say that physical distinctions such as "heavier" or "lighter" are based on an individual's moral valuation. It is the *measurement scale of scientific knowledge* about quantities that is meaningfully calibrated in terms of utility, determined operationally by value distinctions. We shall understand exactly what is meant here when we formally define "prevision".

This characterization of scientific knowledge and the role of personal valuation in its formulation may sound strange. It is expressly antithetical to the received doctrine of scientific method which had been formulated and popularized during the mid-quarters of this century—that scientific method presupposes a "value-free" orientation by the practitioner, and that value assertion is a defining characteristic of nonscientific activity. We are proposing here a major conceptual switch in understanding the difference between science and religion. Scientific experience occurs in the context of assertive activities of belief and valuation, whereas religious experience occurs in the context of the nonassertive activity of stillness. Let us move on to see where this new approach to scientific inference leads.

## 2.4 UNCERTAIN KNOWLEDGE—PREVISION

The most extreme positive assertion of knowledge about a quantity is exhibited by the act of precisely specifying its value. I was born in 1948. My gender is male. My mother bore nine living children. I say I am *certain* of these quantities. Even if you do not know the precise value of a quantity for certain, you can at least be certain that the quantity is a member of a specifiable finite set of possibilities. Minimally, as long as you understand the operational definition of the quantity, you know it to be some member of its realm. With respect to an event, then, you may be certain that the event equals 1; you may be certain that the event equals 0; or you may be certain merely that the event equals 0 or 1. These three situations exhaust the possibilities for the status of anyone's *certain* knowledge of an event.

The operational subjective theory of statistics proposes that your *uncertain* scientific knowledge about a quantity can be specified more precisely than merely by identifying its realm. This precision is achievable operationally by your situating the quantity within an ordering that is based simultaneously upon your opinions and your utility valuations, your desires. Since it is preferable not to confound these two bases for the measure of your knowledge when we communicate with one another, we define a procedure for measuring uncertain knowledge in which the component of utility valuation is tamed.

Consider a monetary unit that is large enough to be divisible into many noticeably distinct units, but not so large that your acquisition or loss of this amount would change your life possibilities inordinately. We are going to refer to this unit as *the scale of your maximum stake* for prevision assessment purposes. For many of us, U.S. dollar could suffice. It is divisible into 100 pennies. You would not ignore a dollar bill if you came upon one on the roadside. Yet the loss of one through carelessness would not reduce you to a state of interminable woe. For some of us, the unit of \$10 or even \$100 may be more appropriate. For although a dollar is divisible into 100 units, many of us do ignore a penny when we pass it in the street. If you ignore a penny, do you ignore a dime? a quarter? (The smallest denomination of currency that is noticeably of interest to you might be called the scale of your *minimum* stake. But we shall ignore this subtlety for the moment.) Perhaps \$10 or even \$100 is the appropriate scale of your maximum stake in order that it be divisible into many noticeably distinct units of currency sizable enough to interest you. This scale typically varies with time as your financial fortune varies. You will understand better how to determine the scale of *your* maximum stake for yourself once you see the precise use that we want to make of it.

The *standard procedure* for measuring your uncertain scientific knowledge, which we now develop, is specified relative to your stipulation of the scale of your maximum stake. The consequences of this relativity are minimal, as long as you are aware of it. The standard procedure is proposed as applicable to the specification of your knowledge about many quantities. We discuss some problematic quantities in due time.

The main presumption behind the standard procedure for measuring your uncertain knowledge is that if $A$ were an event that you were certain to equal 1, then you would be indifferent between owning a redeemable claim to "one U.S. dollar" and owning a redeemable claim to "$A$ U.S. dollars." You would willingly exchange either claim for the other, for you are certain that they represent the same exchange value. But if you were *uncertain* about the numerical value of $A$, you would presumably prefer a claim to one dollar above a claim to $A$ dollars. For once the numerical value of $A$ were known for certain, the redemption value of the latter claim could only be less than or equal to the claim to the dollar. Yet you would presumably always prefer holding the claim to $A$ dollars above holding a claim to nothing at all. For once the numerical value of $A$ were known, the redemption value of the claim could only be greater than or equal to the claim to zero dollars. In brief, we are presuming that you prefer having more money than having less.

In order to utilize the theory of inferential statistics according to the standard procedure, you are asked to quantify your uncertain knowledge about any quantity of interest by means of an introspective measurement whereby you evaluate your preference between having a claim to a specified number of dollars and having a claim to the unknown quantity of dollars. In the extreme, you might assert the precise dollar value that you *equate* with the value of a claim to the unknown quantity of dollars. This asserted valuation is called your *prevision* for the quantity. For example, we define your prevision for an event $A$ as the number $P(A)$ of dollars at which you value a claim to $A$ dollars. That is, you would willingly exchange $A$ dollars for $P(A)$ dollars, and also $P(A)$ for $A$. That is, you would willingly buy a claim to $A$ for $P(A)$, *and* you would also willingly sell a claim to $A$ for $P(A)$. More generally, your prevision for a general quantity, $X$, is the price $P(X)$ dollars at which you value a claim to $X$ dollars, with the understanding that the scale of the transaction is limited so that your (uncertain) net yield from the transaction surely lies within the magnitude of your specified maximum stake.

***Example 1.*** Today a dairy farm will artificially inseminate 50 dairy cows from its herd. Let $X$ equal the number of these who will bear a live calf within 280 days from today. Notice that the realm of $X$ is $\mathcal{R}(X) = \{0, 1, 2, \ldots, 49, 50\}$. Suppose that I specify the scale of my maximum stake as \$100. I can assess my prevision for $X$ in terms of my valuation of a claim to $X$ dollars, as follows.

| Sequentially I would ask myself: | Suppose I would answer: |
|---|---|
| Do I prefer to have \$5 or \$$X$? | \$$X$ |
| \$45 or \$$X$? | \$45 |
| \$15 or \$$X$? | \$$X$ |
| \$25 or \$$X$? | \$$X$ |
| \$35 or \$$X$? | \$$X$ |
| \$43 or \$$X$? | \$43 |
| \$41 or \$$X$? | indifferent |

This sequential self-questioning has allowed me to identify my prevision assertion $P(X)$ as 41. The transaction of buying or selling the claim to $X$ dollars in return for $P(X)$ dollars can net me at most a net gain or loss of \$41, in the case that I buy \$$X$

for \$41 and the value of $X$ turns out to equal 0. Since \$41 is well within the scale of my maximum stake, I can assess my prevision by thinking of my valuation of $X$ in units of dollars. However, if the scale of your maximum stake were only \$5, say, you would need to assess your prevision for $X$ by evaluing $X$ in units of dimes (10 cent pieces). Only in that way could you be assured that you would not be getting into considerations of transactions that exceed the scale of your maximum stake. You will understand this scale completely when we write a formal definition shortly. □

This procedure for determining your prevision for a quantity $X$ in terms of the monetary price at which you value a claim to $X$ dollars is what we call the *standard procedure* for specifying your prevision.

Before proceeding to formal definitions, let us mention three qualifications to the discussion we have engaged to this point. In order to delay further distracting discussion so as to get on with formal definitions, we merely list these qualifications here in cryptic form. A detailed discussion of these three qualifications is appended separately to this section. You will understand the discussion better after you work through the formalities in what follows.

QUALIFICATION 1. The standard procedure for assessing your previsions can be applied to your knowledge only about quantities whose numerical values are not too critical to your welfare. In assessing your previsions for quantities of very great importance, the considered "transactions" must be denominated in units of utility.

QUALIFICATION 2. The standard procedure for prevision determination can calibrate distinguishable prevision assertions only in discrete units. I ended the questioning process in Example 1 with indifference at a value of \$41. Could it be possible operationally to end the story at \$41.56? at \$41.56418?

QUALIFICATION 3. Based on a limited amount of introspection, you might arrive only at the limited assertion that you are willing to pay at least $P_l(X)$ for $X$, and that you are willing to sell a claim to $X$ for $P_u(X)$. This would be to assert the closed interval $[P_l(X), P_u(X)]$ as a partial assertion of your prevision.

It should be evident from our discussion to this point that your prevision for a specified quantity is an assertion that *you* make on the basis of *your* state of knowledge of the quantity. If your state of knowledge were to change, presumably you would arrive at a different value for your prevision of the specified quantity in that new state of knowledge. The value that you assert as your prevision can be evaluated neither as "correct" nor "incorrect" by anyone else. You assert your prevision to be whatever you assess it to be. For it is operationally defined as your measurement of your state of knowledge about a quantity. Your prevision for a quantity *is not* defined as a measurable characteristic of what's happening with regard to the generation of $X$. The quantity itself is measuring that. Your prevision *is* a measurement of what's happening only in the sense that it measures what's happening to you when you consider the value of $X$. Someone else may be surprised by what you think about the values of quantities, and may well think something different about them. After some discussion with that person, you may consider yourself to have been newly informed and come to specify your prevision as closer to your "informant's" prevision. Or again, you may contest this "informant's" knowledge and remain firm in your own quite different prevision assertion. But someone else may not tell you that you do not think what you say you think! You are the only one who can judge that.

Finally, just as we use a special name—events—for quantities whose realm is $\{0, 1\}$, we also use a special name for your prevision assertions regarding events. We call them probabilities. Your prevision for an event is called your probability for the event. All probabilities are previsions, but not all previsions are probabilities. For, as we have already observed, all events are quantities, though not all quantities are events.

We have reached the point at which we should formally define your prevision for a quantity and the scale of your maximum stake.

***Definition 2.9.*** Let $X$ be any quantity with bounded discrete realm $\mathscr{R}(X) = \{x_1, \ldots, x_K\}$. In asserting your state of knowledge about $X$ and your state of utility valuation, your *prevision configuration* for $X$ is the pair of numbers $P(X)$ and $S > 0$ you specify with the understanding you are thereby asserting your willingness to engage *any* transactions that would yield you a net gain of amount $s[X - P(X)]$ as long as the scale factor, $s$, is a number for which $|s[x - P(X)]| \leqslant S$ for every $x \in \mathscr{R}(X)$. Individually, $P(X)$ is called your *prevision* for $X$, and $S$ is called the *scale of your maximum stake*. If the quantity $X$ is an event, then your prevision is also called your *probability* for the event. □

Consistent with our definitions throughout, all of the symbols that appear in the definition, namely $x, X, P(X), s$, and $S$, represent *unitless* numbers. A transaction, and thus your "gain," can be denominated in any units whatsoever—peaches, pimentos, or cash. But typically, we think of your gain in units of currency, the almost universally recognized medium of exchange. For quantities whose values concern us greatly, the considered transactions are to be denominated in units of pleasure, called "utils."

Both mathematically and operationally, the most important aspect of the definition's specifying "your willingness to engage any transaction..." is that the scale factor $s$ may be positive or negative. For positive values of $s$, the "transaction" is interpreted as your buying the unknown return, $sX$, at the specfied price, $sP(X)$. For negative $s$, the interpretation is that you are selling a claim to the unknown quantity, $-sX$, at the specified price, $-sP(X)$. In either case, your net gain from the transaction would be representable as $s[X - P(X)]$. Thus, in asserting your prevision $P(X)$ you are specifying a price at which you are willing to buy and to sell $X$ (at an appropriate scale), and that in doing either you would regard yourself as no better or worse off than in making no transaction at all.

Evidently the realm of $X$ must be bounded in order that this definition be operational. Otherwise there does not exist any number that can perform the role required of $S$, the scale of your maximum stake.

In most of this textbook and in most applications of operational subjective statistical procedures, we refer to your prevision $P(X)$ alone, without explicit reference to the scale of your maximum stake, $S$. Your prevision assertion represents your uncertain knowledge about $X$. That is what interests us, whereas $S$ merely plays a role in the procedure for your eliciting your prevision. Once the scale of your maximum stake is specified in your prevision configuration for any quantity, the same number can be used in the specification of your prevision for any other quantity. Shortly, you can try to determine the scale of your maximum stake as an exercise.

The definition for your prevision of a quantity can be easily extended to a definition of your prevision for a vector of quantities in a way that is consistent with the definition of your prevision for a single quantity. Your prevision for a vector of quantities, $\mathbf{X}_N = (X_1, \ldots, X_N)^{\mathrm{T}}$, equals the vector of your previsions for them individually: $P(\mathbf{X}_N) \equiv (P(X_1), \ldots, P(X_N))^{\mathrm{T}}$. You specify each component of the vector $P(\mathbf{X}_N)$ with the understanding that you are asserting your indifference to any combination of transactions involving all the components of $\mathbf{X}_N$, again just so long as the maximum total net gain or loss from the combination will not exceed your specified scale of maximum stake, $S$. To clinch these ideas, and to conclude this section, let us write the definition of your prevision for a vector.

***Definition 2.10.*** Let $\mathbf{X}_N \equiv (X_1, \ldots, X_N)^{\mathrm{T}}$ be any vector of quantities with bounded discrete realm, $\mathscr{R}(\mathbf{X}_N)$. In asserting your state of knowledge about $\mathbf{X}_N$ and your state of utility valuation, your *prevision configuration* for $\mathbf{X}_N$ is the vector of numbers $P(\mathbf{X}_N) \equiv (P(X_1), \ldots, P(X_N))^{\mathrm{T}}$ and the number $S > 0$, which you specify with the understanding you are thereby asserting your willingness to engage *any* transaction that would yield you a net gain of amount $\mathbf{s}_N^{\mathrm{T}}[\mathbf{X}_N - P(\mathbf{X}_N)]$, as long as the vector of scale factors, $\mathbf{s}_N$, is a vector for which $|\mathbf{s}_N^{\mathrm{T}}[\mathbf{x}_N - P(\mathbf{X}_N)]| \leqslant S$ for every $\mathbf{x}_N \in \mathscr{R}(\mathbf{X}_N)$. Individually, $P(\mathbf{X}_N)$ is called your *prevision* for $\mathbf{X}_N$, and $S$ is called the *scale of your maximum stake.* □

The vector product $\mathbf{s}_N^{\mathrm{T}}[\mathbf{X}_N - P(\mathbf{X}_N)]$, which appears in this definition, can also be written as the summation $\sum_{i=1}^{N} s_i[X_i - P(X_i)]$. This sum represents the net gain from engaging each of the transactions $[X_i - P(X_i)]$ at the scale and direction specified by the coefficient $s_i$. It is worth stressing here again that the stipulated acceptability of "any transaction" amounts to the stipulation that the scale vector $\mathbf{s}_N$ may have any direction. All or any of its components may be positive, and all or any of them may be negative. The total transaction would thus entail your buying claims to some $X_i$ components and selling claims to others.

***Example 2.*** Suppose there are 40 students in your class. Let $D$ equal the number of them who have driven an automobile today. What is your $P(D)$? Suppose you assert your $P(D) = 7.6$. Let us define the quantities $M = (D > 10)$ and

$L = (D < 6)$. If you assert your previsions for $M$ and $L$ as $P(M) = .33$ and $P(L) = .1$, you would have asserted your prevision for a vector: $P(D, M, L) = (7.6, .33, .1)$. What is the realm matrix for this vector? □

***Historical Philosophical Note***

Notice that we have defined your prevision for quantities as an *assertion* that you make. Gottlob Frege (1879) originated the distinction in deductive logic between the *content* of a sentence or its negation, denoted respectively as $A$ or $\tilde{A}$, and the *assertion* by someone of the truth of a sentence or its negation, which he denoted by $\vdash A$ or $\vdash \tilde{A}$, respectively. This is precisely the distinction we make between the numerical value of an event, such as $A = 0$ or $A = 1$, and your asserted prevision for an event, $P(A) = \#$. The only difference is that, whereas in the logic of certainty you can only assert that a sentence is true or is false, $\vdash(A) = 1$ or $\vdash(A) = 0$, in the logic of uncertain knowledge (the theory of prevision) we allow that your assertion value for an event may be any number, even though the event itself can equal only 0 or 1. DeFinetti (1974a) mistakenly attributed the distinction between a contemplated proposition and an asserted proposition to B. O. Koopman (1940), who did use it along with Frege's symbol, $\vdash$. But Frege's distinction was already recognized and used even by Whitehead and Russell (1913). See also the discussions by Jeffreys (1961, Section 1.51) and Levi (1980).

## PROBLEMS

1. Try the following procedure for determining the scale of your maximum stake. Consider the flip of a coin from your pocket, and define the event $H = 1$ if the coin lands heads up and $H = 0$ if tails up. Now suppose you were offered as a gift either the amount of money $\$H$ or $\$T$, which would you prefer to have? (I think, and hope, you would say, "I'd be just as happy to take either.") How about \$.10 or $\$H$? Which would you prefer to have? How about \$.9 or $\$H$? How about \$.30 or $\$H$? \$.70 or $\$H$? This style of questioning is leading you to conclude that you would be just as happy to be offered $\$H$ as \$.50. Suppose that you decide you would be. Now how about considering a gift of $\$2H$ compared to various gifts of specified amounts of cash? Would you again settle at indifference between \$1 and $\$2H$? I know I would. But if you were to consider gifts of greater value, say $\$1000H$ compared to cash, would you still settle at indifference between an offer of \$500 and an offer of $\$1000H$? For my part, I would prefer the gift of \$500. In fact, I would prefer \$450 to $\$1000H$. How about you? If you are a typically poor student, I think you might even prefer \$300 to $\$1000H$. But any of these choices is for you to decide. The point of this exercise is to determine the *largest* scale of a gift, $\$HS$, that you would regard as equally attractive as a definite cash gift of $\$.5S$. When you discover the largest value of $S$ for which this is the case, you would have discovered the scale of your maximum stake for prevision elicitation purposes. Mine is about \$100. Again, if you are a student, I would guess that yours would be much smaller, perhaps \$10, \$5, \$2, or even \$1. In fact, your

$S$ might even be less. If you find that you are indifferent between $\$H$ and $\$T$, but you definitely prefer \$.50 to $\$H$, your $S$ must be less than \$1. In this case, try the exercise again, but reduce the scale of the offer from dollars to 10 cent pieces. Would you also prefer \$.05 to $\$.10H$? If you are indifferent, then the scale of your maximum stake is at least \$.10, though not as great as \$1. Fair enough. If you do definitely prefer \$.05 to \$.10H, that is fair enough too. But I'm suspicious.

**2.** Think about some quantities that interest you, and define them operationally. Will your car need a new battery this year? How many flat tires will your bicycle get this year? How many students in your class drove an automobile to school today? rode in a private car driven by someone else? rode a bicycle? walked? used some form of public transportation? Assess your prevision for these various quantities. After you assess your $P(D)$, assess your probability for the event $M \equiv (D > P(D) + 3)$.

**3.** For classroom applications it will be useful to settle on a small group of quantities in order to provide prevision assessment practice throughout the course. It would be best if you choose some quantities of general interest, whose recorded values are publicly posted in an accessible document such as the newspaper. Some examples would be the measured rainfall in your locale during the course of each week, the temperature recorded at your local weather station at some specified time each week, the official currency exchange rate between your local currency and that of some other specified country on the Friday of each week, or the interest rate charged on a specified type of loan at one of your local banks. Whatever quantities you choose, be sure that you settle on their precise operational definitions. For example, there are several publicly posted exchange rates. Pick any one. Get into the habit of assessing your previsions for various quantities that interest you.

**4.** Determine your previsions for the amount of New Zealand money you would be offered for \$100 U.S. (cash for cash) by your local bank when it opens next Monday, and when it opens the following Monday. Notice that these are two distinct quantities. Your previsions might be the same, or they might be different, depending on what you think.

## ADDENDUM: A DISCUSSION OF THE THREE QUALIFICATIONS TO DEFINITION 2.9

QUALIFICATION 1. The standard procedure for assessing your previsions can be applied to your knowledge only about quantities whose numerical values are not too critical to your welfare. In assessing your previsions for quantities of very great importance, the considered "transactions" must be denominated in units of utility.

Suppose you buy a ticket in a lottery that will pay you \$1 million if you win and nothing if you lose. Further, suppose you have determined the scale of your maximum stake to be \$5. Now let $W$ denote the event that equals 1 if you win and 0 if you do not. How much would you be willing to pay to receive \$5$W$? In assessing your willing payment price for this transaction as \$$P$, we realize that if $W = 1$, your net gain would be \$1 million + \$5 − \$$P$, whereas if $W = 0$, your net gain would be −\$$P$. The maximum change in your personal fortune that could arise as a result of this transaction would far exceed the scale of your maximum stake. No matter how small we scaled down the size of this "side bet" on $W$, the consequences of $W = 1$ would be momentous. There is an evident problem in understanding our definition of prevision in terms of "money-denominated" transactions. The logical problem impinges on our considerations even more strongly with serious events of non-monetary consequences: how much would you pay to receive \$1 if your local nuclear power plant has an accident and releases high-intensity radioactive contaminants into the river? Would you rather be a "winner" in this transaction or a "loser"? At what price $P(A)$ would you value a claim to \$$A$? There is admittedly a difficulty with using the standard procedure in specifying your uncertain knowledge of $A$. For your valuation of the experience involved in your observation of $A$ can hardly be separated from your valuation of the additional dollar you would receive in the case that $A = 1$.

Frank Ramsey (1926) termed events whose consideration did not entail this valuation problem in such an extreme way as "ethically neutral" events. The standard procedure for determining your prevision requires the ethical neutrality of the quantities under consideration. It is possible to assess uncertain knowledge about nonneutral events, but it is more difficult. You may need to use a more complex procedure in which you consider directly your utility valuation of what's happening that generates the event, not merely your utility for a possible yield of \$1. A detailed study of utility is beyond the scope of our investigations here, but we shall alude to utility functions at various places in this text, especially in Chapter 6 in our discussion of scoring rules. The chapter on utility in the text of Savage (1954) remains today the most profound and easily readable exposition of utility theory, though many elementary texts present the basic ideas. Savage's attitudes, which I share, have been challenged, however. The recent work of Machina and Schmeidler (1992) is stimulating for the advanced reader and controversial.

For some events, such as the extreme one defined by your experience of a nuclear accident, another companion method of evaluating probabilities is often easier to use. It is based on an analysis of personally asserted "belief relations" and of conditions sufficient to represent a system of relative beliefs via numbers. DeGroot (1970) presents one popular axiomatization of beliefs, derived from Villegas (1964). It is well worth your examination. An important philosophical discussion of this approach appears in the article of Suppes (1974). A recent challenging work on related topics is the article of Kadane and Winkler (1988). The response of Lad and Dickey (1990) will be understood in the context of developments that we study in Chapters 3 and 4.

QUALIFICATION 2. The standard procedure for prevision determination can calibrate distinguishable prevision assertions only in discrete units.

If you think about any quantity, are you sure that after introspection you could always arrive at a price $P(A)$ at which you value a claim to $A$, as the valuing process is described in Definition 2.9? Consider the following example, supposing that the scale of your maximum stake is \$1, and that you do distinguish value in units of pennies. Let the event $C \equiv 1$ if the postage price of an inland United States postcard delivery on 2 January 1998 exceeds 26 cents, and $C \equiv 0$ if the price is 26 cents or less. For your information, the postage price of such a postcard delivery on 8 February 1994 is 19 cents. Now consider, at what price $P(C)$ do you value a claim to $C$ dollars? Suppose that upon introspection you determine that you would be willing to buy a claim to $C$ for any price up to 32 cents; you also determine that you would be willing to sell such a claim only if you receive at least 33 cents. What is your evaluation of $P(C)$? Operationally, there does not exist any coinage of a number $P(C)$ that corresponds to your prevision for $C$ as we have defined it. Our attempt to measure your uncertainty has come up against the operational limitation of discrete distinguishable measurements.

The way out of this impasse is to allow you to state whatever number you find appropriate between .32 and .33 as your assessment of $P(C)$. For example, $P(C) = .325$, .326, or even .32π. Yes, that is read "point three two pi," meaning .323141596.... Following this tack, we must admit that the *observable* implications of each of these announcements are identical. Since payments in units of less than one penny cannot be enacted, each of these assertions of prevision expresses the same observable willingness to buy a claim to $C$ at any price less than or equal to $P(C)$ and to sell a claim to $C$ at any price greater than or equal to $P(C)$. If the scale of your maximum stake were larger, say \$10, you could distinguish operationally between assertions of $P(C)$ such as .325 and as .326, but this procedure would still leave you unable to distinguish between announcements of prevision that differ only in their fourth decimal place. A scale of \$100 would allow you to distinguish the prevision to a fourth place, but not to a fifth place, and so on. Our suggested way out of the impasse is to admit that the numerical value of prevision assessment is meaningful only to a limited number of decimal places. What measurement isn't?

Increasing the scale of your maximum stake offers less hope for further precision than might be imagined. For as one's financial fortune increases, increasing the scale of your maximum stake, typically you become less able to distinguish very small monetary magnitudes. Whereas the scale of my maximum stake is about \$100, I really do not care at all about the outcome of any financial transaction to the extent of the nearest pennies. The nearest 10 cents or even 25 cents exhausts my interest. If we were to formalize every detail of our theory, we would have to recognize a companion "scale of the minimum stake" that would interest you. Technically, the consequences of the small stakes issues is to require a recognition of the "grainy" character of prevision assessments, which do not meaningfully range over a continuum of possibilities. The work of Nau (1992) formalizes some of this, again at an advanced level. But for practical matters, the effort required for the formal systematization of every detail regarding the fineness of scale at which probabilities are meaningful seems to be very great relative to the value of the achievement.

In practical applications, the difficulty of making very fine distinctions constitutes a difficult problem for the assessment of very small probabilities. These can be of tremendous concern to us if the consequences of an "unlikely occurrence" are very great. For example, the exposure of radioactive waste to the environment from spent nuclear power fuel can be made very unlikely if sufficient costly care is taken in its storage. But if such a leakage does occur, the consequences are of tremendous magnitude. The only way we can deal with such questions is to consider directly our utility valuations for the large-scale consequences that are under consideration.

There are very special problems of practical consequence regarding the theoretical treatment of events assessed with zero probabilities. Their analysis attracted Borel, Fréchet, and de Finetti as well as theoreticians in our own day. But the questions are rather delicate, and their mathematical treatment is quite advanced, well beyond the scope of this text.

QUALIFICATION 3. Based on a limited amount of introspection, you might arrive only at the limited assertion that you are willing to pay at least $P_l(X)$ for $X$, and that you are willing to sell a claim to $X$ for $P_u(X)$. This would be to assert the closed interval $[P_l(X), P_u(X)]$ as a partial assertion of your prevision.

One might object to the proposed operational definition of prevision, saying, "Hey! For any particular quantity (such as the proportion of U.S. homes on January 1, 1998 that will contain a computer with at least 256K memory) I can identify numbers (of dollars) that I value as preferable to $X$ (dollars), and numbers that I value as inferior to $X$. But I cannot yet identify a precise number $P(X)$ at which I value $X$." This objection is perfectly acceptable within the limits of the theory. Specification of $P(X)$ is achieved only through a process of introspection and judgment. The *precise* specification of an asserted $P(X)$ is achieved with effort that is undertaken only if you deem it worthwhile. In practice, we often quit the effort before we reach a precise specification of $P(X)$, your buying and selling price. De Finetti's fundamental theorem of probability, which we study in this chapter, is directly relevant to this idea, stating precisely how and why an interval assertion is related to other statements of precise prevision. Presumably any prevision interval can be shortened via further reflection and sharpened judgment if you decide it is worth the effort.

The practical importance of probability intervals has long been recognized in a tradition which has been developed in the writings of Keynes (1921), Borel (1924), Van Deurhen (1934), Reichenbach (1934), Koopman (1940), Good (1950), C. A. B. Smith (1961), Scott (1964), Fishburn (1964, 1985), Dempster (1967), Suppes (1974, 1981), Shafer (1976), Walley and Fine (1982), Leamer (1986), and Walley (1991), to select merely a few important contributors.

We study the content of interval prevision assertions in detail in Sections 2.10 and 2.11. But for now we begin our construction of the subjective theory by supposing that you do identify a precise value at which you are willing to buy and to sell claims to the unknown quantity of dollars. The mathematical consequences of the weaker, interval assertions will follow easily from the results we achieve based on stronger assertion.

## 2.5 COHERENCY OF PREVISION

In specifying your state of knowledge, you are at complete liberty to assert your previsions for several quantities as any numbers you please. For you are merely identifying a list of prices at which you are willing to engage in transactions. However, there is one requirement of prevision that is necessary for the state of knowledge which motivates the assertion to be considered scientific, or logical. This requirement is that your assertions be *coherent*. Coherency is defined as a global property of *all* the assertions you avow in specifying your prevision for a vector of quantities. It amounts to the condition that you do not thereby avow your willingness to engage in transactions that would surely yield for you a net loss, no matter what the value of the quantities happen to be. For such a willingness would contradict your presumed general preference of more money to less. If you also assert your willingness to engage in an array of transactions that surely would yield you a loss, we must say that we can no longer understand what you mean when you say anything. For you apparently aver that you prefer more to less and that you are willing to exchange more in return for less. Let us make a formal definition and consider some examples.

***Definition 2.11.*** Your prevision for the vector of quantities $\mathbf{X}_N$ with realm $\mathscr{R}(\mathbf{X}_N)$ is *coherent* as long as there exists no vector $\mathbf{s}_N$ with allowable scale for which $\mathbf{s}_N^{\mathrm{T}}[\mathbf{x}_N - P(\mathbf{X}_N)] < 0$ for every $\mathbf{x}_N \in \mathscr{R}(\mathbf{X}_N)$. If such a vector could be found, we would say that your assertions are *incoherent*, since acting upon them would make you a *sure loser*. □

"Allowable scale" to the vector $\mathbf{s}_N$ refers to the qualification that the scale of the net conglomerate transaction, $|\mathbf{s}_N^{\mathrm{T}}[\mathbf{x}_N - P(\mathbf{X}_N)]| = |\sum_{i=1}^{N} s_i[x_i - P(X_i)]|$, not exceed your specified value for the scale of your maximum stake, $S$, no matter what the value of $\mathbf{x}_N \in \mathscr{R}(\mathbf{X}_N)$ is.

***Example 1.*** Suppose that $E$ and $F$ are events for which $F \equiv \tilde{E} = 1 - E$. The event $E$ might represent the recording of measurable rain today at your local weather station, and the event $F$ the recording of no rain. Now suppose that you assert your prevision for these events as $P(E) = .6$ and $P(F) = .9$, stipulating the scale of your maximum stake as \$5. Using row vector notation, we can write $P(E, F) = (.6, .9)$. According to the definitions of $E$ and $F$, the realm $\mathscr{R}[(E, F)]$ contains only two elements: $\mathscr{R}[(E, F)] = \{(0, 1), (1, 0)\}$. The net gain that you achieve by engaging both the transaction of paying $P(E)$ for in return for $E$ *and* paying $P(F)$ in return for $F$ would be

$$NG(E,F) = [E - P(E)] + [F - P(F)] = E + F - P(E) - P(F)$$
$$= 1 - .6 - .9 = -.5,$$

no matter whether $(E,F) = (0,1)$ or $(E,F) = (1,0)$. In either case, $E + F = 1$. If you are really willing to assert these two numbers as your prevision assessment for $E$ and $F$, then you are a sure loser. I can't understand anything you say. In the language of Definition 2.11, by noticing that $(1,1)[(e,f)^{\mathrm{T}} - P(E,F)^{\mathrm{T}}] = -.5 < 0$ for every $(e,f) \in \mathcal{R}[(E,F)]$, we have found a vector $(s_e, s_f) = (1,1)$ whose existence defies the coherency of this prevision assertion. □

The principle of coherency of prevision amounts to a generalization of "the principle of noncontradiction" in two-valued propositional logic. In the first place, this principle motivates the "sure loser" criterion, which does not allow you to say both "I prefer more money to less" and "I prefer less money to more." In the second place, the sure loser criterion identifies operationally the problem with asserting contradictory statements. Consider what would happen to you if you would assert as your previsions both $P(A) = 1$ and $P(\tilde{A}) = 1$. These assertions would defy the principle of noncontradiction, for they would amount to asserting that both of the contradictory propositions associated with the events $A$ and $\tilde{A}$ are true. Operationally, these previsions specify that you would willingly exchange \$1 for \$$A$ and also \$1 for \$$\tilde{A}$. But if you made both exchanges, you would be paying \$2 in exchange for \$$A$ + \$$\tilde{A}$, which equals only \$1, no matter whether $A$ equals 1 or 0. Assertion of the truth of two contradictory statements makes you a sure loser.

***Philosophical Comment***

We noted in Section 1.2 some similarity between the "intuitionist" theory of mathematics and the mathematical foundation for the operational subjective theory of probability. Although there are definite similarities, such as the emphasis on finite mathematics, the operational definition of coherency provides the basis for an extreme difference between the two approaches. Intuitionists reject the law of the excluded middle as not being representative of any intuition, and thus they do not allow the procedure of proof by contradiction as a "constructive" proof of any proposition. Quite the contrary, the operational subjective theory identifies the operational consequences of defying the principle of noncontradiction and uses a natural extension of the meaning of this law (the principle of coherency) to construct the logic of uncertain knowledge. Evidently we allow proof of propositions by appeal to the principle of noncontradiction, and we consider such proof as constructive, with the active meaning associated with "sure loser" status.

Throughout this text we study coherent and incoherent previsions in various contexts. How can we tell whether someone's prevision assertions are coherent? Rather than multiply individual examples here, let us state the following simple algebraic *characterization* of coherent prevision.

**Theorem 2.4.** The requirement that your prevision be coherent is equivalent to the two restrictions that for any quantities $X$ and $Y$ your prevision satisfies

**(i)** $\min \mathscr{R}(X) \leqslant P(X) \leqslant \max \mathscr{R}(X)$,
**(ii)** $P(X+Y) = P(X) + P(Y)$,

where $\min \mathscr{R}(X)$ represents the smallest member of the realm $\mathscr{R}(X)$ and $\max \mathscr{R}(X)$ represents the largest member of the realm $\mathscr{R}(X)$. □

We defer a formal proof of this innocuous-looking theorem until the end of this section, following the problems. But a corollary to the theorem is the easily understandable and very important requirement that coherent prevision assertions operate linearly over any finite linear combination of quantities. Any linear combination of quantities defines another quantity, and thus it is relevant to consider your prevision for it. The linearity of coherent prevision in this form is so important to the theory and its application that we state it formally as a corollary and prove it before we discuss the theorem and some examples.

**Corollary 2.4.1.** For any vector of quantities, $\mathbf{X}_N$ and any vector of real-valued constants, $\mathbf{s}_N$, coherent prevision assertion requires that

$$P(\mathbf{s}_N^{\mathrm{T}}\mathbf{X}_N) = \mathbf{s}_N^{\mathrm{T}} P(\mathbf{X}_N).$$

*Proof.* For integer values of the coefficients $s_i$, the corollary follows simply from restriction (ii) of Theorem 2.4 by finite induction. To establish this result for rational coefficients, note that for any integer $k$, $P(X) = P(kX/k) = kP(X/k)$. Thus, $P(X/k) = P(X)/k$. So for any integers $j$ and $k$, it must be the case that $P[(j/k)X] = jP(X/k) = (j/k)P(X)$. Finally, the boundedness of $P$ (restriction i) then establishes the linearity result for any real scalar components of $\mathbf{s}_N$. □

We noted in Section 2.1 that any function of a quantity also defines a quantity. Conceiving of your prevision assertion as an operation on a space of functions of quantities, Theorem 2.4 and its corollary can be understood to say that a *coherent prevision* operator, $P$, *specifies a bounded linear functional* on the space of linear functions of quantities for which prevision is asserted. The boundedness of $P$ is due to the fact that the realm of the quantity vector $\mathbf{X}_N$ is finite. It still allows that for any value of $K$, $P(KX) = KP(X)$. For the realm of $KX$ is bounded too. In Chapter 4 we remind ourselves of the definition of a linear functional, and we study some more extensive implications of this characterization of coherent prevision.

Any reader familiar with the standard laws of finite probability will recognize that they are all immediate consequences of the linearity of coherent prevision, when applied to events. Let us merely state here their representation using the syntax of quantities and coherent prevision:

**(i)** If $C$ is a constant, then $P(C) = C$. In particular, $P(0) = 0$.
**(ii)** For any event $A, P(\tilde{A}) = P(1 - A) = 1 - P(A)$.
**(iii)** For any events $A$ and $B, P(A \vee B) = P(A + B - AB) = P(A) + P(B) - P(AB)$.
**(iv)** If $A$ and $B$ are events whose definitions *require* that $A \leqslant B$, then $P(A) \leqslant P(B)$.

As an example of statement (iv), define the events $A$ and $B$ by

$A \equiv$ (The president of France on 1 May 2002 is a male aged more than 60 years.)

$B \equiv$ (The president of France on 1 May 2002 is a male.)

Clearly, the definitions of $A$ and $B$ require that $A \leqslant B$, since $A$ cannot equal 1 if $B$ equals 0. Thus, your probabilities, if they are to be coherent, must satisfy $P(A) \leqslant P(B)$. Notice that $B = AB + \tilde{A}B = A + \tilde{A}B$ in this context. Thus, $P(B) \geqslant P(A)$. An interesting example of common incoherencies in this context appears in Kahneman, Slovic, and Tversky (1982, p. 92).

The following example displays an application of the linearity of coherent prevision in a practical problem.

***Example 2.*** Let $U$ denote the event that the New Zealand unemployment rate, measured according to the number of people who register for the unemployment benefit during the first quarter of 1990, exceeded 11%. Let I denote that the rate of increase in the Consumer Price Index (CPI) for New Zealand over the year ending in the first quarter of 1990 did not exceed 7%. Suppose that you assert the probabilities $P(U) = .6$, $P(U \Rightarrow I) = .85$, and $P(U \vee I) = .95$. What does coherency require for your assertion of $P(I)$ and for your assertion of $P(I \Rightarrow U)$?

***Answer.*** First, note that the linearity of coherent prevision requires that your

$$P(U \Rightarrow I) = P(1 - U + UI) = 1 - P(U) + P(UI) = .85,$$

$$P(U \vee I) = P(U + I - UI) = P(U) + P(I) - P(UI) = .95.$$

Solving these two extreme right hand side equations along with your assertion of $P(U) = .6$ yields the solutions $P(UI) = .45$ and $P(I) = .8$. Finally, the linearity of coherent prevision, once again, requires in addition that

$$P(I \Rightarrow U) = P(1 - I + IU) = 1 - P(I) + P(IU) = .65.$$

The mechanical use of the linearity property of coherent prevision in this example should be easily understood. As to the substantive content of the answer, it might be surprising to you that the events $(U \Rightarrow I)$ and $(I \Rightarrow U)$ are both assessed with probabilities greater than .5 in the same problem. Let me only remark here that the notion of logical implication has nothing to do with the supposed physical relation of "causality." With proper deference to David Hume, the operational subjective tradition in probability theory rejects the notion of "cause" as observationally meaningless. Neither does low inflation cause high unemployment, nor does high unemployment *cause* low inflation. While the idea of rejecting causality as a meaningful concept is not old, you may need to struggle with it for some time, since crude notions of causality are commonly presumed both in sophisticated scientific journals as well as the popular press. While we can all observe the value of $U$, the value of $I$, and even the value of $(U \Rightarrow I)$, no one ever observes "$U$ causing $I$."

□

It is time for you to try some problems. If you find yourself solving a lot of linear equations and wondering that surely there must be a more efficient way to be doing all this, just carry on for now and complete the problems as posed. By the end of this chapter we shall have developed more efficient numerical solution procedures as well as richer theoretical conceptions of what we are doing here.

## PROBLEMS

1. Based upon your solutions to problems 3 and 4 of Section 2.2, use the linearity of coherent prevision to establish a formula for the probability that at least one of three events occurs, $P(E_1 \vee E_2 \vee E_3)$; ... your probability that at least one of $N$ events occurs, $P(E_1 \vee E_2 \vee \cdots \vee E_N)$.

2. A climbing expedition has the goal to reach the peak of Mount Dhaulagiri in central Nepal. Suppose that you, as one of the team members, surmising the difficulty of the climbing, the current state of weather conditions on the ice, and the preparation schedule of the crew, assert the following previsions: your prevision for the event $R$, that at least one team member eventually reaches the peak on the expedition, equals .45; your prevision for the event $F$, that at least one member dies in a fatal accident, equals .15; and your prevision that at least one of these two happenings occurs is .50. Determine the cohering previsions for events that

   **a.** there is a fatal accident or no one reaches the peak;
   **b.** someone reaches the peak or there are no fatal accidents;
   **c.** there is a no fatal accident and no one reaches the peak.

3. For your information, this problem was formulated in early April 1989. Let the quantity

   $U$ denote the event that the NZ unemployment rate for the second quarter of 1989 exceeds 13%,

$I$ denote the event that the NZ inflation rate in the CPI for the second quarter of 1989 exceeds 5%,

$C$ denote the event that 1.0 \$NZ > .61\$U.S. at the close of trade in New York on 1 July 1989, and

$R$ denote the event that the Bank of New Zealand interest rate offered for a three-month certificate of deposit on 1 July 1989 exceeds 11%.

An economic forecaster asserts the probabilities

$$P(\tilde{U}) = .75, \quad P(I) = .10, \quad P(C) = .20, \quad P(R) = .30,$$

$$P(U\tilde{I}) = .20, \quad P(UR) = .10, \quad P(\tilde{U}\tilde{C}) = .60, \quad P(\tilde{C}I) = .05.$$

[In fact, these were my (Frank Lad's) probabilities, asserted on 3 April 1989 at 8:27 A.M. New Zealand time.]

**(i)** Determine the forecaster's probabilities for the event $(\tilde{U} \vee \tilde{I})$ and for the event $(\tilde{U} \vee I)$ that are required by coherency.

**(ii)** Use your answer to problem 1 to determine the forecaster's probability for the event $(C \vee \tilde{U} \vee I)$ as required by coherency. Can you find a numerical answer to this problem? (Thinking ahead question: can you determine any bounds on what this answer might be?)

**(iii)** Think about the events specified and the various things you may know that you consider relevant to them. Assert *your* probabilities for the same events assessed by this forecaster. Do you feel any problem on account of the fact that the values of these events have already been determined and that you *could* find their exact values by looking up the relevant statistics in a New Zealand statistics annual? Have you looked them up? In your current state of uncertain knowledge, what are your probabilities for these events?

**(iv)** Adjust the events considered in this problem to identify four events currently relevant to the situation of unemployment, inflation, foreign exchange, and interest rates in whatever country interests you. Assess your probabilities corresponding to the eight probabilities that are "given" in this problem and solve the problems posed in (i) and (ii).

**4.** Suppose you will record the date of each stranding of a group (at least two) of pilot whales on Farewell Spit (along the Cook Straight of New Zealand) during the next 10 years. Then for each observed stranding record the number of days that have passed since the most recent new moon. Thus, for example, a stranding occurring on the day (or night) of a new moon will be summarized by recording 0; a stranding occurring one day after the new moon would specify the recording of 1; and so on. Strandings on the day (night) of a full moon would specify a recording of 14; on the day of the third quarter, 21; and on the day before a new moon, 28. Now define the quantity $T$ to equal the average of all such recordings during the decade. What is your $P(T)$? What is your $P(|T - 14| \leqslant 3)$?

*N.B.* This is mainly a thinking problem. Detailed attention to this problem would require some precise efforts that you need not expend now. What are your main considerations in identifying your prevision here? For your information, the average duration of a lunar month, called the synodic period of the moon, is known quite precisely to equal 29.5306 solar days. A detailed study relevant to this problem appears in Lad and Brabyn (1993).

## PROOF OF THEOREM 2.4

*Proof* ($\Rightarrow$). Suppose that $P$ is coherent and prove restrictions (i) and (ii) by contradiction.

(i) Suppose your $P(X) > \max \mathscr{R}(X)$. Any choice of $s > 0$ would mean that you are buying $s$ units of $X$ for $P(X)$ per unit and would result in a net gain for you of $s(x - P(X)) < 0$ for every $x \in \mathscr{R}(X)$. Such a predicament characterizes you as a sure loser, establishing the incoherency of your assertion. Choosing the scale factor as $s = S/[P(X) - \min \mathscr{R}(X)]$ would ensure that the scale of your maximum stake is not exceeded. A similar result follows if your $P(X) < \min \mathscr{R}(X)$. The sure loss transaction would be identified by a negative value of $s$, involving your selling a claim to $X$ in return for $P(X)$. Thus, any coherent $P(X)$ must be bounded by the extremes of $\mathscr{R}(X)$.

(ii) Suppose your $P(X + Y) = P(X) + P(Y) + \delta > P(X) + P(Y)$, with $\delta > 0$. In this case, suppose we pick scalars $s_{X+Y}, s_X$, and $s_Y$ of permitted scale that satisfy $s_{X+Y} = -s_X = -s_Y > 0$. [In the language of buying and selling, this would mean you buy $s_{X+Y}$ units of $X + Y$ for the price $P(X + Y)$ while you also sell $s_X$ units of $X$ at the price $P(X)$ and $s_Y$ units of $Y$ at the price $P(Y)$.] When it is determined by observation that $X = x \in \mathscr{R}(X)$ and $Y = y \in \mathscr{R}(Y)$, your net gain from the combination of transactions identified by these scalars would be

$$\begin{aligned} NG(x, y{:}\, s_{X+Y} = -s_X = -s_Y) &= s_{X+Y}[x + y - P(X + Y)) \\ &\quad + s_X[x - P(X)] + s_Y[y - P(Y)] \\ &= -s_{X+Y}P(X + Y) - s_X P(X) - s_Y P(Y) \\ &= -s_{X+Y}[P(X) + P(Y) + \delta] - s_X P(X) - s_Y P(Y) \\ &= -s_{X+Y}\delta \\ &< 0 \qquad \text{for every } x \in \mathscr{R}(X) \text{ and } y \in \mathscr{R}(Y). \end{aligned}$$

But this contradicts the coherency of $P$. If you were called on your avowed willingness to act on your asserted previsions by paying $s_{X+Y}P(X + Y)$ for a claim to $X + Y$ and by selling claims to $X$ and to $Y$ in return for $s_X P(X)$ and $s_Y P(Y)$, then you would be forced in net to pay out $s_{X+Y}\delta$ no matter what the values of $X$, $Y$, and $X + Y$ happen to be. You are a sure loser. A similar result follows if your $P(X + Y) = P(X) + P(Y) - \delta < P(X) + P(Y)$. Thus, any coherent assertions of $P$ must specify a finitely additive operator: $P(X + Y) = P(X) + P(Y)$. □

*Proof* ($\Leftarrow$). The general linearity of coherent prevision implies that once you assert your $P(\mathbf{X}_N)$ you must also assert

$$P(\mathbf{s}_N^{\mathrm{T}}[\mathbf{X}_N - P(\mathbf{X}_N)]) = P(\mathbf{s}_N^{\mathrm{T}}\mathbf{X}_N) - P(\mathbf{s}_N^{\mathrm{T}} P(\mathbf{X}_N)] = 0$$

for any vector of scalars $\mathbf{s}_N$. Realizing that $(\mathbf{s}_N^{\mathrm{T}}[\mathbf{X}_N - P(\mathbf{X}_N)])$ denotes another quantity, by restriction (i) we know that for any such choice of scalars the number 0 must lie within the extremes of the realm of the quantity $\mathbf{s}^{\mathrm{T}}[\mathbf{X}_N - P(\mathbf{X}_N)]$. Thus, there are no scalars $\mathbf{s}_N$ that can ensure $\mathbf{s}^{\mathrm{T}}[\mathbf{x}_N - P(\mathbf{X}_N)] < 0$ for every possible vector $\mathbf{x}_N \in \mathscr{R}(\mathbf{X}_N)$. That is, $P$ is coherent. □

## 2.6 A GEOMETRICAL REPRESENTATION OF THE REQUIREMENT OF COHERENCY

A very general understanding of the mathematical implications of coherency can be achieved by studying the geometrical representation of all coherent prevision assessments. If you are a beginner at linear algebra and its geometry, you may need some background help from your teacher in understanding this section. However, the concepts are not difficult. If you are more advanced in the use of these mathematical concepts and notation, this section should be simple and straightforward. In either case, the resulting conclusion is very powerful and important to our further developments: the coherence of an asserted prevision vector is equivalent geometrically to the stipulation that the vector, considered as a point, lies within the convex hull of the realm of the quantity vector in question. Let us develop this result.

The coherency of asserting $P(\mathbf{X}_N) = \mathbf{p}_N \equiv (p_1, \ldots, p_N)^{\mathrm{T}}$ demands minimally that for each component $X_i$, the previsions satisfy $\min \mathscr{R}(X_i) \leqslant p_i \leqslant \max \mathscr{R}(X_i)$. Thus, any coherent prevision vector $P(\mathbf{X}_N) = \mathbf{p}_N$ is identifiable geometrically as a point in $N$-dimensional space, bounded in each of its dimensions by the realm of the quantity defining the corresponding component of $\mathbf{X}_N$. But we can be even more precise; the boundedness of $\mathbf{p}_N$ in every dimension is not sufficient to establish its coherency.

To establish and to understand the necessary and sufficient geometrical conditions for the coherency of the assertion $P(\mathbf{X}_N) = \mathbf{p}_N$, we need to consider two definitions that relate algebraical and geometrical concepts: a supporting hyperplane and the convex hull of a set of points.

***Definition 2.12.*** The vector $\mathbf{s}_N$ and constant $c$ together represent a *supporting hyperplane* for the setof points $\mathscr{R}(\mathbf{X}_N) = \{\mathbf{x}_{N_1}, \ldots, \mathbf{x}_{N_K}\}$ if $\mathbf{s}_N^{\mathrm{T}}\mathbf{x}_N = c$ for some $\mathbf{x}_N \in \mathscr{R}(\mathbf{X}_N)$, and $\mathbf{s}_N^{\mathrm{T}}\mathbf{x}_N \geqslant c$ for every $\mathbf{x}_N \in \mathscr{R}(\mathbf{X}_N)$. If $(\mathbf{s}_N, c)$ represents a supporting hyperplane for $\mathscr{R}(\mathbf{X}_N)$, then the set $\{\mathbf{x}_N | \mathbf{s}_N^{\mathrm{T}}\mathbf{x}_N \geqslant c\}$ is called the *supporting half-space* for $\mathscr{R}(\mathbf{X}_N)$ defined by $\mathbf{s}_N$. □

This nomenclature is motivated by the fact that the linear equation $\mathbf{s}_N^{\mathrm{T}}\mathbf{x}_N = c$ defines a plane in many dimensions (a hyperplane); and the linear inequality $\mathbf{s}_N^{\mathrm{T}}\mathbf{x}_N \geqslant c$

for each $\mathbf{x}_N \in \mathscr{R}(\mathbf{X}_N)$ ensures that no member of $\mathscr{R}(\mathbf{X}_N)$ "falls below" this hyperplane. Thus, the hyperplane is said to "support" the set. The inequality partitions real $N$-dimensional space, $\mathbb{R}^N$, into two half-spaces, one containing all points that satisfy the inequality, the other containing all points that do not.

The following definition states three equivalent identifications of a convex hull. (I hope you remember what it means for a set to be convex.)

***Definition 2.13.*** The *convex hull* of the set of points $\mathscr{R}(\mathbf{X}_N) = \{\mathbf{x}_{N_1}, \ldots, \mathbf{x}_{N_K}\}$, to be denoted by $\mathscr{C}(\mathscr{R}(\mathbf{X}_N))$, is

(i) the intersection of all supporting half-spaces for $\mathscr{R}(\mathbf{X}_N)$;
(ii) the smallest convex set containing the elements of $\mathscr{R}(\mathbf{X}_N)$; and
(iii) the set of all convex combinations of vectors $\mathbf{x}_N \in \mathscr{R}(\mathbf{X}_N)$. □

Specifying $\mathscr{C}(\mathscr{R}(\mathbf{X}_N))$ as the "smallest" convex set in statement (ii), means that no convex proper subset of $\mathscr{C}(\mathscr{R}(\mathbf{X}_N))$ contains all elements of $\mathscr{R}(\mathbf{X}_N)$. Specifying $\mathscr{C}(\mathscr{R}(\mathbf{X}_N))$ as the set of all "convex combinations" in statement (iii) means that

$$\mathscr{C}(\mathscr{R}(\mathbf{X}_N)) = \left\{ \mathbf{p}_N \,|\, \mathbf{p}_N = \sum_{i=1}^{K} q_i \mathbf{x}_{Ni}, \text{ where each } q_i \geqslant 0, \text{ and } \sum_{i=1}^{K} q_i = 1 \right\}.$$

For, in general, a convex combination is a linear combination defined by non-negative coefficients that sum to 1.

Proofs of the equivalence of these three definitions can be found in most any text on linear algebra or real analysis, though you might well think them out yourself. All three definitions will aid your intuitions in understanding the following important theorem. We give a simple proof, based on the linearity of coherent prevision and the characterization (iii) of a convex hull. However, the proof does rely on conceptions of a partition and of partial prevision assertion that we shall not study in detail until later in this chapter.

**Theorem 2.5.** Let $\mathbf{X}_N$ be a vector of quantities with realm $\mathscr{R}(\mathbf{X}_N) = \{\mathbf{x}_{N_1}, \ldots, \mathbf{x}_{N_K}\}$. The assertion $P(\mathbf{X}_N) = \mathbf{p}_N$ is coherent if and only if $\mathbf{p}_N$ lies within the convex hull $\mathscr{C}(\mathscr{R}(\mathbf{X}_N))$.

*Proof*. Remember from Theorem 2.3 that any quantity vector $\mathbf{X}_N$ can be represented as $\mathbf{X}_N = \mathbf{R}(\mathbf{X}_N)\mathbf{Q}(\mathbf{X}_N)$, where $\mathbf{R}(\mathbf{X}_N)$ is the $N \times K$ realm matrix of constants and $\mathbf{Q}(\mathbf{X}_N)$ is a $K \times 1$ vector of events. Thus, the linearity of coherent prevision requires that $P(\mathbf{X}_N) = \mathbf{R}(\mathbf{X}_N)P(\mathbf{Q}(\mathbf{X}_N))$. Now each component of $\mathbf{Q}(\mathbf{X}_N)$ is an event. Thus, Theorem 2.4 would require that any prevision assertion $P(Q_i(\mathbf{X}_N)) \equiv q_i \in [0, 1]$. Moreover, the components of $\mathbf{Q}(\mathbf{X}_N)$ are specific events, i.e., the constituents of the partition generated by $\mathbf{X}_N$. Thus, one of them must equal 1 and the rest 0. So there is one linear restriction governing them, namely that their components sum to 1. Again, Theorem 2.4 would thus require that

$1 = P(\sum_{i=1}^{K} Q_i(\mathbf{X}_N)) = \sum_{i=1}^{K} P(Q_i) \equiv \sum_{i=1}^{K} q_i$. Although there is no requirement that you assert your $P(\mathbf{Q}(\mathbf{X}_N))$ as some specific vector $\mathbf{q}_K$, coherency would not allow you to assert previsions that do not satisfy the constraints that each $q_i \in [0, 1]$ and $\sum_{i=1}^{K} q_i = 1$. Thus, at a minimum, $P(\mathbf{X}_N) = \mathbf{R}(\mathbf{X}_N)\mathbf{q}_K = \sum_{i=1}^{K} q_i \mathbf{x}_{Ni}$, for some vector $\mathbf{q}_K$ that satisfies them. Thus, to be coherent, the vector $P(\mathbf{X}_N)$ must lie within the convex hull $\mathscr{C}(\mathscr{R}(\mathbf{X}_N))$. And if it does, it is. □

The coherency of any prevision vector lying within the convex hull of the realm is corollary to one of the most famous theorems of twentieth-century mathematics, the separating hyperplane theorem: If $\mathbf{x}$ is any member of a closed convex set $\mathscr{C}$, then there exists no hyperplane $\mathbf{s}$ and constant $c$ for which $\mathbf{s}^T\mathbf{x} < c < \mathbf{s}^T\mathbf{y}$ for every other element $\mathbf{y}$ of $\mathscr{C}$. That is, there is no hyperplane that separates any element of a closed convex set from the remaining elements in the set. Such a characteristic is virtually the definition of the coherence of a prevision assertion $P(\mathbf{X}_N) = \mathbf{p}_N$: there is no vector of scalars $\mathbf{s}_N$ (no hyperplane) with the property that $\mathbf{s}_N^T(\mathbf{x}_N - \mathbf{p}_N) < 0$ for every $\mathbf{x}_N \in \mathscr{R}(\mathbf{X}_N)$. That is, there is no hyperplane that will separate a coherent prevision vector from the realm of the quantity vector in question. Any coherent prevision vector must reside in every half-space that contains $\mathscr{R}(\mathbf{X}_N)$. Any coherent prevision vector must reside in the convex hull $\mathscr{C}(\mathscr{R}(\mathbf{X}_N))$. Let's look at two easy examples.

***Example 1.*** Consider the operation of rolling two standard cubic dice. Let $X_1$ and $X_2$ denote the numbers of dots showing on their top faces. The realm of each component is identical, $\mathscr{R}(X_i) = \{1, 2, 3, 4, 5, 6\}$; and the realm of the vector $\mathbf{X}_2$ is the Cartesian product of the realms of its components: $\mathscr{R}(\mathbf{X}_2) = \mathscr{R}(X_1) \otimes \mathscr{R}(X_2) \equiv \{(x_1, x_2) | x_1 \in \mathscr{R}(X_1) \text{ and } x_2 \in \mathscr{R}(X_2)\}$. (We review the definition of a Cartesian product in the next section.) Figure 2.1 depicts the convex hull $\mathscr{C}(\mathscr{R}(\mathbf{X}_2))$ generated by the realm of the quantity vector $\mathbf{X}_2$. Any coherent prevision for $\mathbf{X}_2$ is representable as a vector lying within this square hull. Any vector lying within this square hull represents a coherent prevision for the vector $\mathbf{X}_2$. □

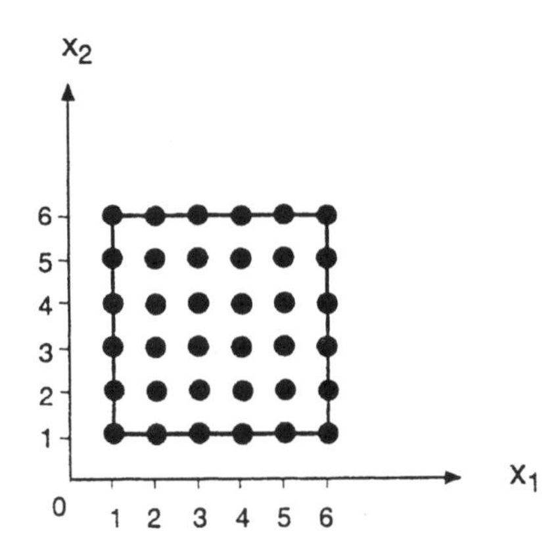


**Figure 2.1.** The convex hull of the realm of a vector of quantities defined by the rolls of two dice.

***Example 2.*** Consider the operation of flipping a coin three times. Denote by $E_1, E_2$, and $E_3$ the events that equal 1 if the subscripted flip yields a head, and otherwise equal 0. Each possible measurement $E_i = 0$ and $E_i = 1$ is compatible with each other possible measurement $E_j = 0$ or $E_j = 1$. Then the realm of the vector of events $\mathbf{E}_3 = (E_1, E_2, E_3)^T$ is the set of the eight possible triples of 0's and 1's that define the vertices of a unit cube in three-dimensional space: $\mathscr{R}(\mathbf{E}_3) = \mathscr{R}(E_1) \otimes \mathscr{R}(E_2) \otimes (E_3)$. Any point within the convex hull of these eight points represents a specification of coherent prevision for $\mathbf{E}_3$. Figure 2.2 depicts these eight elements of $\mathscr{R}(\mathbf{E}_3)$ as points in 3-dimensional space. The convex hull $\mathscr{C}(\mathscr{R}(\mathbf{E}_3))$ is the cube having these eight points as its vertices. Any coherent prevision for $\mathbf{E}_3$ is representable by a vector that lies within this cubic hull. Any vector lying within this hull represents a coherent prevision for the vector of events $\mathbf{E}_3$. □

Our main interest in these two examples just now is to notice explicitly that any vector lying in an appropriate convex hull of points in a realm can represent someone's coherent prevision. Any assertion of prevision that can be so represented must be considered an acceptable assessment based on someone's considered state of information, acceptable in the sense that there is nothing illogical about it, acceptable in the sense that you (as its proponent) cannot be made a sure loser if someone would "call you" on this assertion. An objectivist text on probability would offer these two examples as probability problems in which the dice or the coin would be described as "perfectly balanced," "symmetric," or "fair." You would be required to say that the "correct" probability vector for $\mathbf{E}_3$ is (.5, .5, .5), and that the "correct" expectation of $\mathbf{X}_2$ is (3.5, 3.5). By the way, what are your previsions for $\mathbf{E}_3$ and for $\mathbf{X}_2$?

More interesting examples of convex hulls representing all coherent prevision assertions will arise when we consider vectors whose realms do not equal the Cartesian products of the realms of their component quantities. The usefulness of this geometric interpretation of coherency of prevision develops from a deeper study of possible logical dependencies among distinct quantities. We now turn to a formal characterization of this topic.

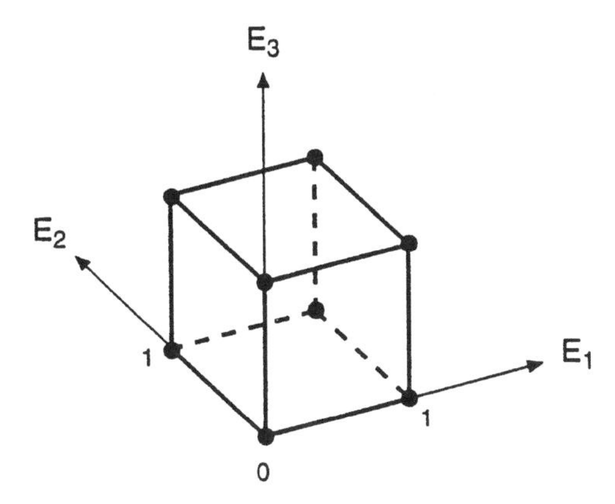


**Figure 2.2.** The convex hull of the realm of a vector of events defined by three flips of a coin.

## 2.7 LOGICAL RELATIONS AMONG QUANTITIES

When discussing the quantities $X_1$ and $X_2$ in Example 1 of Section 2.6, we noticed that every possible outcome of each quantity would be compatible with every possible outcome of the other. The fact that one die turns up a 5, say, places no restriction on whether the other die turns up 1, 2, 3, 4, 5, or 6. Thus, the realm of the vector of the two quantities equals the Cartesian product of the realms of its components. (To remind you, the Cartesian product of two sets is defined as the set of all possible ordered pairs of elements of the two sets. Standard notation is $A \otimes B \equiv \{(a, b) | a \in A \text{ and } b \in B\}$.) While the realms of many quantity vectors have this characteristic, this is not true for every vector of quantities. When it does not hold, we say that the quantities entail some form of logical dependence among themselves. Evidently, in the examples of Section 2.6, the quantities $X_1$ and $X_2$ and the events $E_1$, $E_2$, and $E_3$ are logically independent. Let us now consider some examples of logically dependent quantities. We conclude this Section by stating a formal definition.

***Example 1.*** Suppose that three pieces of output were each rejected in a quality control check of production, on the basis of at least one of several specified criteria. For example, each of three metal bolts might have been rejected for at least one of the following criteria: too long, too short, uneven threads, mangled threads, mangled head, missing head, or missing tail. Let $E_i$ be defined as the event that equals 1 if at least one criterion disqualifying piece $i$ also disqualifies another of these pieces of output—that is, if at least one of the other two pieces was rejected for at least one of the same reasons as was piece $i$. Otherwise, $E_i$ equals 0. Clearly each event has the realm $\mathscr{R}(E_i) = \{0, 1\}$. But the realm for the row vector of events $\mathbf{E}_3 \equiv (E_1, E_2, E_3)$ would be $\mathscr{R}(\mathbf{E}_3) = \{(0, 0, 0), (1, 1, 0), (1, 0, 1), (0, 1, 1), (1, 1, 1)\}$. This is a *proper* subset of $\mathscr{R}(E_1) \otimes (\mathscr{R}(E_2) \otimes \mathscr{R}(E_3)$. The triples $(1, 0, 0)$, $(0, 1, 0)$, and $(0, 0, 1)$ are not elements of $\mathscr{R}(\mathbf{E}_3)$ since it is *impossible* that they represent the value of the event vector. For it is impossible that exactly one piece of output is disqualified by a quality criterion which also disqualifies another piece. If a fault in one piece has a match among the other two pieces, then the piece(s) with a matching fault has (have) a match as well! The operational definitions of the measurements defining the three component events impose a *logical relation* among them. The logical relation stifles the free reign of each event to lie anywhere in its realm without any consideration of the value of the others.

Figure 2.3 depicts the five elements of $\mathscr{R}(\mathbf{E}_3)$ in three-dimensional space for this example. The convex hull of these points is the polytope that has these five points as its vertices. Any coherent prevision asserted for the vector of events $\mathbf{E}_3$ can be represented by a point lying somewhere within this hull. (It may lie on an edge, or even be one of the vertices.) It is worth noticing in the graph that none of the three events is defined as a *function* of the other two. (How can you tell this easily from looking at the graph?) As to terminology, a polygon is a many-sided figure in two dimensions; a polyhedron is a many-sided (multifaced) figure in three dimensions; and generally, a polytope is a many-sided figure in many dimensions. □

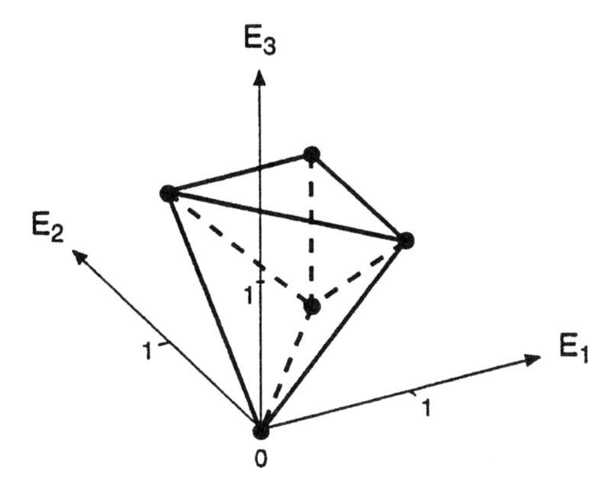


**Figure 2.3.** The convex hull of the realm of a vector of three events whose definitions entail a logical dependency among them. Each event equals 1 only if a fault in a specific piece of output matches a fault in at least one of two other specific pieces of output.

*Example 2.* Events that are defined by logical functions of other events, such as we studied in Section 2.2, can be identified geometrically in the same fashion. Suppose $F_1$ and $F_2$ are logically independent events. It is possible that both, either, or neither of them equal 1. Thus, $\mathscr{R}(\mathbf{F}_2) = \{(0, 0), (0, 1), (1, 0), (1, 1)\}$. However, if we define a third event as $F_3 \equiv F_1 \vee F_2 = F_1 + F_2 - F_1F_2$, and the vector of events as $\mathbf{F}_3 \equiv (F_1, F_2, F_3)$, then $\mathscr{R}(\mathbf{F}_3) = \{(0, 0, 0), (0, 1, 1), (1, 0, 1), (1, 1, 1)\}$. For each triple in $\mathscr{R}(\mathbf{F}_3)$, the third component equals the value required of $F_3$ by its definition as a function of the first two components. You should graph these realm elements yourself, as suggested in the first problem after Example 3. □

In the examples and problems considered thusfar, the values that are possible for each individual event are different depending on what are the values of the other two events. We say that the events are logically dependent on one another. General quantities can be logically dependent, too, as the following example and problems suggest.

*Example 3.* Suppose we have a balance that can measure weights of animals on the hoof in 5-kg units from 0 through 100. Let $X_1 =$ the recorded weight for a particular three-day-old Holstein calf 1, $X_2 =$ the recorded weight for a particular three-day-old Holstein calf 2, and $X_3 \equiv X_1 + X_2$. The realms of the individual quantities are $\mathscr{R}(X_1) = \mathscr{R}(X_2) = \{0, 1, 2, \ldots, 99, 100\}$ and $\mathscr{R}(X_3) = \{0, 1, 2, \ldots, 199, 200\}$. For the vector $\mathbf{X}_3$, $\mathscr{R}(\mathbf{X}_3) = \{(x_1, x_2, x_3) | x_1 \in \mathscr{R}(X_1), x_2 \in \mathscr{R}(X_2)$, and $x_3 = x_1 + x_2\}$. Notice that $\mathscr{R}(\mathbf{X}_3) \neq \mathscr{R}(X_1) \otimes \mathscr{R}(X_2) \otimes \mathscr{R}(X_3)$, since, for example, (26, 22, 50) is not an element of $\mathscr{R}(\mathbf{X}_3)$ even though it is an element of the Cartesian product. □

## PROBLEMS

**1.** In the context of Example 2, identify the four points of $\mathscr{R}(\mathbf{F}_3)$ in three-dimensional space and draw the convex hull of coherent prevision assessments for the three events. (For help in "seeing" your convex hull, follow the standard convention of using unbroken lines to represent any edge that can be seen from your viewing perspective and use dashed lines to represent any edges that cannot be seen.) Notice that although $F_3$ is defined as a function of $F_1$ and $F_2$, neither is $F_1$ a function of $F_2$ and $F_3$, nor is $F_2$ a function of $F_1$ and $F_3$.

**2.** Define $G_1$ as the event that the current quarter's measured unemployment rate for the United States (or for whatever country you choose) exceeds the measured rate for the quarter just completed, and define $G_2$ as the event that the trade deficit for the current quarter exceeds the trade deficit in the most recent quarter. Define $G_3 \equiv (G_1 \Rightarrow G_2) = 1 - G_1 + G_1 G_2$.

**a.** What is the realm of the vector of events, $\mathbf{G}_3 = (G_1, G_2, G_3)$? Identify on a graph in three dimensions the elements of $\mathscr{R}(\mathbf{G}_3)$.

**b.** Define $G_4 \equiv (G_2 \Rightarrow G_1)$. Identify the realm $\mathscr{R}(G_1, G_3, G_4)$ and graph its elements in three-dimensions. (This is done most easily by constructing the realm matrix $\mathbf{R}(\mathbf{G}_4)$ and extracting from it the appropriate three-dimensional column vectors whose components include only the first, third, and fourth row elements of this matrix.) Are these events logically dependent? Is any of these events defined as a function of the other two events?

**c.** Continuing part b, suppose that someone asserts the previsions $P(G_1, G_3, G_4) = (.6, .8, .4)$. Is each of these prevision components within its required limits of $[\min \mathscr{R}(G_i), \max \mathscr{R}(G_i)]$? Is the prevision vector coherent? How can you tell? Draw separately the convex hulls of the realms of the two-dimensional vectors $(G_1, G_3)$, $(G_1, G_4)$, and $(G_3, G_4)$ and determine whether the relevant pairs of prevision assertions would be coherent. (If you were having difficulty determining whether the three-dimensional prevision vector is coherent, it may be due to the difficulty of "seeing the situation" in your hand-sketched graph. If so, look ahead to Figure 2.13 in Section 2.9, where you will find a graphical display of the convex hull you are working at.)

**3.** Let $X_1$ and $X_2$ be defined as in Example 3, and define $X_4 \equiv X_1^2, X_5 \equiv X_2^2$, and $X_6 \equiv X_1 X_2$. Identify the realms $\mathscr{R}(X_4)$, $\mathscr{R}(X_5)$, $\mathscr{R}(X_6)$, and $\mathscr{R}(\mathbf{X}_6)$. Use notation similar to that in Example 3.

**4.** Consider the operation of rolling two standard dice. Let $X_1$ equal the sum of the faces showing, and let $X_2$ equal the product of the faces showing. Draw the convex hull of $\mathscr{R}(\mathbf{X}_2)$. Compare it graphically with the convex hull of the Cartesian product $\mathscr{R}(X_1) \otimes \mathscr{R}(X_2)$. *Hint*: Begin by identifying the realm matrix for $\mathbf{D}_2$ where $D_1$ and $D_2$ are the outcomes of the two individual dice.

**5.** Suppose that $\mathbf{X}_4$ is a vector of quantities, each with the same realm, $\mathscr{R}(X_i) = \{1, 2, 3, 4, 5\}$, and that $\mathscr{R}(\mathbf{X}_4)$ is the Cartesian product $\{1, 2, 3, 4, 5\}^4$. Define the quantities $\bar{X} \equiv \sum_{i=1}^{4} X_i/4$ and $S^2 = \sum_{i=1}^{4}(X_i - \bar{X})^2/4$. Determine the realm matrix $\mathbf{R}(\bar{X}, S^2)$, plot its column vectors as points, and identify their convex hull. Presuming you are using a computer, repeat the problem supposing $\mathbf{X}_6$ is a vector of six such quantities and defining $\bar{X}$ and $S^2$ correspondingly as their average and their average squared difference from their average. Compare your graphs of the convex hulls of $\mathbf{R}(\bar{X}, S^2)$ in the two cases. *Hint*: Review problem 6 of Section 2.1. Adjust your realm.m program to suppress the output of RXN in the loop on values of $i$ by ending that statement with a semi-colon. The commands mean(M) and std(M) produce row vectors of the means and the standard deviations (dividing by $N - 1$) of the columns of the matrix $M$. Read "help plot" to learn the syntax for plotting your desired points.

We have studied several examples of quantities that are logically dependent by their very definition. In some cases the logical dependence was based upon a functional dependence, as when $F_3$ was defined as $F_1 + F_2 - F_1F_2$, or when $G_3$ was defined as $1 - G_1 + G_1G_2$, or even when $X_4 \equiv X_1^2$ and $X_6 \equiv X_1X_2$. But in Example 1, wherein we considered the specimens of rejected production items, no one of the quantities could be written as a *function* of the other two. Similarly in problem 4, the realm of the vector of two quantities does not define a function, even though the two component quantities clearly entail some structure of logical dependence among themselves. The relation of logical dependence among quantities is weaker than the relation of functional dependence. Let us now formally define logical independence, and conclude.

***Definition 2.14.*** If the realm of a vector of quantities equals the Cartesian product of the realms of its components, then the quantities are said to be *completely logically independent*. If the realm of the vector is a proper subset of this Cartesian product, then we say the events are *logically dependent* (*among themselves*). □

We have presumed from the outset of our description of quantities and their realms in Section 2.1 that an investigator knows precisely how the quantities of interest are defined operationally. It is worth reminding ourselves of this point, here, insisting that the realm of a vector of quantities is a matrix of known numbers, irrespective of anyone's knowledge of the quantities. It is possible that someone may not recognize the logical relations that restrict the realm of a vector of quantities. In Example 1 of matching production fault errors, it is possible that someone could learn the definitions of the events without thinking of the fact that the outcomes of $(1, 0, 0)$, $(0, 1, 0)$, and $(0, 0, 1)$ are impossible for the vector $\mathbf{E}_3$. With such a lapse, that person might erroneously think that the convex hull of $\mathbf{E}_3$ is the unit cube, and assert the incoherent prevision $P(\mathbf{E}_3) = (.1, .2, .1)$, thinking that it is coherent. But there is not much of interest to be said about this situation other than a caution to pay attention to the operational definitions of the quantities you are studying, both in designed experimental work and in field observations over which you have no control.

Our next project is to develop the vocabulary needed to distinguish an important special type of logical dependence among events, a partition.

## 2.8 PARTITIONS

One particular relation of logical dependence among events that is regularly found useful in the analysis of uncertain knowledge is the structure we call a partition. This is a collection of events that are both incompatible and exhaustive. You may well be familiar with the definition of a partition as it pertains to a collection of sets. If so, in the present section, your familiarity with the set-theoretic definition of a partition will merely be translated into an arithmetic form of expression. We begin directly with three definitions, and then turn to discussion.

***Definition 2.15.*** *N events are incompatible* if at most one of them can equal 1—if, *necessarily*, their sum cannot exceed 1: $\sum_{i=1}^{N} E_i \leqslant 1$. □

Obviously, any events $E$ and $\tilde{E}$ are incompatible. Also, if $H$ is someone's height measurement in centimeters, then $(H < 130)$ and $(H > 150)$ are incompatible events. If two events are incompatible, their product must equal 0. Furthermore, if $N$ events are incompatible then they are also pairwise incompatible.

**Theorem 2.6.** If $N$ events are incompatible then $E_1 \vee E_2 \vee \cdots \vee E_N = \sum_{i=1}^{N} E_i$.

The event appearing on the left-hand side of this equation is called the *logical sum* of the $N$ events. In this terminology, the theorem states that if $N$ events are incompatible then their logical sum equals their arithmetic sum. The proof is immediate using the equality that you proved as problem 4 of Section 2.2.

***Definition 2.16.*** *N events are exhaustive* if at least one of them must equal 1—if, *necessarily*, their sum must be greater than or equal to 1: $\sum_{i=1}^{N} E_i \geqslant 1$. □

As to someone's height measurement, $H$, the events $(H > 130)$ and $(H < 150)$ are exhaustive.

***Definitions 2.17.*** *N events constitute a partition* if they are incompatible and exhaustive—if their sum necessarily equals 1: $\sum_{i=1}^{N} E_i = 1$. Each of the events that together constitute a partition is called a *constituent* of the partition. The set of constituents is referred to as a *partition*. The number of constituents is called the *size* of the partition. □

The events $(H < 130)$, $(130 \leqslant H \leqslant 150)$, and $(H > 150)$ constitute a partition.

**Warning!** There is syntactic novelty here! Notice that a *partition*, as defined here, is a set whose members (the constituents of the partition) are events—numbers whose values are unknown. You are probably familiar with another definition of the *partition of a set S*, as a set of subsets of $S$ which are disjoint and whose union equals $S$. It may take some concentration on your part to become used to the syntax of a partition as we have defined it here. *A partition is a set of events whose sum equals 1.* There is still another mathematical use for the word "partition," which is more relevant to our usage here. *A partition of a positive integer* is any collection of positive integers that sum to the said integer. For example, the integers 1, 2, 3 and the integers 2, 4 are both said to be partitions of 6. (Repeated integers are allowed in these collections too. For example, 3, 3 and 2, 2, 2 are also both partitions of 6.) In this spirit, it may be helpful to think of partitions characterized by Definition 2.17 to be partitions of the number 1, since the constituents are numbers that sum to 1.

We may also use the word "constituent" in another way, as in the "constituents of an event." Let us merely note the definition and proceed.

***Definition 2.18.*** If an event $F$ equals the sum of $K$ incompatible events, the individual summand events of $F \equiv \sum_{i=1}^{K} E_i$ are called the *constituents of F.* □

The constituents of $F$ need not sum to 1. For their sum is the event $F$, which might equal 0 or 1.

Of course it is common in every field of scientific investigation to be interested in several events that do *not* constitute a partition. For example, the events that the unemployment rate increases this quarter, that net capital expenditure increases, and that consumer debt increases do *not* constitute a partition. These three events are neither exclusive nor exhaustive. In fact, they are completely logically independent.

Nonetheless, with *every* collection of events there is associated a particular specifiable partition which the collection is said to generate. This special partition is identified by our next theorem.

**Theorem 2.7.** Any collection of $N$ events $E_1, E_2, \ldots, E_N$ generates a partition with size $S(N) \leqslant 2^N$ constituents. The constituents of the partition generated by $\mathbf{E}_N$ are the summand product *events* in the multiplicative expansion of the equation

$$1 = \prod_{i=1}^{N} (E_i + \tilde{E}_i) = (E_1 + \tilde{E}_1)(E_2 + \tilde{E}_2) \cdots (E_N + \tilde{E}_N).$$

*Proof.* The multiplicative expansion of this expression yields $2^N$ summands. A typical summand is the *product* of $N$ events, some of the multiplicands being events chosen from among $E_1, E_2, \ldots, E_N$, and the rest of the multiplicands being negations of the remaining events. Two examples would be the product events $E_1 E_2 \tilde{E}_3 E_4 \tilde{E}_5 \cdots \tilde{E}_{N-1} E_N$ and $E_1 E_2 E_3 \cdots E_N$. Depending upon the operational definitions of the original events $E_1, \ldots, E_N$, however, some of these product summands may not be events but the constant 0 on account of some incompatibilities. Thus, the number $S(N)$ of *events* that constitute the partition generated by the multiplicative expansion denoted in the theorem may be less than $2^N$. □

***Example 1.*** Reconsider the case of three completely logically independent events that were defined by the results of three flips of a coin in Example 2 of Section 2.6. In this case, the expansion of the product specified in Theorem 2.7 reads

$$\begin{aligned} 1 &= (E_1 + \tilde{E}_1)(E_2 + \tilde{E}_2)(E_3 + \tilde{E}_3) \\ &= E_1 E_2 E_3 + \tilde{E}_1 E_2 E_3 + E_1 \tilde{E}_2 E_3 + E_1 E_2 \tilde{E}_3 \\ &\quad + E_1 \tilde{E}_2 \tilde{E}_3 + \tilde{E}_1 E_2 \tilde{E}_3 + \tilde{E}_1 \tilde{E}_2 E_3 + \tilde{E}_1 \tilde{E}_2 \tilde{E}_3. \end{aligned}$$

The eight "triple-product" events are the constituents of the partition generated by $E_1, E_2$, and $E_3$, noting that none of them are impossible by definition. It is useful to denote these constituents expressly as events, with denotations $Q_1, Q_2, \ldots$, and $Q_8$. Table 2.2 summarizes the several types of objects we have constructed and the relations among them. □

***Example 2.*** Reconsider Example 1 in Section 2.7 in which events $E_1, E_2$, and $E_3$ were defined in terms of whether a piece of faulty output has a fault that matches a fault found in one of the other pieces. Some of the triple products of events listed as "constituents" in Table 2.2 would necessarily equal 0 in this example. (Which ones are they?) In this case the original three events generate a partition with $S(3) = 5$ constituents, less than $2^3 = 8$. □

The results and discussion of this section should remind you of the general result we stated as Theorem 2.3 long ago, and which we used in our

**Table 2.2. Notation for the Partition Generated by Three Events**

| Vector of Events of Interest | Constituents of the Partition They Generate | Constitutent Denotation | Representation of Constituent by Vector of Events of Interest |
|---|---|---|---|
| $\mathbf{E}_3^T =$ | $E_1 E_2 E_3$ | $Q_1$ | (1, 1, 1) |
| | $\tilde{E}_1 E_2 E_3$ | $Q_2$ | (0, 1, 1) |
| $(E_1, E_2, E_3)$ | $E_1 \tilde{E}_2 E_3$ | $Q_3$ | (1, 0, 1) |
| | $E_1 E_2 \tilde{E}_3$ | $Q_4$ | (1, 1, 0) |
| | $E_1 \tilde{E}_2 \tilde{E}_3$ | $Q_5$ | (1, 0, 0) |
| | $\tilde{E}_1 E_2 \tilde{E}_3$ | $Q_6$ | (0, 1, 0) |
| | $\tilde{E}_1 \tilde{E}_2 E_3$ | $Q_7$ | (0, 0, 1) |
| | $\tilde{E}_1 \tilde{E}_2 \tilde{E}_3$ | $Q_8$ | (0, 0, 0) |

proof of Theorem 2.5. If $\mathbf{X}_N$ is a vector of general quantities with the realm $\mathscr{R}(\mathbf{X}_N) = \{\mathbf{x}_{N_1}, \mathbf{x}_{N_2}, \ldots, \mathbf{x}_{N_K}\}$ and realm matrix $\mathbf{R}(\mathbf{X}_N) \equiv [\mathbf{x}_{N_1} \ \ \mathbf{x}_{N_2} \ \ \cdots \ \ \mathbf{x}_{N_K}]$, then the $K$ components of the event vector $\mathbf{Q}(\mathbf{X}_N) \equiv [(\mathbf{X}_N = \mathbf{x}_{N_1}), \ldots, (\mathbf{X}_N = \mathbf{x}_{N_K})]^{\mathrm{T}}$ constitute the partition generated by $\mathbf{X}_N$. The equation $\mathbf{X}_N = \mathbf{R}(\mathbf{X}_N)\mathbf{Q}(\mathbf{X}_N)$ is a very general statement of the fact that any quantity vector can be written as a linear combination of the constituents of the partition it generates. The linear coefficients are the various row elements of the realm matrix. The following two examples display this equation in the context of an event vector.

***Example 1 continued.*** In this example with $N = 3$, it is easy to see that

$$\mathbf{E}_3 = \begin{pmatrix} E_1 \\ E_2 \\ E_3 \end{pmatrix} = \begin{pmatrix} 1 & 0 & 1 & 1 & 1 & 0 & 0 & 0 \\ 1 & 1 & 0 & 1 & 0 & 1 & 0 & 0 \\ 1 & 1 & 1 & 0 & 0 & 0 & 1 & 0 \end{pmatrix} \begin{pmatrix} Q_1 \\ Q_2 \\ Q_3 \\ Q_4 \\ Q_5 \\ Q_6 \\ Q_7 \\ Q_8 \end{pmatrix} = \mathbf{R}(\mathbf{E}_3)\mathbf{Q}(\mathbf{E}_3).$$

The column vectors of the $3 \times 8$ matrix of 0's and 1's are points in three-dimensional space that represent the eight constituent events of the partition generated by $E_1, E_2$, and $E_3$. For when $Q_i = 1$, the observation vector $(E_1, E_2, E_3)^{\mathrm{T}}$ is identical with column $i$ of this matrix. These points were exhibited in Figure 2.2. □

***Example 2 continued.*** When the three events are defined by the observation of matching faults among three pieces of output, the vector of events $\mathbf{E}_3$ can be expressed as

$$\begin{pmatrix} E_1 \\ E_2 \\ E_3 \end{pmatrix} = \begin{pmatrix} 1 & 0 & 1 & 1 & 0 \\ 1 & 1 & 0 & 1 & 0 \\ 1 & 1 & 1 & 0 & 0 \end{pmatrix} \begin{pmatrix} Q_1 \\ Q_2 \\ Q_3 \\ Q_4 \\ Q_8 \end{pmatrix}.$$

The column vectors of the $3 \times 5$ matrix of 0's and 1's are identical with those in the $3 \times 8$ matrix of Example 1, except that columns 5, 6, and 7 have been deleted. The remaining columns identify points in three-dimensional space that represent the five constituent events of the partition generated by $E_1, E_2$, and $E_3$. They were exhibited in Figure 2.3. □

The main purpose of this section has been to develop your familiarity with the logical structure and the arithmetical syntax of a partition, which will play a central role in our study of the fundamental theorem of prevision in Section 2.10. But it is worth saying right now that if you assert probabilities for all the constitutents of a partition, then we commonly refer to these assertions as specifying your *probability mass function* over the partition. Coherency obviously requires that the sum of these probabilities equals 1. We defer an extensive study of probability distributions until Chapter 5, but we have reasons to refer sometimes to probability distributions before that.

## PROBLEMS

1. Suppose $E_1, E_2$, and $E_3$ are exhaustive events, and that $E_4 \equiv 1 - E_1 + E_1 E_2$ and $E_5 \equiv E_1 + E_2 - E_1 E_2$. What is the size of the partition generated by these five events? Identify $\mathbf{E}_5$ as the product of its appropriate realm matrix and the vector of constituents of the partition it generates.

2. Three friends play two games of three-hand cribbage during lunch. As soon as the winner of a game emerges, each player pays a penny per point to every other player who is ahead of him. Denote $E_i$ as the event equal to 1 if player $i$ concludes the match without incurring a net loss. Otherwise $E_i$ equals 0. Determine the constituents of the partition generated by $\mathbf{E}_3 = (E_1, E_2, E_3)^{\mathrm{T}}$, and graph the convex hull of the realm of the vector $\mathscr{C}(\mathscr{R}(\mathbf{E}_3))$ in three-dimensional space. How would the realm be different if they play only one game? In each case, identify $\mathbf{E}_3$ as the product of its realm matrix and the vector of constituents of the partition it generates.

3. Recognize why the realm matrix for a partition vector $\mathbf{Q}_S$ is the $S \times S$ identity matrix, $\mathbf{I}_S$.

4. Return to problem 2 in Section 2.5 concerning the mountain-climbing expedition. Determine the partition generated by the events $R$ and $F$, and write out the representation of the event vector $(R, F, RF, R \vee F, F \vee \tilde{R}, R \vee \tilde{F}, \tilde{F}\tilde{R})^{\mathrm{T}}$ as the product of its realm matrix with a partition vector. Try building this realm matrix using MATLAB, beginning by creating $\mathbf{R}(R, F)^{\mathrm{T}}$ with the subroutine realm.m you created in problem 6 of Section 2.1. Then generate the subsequent rows of the full realm matrix, using the appropriate arithmetical representations of each row as a function of the first two rows.

5. Return to problem 3 of Section 2.5 involving the economic events $U$, $I$, $C$, and $R$. Represent the event vector $(U, I, C, R, U\tilde{I}, UR, \tilde{U}\tilde{C}, \tilde{C}I, \tilde{U} \vee \tilde{I}, \tilde{U} \vee I, C \vee \tilde{U} \vee I)^{\mathrm{T}}$ as the product of its realm matrix and a partition vector. Build the realm matrix using MATLAB.

## 2.9 A SYSTEMATIC GEOMETRICAL EXAMPLE OF POSSIBLE RELATIONS OF LOGICAL DEPENDENCE ENTAILED AMONG THREE EVENTS

This section is meant to consolidate your understanding of geometrical representations of coherent prevision with the concept of logical relations among quantities. We systematically consider the geometrical aspects of all the various structures of logical dependence that might be entailed among three events. Our focus here is not on practical applications, but on geometrical structures. Actually, it is easy to create practical applications of each of the structures we shall study. Perhaps you can do it yourself as we discuss each example.

The easiest way to construct practical examples of events that are logically related in the ways to be described is to define three events in terms of appropriate functions of a general quantity. For example, suppose that $X$ is the number of correct responses of a subject to a 100-item multiple-choice test. The realm of $X$ is $\mathscr{R}(X) = \{0, 1, 2, \ldots, 99, 100\}$. Suppose we define

$$E_1 \equiv (45 < X < 55) + (X > 90),$$
$$E_2 \equiv (40 < X < 60) + (85 < X < 95),$$
$$E_3 \equiv (35 < X < 65) + (95 < X < 97).$$

You can satisfy yourself that the events $E_1$, $E_2$, and $E_3$ are completely logically independent by creating the realm matrix for the quantity vector $(X\ E_1\ E_2\ E_3)^{\mathrm{T}}$. The first row would be the vector of integers $(0\ 1\ 2\ \cdots\ 98\ 99\ 100)$. Underneath this row write the next three rows specified by the event functions defining $E_1$, $E_2$, and $E_3$. You can then satisfy yourself by inspection that the events are completely logically independent; all eight elements of $\{0, 1\}^3$ appear in the realm of $\mathbf{E}_3$. For each example of this section, these definitions of $E_1$, $E_2$, and $E_3$ can be modified to identify three events that entail the structure of logical dependence described.

In the course of our discussion, we regularly count the numbers of geometrical structures that share various specified characteristics. The presentation presumes that you are familiar with combinatoric enumeration of the number of ways that $R$ distinct objects can be chosen from a group of $N$ objects. We use the standard notation

$${}^{N}C_{R} \equiv N!/R!(N-R)!.$$

If you are not familiar with this counting procedure, or if you would like to solidify your understanding of permutations and combinations at this time you may turn ahead to Section 3.11.1, which reviews the main principles of counting. But if you don't want to get into that right now, just ignore the references in the following discussion to the *counting* of structures.

### 2.9.1 No Logical Restrictions

Three events are completely logically independent if the realm of the vector $\mathbf{E}_3$ equals the Cartesian product $\{0, 1\}^3$. Each event is free to be either 0 or 1, irrespective of the values of the other two events. In representing the vector $\mathbf{E}_3$ as

$$\mathbf{E}_3 = \mathbf{R}_{3,8}\mathbf{Q}_8 = \begin{pmatrix} 1 & 0 & 1 & 1 & 1 & 0 & 0 & 0 \\ 1 & 1 & 0 & 1 & 0 & 1 & 0 & 0 \\ 1 & 1 & 1 & 0 & 0 & 0 & 1 & 0 \end{pmatrix} \mathbf{Q}_8,$$

the columns of the realm matrix $\mathbf{R}_{3,8}$ can be identified geometrically as the eight vertices of the unit cube, which we have already displayed in Figure 2.2. As we continue this section, we study the structures of the degenerating cube that can be achieved by removing successively one, two, three, and more of its vertices.

### 2.9.2 One Logical Restriction

Suppose that the definitions of the component events of $\mathbf{E}_3$ are such that the product quantity $\tilde{E}_1\tilde{E}_2E_3$ necessarily equals 0, but all other products of the events and their complements in the expansion of $\prod_{i=1}^{N}(E_i + \tilde{E}_i)$ are truly events. (That is, they can equal either 0 or 1.) Then we can represent $\mathbf{E}_3$ as

$$\mathbf{E}_3 = \mathbf{R}_{3,7}\mathbf{Q}_7 = \begin{pmatrix} 1 & 0 & 1 & 1 & 1 & 0 & 0 \\ 1 & 1 & 0 & 1 & 0 & 1 & 0 \\ 1 & 1 & 1 & 0 & 0 & 0 & 0 \end{pmatrix} \mathbf{Q}_7.$$

The column $(0, 0, 1)^{\mathrm{T}}$ has been removed from $\mathbf{R}_{3,8}$ to create $\mathbf{R}_{3,7}$, since this column would be an impossible value for the vector $\mathbf{E}_3$. The column vectors of $\mathbf{R}_{3,7}$ are identified by solid balls in Figure 2.4. Evidently, a single vertex, $(0, 0, 1)^{\mathrm{T}}$, has been removed from the unit cube to create the convex hull displayed in Figure 2.4. This

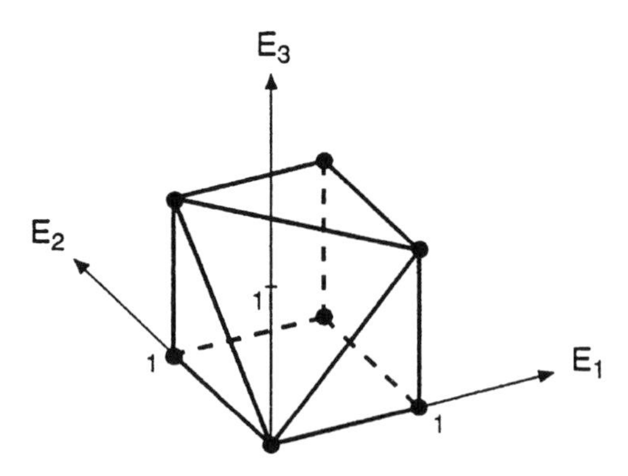


**Figure 2.4.** The convex hull of the realm of $\mathbf{E}_3$ whose component events are logically dependent on account of a single restriction.

particular hull identifies coherent probabilities for $\mathbf{E}_3$ under the logical restriction that $E_1 \vee E_2 = 1$ is a necessary but not sufficient condition for $E_3 = 1$. This is the underlying structure in a famous problem posed by George Boole (1854b), whose several respondents are discussed in a beautiful commentary by Hailperin (1986). There are ${}^8C_1 = 8$ different vertices that could be removed from a cube in order to exemplify a similar relation of logical dependence among three events. The convex hulls deriving from the removal of any other vertex are congruent to this one. They can be obtained by an appropriate rotation of the hull depicted in Figure 2.1. For example, if the three events describe whether the three players in a game of "Oh, Hell!" (a card game) make their bids in one of the hands, only the vector $(1, 1, 1)^T$ would be removed from the unit cube, since the game is explicitly designed so that all players cannot make their bids in the same hand.

### 2.9.3 Two Logical Restrictions

Removing two columns from $\mathbf{R}_{3,8}$ is equivalent to removing two vertices from the unit cube. There are ${}^8C_2 = 28$ different ways that this can be done. But these several ways of removing two vertices yield only three distinct geometric structures that could represent the convex hull of the realm of $\mathbf{E}_3$. For the two vertices to be removed could lie either on an edge of the cube, on a diagonal of a face, or on a central diagonal of the cube. There are 12 different edges, 12 facial diagonals, and 4 central diagonals of a cube. Thus, these three removal strategies generate the 28 possible geometric structures. The convex hulls shown in Figures 2.5, 2.6, and 2.7, each depict one example of a distinct type. The other possible convex hulls could be obtained by appropriately rotating one of these. The algebraic representation of $\mathbf{E}_3$ as $\mathbf{R}_{3,6}\mathbf{Q}_6$ corresponding to each of the displayed figures appears in its caption.

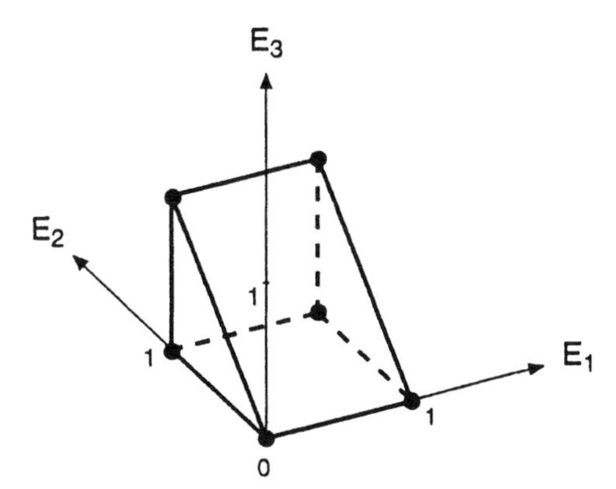


**Figure 2.5.** The convex hull of $\mathscr{R}(\mathbf{E}_3)$ when the two vertices removed from the unit cube lie on an edge:

$$\mathbf{E}_3 = \begin{pmatrix} 1 & 0 & 1 & 1 & 0 & 0 \\ 1 & 1 & 1 & 0 & 1 & 0 \\ 1 & 1 & 0 & 0 & 0 & 0 \end{pmatrix} \mathbf{Q}_6.$$

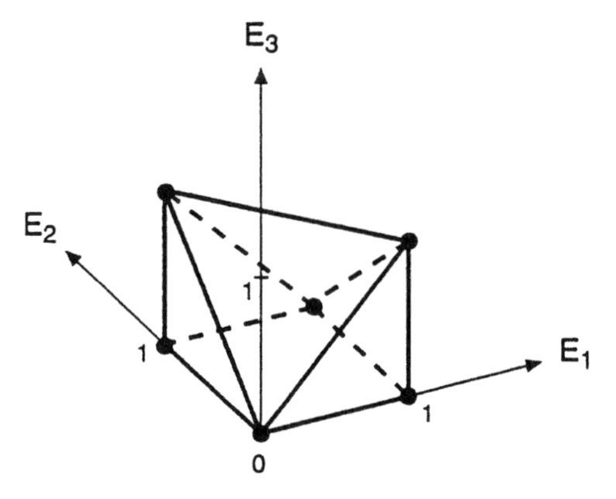


**Figure 2.6.** The convex hull of $\mathscr{R}(\mathbf{E}_3)$ when the two vertices removed from the unit cube lie on the diagonal of a face:

$$\mathbf{E}_3 = \begin{pmatrix} 0 & 1 & 1 & 1 & 0 & 0 \\ 1 & 1 & 0 & 0 & 1 & 0 \\ 1 & 0 & 1 & 0 & 0 & 0 \end{pmatrix} \mathbf{Q}_6.$$

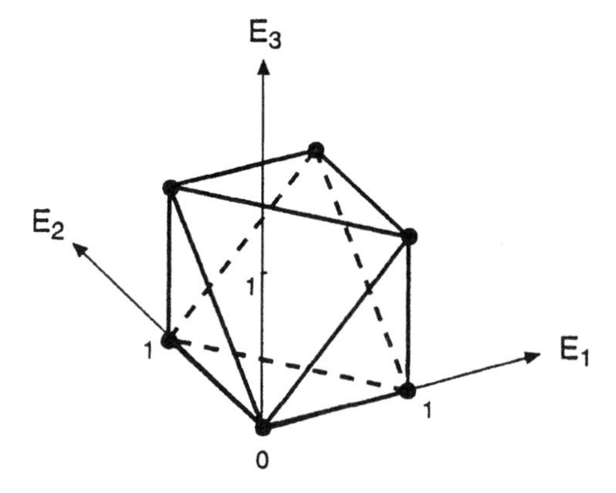


**Figure 2.7.** The convex hull of $\mathscr{R}(\mathbf{E}_3)$ when the two vertices removed from the unit cube lie on a central diagonal:

$$\mathbf{E}_3 = \begin{pmatrix} 1 & 0 & 1 & 1 & 0 & 0 \\ 1 & 1 & 0 & 0 & 1 & 0 \\ 1 & 1 & 1 & 0 & 0 & 0 \end{pmatrix} \mathbf{Q}_6.$$

### 2.9.4 Three Logical Restrictions

There are ${}^8C_3 = 56$ different ways that three vertices can be removed from the eight vertices of a cube. If the joint definitions of the components of $\mathbf{E}_3$ are such that three columns must be removed from $\mathbf{R}_{3,8}$, then again there are only three noncongruent geometrical structures that can represent the convex hull of the realm of $\mathbf{E}_3$. The convex hulls associated with these three structures are exhibited in Figures 2.8, 2.9, and 2.10. The three vertices to be removed could all be selected from the same face; or all three could be selected from a diagonal plane that divides the cube in half; or two could be picked off the diagonal of a face while the third is picked off the perpendicular diagonal of the opposite face, leaving a convex hull that is symmetric about one of the central diagonals of the original

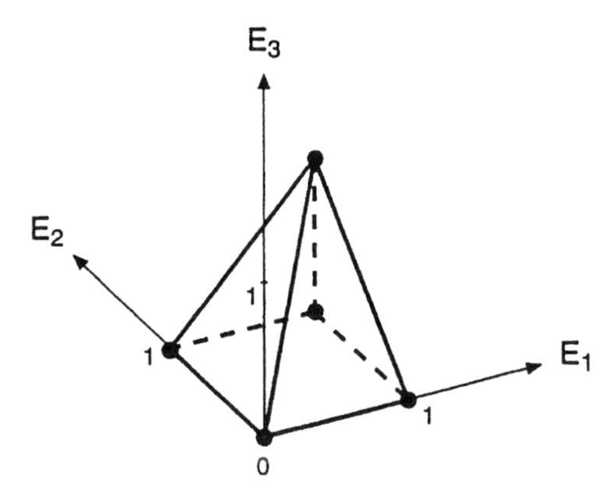


**Figure 2.8.** The convex hull of $\mathscr{R}(\mathbf{E}_3)$ when three vertices are removed, and four of the five remaining vertices constitute one face of the unit cube:

$$\mathbf{E}_3 = \begin{pmatrix} 1 & 1 & 1 & 0 & 0 \\ 1 & 1 & 0 & 1 & 0 \\ 1 & 0 & 0 & 0 & 0 \end{pmatrix} \mathbf{Q}_5.$$

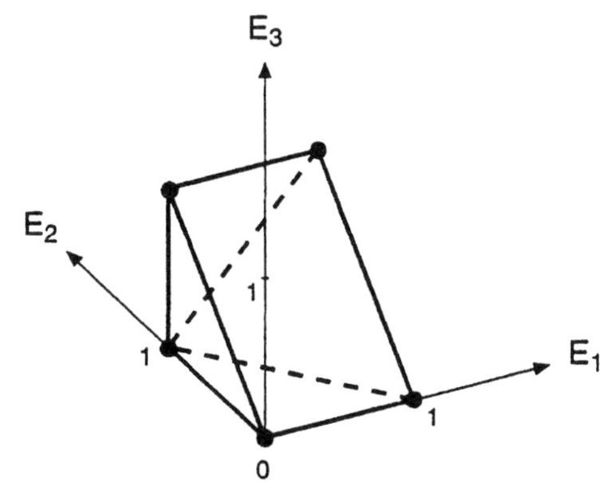


**Figure 2.9.** The convex hull of $\mathscr{R}(\mathbf{E}_3)$ when three vertices are removed, and four of the five remaining vertices lie on a diagonal plane through the unit cube:

$$\mathbf{E}_3 = \begin{pmatrix} 1 & 0 & 1 & 0 & 0 \\ 1 & 1 & 0 & 1 & 0 \\ 1 & 1 & 0 & 0 & 0 \end{pmatrix} \mathbf{Q}_5.$$

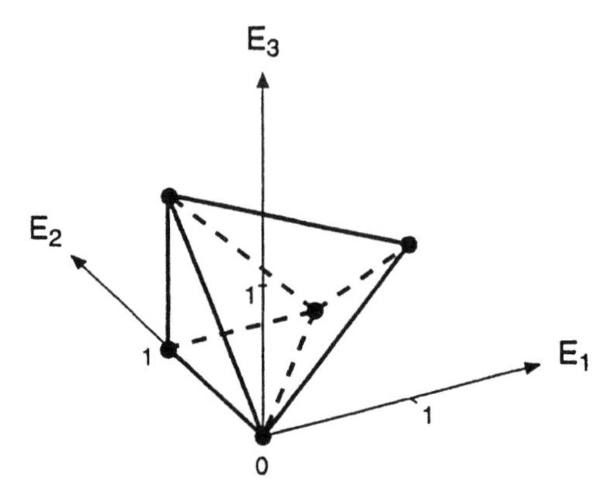


**Figure 2.10.** The convex hull of $\mathscr{R}(\mathbf{E}_3)$ when three vertices are removed from the unit cube, and the five remaining vertices of the hull are symmetric about a central diagonal of the unit cube:

$$\mathbf{E}_3 = \begin{pmatrix} 1 & 0 & 1 & 0 & 0 \\ 0 & 1 & 1 & 1 & 0 \\ 1 & 1 & 0 & 0 & 0 \end{pmatrix} \mathbf{Q}_5.$$

cube. The numbers of distinct but congruent hulls that can be obtained by rotating these are 24, 24, and 8, respectively. Can you count them? Again, the algebraic representation of $\mathbf{E}_3$ as $\mathbf{R}_{3,5}\mathbf{Q}_5$ associated with each figure appears in its caption.

### 2.9.5 Four Logical Restrictions

There are ${}^8C_4 = 70$ distinct choices of four vertices to be removed from a unit cube. For our purposes we can ignore six of these choices—those that would result in the removal of all four vertices from the same face of the cube. For the vectors identified by the remaining vertices would all be constant in one of their components. Thus, they could not represent the realm of a vector of *events*. (It must be possible that an event can equal either 0 or 1.) The remaining 64 ways of removing four vertices from a cube yield six distinct geometrical structures of convex hulls. We discuss their distinctive aspects in pairs. The structures are displayed in Figures 2.11–2.16. In each case, the caption identifies the associated algebraical representation of $\mathbf{E}_3$ as $\mathbf{R}_{3,4}\mathbf{Q}_4$.

A convex hull with the shape displayed in Figure 2.11 is uniquely identified by the corner of the cube in which it is situated. Since there are eight corners on a cube, there are eight distinct but congruent structures that can be obtained by rotating the one depicted. Rotating the unit cube so that the identifying corner of a hull with this shape is $(0, 0, 0)^T$, for example, would depict the convex hull of $\mathscr{R}(\mathbf{E}_3)$ for three incompatible events. The displayed Figure represents the hull of three events whose sum must equal either 2 or 3. The inverses of these three events, $\tilde{E}_1$, $\tilde{E}_2$, and $\tilde{E}_3$, are incompatible.

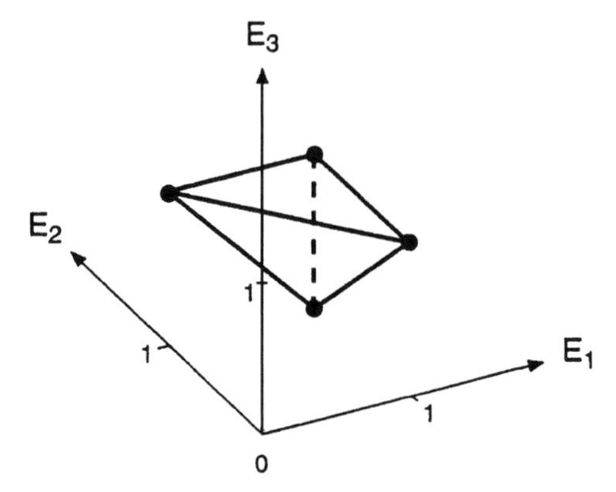


Figure 2.11

$$\mathbf{E}_3 = \begin{pmatrix} 0 & 1 & 1 & 1 \\ 1 & 1 & 0 & 1 \\ 1 & 1 & 1 & 0 \end{pmatrix} \mathbf{Q}_4.$$

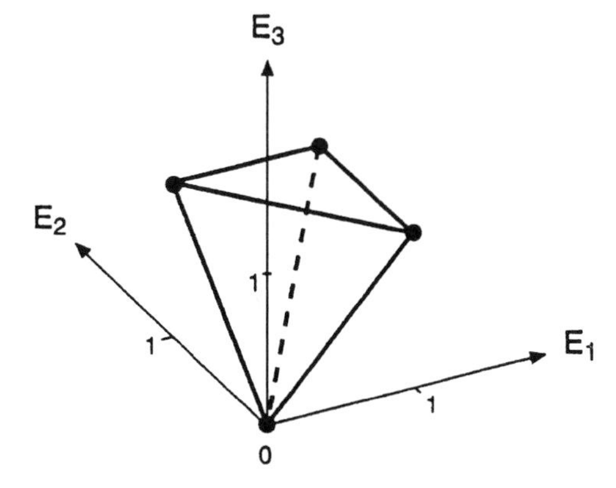


Figure 2.12

$$\mathbf{E}_3 = \begin{pmatrix} 0 & 1 & 1 & 0 \\ 1 & 1 & 0 & 0 \\ 1 & 1 & 1 & 0 \end{pmatrix} \mathbf{Q}_4.$$

Figure 2.12 depicts a convex hull that has one central diagonal and three facial diagonals of the unit cube among its edges. The easiest way to count the number of convex hulls that are congruent to the one depicted is to notice that the facial diagonal on the plane $E_3 = 1$ plays a special role in identifying this geometric figure. It is an edge of the only face that is configured as a right isosceles triangle. This very same facial diagonal plays this same role for one other figure, the one achieved by rotating this convex hull 180° around the line for which

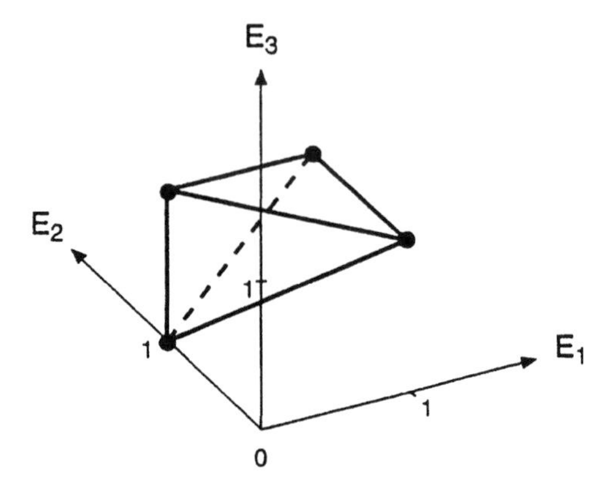


Figure 2.13

$$\mathbf{E}_3 = \begin{pmatrix} 0 & 1 & 1 & 0 \\ 1 & 1 & 0 & 1 \\ 1 & 1 & 1 & 0 \end{pmatrix} \mathbf{Q}_4.$$

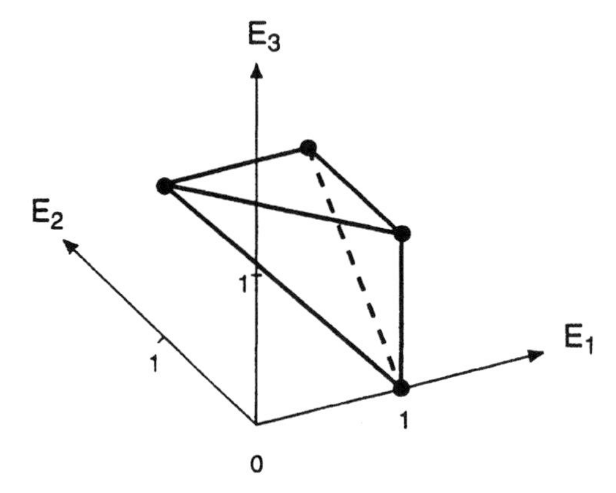


Figure 2.14

$$\mathbf{E}_3 = \begin{pmatrix} 0 & 1 & 1 & 1 \\ 1 & 1 & 0 & 0 \\ 1 & 1 & 1 & 0 \end{pmatrix} \mathbf{Q}_4.$$

$E_1 = E_2 = 1/2$. In the same way, each facial diagonal of a cube is an edge defining two distinct convex hulls that are congruent with this one. Since there are 12 facial diagonals on a cube, there are then 24 distinct hulls congruent to the one depicted. The convex hull displayed in Figure 2.12 represents the logical relations among three events for which $E_1$ and $E_2$ are completely logically independent, while $E_3$ is defined by a logical function of them: $E_3 \equiv E_1 \vee E_2 = E_1 + E_2 - E_1 E_2$.

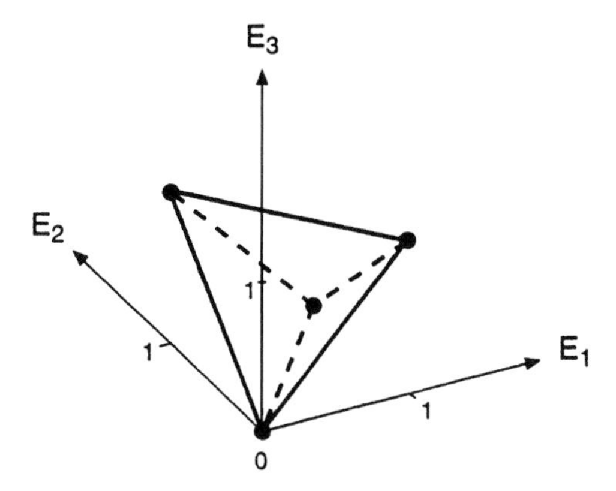


Figure 2.15

$$\mathbf{E}_3 = \begin{pmatrix} 0 & 1 & 1 & 0 \\ 1 & 0 & 1 & 0 \\ 1 & 1 & 0 & 0 \end{pmatrix} \mathbf{Q}_4.$$

Figures 2.13 and 2.14 depict convex hulls that are not congruent to each other, but are mirror images of one another. Each uses three connected edges of the original unit cube to serve as its own edges. A cube of either shape is uniquely identified by the middle one of these edges. Since a cube has 12 edges, there are 12 distinct but congruent convex hulls that can be obtained by appropriately rotating each of these figures.

The convex hull shown in Figure 2.15 exhibits the unusual feature that projecting it onto any face of the unit cube yields the full unit square. Any two component events of $\mathbf{E}_3$ are completely logically independent. Yet each one of the components is representable as a function of the other two! To be precise, each of the component events equals the logical "exclusive or" composition of the other two: $E_i = E_j + E_k - 2E_jE_k$, where $i$, $j$, and $k$ can be replaced by the numbers 1, 2, and 3, in any order. According to the conceptions of two-valued propositional logic, one might be tempted to say that the components of $\mathbf{E}_3$ are completely logically dependent. For the numerical values of any two of them determines the value of the third. However, in the many-valued logic of uncertain knowledge, it is not true that your prevision for any two of these quantities determines your coherent prevision for the third. (That would be the case only if the three events were *linearly* dependent, as in Figure 2.16.) The three events associated with the symmetric tetrahedron depicted in Figure 2.15 have the arithmetical property that their sum equals either 0 or 2. There is only one other convex hull congruent with this one. Its vertices are those that were *removed* from the original cube to create this one. It represents the hull of the realm of three events whose sum equals either 1 or 3.

The obvious feature to notice in Figure 2.16 is that all four vertices of the convex hull lie in a plane. Thus, the three events are so restricted as to be *linearly*

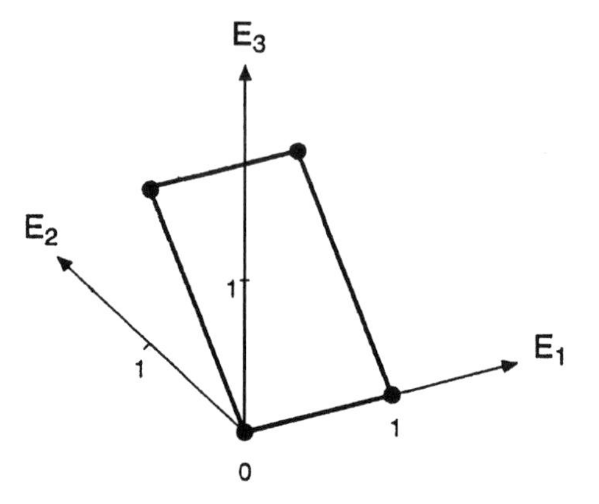


Figure 2.16

$$\mathbf{E}_3 = \begin{pmatrix} 0 & 1 & 1 & 0 \\ 1 & 1 & 0 & 0 \\ 1 & 1 & 0 & 0 \end{pmatrix} \mathbf{Q}_4.$$

dependent. The linear restriction associated with the displayed plane is a simple one: $E_2 = E_3$. Since there are six different diagonal planes in a unit cube, six of the ways of removing four vertices yield a geometrical structure for the hull of $\mathscr{R}(\mathbf{E}_3)$ that is congruent with this one. The associated restrictions that define the other five structures would be $E_1 = E_2$, $E_1 = E_3$, $E_1 = \tilde{E}_2$, $E_1 = \tilde{E}_3$, and $E_2 = \tilde{E}_3$. One might object to the classification scheme that identifies these logical restrictions as characterizing of three events. For if $E_1$ *must* equal $E_2$, for example, then you might want to consider these two events as merely different names for the same event. Fair enough.

We have now accounted for all 70 different ways of removing four of the eight vertices from a unit cube. Six were ignored, since they were not really associated with the convex hulls of the realm of three *events*; and the shapes displayed in our six figures were found to be replicable 8, 24, 12, 12, 2, and 6 times by appropriate rotations. We are ready to exhibit the structures yielded by removing five vertices from a cube.

### 2.9.6 Five Logical Restrictions

When five vertices are removed from the unit cube, only three remain, and thus they must lie in a plane. The three component events of $\mathbf{E}_3$ whose realm is so represented are linearly dependent. There are ${}^8C_5 = 56$ different ways the five vertices can be chosen for removal. However, we may ignore any removal procedure that results in the remaining three vertices all situated on the same face of the cube. For one of the components would be constant in all three vertex vectors that remain. Thus, these vectors could not represent the realm of a vector of *events*. Associated with each of the six faces on a cube are four removal

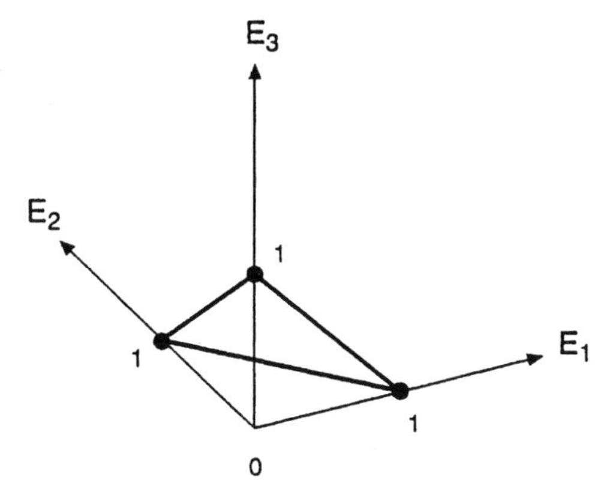


Figure 2.17

$$\mathbf{E}_3 = \begin{pmatrix} 1 & 0 & 0 \\ 0 & 1 & 0 \\ 0 & 0 & 1 \end{pmatrix} \mathbf{Q}_3.$$

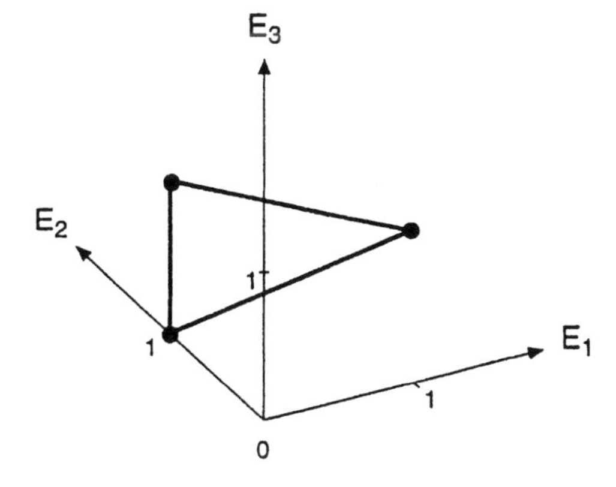


Figure 2.18

$$\mathbf{E}_3 = \begin{pmatrix} 0 & 0 & 1 \\ 1 & 1 & 0 \\ 1 & 0 & 1 \end{pmatrix} \mathbf{Q}_3.$$

procedures for which three vertices remain on that face. Thus, in all there are 24 removal procedures that may be ignored for our purposes. The remaining 32 ways of removing five vertices yield only two distinct structures for the convex hull. They are depicted in Figures 2.17 and 2.18.

Figure 2.17 displays the convex hull of the realm of three events that constitute a partition. (To foreshadow some terminology we use regularly in Chapters 3 and 5, the convex hull of a partition is called a "unit simplex.") A convex hull with this shape is uniquely identified by the single vertex removed from the unit cube that is separated by the triangular hull from the other four removed vertices. Since the cube has eight vertices, there are eight congruent hulls that can be generated from this one by rotating the original cube. As to Figure 2.18, there are 24 convex hulls congruent to the one depicted. Any one of them is uniquely defined by an edge of the cube and a connected central diagonal of the cube. For each of the 12 edges there are two connected central diagonals.

### 2.9.7 Six or Seven Logical Restrictions?

Finally, for the sake of completeness, we should comment on the possibility of removing more than five vertices from the unit cube. There are ${}^{8}C_{6} = 28$ ways that six vertices could be removed. But the two remaining vertices may not define an edge or a facial diagonal of the cube, or else they could not specify the realm of a vector of *events*. There are 12 edges and 12 facial diagonals on a cube. The only four remaining possibilities are the pairs of vertices that define the central diagonals of the cube. Such pairs can define the realm of a vector of three events. For example, if the components of $\mathbf{E}_3$ are defined identically, $E_1 \equiv E_2 \equiv E_3$, then the realm of the vector is $\mathscr{R}(\mathbf{E}_3) = \{(0, 0, 0), (1, 1, 1)\}$. Another example would be definitions that entail $E_1 \equiv \tilde{E}_2 \equiv \tilde{E}_3$. (Which central diagonal of the cube would represent the realm of this vector of events?) In any such case, not only are the three events linearly dependent, but also any two of the three are linearly dependent! In fact, you could very well raise the objection once again that in these situations there are not truly three events but only one event with three different names.

It is impossible to remove seven columns from $\mathbf{R}_{3,8}$ on account of logical dependence among three *events*. If only one column were a possible measurement for the vector $\mathbf{E}_3$, it would be a vector of constants, not events. Geometrically, it would be represented by a point.

We have systematically exhibited all possible ways that three events can entail logical dependence among themselves. As you might guess, there is an algebraic structure that incorporates all the possible geometries of logical relations among events, and more generally among quantities. The structure is that of a ring. If you are familiar with the concepts of abstract algebra, and you are interested by the systematization of logical relations, the rewarding papers of Glynn (1987, 1988) show how the system of geometrical structures constitutes a ring.

Having studied this extensive example, your conceptions of coherent prevision vectors as points lying within the convex hull of a realm are developed well enough that we can study forthwith the fundamental theorem of prevision, which unifies the logic of uncertainty as it applies to most any practical problem.

## 2.10 THE FUNDAMENTAL THEOREM OF PREVISION. PART I

The project of eliciting your prevision for several quantities, or even for a single quantity, requires introspective effort on your part, which may be considerable.

About a quantity such as the number of dots showing after the roll of a die from your backgammon set, you may well easily decide to assert that your $P(X) = 3.5$, and even that your $P(X = 1) = P(X = 2) = \cdots = P(X = 6) = 1/6$. But about the number to be announced by the Bureau of Labor Statistics as the aggregate rate of unemployment in the United States during the third quarter of 1998, you would want to ponder long over numerous characteristics of the United States and world economies before asserting, say, your $P(U_{98/3}) = .073$. Given the compounded difficulty involved in making considered assertions about several quantities, it is of practical interest to be able to specify explicitly all the numerical restrictions that the principle of coherency and your assertion of prevision for any limited number of quantities would imply for your coherent assertion of prevision for any other quantities. This is the content of the theorem we designate the fundamental theorem of prevision. This theorem is a broadening of one designated by de Finetti (1974a, Section 3.10) as the fundamental theorem of probability, which pertains to the special case in which all the quantities considered are events.

The statement of the fundamental theorem is couched in terms of the computational optimization technique known as linear programming. If you are completely unfamiliar with linear programming, your teacher will review for you the basic ideas of the general problem it solves and the basis for its computational solution. There are many textbooks available at both theoretical and applied levels. Walsh (1985) is a good one.

The plan of this section is, first, to state and prove the fundamental theorem and to provide a small practical example of its use. Next, the dual form of the linear program involved in the theorem statement will be identified and interpreted. Third, the theorem will be extended to specify the implications of $N$ prevision assertions for the cohering prevision of any further *vector* of quantities, and some notable features of this extension are discussed. Finally, we discuss two important interpretations of the fundamental theorem. On the one hand, we notice that the theorem expands the well-known closure result of deductive propositional logic, which characterizes *all the propositions deductively implied* by any finite set of propositions accepted as axioms. On the other hand, we find that the theorem also allows us to recognize the operational meaning of a weaker form of uncertainty assertion than prevision, which we have defined as a precise number. The weaker form of assertion about a quantity is represented by an interval. We refer to it as a partial assertion of your prevision. Details will be developed in Section 2.11.

The statement of the theorem involves the following notation:

Let

$\mathbf{X}_{N+1}$ denote a vector of any $N + 1$ quantities that interest you

$S(N + 1)$ denote the size of the realm $\mathscr{R}(\mathbf{X}_{N+1})$

$\mathbf{R}_{(N+1),S(N+1)}$ denote the realm matrix for $\mathbf{X}_{N+1}$

$\mathbf{Q}_{S(N+1)}$ denote the vector of constituents of the partition generated by $\mathbf{X}_{N+1}$

$\mathbf{R}_{N,S(N+1)}$ denote the matrix composed by the first $N$ rows of $\mathbf{R}_{(N+1),S(N+1)}$
$\mathbf{r}_{N+1}$ denote the final *row* of $\mathbf{R}_{(N+1),S(N+1)}$
$\mathbf{0}_{S(N+1)}$ and $\mathbf{1}_{S(N+1)}$ denote $S(N+1) \times 1$ column vectors of 0's and 1's

**Theorem 2.8.** The Fundamental Theorem of Prevision: Part I. Suppose you assert your prevision for $N$ quantities as $P(\mathbf{X}_N) = \mathbf{p}_N$. Let $X_{N+1}$ be any further quantity. Then for the coherency of your prevision over the vector $\mathbf{X}_{N+1}$, it is necessary and sufficient that your further assertion of $P(X_{N+1})$ lie within the interval $[P_l(X_{N+1}), P_u(X_{N+1})]$, where the boundaries of this interval are defined by the solutions to the following two linear programming problems:

Find the $S(N+1)$-tuples $\mathbf{q}_{S(N+1)} = (q_1, \ldots, q_{S(N+1)})^{\mathrm{T}}$ that characterize
$P_l(X_{N+1}) \equiv \text{minimum } \mathbf{r}_{N+1}\mathbf{q}_{S(N+1)}$ and $P_u(\mathbf{X}_{N+1}) \equiv \text{maximum } \mathbf{r}_{N+1}\mathbf{q}_{S(N+1)}$
both subject to the linear constraints that

$$\mathbf{R}_{N,S(N+1)}\mathbf{q}_{S(N+1)} = \mathbf{p}_N,$$
$$\mathbf{1}^{\mathrm{T}}_{S(N+1)}\mathbf{q}_{S(N+1)} = 1,$$
$$\mathbf{q}_{S(N+1)} \geqslant \mathbf{0}_{S(N+1)}.$$

If the feasible set of solutions to these linear programming problems is empty, then the asserted prevision $P(\mathbf{X}_N) = \mathbf{p}_N$ is itself incoherent.

*Proof* ($\Rightarrow$). We presume your prevision is coherent over the components of $\mathbf{X}_{N+1}$, and show that the stated linear programming problems are appropriate. From Section 2.6 we know that the coherency of $P$ over $\mathbf{X}_{N+1}$ is equivalent to the statement that $P(\mathbf{X}_{N+1})$ lies within the convex hull of the vectors in $\mathscr{R}(\mathbf{X}_{N+1})$. That is, $P(\mathbf{X}_{N+1}) = \mathbf{R}_{(N+1),S(N+1)}\mathbf{q}_{S(N+1)}$ for some vector $\mathbf{q}_{S(N+1)}$ with nonnegative components that sum to 1. The conditions of the linear programming problems stated in our theorem include these two requirements on the vector $\mathbf{q}_{S(N+1)}$. The primary condition that $\mathbf{R}_{N,S(N+1)}\mathbf{q}_{S(N+1)} = \mathbf{p}_N$ applies the convexity requirement to the components of $P(\mathbf{X}_{N+1})$ that the theorem presumes you have specified, $P(\mathbf{X}_N) = \mathbf{p}_N$. The objective functions of the two linear programming problems merely formalize the extreme values of your possible assertion for $P(X_{N+1})$ that would allow the full vector $P(\mathbf{X}_{N+1})$ to lie within the required convex hull. If there were no vector $\mathbf{q}_{S(N+1)}$ that meets the conditions of the linear programming problems, your asserted vector $P(\mathbf{X}_N) = \mathbf{p}_N$ would not lie in the convex hull of $\mathscr{R}(\mathbf{X}_N)$ and thus would itself be incoherent. □

*Proof* ($\Leftarrow$). We presume that $P(X_{N+1})$ lies within the interval specified by the solutions to the two linear programming problems, and show that the asserted $P(\mathbf{X}_{N+1})$ is coherent. The convex hull of the realm $\mathscr{R}(\mathbf{X}_{N+1})$ is characterized by the realm matrix $\mathbf{R}_{(N+1),S(N+1)}$ along with the vectors $\mathbf{q}_{S(N+1)}$ that meet both the summation constraint, $\mathbf{1}^{\mathrm{T}}_{S(N+1)}\mathbf{q}_{S(N+1)} = 1$, and the nonnegativity constraints $\mathbf{q}_{S(N+1)} \geqslant \mathbf{0}_{S(N+1)}$. Any asserted prevision vector $P(\mathbf{X}_{N+1})$ that equals

$\mathbf{R}_{(N+1),S(N+1)}\mathbf{q}_{S(N+1)}$ for some vector $\mathbf{q}_{S(N+1)}$ satisfying these conditions must be coherent. The two specified linear programming problems ensure that these conditions are met. Thus, any prevision assertion $P(\mathbf{X}_N) = \mathbf{p}_N$ that allows for feasible solutions to the linear programming problems, appended by an assertion of $P(X_{N+1})$ lying within the interval $[P_l(X_{N+1}), P_u(X_{N+1})]$ defined by the solutions to the two linear programming problems, must be coherent. □

At one extreme, if $X_{N+1}$ is logically related to $\mathbf{X}_N$ by a *linear* function, $X_{N+1} \equiv \mathbf{c}_N^{\mathrm{T}}\mathbf{X}_N$, then $P(X_{N+1})$ is determined precisely on account of the linearity of a coherent prevision functional, $P$. In this case, every feasible solution to the linear programming problems would yield this same number for the objective function value: $P_l(X_{N+1}) = P_u(X_{N+1}) = P(X_{N+1}) = \mathbf{c}_N^{\mathrm{T}}P(\mathbf{X}_N) = \mathbf{r}_{N+1}\mathbf{q}_{S(N+1)}$ for every feasible solution vector $\mathbf{q}_{S(N+1)}$.

At the other extreme, if $X_{N+1}$ happens to be completely logically independent of $\mathbf{X}_N$ (that is, if every possible value of $X_{N+1}$ is compatible with every constituent of the partition generated by $\mathbf{X}_N$), then your assertion of $P(\mathbf{X}_N)$ has no implications for the range of coherent assessment of $P(X_{N+1})$, which remains as the interval $[\min \mathscr{R}(X_{N+1}), \max \mathscr{R}(X_{N+1})]$. Numerical computation of the linear programming solutions would yield $P_l(X_{N+1}) = \min \mathscr{R}(X_{N+1})$ and $P_u(X_{N+1}) = \max \mathscr{R}(X_{N+1})$.

Between these two extremes are all the intermediate structures of logical dependence possible among the quantities $X_1, \ldots, X_{N+1}$. In these cases, the tightness of the bound depends on the specific value of your prevision assertion $P(\mathbf{X}_N)$ and on the structure of logical dependence entailed among the components of $\mathbf{X}_{N+1}$. We shall examine a suggestive numerical example shortly.

The usefulness of this theorem is found in the fact that the logical relations among the quantities $\mathbf{X}_{N+1}$ and the constituents of the partition they generate, $\mathbf{Q}_{S(N+1)}$, can be exploited in your evaluation of $P(X_{N+1})$ *without* the necessity that you evaluate your prevision for every detailed constituent.

Our first example is based upon a realistic and practical problem. We discuss both its geometrical and computational aspects. It makes only a simple use of the fundamental theorem of prevision, in that each of the three quantities involved are events. The advantage of this simplification is that every aspect of the example can be easily understood.

***Example 1.*** This example was designed on 31 March 1987, following a day when the Dow Jones average of industrial stock prices on the New York Stock Exchange dropped 57.39 points, amidst investors' concerns over a possible impending "trade war" between the United States and Japan. Similar concerns remain in the air today, 16 February 1994, as I review this.

Let $E_1$ be the event equal to 1 if Japan refuses to honor a trade agreement, made in the summer of 1986, to import U.S.-produced semiconductors before 12 April 1987. Let $E_2$ be the event equal to 1 if between 1 April and 12 April 1987, the United States imposes special tariffs on imported Japanese electronic products. Finally, let $E_3$ be the event equal to 1 if either, but not both, of $E_1$ and $E_2$ equal 1:

$E_3 \equiv E_1 + E_2 - 2E_1E_2$. This event is substantively interesting, since it represents the possibility of some formal, nonreciprocal, antagonistic behavior between the two countries. The event vector $\mathbf{E}_3$ is represented algebraically as

$$\mathbf{E}_3 = \begin{pmatrix} 1 & 1 & 0 & 0 \\ 1 & 0 & 1 & 0 \\ 0 & 1 & 1 & 0 \end{pmatrix} \mathbf{Q}_4 = \mathbf{R}_{3,4}\mathbf{Q}_4,$$

and the convex hull of $\mathscr{R}(\mathbf{E}_3)$ defines the symmetric tetrahedron displayed in Figure 2.19.

Now, suppose you assert your $P(E_1) = .1$ and your $P(E_2) = .75$. What range does the coherency of your prevision allow for your assertion of $P(E_3)$? Any coherent assertion vector $P(\mathbf{E}_3)$ must lie inside the polytope shown in Figure 2.19. This is expressed algebraically by the requirement that $P(\mathbf{E}_3) = \mathbf{R}_{3,4}\mathbf{q}_4$ for some vector $\mathbf{q}_4$ meeting the conditions that $\mathbf{1}_4^{\mathrm{T}}\mathbf{q}_4 = 1$ and $\mathbf{q}_4 \geqslant \mathbf{0}_4$. That is, $P(\mathbf{E}_3)$ must be some convex combination of vectors in $\mathscr{R}(\mathbf{E}_3)$. These requirements are expressed explicitly in the conditions of the linear programming problems defined in the statement of the theorem:

Find the vectors $\mathbf{q}_4$ that characterize $P_l(E_3) = \text{minimum } (0 \quad 1 \quad 1 \quad 0)\mathbf{q}_4$ and $P_u(E_3) = \text{maximum } (0 \quad 1 \quad 1 \quad 0)\mathbf{q}_4$ both subject to the linear constraints

$$\begin{pmatrix} 1 & 1 & 0 & 0 \\ 1 & 0 & 1 & 0 \\ 1 & 1 & 1 & 1 \end{pmatrix} \mathbf{q}_4 = \begin{pmatrix} .10 \\ .75 \\ 1 \end{pmatrix},$$

and the nonnegativity constraint $\mathbf{q}_4 \geqslant \mathbf{0}_4$.

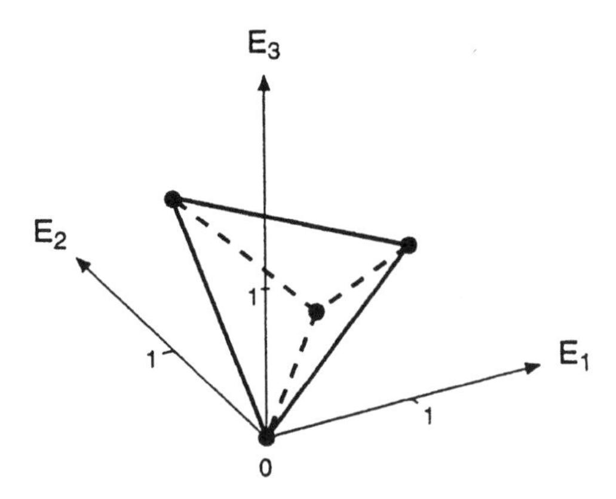


**Figure 2.19.** The convex hull of the realm of $\mathbf{E}_3$:

$$\mathbf{E}_3 = \begin{pmatrix} 1 & 1 & 0 & 0 \\ 1 & 0 & 1 & 0 \\ 0 & 1 & 1 & 0 \end{pmatrix} \mathbf{Q}_4 = \mathbf{R}_{3,4}\mathbf{Q}_4.$$

The first two rows of the constraint matrix derive from the assertion values for $P(E_1)$ and $P(E_2)$, and the third row ensures that the components of $\mathbf{q}_4$ sum to 1. The following solution vectors to the minimum and maximum problems are yielded by a linear programming routine:

$$\mathbf{q}_4(\min) = (.1 \quad 0 \quad .65 \quad .25)^{\mathrm{T}} \qquad \text{and} \qquad \mathbf{q}_4(\max) = (0 \quad .1 \quad .75 \quad .15)^{\mathrm{T}}.$$

These vectors identify the bounds on $P(E_3)$ as $P_l(E_3) = .65$ and $P_u(E_3) = .85$.

Geometrically, each of your assertions, $P(E_1)$ and $P(E_2)$, stipulates that your prevision vector $P(\mathbf{E}_3)$ must lie on a particular plane. The first requires that $P(\mathbf{E}_3)$ lie on the plane $(.1, P(E_2), P(E_3))$, and the second requires it to lie on $(P(E_1), .75, P(E_3))$. Let us study sequentially the implications of imposing these requirements.

Figure 2.20 depicts the intersections of various planes of the form $P(E_1) = C$ with the convex hull of coherent prevision vectors depicted in Figure 2.19. The $E_2$- and $E_3$-axes are directed in Figure 2.20 so that you can easily visualize these intersections in the octant shown in Figure 2.19. For example, the assertion $P(E_1) = 0$ would require that $P(E_2) = P(E_3)$. This configuration of prevision defines the line that is the edge of the convex hull in the $(E_2, E_3)$-axis plane. At the opposite extreme, the assertion $P(E_1) = 1$ would require that $(P(E_2), P(E_3))$ lie on the line $P(E_3) = 1 - P(E_2)$, which defines the opposite (and perpendicular) edge of the tetrahedron. For values of $C$ within (0, 1), the assertion $P(E_1) = C$ requires that any point representing a coherent assertion of $P(E_2)$ and $P(E_3)$ must lie within a rectangle, tilted at 45°. As shown in Figure 2.20, these rectangles expand in one direction and contract in the other as the value of $C$ increases from 0 to 1. At these extremes, the rectangles reduce to the lines mentioned. When $C = .5$, for example, the rectangle is a square. Referring to Figure 2.19, you should be able to visualize this plane of intersection. It touches the top edge of the tetrahedron at (.5, .5, 1) and the bottom edge at (.5, .5, 0). It touches the side edges at (.5, 0, .5) and at (.5, 1, .5).

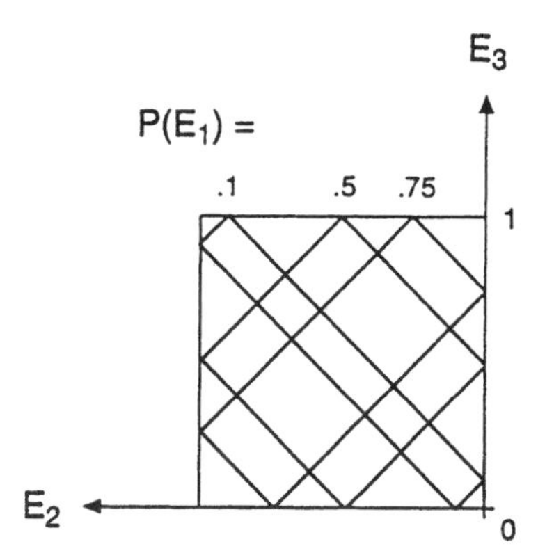


**Figure 2.20.** Intersections of various planes $P(E_1) = C$ with the convex hull of coherent prevision assertions. The number at the apex of each rectangle denotes the relevant value of $C$.

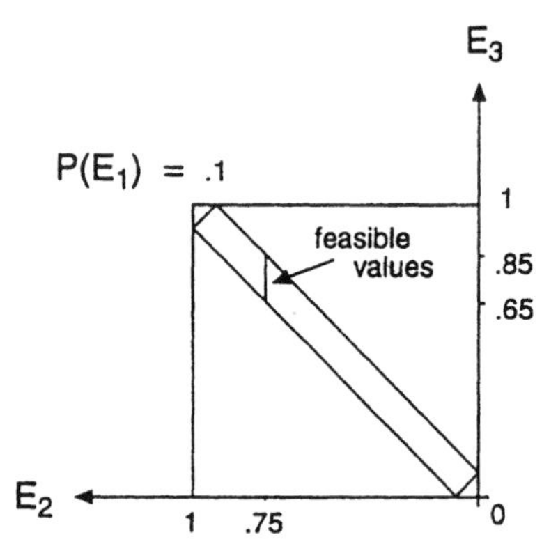


**Figure 2.21.** The range of feasible objective function values for the linear programming problems Max and Min $P(E_3)$ subject to the restrictions that $P(E_1) = .1$, $P(E_2) = .75$, and that $P$ be coherent.

In order to complete our study of the example assertion of $P(E_1) = .1$ and $P(E_2) = .75$, we should isolate for special consideration the plane in $(E_2, E_3)$-space associated with $P(E_1) = .1$. This plane is depicted in Figure 2.21.

The rectangle inscribed within the unit square in Figure 2.21 contains all pairs $(P(E_2), P(E_3))$ that would cohere with the mere assertion of $P(E_1) = .1$. Asserting further that $P(E_2) = .75$ stipulates that only points within this rectangle having .75 as their $E_2$ coordinate would meet all the requirements of the linear programming problems defining the bounds on the coherent assertion of $P(E_3)$. The range of cohering assertion values for $P(E_3)$ can be recognized as the interval [.65, .85].

The significance of the extreme $\mathbf{q}_4$ vectors that solve the linear programming problems can also be understood geometrically. Recall that

$$\mathbf{E}_3 = \begin{pmatrix} 1 & 1 & 0 & 0 \\ 1 & 0 & 1 & 0 \\ 0 & 1 & 1 & 0 \end{pmatrix} \mathbf{Q}_4 \equiv \mathbf{R}_{3,4}\mathbf{Q}_4.$$

Every coherent prevision vector $P(\mathbf{E}_3)$ that agrees with your asserted first two components, $P(E_1) = .1$ and $P(E_2) = .75$, must be some convex combination of the column vectors defining the matrix $\mathbf{R}_{3,4}$. The four components of each extreme $\mathbf{q}_4$ vector found by the linear programming algorithm identify the convex weight coefficients associated with the corresponding columns of $\mathbf{R}_{3,4}$ (each being a vertex of the convex hull). Thus, the vector $\mathbf{q}_4(\text{min}) = (.1 \quad 0 \quad .65 \quad .25)^T$ identifies the lower extreme point on the line segment of cohering prevision vectors shown in Figure 2.22. This point is situated on the left rear face of the tetrahedron as you are viewing it. For it involves a weight of 0 on the vertex defined by the second column vector, $(1 \quad 0 \quad 1)^T$. Similarly, the vector $\mathbf{q}_4(\text{max}) = (0 \quad .1 \quad .75 \quad .15)^T$ identifies the upper extreme point on the line segment, situated on the front face of the tetrahedron as you are viewing it. For $\mathbf{q}_4(\text{max})$ gives a weight of 0 to the vertex $(1 \quad 1 \quad 0)^T$. □

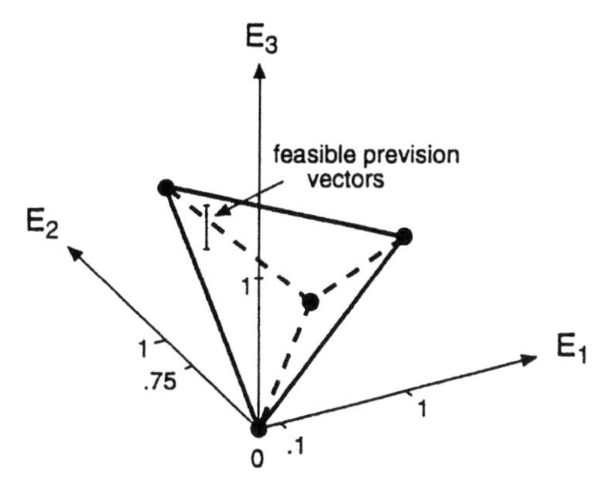


**Figure 2.22.** The range of cohering prevision vectors involving $P(E_1) = .1$ and $P(E_2) = .75$, shown within the convex hull of all coherent prevision vectors.

***Historical Note***

The logic behind the fundamental theorem of probability was presented by de Finetti in his lectures in Paris in 1935. But he named the result only in his text of 1970. Oddly, de Finetti's formulation of the theorem in the text was couched in the syntax of a set-theoretic conception of events, as was used in Kolmogorov's axiomatization of probability. The statement of de Finetti's theorem in the context of a linear programming problem was first presented by Bruno and Gilio (1980). An extensive discussion of the fundamental theorem of prevision, with several computational and geometrical examples, can be found in Lad, Dickey, and Rahman (1990, 1992). The applicability of linear programming to related questions had already been recognized by Fishburn (1964), Scott (1964), and Hailperin (1965). Fittingly, these papers appeared in the centenary year of the death of George Boole, who was grappling with similar ideas during the last 10 years of his life, following the publication of his *Laws of Thought* (1854a). Particularly relevant are Boole's papers "On a General Method in the Theory of Probabilities" (1854b) and "On the Theory of Probabilities" (1862). This aspect of Boole's writings has been studied in detail in the book by Hailperin (1986). A formalist development of a dual form of the fundamental theorem of prevision, part I, to be studied shortly, was presented by Whittle (1970, 1971), whose work develops an axiomatic treatment of "expectation" which is formally related to de Finetti's theory of prevision.

Computational procedures for solving linear programming problems are available in many computing packages. The algebra behind the procedures is presented in several textbooks, such as Walsh's *Introduction to Linear Programming* (1985). But you can learn to set up programming problems for practical solution using MATLAB or TURBOLP or whatever, even if you have not studied the theoretical basis for the simplex method. To solidify your understanding at this point, you should try the following problems. Some of them involve general quantities rather than events, and require a computer.

## PROBLEMS

1. For the various structures of logical dependence possible among three events that were displayed in Section 2.9, conduct a visual analysis of the implication of your specifying your prevision for each pair of events on the range of your possible coherent assertion for the third event. Check the implications of asserting your prevision for *each* of the three pairs of events on the range of coherent assertion of prevision for the third. Notice that the analysis for some figures, such as Figure 2.4, is essentially the same no matter which pair of events are initially assessed. For others, such as Figure 2.5, this is not the case.

2. Return to the economic forecasting problem 3 of Section 2.5, which we revisited as problem 5 of Section 2.8. Remember that the solution to part (ii), identifying coherency requirements on an assertion of $P(C \vee \tilde{U} \vee I)$, yielded not a precise number for an answer but an interval. Recognize this problem now as a linear programming problem, and feed the necessary input into a linear programming computer algorithm to generate the coherent solution bounds.

3. Let $X$ denote the live birthweight of the first registered live human birth in your city or town today. Record the weight in kilograms, measured to the nearest 100 grams, and truncate the extreme measurements so that if the weight is less than 2 kg, you record $X = 2.0$, and if the weight is more than 5 kg, you record $X = 5.0$. Assess your $P(3.0 < X < 3.4)$, your $P(X \leqslant 2.7)$, and your $P(X \geqslant 3.7)$. Use the FTP to determine the cohering bounds on your $P(X)$ and your $P(X^2)$.

4. Suppose the components of $\mathbf{X}_4$ are logically independent quantities with the identical realm, $\mathscr{R}(X_i) = \{1, 2, 3, 4, 5\}$. Let $\bar{X}_4$ denote their numerical average, let $S^2$ denote the average squared difference between each component and $\bar{X}_4$, and let $\max(\mathbf{X}_4)$ denote the largest component value of $\mathbf{X}_4$. Suppose you assert your $P(\mathbf{X}_4) = (2.2)\mathbf{1}_4$, your $P(S^2)$ as 1.6, your $P(X_i \geqslant 4) = P(X_j \geqslant 4)$ for each pair of components $(X_i, X_j)$, and your $P(\max(\mathbf{X}_4) \geqslant 4) = .3$. Determine the cohering bounds on your assertion of $P[(X_4 = 2) + (X_4 = 3)]$. (*Hint*: Asserting $P(X_i \geqslant 4) = P(X_j \geqslant 4)$ is equivalent to asserting $P[(X_i \geqslant 4) - (X_j \geqslant 4)] = 0$.) What happens to the bounds if $P(\max(\mathbf{X}_4) \geqslant 4)$ is reduced to .2? to .1?

 **N.B.** The following two problems are suggested here only for their computational interest. If you have some background in probability theory, you should recognize their relevance to the Bienaymé-Chebyshev inequality and to Kolmogorov's inequality.

5. Suppose that $E_1, \ldots, E_5$ are completely logically independent events for which you assert the previsions $P(E_1) = P(E_2) = P(E_3) = P(E_4) = P(E_5) = .63$. Define the quantities $\bar{E}_J = J^{-1}\sum_{i=1}^{J} E_i$, for each $J = 2, \ldots, 5$. That is, $\bar{E}_J$ is the average of the first $J$ events. What does coherency of your prevision imply

about your previsions $P(\bar{E}_J)$? Now for each value of $M = 2, \ldots, 5$, and for any specified value of $\varepsilon$, define the events $E(M, \varepsilon) \equiv (|\bar{E}_M - P(\bar{E}_M)| > \varepsilon)$ and $E(M, K, \varepsilon) \equiv (\max_{0 \leqslant k \leqslant K} |\bar{E}_{M+k} - P(\bar{E}_{M+k})| > \varepsilon)$, stipulating that $K = 5 - M$. Can you interpret what these defined events are measuring? Set up the linear programming problems to determine the bounds on your prevision for events of the form $E(M, \varepsilon)$ and $E(M, K, \varepsilon)$ in order that they cohere with the previsions you have asserted for $E_1$ through $E_5$. Compute these bounds for some interesting values of $M$, $K$, and $\varepsilon$, suggested below. If you are in a sizable class of students, you may want to divide among yourselves these various configurations of $M$, $K$, and $\varepsilon$ in order to share the results of your computational efforts:

**(i)** For the values of $M = 2$, 3, 4, and 5 and $\varepsilon = .01$, .03, .07, .15, and .25, compute bounds on cohering previsions $P[E(M, \varepsilon)]$.

**(ii)** For each $M$ and $\varepsilon$ above, and appropriate values of $K$, compute bounds on cohering previsions $P[E(M, K, \varepsilon)]$.

**6.** Suppose that $X_1, \ldots, X_5$ are completely logically independent *quantities*, each with the identical realm $\mathscr{R}(X_i) = \{0, 1, 2, 3\}$. Define $X_6, \ldots, X_{10}$ as the squares of these quantities. That is, $X_6 = X_1^2$, $X_7 = X_2^2, \ldots, X_{10} = X_5^2$. Suppose you assert that your prevision $P(X_1) = P(X_2) = \cdots P(X_5) = 1.1$, and $P(X_6) = P(X_7) = \cdots = P(X_{10}) = 1.6$. Define $\bar{X}_J$ as the average of the first $J$ quantities. What does coherency of your prevision imply about your previsions $P(\bar{X}_J)$? For each value of $M = 1, \ldots, 5$ and for any specified value of $\varepsilon$ define the events $E(M, \varepsilon) \equiv (|\bar{X}_M - P(\bar{X}_M)| > \varepsilon)$ and $E(M, K, \varepsilon) \equiv (\max_{0 \leqslant k \leqslant K} |\bar{X}_{M+k} - P(\bar{X}_{M+k})| > \varepsilon)$, again stipulating that $K = 5 - M$. Can you interpret what these defined events are measuring? Set up the linear programming problems to determine the bounds on your prevision for events of the form $E(M, \varepsilon)$ and $E(M, K, \varepsilon)$ in order that they cohere with the previsions you have asserted for $X_1, \ldots, X_{10}$.

**(i)** For the values of $M = 3, 4$, and 5 and $\varepsilon = .1, .2, .5, .75$, and 1.5, compute bounds on cohering previsions $P[E(M, \varepsilon)]$.

**(ii)** For each $M$ and $\varepsilon$ above, and appropriate values of $K$, compute bounds on cohering previsions $P[E(M, K, \varepsilon)]$.

**7.** Notice that the square of $\bar{X}_J$ is $(\bar{X}_J)^2 = J^{-2}[\sum_{i=1}^{J} X_i^2 + 2\sum_{j>i=1}^{J} X_i X_j]$. Continuing the setup of problem 6, compute bounds on your prevision for $X_1 X_2$ for $\max(X_1 X_2, X_1 X_3, X_2 X_3)$ and for $(\bar{X}_J)^2$, for $J = 2$, and 3. Can you identify the simultaneous restrictions of coherency on your assertion of $P(X_1 X_2)$, $P(X_1 X_3)$, $P(X_2 X_3)$, and $P[(\bar{\mathbf{X}}_3)^2]$?

### 2.10.1 An Extension to Higher Dimensions

The conclusion to the fundamental theorem can be generalized easily to many dimensions. If $K$ further quantities are considered after you have asserted your

$P(\mathbf{X}_N)$, then the conclusion to the theorem would be stated in terms of the realm matrix for all $N+K$ quantities, $\mathbf{R}(\mathbf{X}_{N+K})$, which is partitioned into its first $N$ rows, denoted by $\mathbf{R}_{N,S(N+K)}$, and its last $K$ rows, denoted by $\mathbf{R}_{K,S(N+K)}$. Then for the coherency of your prevision over the entire vector $\mathbf{X}_{N+K}$, it is necessary and sufficient that the vector of your further assertions $[P(X_{N+1}),\ldots,P(X_{N+K})]^{\mathrm{T}}$ be representable as $\mathbf{R}_{K,S(N+K)}\mathbf{q}_{S(N+K)}$, where $\mathbf{q}_{S(N+K)}$ satisfies the linear constraints that

$$\mathbf{R}_{N,S(N+K)}\mathbf{q}_{S(N+K)} = \mathbf{p}_N,$$

$$\mathbf{1}^{\mathrm{T}}_{S(N+K)}\mathbf{q}_{S(N+K)} = 1,$$

$$\mathbf{q}_{S(N+K)} \geqslant \mathbf{0}_{S(N+K)}.$$

Notice that the coherency of $P$ over $\mathbf{X}_{N+K}$ demands not only that *each* $P(X_{N+i})$ lie within the extreme bounds on a corresponding objective function $\mathbf{r}_{N+i}\,\mathbf{q}_{S(N+K)}$, but more stringently, that the vector of all assertions about the remaining quantities, $[P(X_{N+1}),\ldots,P(X_{N+K})]$, must be representable as a linear function of a vector $\mathbf{q}_{S(N+K)}$ lying within the feasible set of solutions to the linear programming problems specified in the fundamental theorem. To be precise, the linear function is $\mathbf{R}_{K,S(N+K)}\mathbf{q}_{S(N+K)}$. The important catch in this statement can be understood in the context of an example.

***Example 1 (continuing 2.10.1).*** Suppose that you have asserted merely your $P(E_1)=.1$. The extension to the FTP states the joint conditions on the values of $P(E_2)$ and $P(E_3)$ that are required in order for both these assertions to cohere with your specified $P(E_1)$. The *entire rectangle* of points inscribed within the unit square, shown in Figure 2.21, represents feasible solutions to cohering vectors $P(E_2,E_3)$. Now the extreme values of cohering previsions for the two events, considered individually, allow that $\max P(E_2) = \max P(E_3) = 1$, and $\min P(E_2) = \min P(E_3) = 0$. But it is not true that every vector $[P(E_2),P(E_3)]$ meeting merely the conditions $0 \leqslant P(E_1) \leqslant 1$ and $0 \leqslant P(E_2) \leqslant 1$ would cohere with your assertion of $P(E_1)=.1$. The condition that $[P(E_2),P(E_3)]$ lie within the inscribed rectangle is clearly a stronger condition.

A second point of interest is that the events $E_2$ and $E_3$ in this example are *completely* logically independent of one another. Your mere assertion of prevision for either of them would place no logical restrictions on your coherent assertion about the other. Nonetheless, once you assert your prevision for $E_1$, which is logically related to each of them, the mutual logical relation shared among the three events comes into play and forces some structure onto your cohering assessment of these two events. This feature of the example highlights the importance of recognizing the specific structure of the logical dependence among the several quantities that interest you in a particular problem. □

### 2.10.2 A Dual Form of the Fundamental Theorem

The concept of duality has long been a rich source of mathematical understanding, allowing us to see that sometimes seemingly very different results are actually complementary expressions of the same result. You may be familiar with the geometrical example that virtually any true statement concerning points and lines is also true when these words exchange their positions. For example, in Euclidean geometry two lines determine a unique point, and two points determine a unique line. The principle of duality can be applied to linear programming, yielding a similarly fruitful recognition. In this context, the dual objects are the vector of the constraint equation values and the vector of the optimizing coefficients. Without describing all the operational details of the duality principle, which can be found in Walsh (1985, Chap. 3) and elsewhere, let us merely state here the dual forms of the linear programming problems specified in the fundamental theorem of prevision, and discuss their meaning.

The notation and the setup for the dual problem are identical with the Fundamental Theorem 2.8. The only difference is that the boundaries of the prevision interval $[P_l(X_{N+1}), P_u(X_{N+1})]$ are defined in dual form by the solutions to the following two linear programming problems:

Find the $(N+1)$-tuples $\mathbf{a}_{N+1} = (a_0, a_1, \ldots, a_N)^{\mathrm{T}}$ and $\mathbf{b}_{N+1} = (b_0, b_1, \ldots, b_N)^{\mathrm{T}}$ that characterize

$$P_l(X_{N+1}) \equiv \text{maximum } (1 \quad \mathbf{p}_N^{\mathrm{T}})\mathbf{a}_{N+1}$$

subject to the linear inequality constraint

$$[\mathbf{1}_{S(N+1)} \quad \mathbf{R}_{N,S(N+1)}^{T}]\mathbf{a}_{N+1} \leqslant \mathbf{r}_{N+1}^{\mathrm{T}}$$

and

$$P_u(X_{N+1}) \equiv \text{minimum } (1 \quad \mathbf{p}_N^{\mathrm{T}})\mathbf{b}_{N+1}$$

subject to the linear inequality constraint

$$[\mathbf{1}_{S(N+1)} \quad \mathbf{R}_{N,S(N+1)}^{\mathrm{T}}]\mathbf{b}_{N+1} \geqslant \mathbf{r}_{N+1}^{\mathrm{T}}.$$

The only idiosyncracy in the notation here is that the vector denoted by $\mathbf{r}_{N+1}^{\mathrm{T}}$ has the dimension $S(N+1) \times 1$, since $\mathbf{r}_{N+1}$ was defined as the $(N+1)$st *row* of the realm matrix $\mathbf{R}_{N+1,S(N+1)}$. You should notice that the minimization problem in the primal form has become a maximization problem in the dual, and the maximization problem in the primal form has become a minimization problem in the dual. Also, constraining equalities have become inequalities. But the most important feature of the dual form is that the objective function coefficient vector in the primal form, $\mathbf{r}_{N+1}$, has exchanged places with the vector of constraining equation values, $(1 \quad \mathbf{p}_N^{\mathrm{T}})$.

The meaning of the dual formulation is understood by thinking about the class of all quantities that are defined by linear transformations of the quantity vector $\mathbf{X}_N$ for which you have asserted $P(\mathbf{X}_N) = \mathbf{p}_N$. We can specify this class as the set of quantities $\{Y | Y = (1 \quad \mathbf{X}_N^{\mathrm{T}})\mathbf{a}_{N+1}$, for $\mathbf{a}_{N+1} \in \mathbb{R}^{N+1}\}$. Now it is impossible for some of the quantities in this set to exceed the "further quantity" $X_{N+1}$ that is the focus of the fundamental theorem. The coefficient vectors $\mathbf{a}_{N+1}$ defining this subset of quantities are identified by the restricting inequalities of the maximization problem characterizing $P_l(X_{N+1})$: $[\mathbf{1}_{S(N+1)} \quad \mathbf{R}^{\mathrm{T}}_{N,S(N+1)}]\mathbf{a}_{N+1} \leqslant \mathbf{r}^{\mathrm{T}}_{N+1}$. Each row of the matrix multiplying $\mathbf{a}_{N+1}$ from the left includes the first $N$ components of a different vector element of the realm $\mathscr{R}(\mathbf{X}_{N+1})$. The corresponding row value in the right-side column vector, $\mathbf{r}^{\mathrm{T}}_{N+1}$, is the $(N+1)$st component of that same realm element vector. Thus, the inequality is satisfied only by those coefficient vectors $\mathbf{a}_{N+1}$ defining quantities $Y$ that cannot exceed $X_{N+1}$. Since $X_{N+1}$ necessarily exceeds all these (linear combination) quantities, your $P(X_{N+1})$ must exceed your previsions for each of them. Thus, the lower bound for $P(X_{N+1})$ is the largest of your previsions for any quantity within this subset—the largest of all the linear combinations $(1 \quad \mathbf{p}_N^{\mathrm{T}})\mathbf{a}_{N+1}$ from among all qualifying coefficient vectors $\mathbf{a}_{N+1}$. The upper bound for $P(X_{N+1})$ derives from the parallel argument regarding quantities that *surely exceed* $X_{N+1}$.

***Example 1 (continuing 2.10.1).*** In the example of U.S.–Japan trade relations, you asserted your $P(E_1) = .1$ and $P(E_2) = .75$, where the event vector $\mathbf{E}_3$ is represented by

$$\mathbf{E}_3 = \begin{pmatrix} 1 & 1 & 0 & 0 \\ 1 & 0 & 1 & 0 \\ 0 & 1 & 1 & 0 \end{pmatrix} \mathbf{Q}_4 \equiv \mathbf{R}_{3,4}\mathbf{Q}_4.$$

Now consider the following three quantities defined by different affine transformations of $(E_1, E_2)$:

$$\begin{aligned} Y_L &= -1 + E_1 - 1.5E_2 = (1 \quad E_1 \quad E_2)(-1 \quad 1 \quad -1.5)^{\mathrm{T}}, \\ Y &= -2 + 6E_1 - 3.1E_2 = (1 \quad E_1 \quad E_2)(-2 \quad 6 \quad -3.1)^{\mathrm{T}}, \\ Y_U &= 2.5 - .5E_1 - .75E_2 = (1 \quad E_1 \quad E_2)(2.5 \quad -.5 \quad -.75)^{\mathrm{T}}. \end{aligned}$$

Appending an appropriate row to the $\mathbf{R}(\mathbf{E}_3)$ matrix for each of these quantities would mean appending the matrix

$$\begin{pmatrix} -1.5 & 0 & -2.5 & -1 \\ .9 & 4 & -5.1 & -2 \\ 1.25 & 2 & 1.75 & 2.5 \end{pmatrix}.$$

Comparing these rows with the third row of $\mathbf{R}(\mathbf{E}_3)$ shows that $Y_L$ cannot exceed $E_3$ and that $Y_U$ must exceed $E_3$, but that $Y$ may or may not exceed $E_3$ depending on what are the values of $E_1$ and $E_2$. Thus, the coefficient vector defining $Y_L$, $\mathbf{a}_3 = (-1 \quad 1 \quad -1.5)^{\mathrm{T}}$, satisfies the dual-form inequality constraint relevant to $P_l(E_3)$; and thus $P(E_3)$ must exceed $(1 \quad p_1 \quad p_2)\mathbf{a}_3 = (1 \quad .1 \quad .75)(-1 \quad 1 \quad -1.5)^{\mathrm{T}} = -.5$.

The second appended row of the realm matrix exceeds the row for $E_3$ in its first two columns and falls short in the last two. Thus, the associated quantity $Y$ is irrelevant to considerations of the dual form of the fundamental theorem. But the final appended row always exceeds the row corresponding to $E_3$, and thus its coefficient vector $\mathbf{b}_3 = (2.5 \quad -.5 \quad -.75)^{\mathrm{T}}$ satisfies the dual-form inequality constraint relevant to $P_u(E_3)$. Thus, $P(E_3)$ must be less than $(1 \quad p_1 \quad p_2)\mathbf{b}_3 = (1 \quad .1 \quad .75)(2.5 \quad -.5 \quad -.75)^{\mathrm{T}} = 1.8875$.

Solving the two linear programming problems for this example yields the optimal solutions $\mathbf{a}_3^* = (0 \quad -1 \quad 1)$, and $\mathbf{b}_3^* = (0 \quad 1 \quad 1)$. Thus, $P_l(E_3) = (1 \quad .1 \quad .75)\mathbf{a}_3^* = .65$, and $P_u(E_3) = (1 \quad .1 \quad .75)\mathbf{b}_3^* = .85$. □

In considering the fundamental theorem, of course either statement of the programming problem may be considered the primal form and the other the dual. Primal and dual are symmetric words relative to one another in this context. Both forms have sensible interpretations for our understanding. Computationally, the faster form to use is typically the one that specifies the smaller number of constraining equalities or inequalities. In most cases, this will be the form we have specified in Theorem 2.8 as the primal form. Nonetheless, de Finetti's presentation of the theorem in his text is in terms we have identified as the dual form. Whittle's (1970, 1971) presentation of a similar theorem also proceeds according to our dual form.

### 2.10.3 What Is so Fundamental about the FTP?

We have understood the principle of coherency to require the linearity property that for any coefficient vector $\mathbf{s}_N$, your $P(\mathbf{s}_N^{\mathrm{T}}\mathbf{X}_N) = \mathbf{s}_N^{\mathrm{T}}P(X_N)$. The fundamental theorem of prevision takes us another step. It characterizes the restrictions that the coherency of your prevision and your assertion of $P(\mathbf{X}_N)$ place on your assessment of any other quantities $X_{N+1}, \ldots, X_{N+K}$, whether they are linear combinations of $\mathbf{X}_N$ or not. That is, the fundamental theorem characterizes the coherent implications of every prevision you have asserted for every other vector of quantities that you have not yet directly assessed. Seen in this way, the fundamental theorem of prevision is an extension of the famous closure result of deductive logic, described by Hilbert and Ackermann in their text *Mathematical Logic* (1950, Section 1.9) as "the systematic survey of all propositions that can be deduced from any set of axioms." The fundamental theorem of prevision is a truly comprehensive statement of the coherent logic of uncertain knowledge.

Despite the exhaustive scope of this result, it is not widely known among probabilists and statisticians today. Indeed, de Finetti's (1974a, Chap. 3.10) designation of his theorem as "the fundamental theorem of probability" has appeared absurd to many who have bothered to read it without really studying his larger views. As far as I can tell, no one else ever has designated any theorem of probability as "*the* fundamental theorem." If pressed for a response today, I think that most professionals would designate as the fundamental theorem of probability either the law of large numbers or possibly the central limit theorem. There would be a few other contenders mentioned, and many respondents would deny that there is any fundamental theorem. For now, let me merely state that the two famous inequalities of probability theory which provide the basis for the laws of large numbers, the Bienaymé-Chebyshev inequality and the Kolmogorov inequality, are merely special cases of the fundamental theorem of prevision. Similar in content to the problems we have proposed as 5 and 6 of this section, they merely specify an upper bound on the probability you may assert for a large deviation of $\bar{X}$ from your prevision $P(\bar{X})$ if it is to cohere with your concomitant assertion of $P(\bar{X}^2)$. We can think about this again after we study the concept of exchangeability in the next chapter.

But modulo some extensions and developments to be considered in Section 2.11 and Chapter 3, we can make an even bolder statement. *Every inequality of probability theory is a special case of the fundamental theorem of prevision.* For every inequality amounts to the determination of bounds on the prevision for some quantity if it is to cohere with some other specified prevision assertions, which are taken as given.

***Historical Note***

Although de Finetti presented the logic behind his theorem in the Paris lectures of 1935, a time when probability inequalities were of central importance among the cognoscenti of Europe, his result was virtually ignored. Maurice Fréchet highlighted the importance of some particular inequalities in Chapter IV of Volume I of his text on probability theory (Fréchet, 1937) and again in his important two-volume treatise on probabilities associated with a system of compatible and dependent events (Fréchet, 1940, 1943) without mentioning de Finetti's argument, which rendered the particular inequalities as special cases.

The relationship between de Finetti and Fréchet was important, and is interesting. Fréchet was already one of the *grandes hommes* of European mathematics when their exchanges began in 1930. He had presented his thesis "On Several Points of Functional Calculus" at the University of Paris in 1906, the year de Finetti was born. Their exchange began with Fréchet responding negatively to de Finetti's critique of the *complete* additivity of probability functions. The five relevant articles (three by de Finetti, two by Fréchet ) appear in the volume of de Finetti's early papers, *Scritti (1926–1930)* . Subsequently, Fréchet, as the first director of the Institute Henri Poincaré, invited de Finetti to deliver in 1935 the series of lectures, published in 1937 under the title "Prevision: Its Logical Laws, Its Subjective Sources." However, relations between the two men appear to have been somewhat strained, at least around the 1940s. This can be felt in several allusions to de Finetti's results in Fréchet's (1994, 1943) important summary work on compatible and dependent events, especially by a demeaning reference in

Volume II (1943) to a distinction that is central to the understanding of de Finetti's point of view.

However, the two men corresponded regularly over the course of 30 years. Copies of many of de Finetti's papers, annotated in the margins by Fréchet, can be found in the depository of Fréchet's professional effects at the University of Paris VI. They debated publicly at the International Congress on the Philosophy of Science in Paris in 1949. As the presiding figure in the section on probability, Fréchet wrote the general report describing the proceedings, which were evidently quite contentious. Aware particularly of their personal disagreements, Fréchet invited de Finetti to append his own comments to the report. The comments of both are included in the collection of papers by Fréchet, *Les Mathématiques et le Concret* (1955).

In his preface to the English translation of the *Theory of Probability*, de Finetti lauded Fréchet along with Jerzy Neyman and Guido Castelnuovo as "three great men, who, although they all shared an opposed view about our common subject, were always willing to discuss, and were extraordinarily friendly and helpful on every occasion."

## 2.11 BOUNDARY SPECIFICATIONS AS PARTIAL PREVISION ASSERTIONS

The fundamental theorem of prevision introduces naturally the notion that your limited introspection regarding your knowledge about several quantities which interest you may yield only a lower and upper bound for your prevision of each of them. We had alluded to this idea in our discussion of three qualifications to the definition of prevision as we concluded Section 2.4.

Your prevision for a quantity $X$, remember, is the price at which you avow your indifference to an exchange of $\$X$ for $\$P(X)$, and vice versa. According to the assumed principle that you prefer more money to less, this is to say that for any value of $P < P(X)$ you prefer $\$X$ to $\$P$—despite the fact that you may not be certain of the numerical value of $X$. Similarly for any value of $P > P(X)$, you prefer $\$P$ to $\$X$. In the course of your eliciting your $P(X)$, you may consider your prevision for several quantities that are logically related to $X$ and are easier for you to assess than is $X$ itself. For example, you may find it easier to assert your probabilities for events that designate extreme values of $X$ or central values of $X$, such as $(X < a), (X > b)$, or $(c < X < d)$. (Recall problem 3 of Section 2.10.) The fundamental theorem of prevision then specifies that coherency requires your $P(X)$ to lie within the computable interval $[P_l(X), P_u(X)]$ identified by the relevant linear programming problems.

On the one hand, this result can be interpreted as a *cautionary* prescription: if you now proceed to assert your $P(X)$, it had better lie within the specified interval, or else your prevision would be incoherent over all the quantities you have assessed. On the other hand, the fundamental theorem of prevision already specifies an operational *behavioral* implication for you: your assertion of $P(\mathbf{X}_N)$

already implies that you are willing to sell $\$X_{N+1}$ for $\$P$ if $P > P_u(X_{N+1})$, and that you are willing to buy $\$X_{N+1}$ for $\$P$ if $P < P_l(X_{N+1})$. [In the latter case, for example, the dual form of the FTP assures us that a quantity can be defined, as a linear combination of $\mathbf{X}_N$, which cannot exceed $X_{N+1}$, and which you have already avowed your willingness to buy for $P_l(X_{N+1})$.] In this context, your specification of an interval $[P_l(X), P_u(X)]$ can be interpreted as your assertion of a 'bid-ask spread" for transactions involving $\$X$. This terminology, suggested for use in this context by Leamer (1986), is typically used by dealers in the sale of stock options.

Even without recourse to consideration of different quantities, we can imagine your direct specification of a prevision interval as the outcome of your introspective efforts regarding your knowledge of a quantity. In the course of your assessing your $P(X)$, you presumably consider a converging sequence of numbers, bounded by the extremes, $\min \mathscr{R}(X)$ and $\max \mathscr{R}(X)$. For each number in the sequence, you need to decide whether you prefer $\$X$ to that number of dollars, or vice versa. We can represent the former conclusion by the statement that your $P(X) \geqslant K_1$, and the latter by $P(X) \leqslant K_2$. The sequence of numbers will be converging because once you specify that your $P(X) \geqslant K > 0$, say, you can be sure that your $P(X)$ exceeds any number less than $K$. To specify further that your $P(X) > K/2$ in this case would not be contradictory but merely redundant. Thus, after any finite number of considerations, you can identify the largest number, $P_l(X)$, for which you have identified your $P(X) \geqslant P_l(X)$ and the smallest number, $P_u(X)$, for which you have determined your $P(X) \leqslant P_u(X)$. We refer to the numbers $P_l(X)$ and $P_u(X)$, respectively, as the lower and upper bounds on your prevision.

Suppose we regard your specification of an interval as the operational definition of your assertion of your knowledge about a quantity. That is, for any quantity $X$ you specify a pair of numbers $[P_l(X), P_u(X)]$ with the understanding that you are committing yourself thereby to "buy" $X$ at any price below $P_l(X)$, and to "sell" $X$ at any price above $P_u(X)$. Based on your limited inspection of what's happening as far as $X$ is concerned and your limited introspection regarding your assessment of this evidence, specification of this pair may be the best you can do. Greater efforts at introspection may reduce the difference between the upper and lower bounds. But you may quit the effort when you regard the bounds as "close enough" for your purposes relative to the difficulty involved in reducing their difference any further. Evidently, if your $P_l(X) = P_u(X)$, you would have specified the number we have defined heretofore as your prevision, $P(X)$.

In the sense that we have recognized your specification of $P(X)$ as your *assertion* of your prevision, we can refer to your specification of an interval $[P_l(X), P_u(X)]$ as your *partial assertion* of your prevision. At one extreme, if you assert only $P_l(X) = \min \mathscr{R}(X)$ and $P_u(X) = \max \mathscr{R}(X)$, you are making no assertion at all about $X$. You are only avowing your desire to be coherent, which we presume generally in any case. At the other extreme, if you assert an interval so small as to be a single number, $P_l(X) = P_u(X) = P(X)$, you are making a complete assertion of your valuation for $X$. To recognize the meaningful activity of asserting a bid-ask spread for $X$ as constituting your partial assertion of your prevision is merely to recognize that there are meaningful intermediate degrees of assertion available to you, between these two extremes. No one can force you to assert anything. Assertion, by definition, is an act of your own volition.

When you make the partial assertion $[P_l(X), P_u(X)]$, you are giving precise meaning to the oft-heard challenge, "Make me an offer I can't refuse." You are announcing that you will refuse no offer that allows you to sell $X$ for more than $P_u(X)$, nor will you refuse an offer that allows you to buy $X$ for less than $P_l(X)$. Of course, you may still refuse (or accept) offers to buy or sell at prices within the interval. You have just not as yet asserted your position on such an offer. Once you make a decision on any number within the interval, your asserted interval is narrowed.

As a practical matter, the consequence of this discussion is that your assertion of prevision intervals for quantities and your mere assertion of the relative sizes of previsions can specify input restrictions as linear inequalities in a linear programming framework to represent your uncertain knowledge in a problem. The standard extreme value computations would yield your implied prevision bounds on whatever further quantities might interest you. Thus, the statement of the Fundamental Theorem 2.8 can be extended to allow not merely equality restrictions on linear combinations of the $\mathbf{q}$ vector but inequality restrictions as well. At an introductory reading of this text, this statement of our conclusion should suffice. The remainder of this section presents a careful formulation of the logic that develops this conclusion. There are some tricky details. Unless you are an advanced reader, you would probably do well to skip the following details, and turn to the philosophical and historical comments that conclude this section after the problems, if they interest you.

The requirement of coherency can be formulated in a way relevant to your specification of prevision bounds as a partial assertion of your prevision, even when several quantities are under consideration. To this point, we have stated the requirement of coherency as demanding simply that you not commit yourself to transactions that would make you a sure loser. Otherwise you are regarded to be talking nonsense. One obvious implication of coherency would be that, for any quantity $X$, your $P_l(X) \leqslant P_u(X)$. But in order to characterize the coherency of prevision boundaries, we need to append to this requirement a statement that you would not avoid engaging in transactions that would yield you a sure gain. This requirement could remain unvoiced when only the full assertion of prevision was under consideration. We presume once again that your asserted willingness to engage in several individual transactions implies your willingness to engage in any combination of them, subject only to the stipulation that the net transaction cannot exceed the scale of your maximum stake.

The following definitions and theorems, interspersed with brief comments, formalize the ideas we have been discussing.

***Definition 2.19.*** Let $\mathbf{X}_N$ be any vector of quantities, with realm $\mathscr{R}(\mathbf{X}_N)$, and let $\mathbf{a}_N$ be any specified vector of constants. *Your assertion of the inequality*

$\mathbf{a}_N^{\mathrm{T}} P(\mathbf{X}_N) \leqslant b = b(\mathbf{a}_N, \mathbf{X}_N)$ means that you are expressly willing to engage in any transaction that yields you a net gain of $s(b - \mathbf{a}_N^{\mathrm{T}}\mathbf{X}_N)$ as long as the scalar $s$ is nonnegative, and $|s(b - \mathbf{a}_N^{\mathrm{T}}\mathbf{x}_N) \leqslant S$ for every $\mathbf{x}_N \in \mathscr{R}(\mathbf{X}_N)$, where $S$ is the scale of your maximum stake. In the sense that your prevision for any quantity, say, $P(X) = b$, is an assertion of both the inequalities $P(X) \leqslant b$ and $P(X) \geqslant b$, we say that your assertion of an inequality is a *partial assertion of your prevision* for the vector of quantities $\mathbf{X}_N$. □

Interpreting the "transaction" mentioned here, you would be "selling" $\mathbf{a}_N^{\mathrm{T}}\mathbf{X}_N$ in return for $b$, if $b$ is positive. Formally, this definition incorporates a meaning to the partial assertion of $\mathbf{a}_N^{\mathrm{T}} P(\mathbf{X}_N) \geqslant b$, too, since this is equivalent to asserting that $-\mathbf{a}_N^{\mathrm{T}} P(\mathbf{X}_N) \leqslant -b$. (Notice there are no restrictions on the signs of the components of $\mathbf{a}_N$ or $b$.) Thus, the "transaction" specified in the definition would amount to your "buying" $\mathbf{a}_N^{\mathrm{T}}\mathbf{X}_N$ at the price $|b|$ if $b$ is negative. Thus, this definition formalizes the meaning of the statement about any quantity that $P(X) \in [b_1, b_2]$, or, equivalently, the two statements that $P(X) \geqslant b_1$ and $P(X) \leqslant b_2$.

Definition 2.19 implies that $\mathbf{a}_N^{\mathrm{T}} P(\mathbf{X}_N) \leqslant b$ if and only if $P(\mathbf{a}_N^{\mathrm{T}}\mathbf{X}_N) \leqslant b$. Your asserting merely an ordering to your previsions for various quantities, such as $P(X) \leqslant P(Y)$, is equivalent to your asserting a bound on a linear combination of the quantities, $P(X - Y) \leqslant 0$.

Geometrically, your assertion of the inequality $\mathbf{a}_N^{\mathrm{T}} P(\mathbf{X}_N) \leqslant b$ amounts to your specification of a half-space of acceptable transaction prices for $\mathbf{X}_N$, defined by the hyperplane $\mathbf{a}_N^{\mathrm{T}}\mathbf{p}_N \geqslant b$. For any complying vector of prices $\mathbf{p}_N$, you would be willing to exchange $\mathbf{a}_N^{\mathrm{T}}\mathbf{X}_N$ in return for $\mathbf{a}_N^{\mathrm{T}}\mathbf{p}_N$. For any noncomplying vector $\mathbf{p}_N$ for which $\mathbf{a}_N^{\mathrm{T}}\mathbf{p}_N < b$, you have not yet asserted your position on a choice between $\mathbf{a}_N^{\mathrm{T}}\mathbf{p}_N$ and $\mathbf{a}_N^{\mathrm{T}}\mathbf{X}_N$.

***Definition 2.20.*** Let $\mathbf{X}_N$ be any vector of quantities, with realm $\mathscr{R}(\mathbf{X}_N)$. In the process of eliciting your state of uncertain knowledge about $\mathbf{X}_N$, your *partially asserted prevision configuration* for $\mathbf{X}_N$ is the set of $N$-dimensional vectors that satisfy the specifications of all your partial prevision assertions. This set of vectors, denoted by $\mathscr{P}(\mathbf{X}_N)$, is also called your *prevision boundary* for $\mathbf{X}_N$. It is a set specified as the net result of all your partial assertions about $\mathbf{X}_N$. We denote your assertion of this boundary by writing $P(\mathbf{X}_N) \in \mathscr{P}(\mathbf{X}_N)$. □

Evidently, if your prevision boundary contains only a single vector, then that vector is your prevision for $\mathbf{X}_N$ as defined in Section 2.4. In this case, $\mathscr{P}(\mathbf{X}_N) = \{P(\mathbf{X}_N)\}$.

***Definition 2.21.*** *Your partially asserted prevision configuration* for the vector of quantities $\mathbf{X}_N$ is said to be *coherent* as long as

**(i)** you do not thereby avow your willingness to engage in any transaction for which $s(b - \mathbf{a}_N^{\mathrm{T}}\mathbf{x}_N) < 0$ for every $\mathbf{x}_N \in \mathscr{R}(\mathbf{X}_N)$ (remember $s \geqslant 0$),

**(ii)** you do avow your willingness to engage in any transaction that would yield you $s(b - \mathbf{a}_N^{\mathrm{T}}\mathbf{x}_N) > 0$ for every $\mathbf{x}_N \in \mathscr{R}(\mathbf{X}_N)$, and

**(iii)** if you assert the inequality $\mathbf{a}_N^{\mathrm{T}} P(\mathbf{X}_N) \leqslant b$, then for any $b^* > b$ you also assert the inequality $\mathbf{a}_N^{\mathrm{T}} P(\mathbf{X}_N) \leqslant b^*$. □

Coherency of your partial assertions thus ensures [via (i) and (ii)] that neither can you be made a sure loser on the basis of your avowed willingness to engage in transactions, nor would you forgo any transaction that would make you a sure winner. Moreover, if you avow your willingness to engage in any particular transaction, then you will also engage [via (iii)] in any other transaction that will surely yield you a preferable result (even though you may possibly net a loss in the former or in both). We can now characterize the coherency of a partially asserted prevision configuration quite simply.

**Theorem 2.9.** The necessary and sufficient condition for a prevision boundary $\mathscr{P}(\mathbf{X}_N)$ to be coherent is that it be a convex subset of the convex hull $\mathscr{C}(\mathscr{R}(\mathbf{X}_N))$.

*Proof* ($\Rightarrow$). We presume that $\mathscr{P}(\mathbf{X}_N)$ is coherent, and prove that (1) it must be a subset of $\mathscr{C}(\mathscr{R}(\mathbf{X}_n))$, (2) it must be nonempty, and (3) it must be convex.

(1) For any vector $\mathbf{p} \notin \mathscr{C}(\mathscr{R}(\mathbf{X}_N))$, there is a vector $\mathbf{s}$ for which $\mathbf{s}^{\mathrm{T}}(\mathbf{x}_N - \mathbf{p}) < 0$ for every $\mathbf{x}_N \in \mathscr{R}(\mathbf{X}_N)$. Thus, your presumed coherency implies your willingness to assert the inequality $\mathbf{s}^{\mathrm{T}} P(\mathbf{X}_N) < \mathbf{s}^{\mathrm{T}}\mathbf{p}$. For engaging a transaction yielding you $\mathbf{s}^{\mathrm{T}}\mathbf{p} - \mathbf{s}^{\mathrm{T}}\mathbf{x}_N > 0$ for every $\mathbf{x}_N \in \mathscr{R}(\mathbf{X}_N)$ would make you a sure winner. Thus, $\mathbf{p} \notin \mathscr{P}(\mathbf{X}_N)$. Thus, $\mathscr{P}(\mathbf{X}_N) \subseteq \mathscr{C}(\mathscr{R}(\mathbf{X}_N))$.

(2) If $\mathscr{P}(\mathbf{X}_N)$ were empty, then for some vector $\mathbf{a}_N$ there must be numbers $b_1 < b_2$ for which you assert the inequalities $\mathbf{a}_N^{\mathrm{T}} P(\mathbf{X}_N) \leqslant b_1$ and $\mathbf{a}_N^{\mathrm{T}} P(\mathbf{X}_N) \geqslant b_2$. But if you engage in both the associated transactions, yielding you the net gain of $s[b_1 - \mathbf{a}_N^{\mathrm{T}}\mathbf{X}_N] + s[-b_2 - (-\mathbf{a}_N^{\mathrm{T}}\mathbf{X}_N)] = s(b_1 - b_2) < 0$ for every vector $\mathbf{x}_N \in \mathscr{R}(\mathbf{X}_N)$, you would be a sure loser, which contradicts the presumed coherency of $(\mathbf{X}_N)$.

(3) In order for a vector $\mathbf{p}$ to be a member of $\mathscr{P}(\mathbf{X}_N)$, it must be the case that $\mathbf{p}$ satisfies every inequality you have asserted for $P(\mathbf{X}_N)$. Suppose $\mathbf{p}$ and $\mathbf{q}$ are both members of $\mathscr{P}(\mathbf{X}_N)$. If $\mathbf{a}_N$ is a vector for which $\mathbf{a}_N^{\mathrm{T}}\mathbf{p} \leqslant b$ and $\mathbf{a}_N^{\mathrm{T}}\mathbf{q} \leqslant b$, then for any $\alpha \in [0, 1]$ it must be the case that $\mathbf{a}_N^{\mathrm{T}}[\alpha\mathbf{p} + (1 - \alpha)\mathbf{q}] \leqslant b$. Thus, $[\alpha\mathbf{p} + (1 - \alpha)\mathbf{q}]$ must also be member of $\mathscr{P}(\mathbf{X}_N)$. For it satisfies every inequality satisfied by $\mathbf{p}$ and $\mathbf{q}$. Thus, $\mathscr{P}(\mathbf{X}_N)$ is convex.

*Proof* ($\Leftarrow$). We presume that $\mathscr{P}(\mathbf{X}_N)$ is a nonempty convex subset of $\mathscr{C}(\mathscr{R}(\mathbf{X}_N))$, and prove that it must be coherent. There are three parts to the stipulation of a prevision boundary as coherent. We address them in turn.

(1) Any vector $\mathbf{p}$ for which $(\mathbf{a}_N^{\mathrm{T}}\mathbf{p} - \mathbf{a}_N^{\mathrm{T}}\mathbf{x}_N) > 0$ for every $\mathbf{x}_N \in \mathscr{R}(\mathbf{X}_N)$ must lie outside the convex hull $\mathscr{C}(\mathscr{R}(\mathbf{X}_N))$. Since you have specified $P(\mathbf{X}_N) \in \mathscr{P}(\mathbf{X}_N)$, which is a subset of $\mathscr{C}(\mathscr{R}(\mathbf{X}_N))$, you must have asserted that $\mathbf{a}_N^{\mathrm{T}} P(\mathbf{X}_N) \leqslant \mathbf{a}_N^{\mathrm{T}} p$. Thus, you have avowed your willingness to engage any transaction that would make you a sure winner.

(2) If for any vector $\mathbf{p}$ you have asserted that $\mathbf{a}_N^{\mathrm{T}} P(\mathbf{X}_N) \leqslant \mathbf{a}_N^{\mathrm{T}} p$, you cannot at the same time have asserted that $\mathbf{a}_N^{\mathrm{T}} P(\mathbf{X}_N) \geqslant \mathbf{a}_N^{\mathrm{T}} \mathbf{p} + \delta$, where $\delta > 0$. Otherwise $\mathscr{P}(\mathbf{X}_N)$ would be empty. Thus, you cannot be made a sure loser.

(3) Finally, if the vector $\mathbf{p}$ is a member of $\mathscr{P}(\mathbf{X}_N)$ by virtue of its satisfying the inequality $\mathbf{a}_N^{\mathrm{T}} \mathbf{p} \leqslant b$, then it also satisfies $\mathbf{a}_N^{\mathrm{T}} \mathbf{p} \leqslant b^*$ for any $b^* > b$. Thus, if you are willing to engage in any specified transaction, then you are also willing to engage any other transaction that will surely yield you more, even though you may possibly net a loss. In that case, your loss will surely be less. □

Theorem 2.9 is a straightforward generalization of Theorem 2.5. Whereas any coherent assertion of prevision for a quantity vector corresponds to a point within the convex hull of the realm of the quantity vector, any coherent partial assertion of prevision is representable by a nonempty convex subset of this hull.

For practical purposes, it is much more useful to state a constructive result, even if it is not as general as Theorem 2.9. Our next theorem extends the fundamental theorem of prevision to a statement that makes full use of the linear programming technique, which allows linear inequality restrictions as well as equality restrictions. Each equality constraint in a linear programming problem reduces by 1 the dimension of the feasible set of solutions. Inequalities reduce the volume of the feasible set, but do not reduce its dimension.

**Theorem 2.10.** The Fundamental Theorem of Prevision, Part I, Extended to Allow the Partial Assertion of a Prevision Boundary. Let $\mathbf{X}_N$ be any vector of quantities that interest you. Your assertion of any $K$ linear inequalities of the form $\mathbf{A}_{K,N} P(\mathbf{X}_N) \leqslant \mathbf{b}_K$ specifies your partial prevision boundary for $\mathbf{X}_N$ as $\mathscr{P}(\mathbf{X}_N) = \{\mathbf{p}_N \mid \mathbf{p}_N = \mathbf{R}(\mathbf{X}_N)\mathbf{q}$ for some $\mathbf{q} \geqslant \mathbf{0}$ satisfying $\mathbf{A}_{K,N}\mathbf{R}(\mathbf{X}_N)\mathbf{q} \leqslant \mathbf{b}_K$ and $\mathbf{1}^{\mathrm{T}}\mathbf{q} = 1\}$. If this set is empty, then your assertion of $\mathbf{A}_{K,N} P(\mathbf{X}_N) \leqslant \mathbf{b}_K$ is itself incoherent. □

Making use of our understanding of the coherency of a prevision boundary, we can see that this theorem proves itself by its very construction. The requirements that $1 = \mathbf{1}^{\mathrm{T}}\mathbf{q}$ and $\mathbf{q} \geqslant \mathbf{0}$ ensure that the prevision boundary $\mathscr{P}(\mathbf{X}_N)$ lies within $\mathscr{C}(\mathscr{R}(\mathbf{X}_N))$, and the requirements that $\mathbf{A}_{K,N}\mathbf{R}(\mathbf{X}_N)\mathbf{q} \leqslant \mathbf{b}_K$ exhaust the stipulations of your partially asserted prevision configuration for $\mathbf{X}_N$. If all of these latter stipulations were actually equalities, then the theorem would reduce to the fundamental theorem of prevision as stated in Theorem 2.8. The only development would be to offer the meaningful name *prevision boundary* to the feasible set of solutions to the linear programming problems, transformed into a convex subset of $\mathscr{C}(\mathscr{R}(\mathbf{X}_N))$.

## PROBLEMS

1. Consider the vector of events $\mathbf{E}_3$, whose realm elements are displayed graphically in Figure 2.6. Suppose you make the partial assertions of prevision

represented by $.7 \leqslant P(E_1) \leqslant .9$, $.5 \leqslant P(E_2) \leqslant .8$, and $.4 \leqslant P(E_3) \leqslant .9$. Theorem 2.10 then specifies your coherent prevision boundary $\mathscr{P}(\mathbf{E}_3)$ as the set of vectors that lie within the convex hull $\mathscr{C}(\mathscr{R}(\mathbf{E}_3))$ depicted in the figure and meet these three constraints. First try to imagine $\mathscr{P}(\mathbf{E}_3)$ as you view Figure 2.6 by cutting away those parts of the convex hull that are ruled out by the various asserted inequalities. Then use a linear programming algorithm to solve the six linear programming problems specifying the upper and lower coherent bounds on prevision for individual components of $\mathbf{E}_3$. (Notice that some of the six asserted inequality constraints are binding and some are not. There is nothing incoherent about asserting a nonbinding constraint. It is just that your universal desire for coherency characterizes the associated assertion as redundant.) Does $\mathscr{P}(\mathbf{E}_3)$ equal the convex hull of the six solution points?

**2.** Suppose that $E_1$ and $E_2$ are completely logically independent events. Suppose that you assert $.5 \leqslant P(E_1 + 2E_2) \leqslant 1.5$, $.5 \leqslant P(2E_1 + E_2) \leqslant 2.0$, $-1.5 \leqslant P(E_1 - 2E_2) \leqslant .5$, and $-1.5 \leqslant P(-2E_1 + E_2) \leqslant .5$. Graph the coherent prevision boundary characterized by these partial prevision assertions. On the basis of this graph, identify the solutions to the linear programming problems that maximize and minimize the values of $P(E_1)$ and $P(E_2)$ that cohere with these assertions. Does $\mathscr{P}(\mathbf{E}_2)$ equal the convex hull of the four solution points?

**Historical and Philosophical Note**

The extension of the fundamental theorem of prevision to allow the partial assertion of a prevision boundary is the most general extension of the Hilbert–Ackermann closure theorem for deductive logic. Whatever you want to assert about a vector of quantities, no matter how meager an assertion it may be, the fundamental theorem codifies all the implications that coherency of your assertions requires you to assert about any other quantities. Conceived in such a light, this theorem provides a foundation for the construction of a computational system of artificial intelligence. Because there are so many quantities that interest us, it is very difficult to be aware of the coherent implications of our many separate thoughts about them. It is impossible to be thinking about everything at once. The linear programming context of the fundamental theorem of prevision allows you to be aware of all the coherent implications of any particular partial assertions that you want to make. The computing system that routinely derives these implications is well termed a system of artificial intelligence.

The past 20 years has seen great activity in the investigation of such a system by many working groups that have not always communicated well. The eagerness of activity in this area has stemmed partly from a frustration felt by many that the restrictive framework of axiomatic probability theory is not conducive to the representation of the problem or its solution as it arises for someone who could use such a system in a practical instance. But to some extent, developments in this area have not received wholehearted support among proponents of the *operational* view of statistical theory, since many advocates of ideas associated with the names of artificial intelligence, expert systems, and fuzzy logic have not concerned themselves with any operational meaning to the claims they make. They appear willing to entertain probability intervals that have no meaning. When an operational meaning is attributed to some of their proposed "inferential rules," the rules have been found to be incoherent. See, for example, the commentaries by Good, Hill, and Lindley following the article of Shafer (1982). In this context, the results of this section provide an operational subjective support for the popular program of constructing an artificial intelligence system, discussed by Gilio and Scozzafava (1988) and Nilsson (1986). The bid-ask spread interpretation of the meaning of an asserted inequality allows the meaningful formulation of the notion of a probability boundary or interval; and the fundamental theorem of prevision provides the computational structure of the logic, applied to the partial assertion of prevision boundaries.

Long a controversial issue in the theory of subjective probability has been the question of how to represent the complete lack of knowledge about a quantity. It has been argued by some that asserting a uniform probability distribution over every constituent of the partition generated by $X$ represents such knowledge, since it allows that no possible value of $X$ is considered to be more likely than any other possible value. This is the principle of indifference that was criticized by Keynes. Careful analysis shows that in many meaningful senses, such a prevision assessment represents positive knowledge with specific content. Geometrically, it is represented by a unique point in the convex hull of of the partition vector, $\mathbf{Q}(X)$. We can now understand that the logic of partial prevision boundaries is needed to represent the lack of knowledge about a quantity. Only if your prevision interval for a quantity is coextensive with the realm of the quantity can we say that you assert no knowledge of a quantity. This intriguing idea has been advanced spiritedly by Shafer (1976) and in much of his subsequent work.

The important work of Peter Walley (1991) has highlighted the importance of interval probability assertions to the practical representation of uncertain knowledge, and the book is well worth examining on many matters. However, the complicated axiom system introduced to generate his logic of probability intervals is so involved that it makes the reasoning appear more intricate than it need be. The logic of partially asserted prevision boundaries agrees with the spirit of his work and has the capacity to resolve some of the conundrums that are posed in his structures. The philosophical work of Levi (1980) should be mentioned too as a leading exposition of many ideas concerning weaker forms of logical assertions, although its unusual notation and specialized concerns have made it difficult reading for most statisticians.

We have studied virtually all there is to study about probability per se by means of analyzing the structure of coherent prevision assessments. In the next chapter we expand our considerations to assertions of conditional prevision and statistical inference. The analysis will proceed in a way that unifies the theory with the theory of coherent prevision, both algebraically and geometrically.

CHAPTER 3

# Coherent Statistical Inference

*For the service to be considered was not the service of one servant, but of two servants, and even of three servants, and even of an infinity of servants, of whom the first could not out till the second up, nor the second up till the third in, nor the third in till the second up, nor the second up till the first out, every going, every being, every coming consisting with a being and a coming, a coming and a going, a going and a being, nay with all the beings and all the comings, with all the comings and all the goings, with all the goings and all the beings, of all the servants that had ever served Mr Knott, of all the servants that would serve Mr Knott. And in this long chain of consistence, a chain stretching from the long dead to the far unborn, the notion of the arbitrary could only survive as the notion of a pre-established arbitrary.*

*Watt*, SAMUEL BECKETT

## 3.1 INTRODUCTION

In formulating the fundamental theorem of prevision, we have begun to address the question of how your knowledge of some quantities is related to your knowledge of others. The fundamental theorem specifies the weakest relations required by coherency among your previsions (or partial prevision assertions) for several quantities.

It is possible to specify directly a relationship in your knowledge about several quantities by asserting a number that we call your *conditional prevision for a quantity given an event*. Such an assertion will be used to define a new type of quantity, called a *conditional quantity*. The introspective activities behind such assertions provide the basis for procedures of scientific statistical inference, assessing the evidence that the observation of some quantities provides for our knowledge of others.

In this chapter we construct the formal apparatus of inferential statistical procedures. For simplicity, in our theoretical formalities we usually presume that you specify your knowledge of relevant quantities via the full assertion of your prevision. The applicability of partial assertions of prevision boundaries will be mentioned occasionally in examples and problems associated with the second part of the fundamental theorem of prevision.

## 3.2 COHERENT CONDITIONAL PREVISION AND CONDITIONAL QUANTITIES

The definitions of conditional prevision and of a conditional quantity are motivated by the notion of a contingent transaction, which can be explained in the context of a generic event, $E$, and a generic quantity, $X$. These two quantities may be logically related in any way whatsoever.

Consider a transaction in which you pay the amount denoted as $P(X|E)$ for the right to receive $X$ if $E = 1$, and instead, to receive back exactly what you paid if $E = 0$. This is called a *contingent transaction*. It is equivalent to a contract specifying that contingent on $E = 1$, the transaction of $P(X|E)$ for $X$ is made. However, if $E = 0$, then no transaction is made. Your assertion of a conditional prevision $P(X|E)$ is your answer to the question, "At what price $P(X|E)$ would you be indifferent as to buying or selling $X$ contingent on $E$?"

***Example 1.*** Suppose you consider your prevision for the measured height, $H$, (to the nearest centimeter) of a person selected from your class roll by drawing a name from a hat. You may find it hard to think about your assessment of $H$ directly, since you "know" that the males in the class are generally taller than the females, but you don't know whether the person to be selected is male or female. You may not even know how many students are listed on the class role, nor how many are of each sex. Nonetheless, the concept of conditional prevision would allow you meaningfully to assert, for example, $P(H|M) = 177$ and $P(H|F) = 169$. □

***Example 2.*** Let the quantity $X$ be the national unemployment rate in New Zealand measured for the third quarter 1993, recorded to the nearest tenth of a percent, and let $E$ denote the event that equals 1 if the interest rate charged for first home mortgages by the Bank of New Zealand drops below 9.3 percent before 1 July 1993. As I am typing, I read in the paper that the unemployment rate for the fourth quarter of 1992 was 10.3 percent, and the BNZ mortgage interest rate equals 9.9 percent. Although I, Frank Lad, currently assert my $P(X)$ as 10.2 percent, I also assert my $P(X|E)$ as 9.8 percent. Asserting my $P(X)$ specifies my willingness to engage in transactions of $X$ for $P(X)$. Asserting my $P(X|E)$ specifies my willingness to engage in transactions that will be "called off" unless in fact $E$ is observed to equal 1, that is, unless the BNZ reduces its mortgage interest rate below 9.3 percent before July 1. Part of the reason my prevision for the unemployment rate is as high as it is that I think interest rates may well remain quite high. If I condition my transaction commitments on $X$ to be engaged only if the interest rate does fall below 9.3 percent, I would commit myself to a bolder statement regarding a decline in unemployment. □

From the very beginning of our study, we have characterized your assessment of your prevision for a quantity $X$ as a judgment you make about the value of $X$ on the basis of your *current state* of uncertain knowledge. In these two examples, you are uncertain about the values of two quantities, both $X$ and $E$. Because of the operational meaning of asserting a conditional prevision $P(X|E)$, we can interpret such an assertion as your current assessment of the value of $X$ on the basis of your current state of uncertain knowledge supplemented exclusively by the supposed precise summary observation that the event $E$ equals 1. For in asserting $P(X|E)$, you are expressing your willingness to engage transactions regarding $X$ that will have real consequences only under this condition that $E=1$. Thus, in asserting $P(X|E)$ as distinct from your $P(X)$, you would be *asserting the inference* that you would currently make about $X$ on the basis only of the observation that $E=1$. You are asserting the content of the information that you regard the observation $E=1$ would convey regarding $X$.

Formally, the concept of a contingent transaction provides the basis for the following definitions of a conditional prevision and of a conditional quantity given an event. We consider a general quantity and an event, $X$ and $E$, which may be logically related in any way: $\mathscr{R}(X,E) \subseteq \mathscr{R}(X) \otimes \{0,1\}$. The product $XE$ is also a quantity, logically related to $X$ and $E$ by $\mathscr{R}(X,E,XE) = \{(x,e,xe) | (x,e) \in \mathscr{R}(X,E)\}$, which is necessarily a proper subset of $\mathscr{R}(X,E) \otimes \mathscr{R}(XE)$.

***Definition 3.1.*** Let $X$ be any quantity, and $E$ be any event. Your *conditional prevision for $X$ given $E$*, denoted by $P(X|E)$, is the number you specify with the understanding that you are thereby asserting your willingness to engage any transaction that would yield you a net gain of the amount $s[XE - EP(X|E)]$, as long as $|s[xe - eP(X|E)]| \leqslant S$ for every pair of numbers $(e, xe) \in \mathscr{R}(E, XE)$. (Again we denote by $S$ the scale of your maximum stake.) □

It should be evident numerically that your conditional prevision $P(X|E)$ is defined as the price at which you are indifferent to engaging in a contingent transaction for $X$, contingent on $E$. For if $E=1$, your net gain from asserting $P(X|E)$ would equal $s[X - P(X|E)]$, whereas if $E=0$ your net gain would equal 0. In this latter contingency, you would "receive your money back."

***Definition 3.2.*** Having specified your conditional prevision, $P(X|E)$, the *conditional quantity $X$ given $E$*, denoted by $(X|E)$, is defined as

$$(X|E) \equiv XE + (1-E)P(X|E). \qquad \square$$

Why such a definition?

Remember that when we defined your prevision for a quantity $X$, we defined your $P(X)$ as a number you specify with the understanding that you assert thereby your willingness to engage any transaction that would yield you a net gain of $s[X - P(X)]$. Notice that the term in brackets equals the quantity $X$ less your prevision for the quantity. Our definition of a conditional quantity, $(X|E)$, is based upon a parallel interpretation of the structure inside the corresponding brackets in the definition of your conditional prevision, $P(X|E)$. In this definition, the net gain from the contingent transaction is represented by $s[XE - EP(X|E)]$. Once you assert your conditional prevision $P(X|E)$ as an operationally meaningful number, we can identify the expression inside these brackets as *defining a conditional quantity* in terms of the difference between a conditional quantity and your prevision for the conditional quantity

$$(X|E) - P(X|E) \equiv XE - EP(X|E).$$

After algebraic simplification, this amounts to defining the conditional quantity

$$(X|E) \equiv XE + (1 - E)P(X|E).$$

Thus, we have defined a terminology in which it is completely equivalent to speak of your "conditional prevision for the quantity $X$, given $E$" and to speak of your prevision for "the conditional quantity $X$ given $E$." For the operational consequences of the two statements are identical. In this lingo, a conditional quantity is merely a quantity, not categorically different from any other quantity we have discussed. In particular, this quantity is defined as a *linear combination* of the two quantities $XE$ and $\tilde{E}$. In one sense, it is an unusual quantity relative to those we have discussed heretofore: the conditional quantity $(X|E)$ is defined in terms of your prevision for it! The coefficient on $(1 - E)$ in the linear expression defining $(X|E)$ is precisely the number that you have asserted as your $P(X|E)$.

A consequence of this definition of the conditional quantity $(X|E)$ is that it allows a simple derivation of the conditions under which someone's assertions of $P(X|E)$, $P(XE)$, and $P(E)$ cohere with one another. Let us state this very important result as a theorem, often called *the product rule* for coherent previsions, and follow it with a corollary.

**Theorem 3.1.** Let $X$ be any quantity and let $E$ be any event. Then coherency requires that any asserted previsions must satisfy

$$P(XE) = P(X|E)P(E).$$

*Proof.* Since $(X|E) \equiv XE + (1 - E)P(X|E)$, the linearity of coherent prevision requires that $P(X|E) = P(XE) + P(1 - E)P(X|E)$. The stated requirement follows from algebraic simplification. □

**Corollary 3.1.1.** If you assert your $P(E) > 0$, then coherency requires that your further assertions of $P(X|E)$ and $P(X|E)$ satisfy $P(X|E) = P(XE)/P(E)$.

The greater is your background in standard probability theory, the more surely will you recognize that something unusual is happening here. The usual formalist *definition* of conditional probability specifies that "if $P(E_2) > 0$, then the conditional probability of $E_1$ given $E_2$ is defined by $P(E_1|E_2) \equiv P(E_1E_2)/P(E_2)$." Our operational subjective development recognizes this statement in a different light. We have begun our construction with an *operational definition* of your meaningful conditional prevision *assertion* as attesting to some aspect of your uncertain knowledge. We then recognized as a *theorem* that coherency requires your prevision assertions to satisfy the equation $P(XE) = P(X|E)P(E)$. The product rule for conditional prevision is a theorem, not a definition. It is important that you recognize this distinction if you are to understand fully the logical status of coherent statistical inferences. This understanding of coherent conditional prevision was first presented by de Finetti in his famous Paris lectures. The method of proof he used for the product rule in those lectures will be displayed shortly in an example.

Our definition of a conditional quantity will also appear as something quite extraordinary relative to the Kolmogorov-style construction of probability in terms of a measure space, in which there is no such thing as a conditional quantity or a conditional event. Goodman, Nguyen, and Walker (1991) have proposed such an algebraic structure of conditional events to extend the usual Boolean structure of events, interpreting conditional events in the syntax of a particular three-valued logic. A critical commentary on their work, in the spirit of the approach we follow here, appears in Lad (1995).

It is worthwhile to state formally a second corollary to Theorem 3.1, which you can prove yourself. It points out coherent conditional previsions are linear, just as coherent previsions are.

**Corollary 3.1.2.** If you assert your $P(E) > 0$, coherency requires that for any quantities $X$ and $Y$ your $P(X + Y|E) = P(X|E) + P(Y|E)$. Furthermore, the associated conditional quantities respect $(X + Y|E) = (X|E) + (Y|E)$. If $\mathbf{X}_N$ is a vector of quantities and $P(\mathbf{X}_N|E)$ is a vector of your associated conditional previsions, then $P(\mathbf{s}_N^T\mathbf{X}_N|E) = \mathbf{s}_N^T P(\mathbf{X}_N|E)$.

To ward off confusion, we will briefly discuss how the concise notation of $P(X|E)$ expands to more complex situations of application. In the case of economic statistics, for example, you might wish to assert conditional previsions for tomorrow's closing stock market average, $A$, conditional on an array of events defined by possible values of the recent quarter's unemployment statistics, to be announced in the morning. Each of the assertions $P(A|(U = 6.6))$, $P(A|(U = 6.7))$, and $P(A|(U = 6.8))$ provides a different instantiation of a conditional prevision of the generic form $P(X|E)$, where $E$ is an event. We do not need to expand the conditioning event notation to read $P(X|(E = 1))$, since the event $E$ is identical to the event $(E = 1)$. (The latter equals 1 if and only if the former equals 1. Otherwise both are 0.) Similarly, the notation for the assertion $P(X|\tilde{E})$ is identical to $P(X|(E = 0))$.

We are ready to study three detailed examples. The first displays a geometrical representation for coherent conditional prevision and a conditional quantity. It is meant to provoke your mathematical understanding of these concepts. The second example presents an application of conditional prevision assertions in the context of empirical questions in field ecology. It provokes applied scientific understanding. The third example shows how conditional prevision assertions are used in computing the cohering values of further previsions. Each example is followed by some appropriately related problems.

***Example 3.*** Suppose that $X$ is a quantity with realm $\mathscr{R}(X) = \{1, 2, 3, 4, 5\}$, and let $E$ be the event $E \equiv (X > 2)$. Thus, the realm of the vector of quantities $(X, E)$ is $\mathscr{R}(X, E) = \{(1, 0), (2, 0), (3, 1), (4, 1), (5, 1)\}$. Suppose you assert your conditional prevision $P(X|E) = 3.6$. This assertion defines the conditional quantity $(X|E)$ as $(X|E) \equiv XE + 3.6(1 - E)$. So the realm of the vector of four relevant quantities is

$$\mathscr{R}\begin{pmatrix} X \\ E \\ XE \\ X|E \end{pmatrix} = \left\{ \begin{pmatrix} 1 \\ 0 \\ 0 \\ 3.6 \end{pmatrix}, \begin{pmatrix} 2 \\ 0 \\ 0 \\ 3.6 \end{pmatrix}, \begin{pmatrix} 3 \\ 1 \\ 3 \\ 3 \end{pmatrix}, \begin{pmatrix} 4 \\ 1 \\ 4 \\ 4 \end{pmatrix}, \begin{pmatrix} 5 \\ 1 \\ 5 \\ 5 \end{pmatrix} \right\}.$$

Figure 3.1 displays the five elements of the realm of the vector of three quantities $X$, $XE$, and $E$. We know well that any coherent prevision vector $(P(X), P(XE), P(E))$ must lie within the convex hull of these five points.

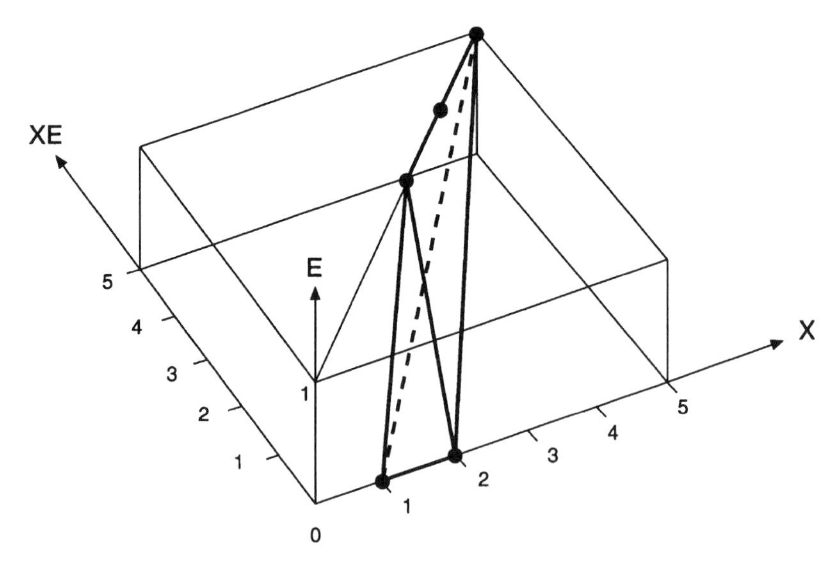


**Figure 3.1.** The dark lines enclose the convex hull $\mathscr{R}(X, XE, E)$, where $\mathscr{R}(X) = \{1, 2, 3, 4, 5\}$ and $E \equiv (X > 2)$. The outlining box and its top diagonal are shown in fine lines merely to allow you to get your bearings for viewing.

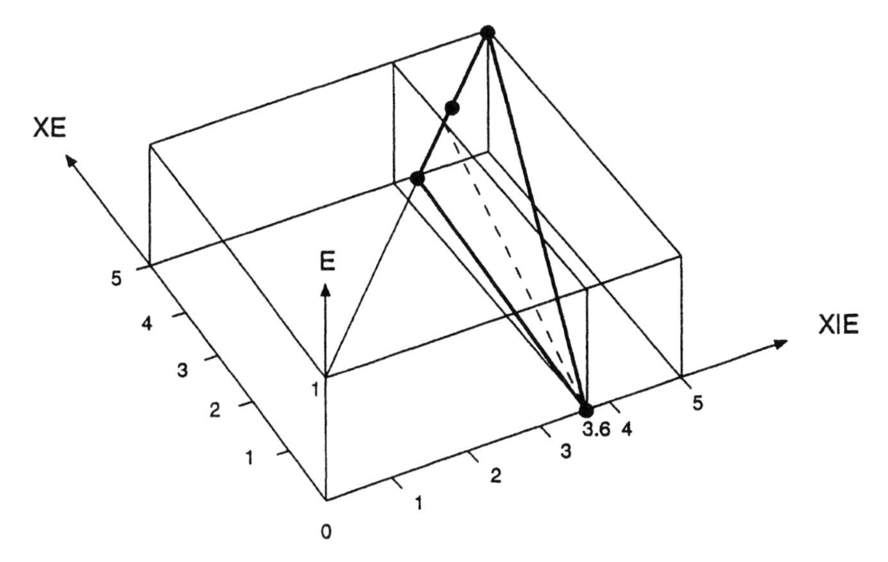


**Figure 3.2.** The convex hull of $\mathscr{R}((X|E), XE, E)$ is the dark-bordered triangle, since $(X|E)$ is defined by a linear function of $XE$ and $E$. The rectangle dividing the outlining box contains only vectors whose first component equals 3.6. The dashed line depicts the intersection of this plane with the triangular convex hull. It characterizes all vectors $(P(X|E), P(XE), P(E))$ that cohere with the assertion $P(X|E) = 3.6$. The equation for this line is $P(XE) = 3.6\,P(E)$.

Figure 3.2 shows the four elements of the realm of the vector of three quantities $(X|E)$, $XE$, and $E$. Notice that the convex hull of these four points constitutes a triangle in a plane, since $(X|E)$ is defined as a linear function of $XE$ and $E$ via $XE + 3.6(1 - E)$. However, not every vector within this convex hull can represent a coherent prevision vector for "you," the person who has asserted $P(X|E) = 3.6$, which was used to define the conditional quantity $(X|E)$. For you, having asserted $P(X|E) = 3.6$ confines the set of possible prevision *vectors* to those that also lie on the plane defined by this assertion! This plane, also shown in Figure 3.2, is represented by the rectangular divider of the outlining box. The dotted line in the figure is the intersection of the triangle specified by the convex hull of the realm $\mathscr{R}((X|E), XE, E)$ and the rectangle specified by $P(X|E) = 3.6$. It identifies the domain of possible coherent prevision vectors for you, the person who has specified this conditional prevision. You should recognize this line of coherent prevision vectors as representing the equation specified in Theorem 3.1:

$$P(XE) = P(X|E)P(E) = 3.6P(E).$$

Finally, Figure 3.3 merges the information portrayed in the previous two. Again, the boldly outlined tetrahedron is the convex hull of the realm $\mathscr{R}(X, XE, E)$. The lightly graveled plane connecting the four hollow balls depicts

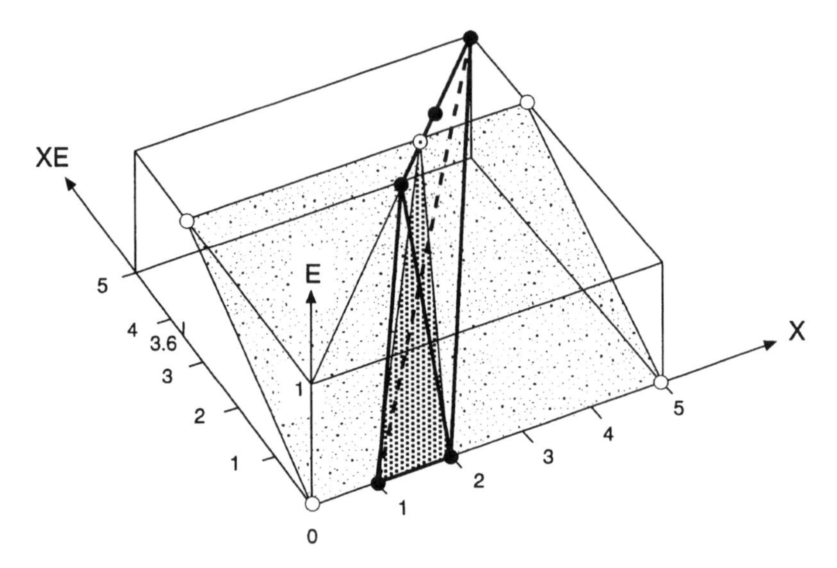


**Figure 3.3.** The convex hull of coherent prevision assertions $P(X, XE, E)$ is the boldly outlined irregular tetrahedron. The lightly graveled plane depicts vectors $(y_1, y_2, y_3)$ for which $y_2 = 3.6\mathbf{y}_3$. The densely dotted triangle, being the intersection points of the tetrahedron and the plane, represents all assertions $P(X, XE, E)$ that cohere with the assertion $P(X|E) = 3.6$.

the information regarding the quantities $X$, $XE$, and $E$ that is specified by your assertion of $P(X|E) = 3.6$. Coherency requires that $P(XE) = P(X|E)P(E)$, here equal to $3.6P(E)$. This equation is respected by all vectors $(y_1, y_2, y_3)$ lying on the lightly graveled plane, which is defined by the equation $y_2 = 3.6y_3$. Any prevision vector $P(X, XE, E)$ that coheres with the assertion $P(X|E) = 3.6$ must both lie within the convex hull of the solid balls and lie on this plane connecting the hollow balls. The densely dotted triangle then denotes all vectors that satisfy both of these restrictions. Examining Figure 3.3 you should see that the assertion of $P(X|E) = 3.6$ places the following individual coherency restrictions on your further assertions:

$$1 \leqslant P(X) \leqslant 3.6, \quad 0 \leqslant P(XE) \leqslant 3.6, \quad \text{and} \quad 0 \leqslant P(E) \leqslant 1.$$

Thus, your assertion of $P(X|E) = 3.6$ restricts your further cohering assertion of $P(X)$ and of $P(XE)$, but it places no restriction on $P(E)$ beyond its standard coherency requirement of lying within the interval $[0, 1]$. □

Before we discuss a substantive scientific problem, you should try the following problems, ordered in increasing degree of difficulty.

## PROBLEMS

1. Repeat the graphical analysis of Example 3, supposing again that $\mathcal{R}(X) = \{1, 2, 3, 4, 5\}$ but that $E$ is an event completely logically independent of $X$. How does the assertion $P(X|E) = 3.6$ restrict cohering values of $P(X)$, $P(XE)$, and $P(E)$ in this case? *Hint*: Be sure to look carefully so that you "see" all the edges of the convex hull. Use dashed lines in drawing any edges that would not be visible from your perspective.

2. Although $(X|E)$ is linearly related to $XE$ and $E$, it is not linearly related $X$ itself. Make a three-dimensional figure of the convex hull of the vector of quantities $((X|E), X, E)$ and examine it to see the liberty with which you may coherently assert previsions $P(X)$ and $P(E)$ concomitantly with your assertion of $P(X|E) = 3.6$. *Hint*: Construct your figure within a framing box exactly as in Figure 3.2, but relable the $XE$-axis as the $X$-axis and make the appropriate adjustments in plotting the points in $\mathcal{R}((X|E), X, E)$.

3. Continue problem 1, supposing that you assert $P(X|\tilde{E}) = 1.2$ in addition to $P(X|E) = 3.6$. First determine algebraically the coherency conditions on $P(X)$, $P(XE)$, and $P(E)$ imposed by asserting $P(X|\tilde{E}) = 1.2$. Then identify three points that lie on the plane defined by this condition. [Check the values of $P(XE)$ and $P(X)$ that would agree with the assertions of $P(E) = 0$ and $P(E) = 1$.] Sketch this plane into your figure. The intersection of the planes associated with these values of $P(X|E)$ and $P(X|\tilde{E})$ determine a line of vectors $(P(X), P(XE), P(E))$ that would cohere with this pair of conditional prevision assertions.

The following extensive example concerns a practical matter of great importance to the future of life on earth. Specific details have been somewhat stylized for simplicity of presentation. Relevant scientific measurements and zoological discussion can be found in articles by Wyman and Hawksley-Lescault (1987) and Wyman (1988, 1990).

***Example 4.*** This example applies the entire development of this textbook to this point. It should lead to your understanding the practicalities of assessing a conditional prevision, $P(X|E)$.

Suppose you are a naturalist, walking through a forest in upstate New York on an expedition to survey the red-backed salamander (*Plethodon cinereus*) population. It is 11:00 A.M., Monday, 11 May 1987. The temperature recorded at the Albany airport is 20°C = 67°F. The ground has been thawed since early April. Your long-term study of recent years has been to analyze the changes in this salamander population in the context of concomitant changes in the acidity of forest soil. In pursuing this study over the years, you have personally examined numerous meter-square "quadrats" of forest soil, recording the number of salamanders discovered within the top 5 cm of soil beneath the loose ground cover. Along with your deliberate search for salamanders, you have been making measurements of several aspects of the quadrats in which you search. At each site the measured quantities include

pH = pH reading of the soil using an instrumentation process calibrated in tenths of acid-base units over the range 3.0 to 5.0;

$T$ = temperature of the soil measured with a ground thermometer calibrated in single degrees centigrade over the range 0 to 40;

$W$ = water content of the soil, recorded on a hydrometer calibrated in units of hundredths between 0 and 1.

Moreover, you have participated in several group expeditions with other people engaging in such a search together, in various types of forest. For example, you have been in virgin hardwood forest and virgin pine forest; in 80-year-old, second-growth forest populated by maples, beach, and oak; and in 60-year-old plantations of pine and of spruce. The second-growth areas have been reforestations of agricultural cropland, of dairy pasture, and of grazing land for sheep. You have also examined creek beds and ponds located within each of these forest types. You have been out looking during the months of April through September, over a period of several years.

Furthermore, you have read research articles by other naturalists, reporting on the results of their similar investigations and their assessments of their experiences. In controlled conditions of a terrarium, you have done extensive experimental work on the behavior and life duration of red-backed salamanders under an extreme array of conditions of water, acidity, and temperature. In a word, "you" are a recognized, experienced salamander expert.

(Informative Note: A pH measurement is a reading of the hydrogen ion concentration in a substance. Each reversible chemical process exhibits its own equilibrium ion concentration at standard temperature and pressure. In particular, the process yielding distilled water is characterized in equilibrium by the ion concentration $[H^+][OH^-] = 10^{-14}$, where brackets around the ion symbol denotes that ion concentration relative to water. A pH reading is based upon a negative logarithmic scale, so that 7 represents distilled water. Higher numbers represent basic substances; lower numbers represent acidic substances. Remember, for example, that dissolving hydrogen chloride in water yields hydrochloric acid, $H^+Cl^-$, increasing the hydrogen ion concentration, whereas dissolving sodium hydroxide yields the basic solution $Na^+OH^-$. For a sense of measurement scale, know that battery acid yields a measurement near to 1, while your stomach acid yields a measurement near 3.)

Now suppose you come upon a specified quadrat in a second-growth forest. It lies well within the ring of the nearest trees in every direction, and there are no protruding roots apparent. In fact, the surface of the meter-square is mainly covered with matted damp leaves, though you can see two flat stones, with surface areas of roughly 150 and 75 cm$^2$. Your examination of this quadrat will yield four operationally defined measurements, denoted by

$$\begin{aligned} \mathrm{S} &= \text{ number of salamanders found,} \\ \mathrm{pH} &= \text{pH reading,} \\ T &= \text{temperature recorded with a ground thermometer,} \\ W &= \text{recorded water content of the soil.} \end{aligned}$$

These quantities recorded are the averages of the measurements at places within the square where salamanders are found. If no salamander is found, the recorded values of these quantities are the averages of measurements at the four corners of the half-meter-square centered within the quadrat. Averages are recorded to the nearest unit of calibration of the measuring instrument. The realms of these quantities, determined by the calibration of the measuring devices being used, are

$$\begin{aligned} \mathscr{R}(S) &= \{0, 1, 2, \ldots, 19, 20\}, \qquad \text{where 20 denotes 20 or more salamanders,} \\ \mathscr{R}(\mathrm{pH}) &= \{3.0, 3.1, 3.2, \ldots, 3.6, 3.7, 3.8, \ldots, 4.9, 5.0\}, \\ \mathscr{R}(T) &= \{0, 1, 2, \ldots, 15, 16, 17, \ldots, 39, 40\}, \\ \mathscr{R}(W) &= \{.00, .01, .02, \ldots, .16, .17, \ldots, .98, .99, 1.00\}. \end{aligned}$$

Although there is no formal logical dependence among these quantities, you would regard some of the constituent events of the partition they generate as quite unlikely, asserting your probability over them as 0. For example, consider the event $((S, \mathrm{pH}, T, W) = (15, 3.0, 39, .02))$. You would regard it as virtually impossible that even one salamander could live under these conditions, and really unbelievable that such a "possible" measurement would summarize this quadrat on this fine day. Upon seeing such an observation recorded in a notebook, you would tend to believe that there had been some carelessness in recording rather than that 15 salamanders had been found under such conditions. Given your sense of your own care in making observations and recording them in your notebook, you assert $P((S, \mathrm{pH}, T, W) = (15, 3.0, 39, .02)) = 0$. Upon being further grilled by a statistician (me, for example), you can be led to assert your prevision for the vector of quantities $(S, \mathrm{pH}, T, W)$. Suppose you do so, asserting your $P(S, \mathrm{pH}, T, W) = (.8, 3.9, 12, .58)$. These numbers seem to make some sense. As we approach the quadrat, you tell me that you regard this to be almost ideal conditions for finding red-backed salamanders in forest litter, though it is perhaps more acidic than would be optimal. Notice that the calendar date and the time of day of the recording are also relevant quantities.

(Informative Note: If you have never searched for salamanders, you may find the assertion $P(S) = .8$ to be rather amazing. You may even wish to make a bet with this naturalist (the "you" of this example). You are welcome to do so. But think for a minute. First, be sure to consider the fact that if you find one salamander it is quite possible that you may find more than one, perhaps two who are mating! It is May 11, early spring in New York after all. Then again you might ponder whether these salamanders congregate as does a den of snakes, or are they solitary creatures? You, the reader, may not know, but I can assure you that a salamander expert knows. Mating habits and social habits aside, suppose I tell you that these salamanders are regarded by naturalists to be the most abundant terrestrial vertebrate species in the northeast United States. Furthermore they play a central role in the food chain, particularly with respect to the decomposition of forest litter. A healthy adult typically measures about 5 cm in length. Are you still so sure you would like to bet? Perhaps now you may regard the naturalist's $P(S) = .8$ as a rather low assessment.)

We study the formalities of relations between different people's prevision assessments in Chapter 6. Let me mention that I have characterized "you" in this example as a salamander expert in order to give some "meat" to the example. We could ask anyone to assert a personal prevision for the quantities mentioned, even someone who is relatively uninformed about matters of red-backed salamanders in northeast North America. The person's prevision assertions, perhaps merely expressed as a prevision boundary, may well be quite different from those of our "expert." This is not to say that a general person's previsions are "wrong" while the expert's are "right." They merely identify the person's avowed state of knowledge as different from that of the expert. But as I said, this is an issue to be taken up in Chapter 6. Let's get back to our story.

On the basis of your laboratory experience and your historical field experience with measurements in northeast forests, you tell me that the red-backed salamander finds ideal a location where the pH reading is in the vicinity of 3.8 to 4.4. The more acidic the soil, you tell me, the rarer would it be for a red-backed salamander to hang around, supposing it has an alternative location to choose. Pressing you to be precise, I ask you to assert your $P(S|(\text{pH} = 3.6))$ and your $P(S|(\text{pH} = 3.9))$. Thinking it over, you respond with the numbers .13 and .86, knowing precisely what this means. Pressing you further, I ask you to identify the (pH, $T$, $W$) measurements associated with the ideal location for finding this type of salamander today. You respond with (4.2, 15, .67), adding, gratis, that your $P(S|(\text{pH}, T, W) = (4.2, 15, .67)) = 1.03$. By saying that these measurements would summarize ideal conditions, you mean that this conditional prevision is greater than that you would assert if the conditioning event were specified by any other vector of measurement values. Moreover, you add, conditioning also on the further specification that the plot with this measurement lies in the forest within 5 meters of a creek bed and that it has a large flat rock within it, your $P[S|((\text{pH}, T, W) = (3.6, 15, .20))CR] = 1.63$. Here $C$ denotes the event that the selected plot lies within 5 meters of a creek bed, and $R$ denotes that a large flat rock is found on the surface of the quadrat. Do you see how multiplicative syntax is used in the conditioning expression so the full symbolic expression represents a particular form of $P(X|E)$?

This example is meant to satisfy you that the operational definition of conditional prevision is meaningful and that the operation involved in assessing it can be performed even in very complicated situations.

(Informative Note: For your intellectual appreciation, if not for your moral appreciation, you may like to know that pH readings of rainwater taken in the Adirondack mountains of upstate New York typically hovered around 4.0 to 4.9 in the 1940s and 1950s. Today (1987 when written) measurements typically range between 3.6 and 4.4. Rainwater is typically more basic than groundsoil, though, obviously, this depends on the type of soil.) □

## PROBLEM

**4.** You should practice making conditional prevision assessments in problems such as the following. Feel free to change the nation or even the subject matter to make a problem relevant to your own interests. Let $X$ be the event that your country's exports to Australia for this year exceed those of last year, and $E$ the event that the exchange value of your currency for an Australian dollar at the end of this year is greater than at the end of last year. Assess your $P(X)$, $P(E)$, and $P(X|E)$.

A convenient feature of the way we have characterized conditional prevision is that assertions of conditional probability can be used to solve probability problems using the same computational technique we developed in Chapter 2, based upon the fundamental theorem of prevision, part I. If any of the asserted previsions used as input in a linear programming problem are conditional previsions, we can merely include the associated conditional quantities in the vector of the first $N$ quantities whose previsions are stipulated. The following example displays this method in solving a problem that was posed in the introductory text of Blackwell (1969).

***Example 5.*** A Post Office Problem. You give your friend a letter to mail. Your probability that your friend forgets to mail it equals .1; your probability that the letter is not delivered given that it is mailed equals .1; and finally, your probability that the addressee does not receive the letter given that it has been delivered equals .1. Presuming you are coherent, what is your probability that the addressee receives the letter, $P(R)$? What is your conditional probability that your friend forgets to mail the letter given that the addressee does not receive it, $P(\tilde{M}|\tilde{R})$?

Let us begin by establishing notation. Let $M$ denote the event that your friend mails the letter, $D$ the event that the post office delivers the letter, and $R$ the event that the addressee receives the letter. In these terms the problem is to determine what the assertions $P(\tilde{M}) = .1$, $P(\tilde{D}|M) = .1$, and $P(\tilde{R}|D) = .1$ imply (via coherency) for an assertion of $P(R)$ and for $P(\tilde{M}|\tilde{R})$.

To begin the solution, remember that if $P(\tilde{R}) > 0$ coherency requires $P(\tilde{M}|\tilde{R}) = P(\tilde{M}\tilde{R})/P(\tilde{R})$. So determining the values of the numerator and the denominator will yield the solution as long as the denominator is positive.

Now examining the columns of possible observations for the vector of events $(M, D, R)^{\mathrm{T}}$ if they were logically independent, viz.,

$$\begin{matrix} M & 1 & 1 & 1 & 1 & 0 & 0 & 0 & 0 \\ D & 1 & 1 & 0 & 0 & 1 & 1 & 0 & 0 \\ R & 1 & 0 & 1 & 0 & 1 & 0 & 1 & 0 \end{matrix}$$

you notice that the third, fifth, sixth, and seventh columns represent impossible observations. For if the post office does not deliver the letter, then the addressee cannot receive it, a situation contradicted in columns 3 and 7. And if your friend does not mail the letter then the post office cannot deliver it, nor can the addressee cannot receive it, situations contradicted in columns 5, 6, and 7. Thus, the realm matrix for the event vector $(M, D, R)^{\mathrm{T}}$ is constituted only by columns 1, 2, 4, and 8, which make up the first three rows of the realm matrix

$$\mathbf{R}\begin{pmatrix} M \\ D \\ R \\ \tilde{M} \\ \tilde{D}|M \\ \tilde{R}|D \\ \tilde{R} \\ \tilde{M}\tilde{R} \end{pmatrix} = \begin{pmatrix} 1 & 1 & 1 & 0 \\ 1 & 1 & 0 & 0 \\ 1 & 0 & 0 & 0 \\ 0 & 0 & 0 & 1 \\ 0 & 0 & 1 & .1 \\ 0 & 1 & .1 & .1 \\ 0 & 1 & 1 & 1 \\ 0 & 0 & 0 & 1 \end{pmatrix}.$$

The remaining five rows of this realm matrix are determined from the fact that their associated quantities are defined functionally in terms of the first three events and by your asserted conditional probabilities. For example, the conditional quantity $(\tilde{D}|M) \equiv \tilde{D}M + (1 - M)P(\tilde{D}|M) = \tilde{D}M + .1(1 - M)$, which yields the fifth line of the realm matrix $\mathbf{R}$.

Since you have asserted previsions for the fourth, fifth, and sixth of these quantities, the fundamental theorem of prevision requires that your prevision for the quantity in any other row must equal $\mathbf{r}_i\,\mathbf{q}_4$, where $\mathbf{r}_i$ is the corresponding row of the realm matrix $\mathbf{R}$, and $\mathbf{q}_4$ is the solution to the matrix equation

$$\begin{pmatrix} .1 \\ .1 \\ .1 \\ 1 \end{pmatrix} = \begin{pmatrix} 0 & 0 & 0 & 1 \\ 0 & 0 & 1 & .1 \\ 0 & 1 & .1 & .1 \\ 1 & 1 & 1 & 1 \end{pmatrix} \begin{pmatrix} q_1 \\ q_2 \\ q_3 \\ q_4 \end{pmatrix} \equiv \mathbf{M}\mathbf{q}_4.$$

The first three rows of this equation represent your specified prevision assertions, the first three rows of $\mathbf{M}$ being appropriately chosen as rows 4 through 6 of $\mathbf{R}$. The final row of $\mathbf{M}$ represents the convexity requirement that $\sum_{i=1}^{4} q_i = 1$. Now the inverse matrix of $\mathbf{M}$ does exist, and the solution to this equation is $\mathbf{q}_4 = \mathbf{M}^{-1}(.1 \quad .1 \quad .1 \quad 1)^{\mathrm{T}} = (.729 \quad .081 \quad .09 \quad .1)^{\mathrm{T}}$, a vector with all nonnegative components as required. The fundamental theorem also tells us that if any components of the solution for $\mathbf{q}_4$ were negative, then the three asserted previsions would be incoherent!

The answer to the first question posed is $P(R) = (1 \quad 0 \quad 0 \quad 0)\,\mathbf{q}_4 = .729$. As to the second question, $P(\tilde{M}|\tilde{R}) = P(\tilde{M}\tilde{R})/P(\tilde{R})$ is required by coherency. Thus, since $P(\tilde{M}\tilde{R}) = (0 \quad 0 \quad 0 \quad 1)\,\mathbf{q}_4 = .1$ and $P(\tilde{R}) = (0 \quad 1 \quad 1 \quad 1)\mathbf{q}_4 = .271$, the numerical solution is $P(\tilde{M}|\tilde{R}) = .1/.271 \approx .369$. You should be able to set up this problem on a computer, using MATLAB. □

If you merely assert an interval bound on a conditional prevision, say $P(X|E) \in [P_l(X|E), P_u(X|E)]$, then you would need to include two quantities in the vector of quantities whose previsions are at least partially asserted by the FTP part I. One would be defined by $L(X,E) \equiv XE + (1-E)P_l(X|E)$, and the other by $U(X,E)) \equiv XE + (1-E)P_u(X|E)$. The associated constraints in the linear programming problems would then be specified by the inequalities $P(L(X,E)) \geqslant P_l(X|E)$ and $P(U(X,E)) \leqslant P_u(X|E)$ respectively. You will have a chance to try this procedure in problem 5d.

## PROBLEMS

**5.** Consider again the mountain-climbing problem, which you solved as problem 2 of Section 2.5 and which you set up in matrix form on a computer in problem 4 of Section 2.8. The following sequence of problems should firm up your computational and conceptual skills regarding conditional prevision and the fundamental theorem of prevision. Together they exhibit the way the FTP can be used interactively to help someone assess realistic coherent previsions in an applied problem.

- **a.** Determine the values of your conditional previsions $P(R|F)$ and $P(F|R)$ that cohere with the assertions $P(R) = .45$, $P(F) = .25$, and $P(R \text{ or } F) = .50$. (You should find $P(R|F) = 4/5$ and $P(F|R) = 4/9$.)
- **b.** Now suppose you think about these solutions, and you find that they do not seem appropriate, thinking you would prefer to assert $P(R|F) = .2$ and $P(F|R) = .6$, along with $P(R) = .45$. Presuming only these three probabilities, determine the cohering assertion value for $P(F)$. (You should find that these three assertions are incoherent!)
- **c.** Reacting to the conclusion in part b, set up a linear programming problem to determine the bounds on $P(R)$ that would be necessary and sufficient for its cohering with the conditional probabilities $P(R|F) = .2$ and $P(F|R) = .6$. Interpret your solution graphically.
- **d.** Repeat part c, again determining coherent bounds on $P(R)$, but presuming only the partial assertion of previsions: $P(R|F) \in [.15, .25]$ and $P(F|R) \in [.55, .65]$. Presume also $P(RF) \geqslant .02$.

6. The following problem is adapted from a "challenge problem" posed by George Boole in the *Cambridge and Dublin Mathematical Journal* in 1851. A detailed discussion of various solutions proposed by Cayley, Boole, Dedekind, and Wilbraham appears in the work of Hailperin (1986).

   Suppose that the events $\mathbf{E}_3$ are logically related by the fact that $E_3$ can equal 1 only if either (or both) $E_1$ and $E_2$ equal 1. Thus, the outcome vector $\mathbf{E}_3 = (0, 0, 1)$ is impossible. (The convex hull of the realm of $\mathbf{E}_3$ appeared as Figure 2.4 in Section 2.9.) Now suppose you assert $P(E_1) = .1$, $P(E_2) = .2$, $P(E_3|E_1) = .8$, and $P(E_3|E_2) = .6$. Find the coherent bounds on a further assertion of $P(E_3)$.

   Boole's problem was posed more generally, in a way that allowed variable values for these four assertion values, and required a solution to be expressed as a function of the four variable assertion values. The very interesting discussion among the several participants is couched in language that considers $E_1$ and $E_2$ as necessary (but not sufficient) "causes" of $E_3$.

We conclude this section with two philosophical comments on the meaning of conditional prevision, and a final example which relates the coherency requirement on conditional prevision assertions to the sure loser criterion.

First, we recognize explicitly that the information about $X$ conveyed by the observation of $E$ is *not* an objective characteristic of $X$ and $E$, but rather the result of a judgment you make in asserting your $P(X)$ and $P(X|E)$. Someone else (or even you) in another state of knowledge, could coherently make quite a different judgment. Thus, making inference about $X$ on the basis of observing $E = 1$ is a free action that you make. It is not a necessary reaction that you make in deference to the rules of logic.

Second, your assertions of $P(X)$ and $P(X|E)$, and thus your inference about $X$ on the basis of the supposed observation $E = 1$, are made before you observe the value of $E$, if in fact you ever observe it. Notice, for example, that $P(X|E)$ and $P(X|\tilde{E})$ are both meaningful assertions, but only one of the observations $E = 1$ and $\tilde{E} = 1$ can possibly be made. By the time you observe the value of $E$ you could well be in a very different state of mind about $X$ for many reasons, based either on subsequent observations, reflections, rememberings, or forgettings. In explicit notation, today you assert your $P_{\text{today}}(X)$ and $P_{\text{today}}(X|E)$. Suppose that tomorrow you do observe that $E = 1$, among whatever other experiences fill your day. In your new state of mind tomorrow, you may assert your $P_{\text{tomorrow}}(X)$ as whatever number you please. Coherency does not require your assertion $P_{\text{tomorrow}}(X)$ in this scenario to bear any necessary relation to your $P_{\text{today}}(X|E)$, even if between today and tomorrow you do observe that $E = 1$.

We address this issue in greater detail in Section 3.6 when we discuss the matter of "learning in the large" and "learning in the small." For now, the bottom line for you to recognize is that the assertions $P(X)$ and $P(X|E)$ are commitments you make at the same point in time, typically without precise knowledge of the values of $X$ and $E$.

The final example of this section determines algebraically how someone who asserts previsions that do not satisfy the coherency requirement $P(XE) = P(X|E)P(E)$ can be made a sure loser on the basis of transactions that these assertions would identify as acceptable. To some extent, this example relies on advanced concepts in linear algebra, so it is not appropriate for the introductory-level reader.

***Example 6.*** An Example of Incoherent Conditional Prevision, $P(X|E, XE, E)$, and a Method for Determining How Its Proponent Can Be Made a Sure Loser. Again, let $X$ be a quantity with a realm $\mathscr{R}(X) = \{1, 2, 3, 4, 5\}$ and let $E$ be the event $E \equiv (X > 2)$. Suppose someone asserts the previsions $P(X|E) = 3.6$, $P(XE) = 4.1$, and $P(E) = .6$. Since these specifications do not honor the multiplication rule for coherent prevision, their proponent is incoherent and can be made to be a sure loser. How would this be done?

If the proponent of these assertions were "called" on the expressed willingness to engage in all three of the types of transactions identified by these prevision values, at scales specified by coefficients $s_{X|E}$, $s_{XE}$, and $s_E$, the net gain to the proponent would be

$$NG((X|E), XE, E) = s_{X|E}[(X|E) - P(X|E)] + s_{XE}[XE - P(XE)] + s_E[E - P(E)]$$

$$= ([X|E) - P(X|E)] \quad [XE - P(XE)] \quad [E - P(E)]) \begin{pmatrix} s_{X|E} \\ s_{XE} \\ s_E \end{pmatrix}.$$

Notice that the net gain is defined as a new quantity, which is a linear function of the quantities $(X|E)$, $XE$, and $E$, with the linear coefficients being specified by the scale factors $s_{X|E}$, $s_{XE}$, and $s_E$. Our question regarding sure loss translates mathematically into the following question: for what specification of these three scale factors would $NG((X|E), XE, E)$ be negative for every possible observation $((x|e), xe, e) \in \mathscr{R}((X|E), XE, E)$?

Now the realm for this vector of three quantities is

$$\mathscr{R}((X|E), XE, E) = \{(3.6, 0, 0), (3, 3, 1), (4, 4, 1), (5, 5, 1)\}.$$

Thus, the net gains deriving from these four possible observation vectors, written as a column vector, can be expressed as the product of a matrix and a vector:

$$\begin{pmatrix} \mathrm{NG}(3.6,\ 0,\ 0) \\ \mathrm{NG}(3,\ 3,\ 1) \\ \mathrm{NG}(4,\ 4,\ 1) \\ \mathrm{NG}(5,\ 5,\ 1) \end{pmatrix} = \begin{pmatrix} 0 & -4.1 & -.6 \\ -.6 & -1.1 & .4 \\ .4 & -.1 & .4 \\ 1.4 & .9 & .4 \end{pmatrix} \begin{pmatrix} s_{X|E} \\ s_{XE} \\ s_E \end{pmatrix}.$$

This is a linear equation, described in obvious notation by

$$\mathbf{NG}_{4,1} = \mathbf{M}_{4,3}\,\mathbf{S}_{3,1}.$$

The rows of $\mathbf{M}$ are obtained by subtracting the prevision vector (3.6, 4.1, .6) from each of the four vector elements of $\mathscr{R}((X|E), XE, E)$. The fact that each column of $\mathbf{M}$ contains positive and negative elements means that, individually, each of the presumed prevision assertions might yield a gain or a loss, depending on the observed values of $X$ and $E$.

However, if this matrix $\mathbf{M}$ has a generalized inverse, $\mathbf{M}^-$, then you can be sure that the vector $\mathbf{NG}$ of possible net gain values is composed exclusively of negative components, say $\mathbf{NG} = (-1, -1, -1, -1)^{\mathrm{T}}$, by selecting the vector $\mathbf{S} = \mathbf{M}^-\mathbf{NG}$ to represent the scale factors $s_{X|E}$, $s_{XE}$, and $s_E$ for the three avowed acceptable transactions. In fact, this matrix $\mathbf{M}$ does have a generalized inverse, displayed to the nearest $10^{-4}$ as

$$\mathbf{M}^- = \begin{pmatrix} .2062 & -.8763 & 1.1856 & 0 \\ -.2062 & -.1237 & -.1856 & 0 \\ -.2577 & .8454 & 1.2680 & 0 \end{pmatrix}.$$

If you are not familiar with the generalized inverse of a matrix, just check for yourself that $\mathbf{M}^-\mathbf{M} = \mathbf{I}_3$, the $3 \times 3$ identity matrix. (The first three columns of $\mathbf{M}^-$ constitute the inverse matrix for the first three rows of the matrix $\mathbf{M}$. You can check it on MATLAB.) Thus, the solution to our problem is that the choice of

$$\mathbf{S} = \mathbf{M}^-(-1, -1, -1, -1)^{\mathrm{T}} = (-.5155 \quad .5155 \quad -1.8557)^{\mathrm{T}} = (s_{S|E}, s_{XE}, s_E)^{\mathrm{T}}$$

would make the proponent of this incoherent prevision assertion a sure loser of 1 unit, no matter what value of $X$ is observed.

If you are familiar with the theory of the generalized inverse of a matrix, you can go on to prove for yourself that in general, the matrix constructed by rows denoting the values of $([XE - EP(X|E)], [XE - P(XE)], [E - P(E)])$ for the various vectors in $\mathscr{R}((X|E), XE, E)$, does have a generalized inverse unless $P(XE) = P(X|E)P(E)$. If the matrix has a generalized inverse, then the formula $\mathbf{S} = \mathbf{M}^-\mathbf{NG} = (s_{X|E}, s_{XE}, s_E)^{\mathrm{T}}$ can be used to determine the combination of transactions that would result in the proponent of an incoherent previson $(P(X|E), P(XE), P(E))$ being a sure loser. This was the original proof scheme used by de Finetti (1937) to establish the general coherency requirement for conditional prevision. □

## 3.3 THE FUNDAMENTAL THEOREM OF PREVISION. PART II

We have introduced conditional prevision in such a way that asserted conditional previsions can be introduced as input into the linear programming problems specified by the fundamental theorem of prevision in the same way that simple previsions are input. We merely include the relevant conditional quantities $(X|E)$ among the quantities whose previsions are presumed asserted in the problem setup. The one problem left unformalized is to determine the *bounds on a further conditional prevision* if it is to cohere with a list of specified previsions and conditional previsions. Since we have determined the necessary and sufficient condition for coherent conditional previsions as the requirement that $P(XE) = P(X|E)P(E)$, we can formulate the solution to this problem directly, calling it the fundamental theorem of prevision, Part II. After stating the theorem we discuss its computational resolution and provide an example with further discussion.

The statement of the theorem involves the following notation:

Let

$\mathbf{X}_{N+3} =$ vector of quantities whose first $N + 1$ components are any quantities that interest you, whose $(N + 2)$nd component is any event, and whose $(N + 3)$rd component is the product quantity $X_{N+3} \equiv X_{N+1}E_{N+2}$;

$S(N + 3) =$ size of the realm $\mathscr{R}(\mathbf{X}_{N+3})$;

$\mathbf{R}_{(N+3),S(N+3)} =$ realm matrix for $\mathbf{X}_{N+3}$;

$\mathbf{Q}_{S(N+3)} =$ vector of constituents of the partition generated by $\mathbf{X}_{N+3}$;

$\mathbf{R}_{N,S(N+3)} =$ first $N$ rows of $\mathbf{R}_{(N+3),S(N+3)}$;

$\mathbf{r}_{N+1} = (N + 1)$st row of $\mathbf{R}_{(N+3),S(N+3)}$;

$\mathbf{r}_{N+2} = (N + 2)$nd row of $\mathbf{R}_{(N+3),S(N+3)}$;

$\mathbf{r}_{N+3} = (N + 3)$rd row of $\mathbf{R}_{(N+3),S(N+3)}$;

$\mathbf{0}_{S(N+3)}$ and $\mathbf{1}_{S(N+3)}$ denote $S(N + 3) \times 1$ column vectors of 0's and 1's.

**Theorem 3.2.** The Fundamental Theorem of Prevision. Part II. Suppose you assert your prevision for $N$ quantities as $P(\mathbf{X}_N) = \mathbf{p}_N$. Let $X_{N+1}$ be any further quantity, let $X_{N+2} \equiv E_{N+2}$ be any further event, and let $X_{N+3} \equiv X_{N+1}E_{N+2}$. Then as long as cohering values of $P(E_{N+2})$ are bounded away from 0 by the FTP part I, for the coherence of any further assertion of conditional prevision $P(X_{N+1}|E_{N+2})$ with $P(\mathbf{X}_N)$ it is necessary and sufficient that $P(X_{N+1}|E_{N+2})$ lies within the interval $[P_l(X_{N+1}|E_{N+2}), P_u(X_{N+1}|E_{N+2})]$, where the boundaries of this interval are defined by the solutions to the following two nonlinear programming problems:

Find the $S(N + 3)$-tuples $\mathbf{q}_{S(N+3)} = (q_1, \ldots, q_{S(N+3)})^{\mathrm{T}}$ that characterize

$$P_l(X_{N+1}|E_{N+2}) \equiv \text{minimum } \mathbf{r}_{N+3}\mathbf{q}_{S(N+3)}/\mathbf{r}_{N+2}\mathbf{q}_{S(N+3)},$$

$$P_u(X_{N+1}|E_{N+2}) \equiv \text{maximum } \mathbf{r}_{N+3}\mathbf{q}_{S(N+3)}/\mathbf{r}_{N+2}\mathbf{q}_{S(N+3)}$$

both subject to the linear constraints that

$$\mathbf{R}_{N,S(N+3)}\mathbf{q}_{S(N+3)} = \mathbf{p}_N,$$

$$\mathbf{1}^{\mathrm{T}}_{S(N+3)}\mathbf{q}_{S(N+3)} = 1,$$

$$\mathbf{q}_{S(N+3)} \geqslant \mathbf{0}_{S(N+3)}.$$

If the feasible set of solutions to these nonlinear programming problems is empty, then the asserted prevision $P(\mathbf{X}_N) = \mathbf{p}_N$ is itself incoherent.

*Proof*. This result follows immediately from Corollary 3.1.1 in Section 3.2, that if $P(E_{N+2}) > 0$ then $P(X_{N+1}|E_{N+2}) = P(X_{N+1}E_{N+2})/P(E_{N+2})$. The objective functions in the two nonlinear programming problems identify the extreme values of this quotient that would cohere with the presumed values of $P(\mathbf{X}_N)$. Since linearity is a necessary and sufficient condition for coherency of prevision, the linear conditions on $\mathbf{q}_{S(N+3)}$ specified in this present theorem are both necessary and sufficient for the coherence of $P(X_{N+1}|E_{N+2})$ with $P(\mathbf{X}_N)$ as well. □

The programming problem specified by the theorem is nonlinear only on account of the objective function, which is a *quotient of two linear functions* of the vector of variables $\mathbf{q}_{S(N+3)}$. The constraining equations (or inequalities in an obvious generalization when only prevision boundaries are asserted for components of $\mathbf{X}_N$) remain linear functions of $\mathbf{q}_{S(N+3)}$, just as in the FTP part I. A nonlinear problem of this type is called a *fractional programming problem* for this reason. Happily, for its computational resolution it can be transformed into a linear programming problem, as described by Whittle (1982). Using generic notation, the transformation argument is as follows.

Suppose $\mathbf{n}$, $\mathbf{d}$, and $\mathbf{x}$ are column vectors of the same dimension, and $\mathbf{A}$ and $\mathbf{b}$ are a matrix and a vector appropriately conformable so that $\mathbf{Ax} = \mathbf{b}$. The fractional programming problem is:

Find the vector $\mathbf{x}$ that maximizes the quotient $f(\mathbf{x}) = \mathbf{n}^{\mathrm{T}}\mathbf{x}/\mathbf{d}^{\mathrm{T}}\mathbf{x}$ subject to the constraints $\mathbf{Ax} = \mathbf{b}$ with $\mathbf{x} \geqslant \mathbf{0}$, where $\mathbf{d}^{\mathrm{T}}\mathbf{x} > 0$ for every $\mathbf{x}$ satisfying these constraints.

This is precisely the form of the programming problem stated in the FTP part II, with $\mathbf{r}_{N+3} = \mathbf{n}^{\mathrm{T}}$, $\mathbf{r}_{N+2} = \mathbf{d}^{\mathrm{T}}$, $\mathbf{q}_{S(N+3)} = \mathbf{x}$,

$$\begin{pmatrix} \mathbf{R}_{N,S(N+3)} \\ \mathbf{1}^{\mathrm{T}}_{S(N+3)} \end{pmatrix} = \mathbf{A}, \qquad \text{and} \qquad \begin{pmatrix} \mathbf{p}_N \\ 1 \end{pmatrix} = \mathbf{b}.$$

Now consider the transformations $y(\mathbf{x}) = 1/\mathbf{d}^{\mathrm{T}}\mathbf{x}$ and $\mathbf{z}(\mathbf{x}) = \mathbf{x}y(\mathbf{x})$. In the context of this problem, there are no special conditions placed on the scalar $y = y(\mathbf{x})$ beyond those stated in $\mathbf{Ax} = \mathbf{b}$ and $\mathbf{x} \geqslant \mathbf{0}$. However, these definitions do imply a linear constraint on the vector $\mathbf{z}$ that $\mathbf{d}^{\mathrm{T}}\mathbf{z} = 1$, since $\mathbf{d}^{\mathrm{T}}\mathbf{z}(\mathbf{x}) = \mathbf{d}^{\mathrm{T}}\mathbf{x}y(\mathbf{x}) = \mathbf{d}^{\mathrm{T}}\mathbf{x}/\mathbf{d}^{\mathrm{T}}\mathbf{x} = 1$. Now the fractional programming problem can be rewritten as a linear programming problem in terms of $\mathbf{z}$ and $y$ as follows:

Find the vector $(\mathbf{z}^{\mathrm{T}} \quad y)^{\mathrm{T}}$ that maximizes $f(\mathbf{z}, y) = (\mathbf{n}^{\mathrm{T}} \quad 0)(\mathbf{z}^{\mathrm{T}} \quad y)^{\mathrm{T}}$ subject to the constraints $\mathbf{Az} = \mathbf{b}y$ and $\mathbf{d}^{\mathrm{T}}\mathbf{z} = 1$, with $\mathbf{z} \geqslant \mathbf{0}$ and $y > 0$.

In matrix form these linear constraints are written

$$\begin{pmatrix} \mathbf{A} & -\mathbf{b} \\ \mathbf{d}^{\mathrm{T}} & 0 \end{pmatrix} \begin{pmatrix} \mathbf{z} \\ y \end{pmatrix} = \begin{pmatrix} \mathbf{0} \\ 1 \end{pmatrix}.$$

This linear programming problem is identical to the fractional problem specified in terms of **A**, **b**, **n**, and **d**, since the objective functions are identical and the constraints are identical, as seen by direct substitution. If $(\mathbf{z}^{\mathrm{T}} \quad y)^{\mathrm{T}}$ is the solution to the transformed problem, then $\mathbf{x} = \mathbf{z}/y$ is the solution to the original problem.

Computationally, then, the programming problems specified by the FTP part II are representable in the standard linear programming format of finding the extreme values of $\mathbf{c}^{\mathrm{T}}\mathbf{x}$ subject to the restrictions $\mathbf{A}\mathbf{x} = \mathbf{b}$ with $\mathbf{x} \geqslant \mathbf{0}$, where the input components **A**, **b**, and **c** are defined as

$$\mathbf{A} = \begin{pmatrix} \mathbf{R}_{N,S(N+3)} & -\mathbf{p}_N \\ \mathbf{1}^{\mathrm{T}}_{S(N+3)} & -1 \\ \mathbf{r}_{N+2} & 0 \end{pmatrix}, \qquad \mathbf{b} = \begin{pmatrix} \mathbf{0}_N \\ 0 \\ 1 \end{pmatrix}, \qquad \text{and} \qquad \mathbf{c}^{\mathrm{T}} = (\mathbf{r}_{N+3} \quad 0).$$

Before a brief discussion, let's compute a numerical example.

***Example 1.*** We focus merely on the computational setup of this problem, since a complete substantive and geometrical exposition appears in Lad, Dickey, and Rahman (1992). Suppose $X$ is a quantity with realm $\{1, 2, 3, 4, 5\}$, and you assert the previsions $P(X) = 2.2$ and $P(X^2) = 6.0$. Define the event $E_{1.2}$ by $E_{1.2} \equiv (|X - 2.2| \geqslant 1.2)$. You should be able to show, using the FTP part I, that coherency requires $P(E_{1.2}) \in [.2\bar{6}, .5]$, so $P(E_{1.2}) > 0$. What does coherency allow as assertion values for the conditional prevision $P(X|E_{1.2})$?

To begin, the realm matrix corresponding to $\mathbf{R}_{(N+3),S(N+3)}$ is

$$\mathbf{R}\begin{pmatrix} X \\ X^2 \\ X \\ E_{1.2} \\ XE_{1.2} \end{pmatrix} = \begin{pmatrix} 1 & 2 & 3 & 4 & 5 \\ 1 & 4 & 9 & 16 & 25 \\ 1 & 2 & 3 & 4 & 5 \\ 1 & 0 & 0 & 1 & 1 \\ 1 & 0 & 0 & 4 & 5 \end{pmatrix}.$$

The first two rows pertain to the two quantities whose previsions are asserted initially in the statement of the problem. So $N = 2$ and $\mathbf{X}_2 = (X, X^2)^{\mathrm{T}}$. The third row appears redundant, but it is included to represent the formalities of the theorem as stated. It exemplifies an important point in the theorem that the bounds can be computed for any further conditional prevision $P(X_{N+1}|E_{N+2})$, even even if $X_{N+1}$ is a quantity whose unconditional prevision has already been specified.

Now the appropriate linear programming problem that determines coherent bounds on $P(X|E_{1.2})$ is specified by the objective function coefficients vector $\mathbf{c}^{\mathrm{T}} = (1 \quad 0 \quad 0 \quad 4 \quad 5 \quad 0)$, which is $\mathbf{r}_5$ $(= \mathbf{r}_{N+3})$ with a zero appended; the constraint coefficients matrix is

$$\mathbf{A} = \begin{pmatrix} 1 & 2 & 3 & 4 & 5 & -2.2 \\ 1 & 4 & 9 & 16 & 25 & -6.0 \\ 1 & 1 & 1 & 1 & 1 & -1 \\ 1 & 0 & 0 & 1 & 1 & 0 \end{pmatrix};$$

and the constraint values vector is $\mathbf{b}^{\mathrm{T}} = (0 \quad 0 \quad 0 \quad 1)^{\mathrm{T}}$. The first two rows of $\mathbf{A}$ are the first two rows of $\mathbf{R}$ with an appended column of $-(P(X) \quad P(X^2))^{\mathrm{T}}$. The third row is the appropriately dimensioned vector of 1's with $-1$ appended, and the fourth row is $\mathbf{r}_4$ $(=\mathbf{r}_{N+2})$ with a 0 appended.

The solution vector (to four decimal places) for the minimization problem is $\mathbf{x} = (.9444 \quad 0 \quad 1.2222 \quad 0 \quad .0556 \quad 2.2222)^{\mathrm{T}}$, yielding the lower extreme value of $P_l(X|E_{1.2}) = 1.2222$. Dividing $\mathbf{x}$ by its last component yields the solution to the fractional problem as posed, $\mathbf{q}_5 = (.425 \quad 0 \quad .55 \quad 0 \quad .025)$, a nonnegative vector whose components sum to 1.

The solution vector for the maximization problem is $\mathbf{x} = (.5625 \quad 2.75 \quad 0 \quad 0 \quad .4375 \quad 3.75)^{\mathrm{T}}$, yielding the upper extreme value of $P_u(X|E_{1.2}) = 2.75$. Dividing $\mathbf{x}$ by its final component yields the solution $\mathbf{q}_5 = (.15 \quad .7\overline{3} \quad 0 \quad 0 \quad .11\overline{6})$, again a nonnegative vector whose components sum to 1. □

You might have noticed that the quantity $X_{N+3} \equiv X_{N+1}E_{N+2}$ involved in the statement of the FTP part II would be a difficult one to contemplate. It is worth mentioning explicitly that you do not have to think about $X_{N+3}$ to do any probability problem. You are completely free to assess your opinion only about quantities that are natural and easy for you to think about in any specific application. The role of $X_{N+3}$ is merely computational in bringing the requirements of coherency to bear on your solution.

You can visualize the logic of the FTP part II by examining Figure 3.3 of Section 3.2 in the following light. Whatever initial assertions you make regarding quantities that are logically related to any $X$ and $E$ will identify your partial prevision polytope for the vector $(X, XE, E)$ as some convex polytope within the convex hull of a realm such as shown in the figure. [Imagine, for example, reducing the displayed tetrahedron further by slicing sections off of it by using planes, at least cutting it away from the floor on which $P(E) = 0$.] Since coherency requires $P(XE) = P(X|E)P(E)$, any assertion vector $P(X, XE, E)$ that coheres with the initial assertions must lie on a plane that intersects the resulting polytope and contains the $X$-axis, as does the lightly dotted plane shown in the figure. The value of $P(X|E)$ corresponding to any plane so defined equals the slope of the line cut by this plane in $(P(E), P(XE))$-space. Solutions to the fractional programming problems amount to the extreme points on the polytope that are identified by *swinging* a plane through the polytope, using the $X$-axis as a hinge. Of all such planes that intersect the polytope in at least one point, $P_u(X|E)$ is defined by the slope of the steepest line in $(P(E), P(XE))$-space, and $P_l(X|E)$ by the slope of the flattest line. You can see in Figure 3.3 that if $P(E)$ is not bounded away from 0, then the $X$-axis itself (the hinge) will always intersect the polytope, so no coherent bound on $P(X|E)$ can be computed. This situation can be interesting, but we will not address it further.

In scientific and commercial applications, the FTP part II would be useful computationally whenever you are unable to assess all your probabilities regarding the possible measurements deriving from a sequence of experiments or historical observations. Merely assert whatever you can about whatever quantities you can assess easily. Then, when you observe the measured values of some specific quantities, use the theorem to compute bounds on your conditional previsions for whatever other quantities interest you, conditioned on the product event that is specified by the measurements you have made. The following problems, requiring small-scale computations, will suggest some ideas in this regard. More interesting examples will arise after we have studied the judgment to regard a sequence of quantities exchangeably.

## PROBLEMS

The first problem is easy, just to solidify your understanding of the computational procedure involved in applying the FTP part II to a practical problem. The second is a somewhat larger-scaled problem and will suggest the usefulness of part II to problems of inference. The remaining problems, adapted from University of Canterbury final examinations, review your understanding of the entire text to this point.

**1.** Return again to the mountain-climbing problem 5 of Section 3.2. Using the FTP part II and the transformation of its fractional programming problem into a linear programming problem, compute bounds on $P(F|R)$ that would be necessary and sufficient for its cohering with the conditional probabilities $P(R|F) = .2$ and $P(R) = .45$. Compute the bounds, supposing only $P(R|F) \in [.15, .25]$ and $P(R) \in [.4, .5]$ along with $P(RF) \geqslant .02$.

**2.** The components of $\mathbf{X}_5$ are logically independent quantities, with the same realm $\mathscr{R}(X_i) = \{1, 2, 3, 4, 5\}$. Let $\bar{X}_5 = \sum_{i=1}^{5} X_i/5$ and $S_5^2 \equiv \sum_{i=1}^{5} (X_i - \bar{X}_5)^2/5$. Suppose you assert

**(i)** Each $P(X_i) = 2.2$
**(ii)** $P(X_i > 3)$ is nondecreasing in $i$.
**(iii)** $P(X_1 \leqslant X_2 \leqslant X_3 \leqslant X_4 \leqslant X_5) \geqslant .2$.
**(iv)** $P(\max \mathbf{X}_5 > 3) \geqslant .4$.
**(v)** $P(S_5^2 > 2.25) \geqslant .3$.

Further, define $M \equiv (X_4 + X_5)/2$ and $V = [(X_4 - M)^2 + (X_5 - M)^2]/2$.

**a.** Compute cohering bounds on $P(X_4)$, $P(M)$, and $P(V)$.
**b.** Compute cohering bounds on $P(X_4|E)$, $P(M|E)$, and $P(V|E)$, when the event $E$ is defined as $E \equiv (\mathbf{X}_3 = (2, 2, 2))$, as $E \equiv (\mathbf{X}_3 = (2, 5, 4))$, and as $E \equiv (\mathbf{X}_3 = (3, 5, 1))$. Presume each $P(E) \geqslant 10^{-6}$.

**3.** A Problem Regarding Export Statistics. Let $XA$, $XJ$, and $XU$ denote the monetary value of New Zealand's exports during 1989 to Australia, Japan, and the United States, respectively. The *NZ Statistical Yearbook* for 1990 lists the total world value of NZ exports during 1989 as NZ$14,905 millions.

**a.** Describe the realm of the vector of quantities $(XA, XJ, XU)$ as simply as you can, paying attention to any logical relations among them.

**b.** Let $RXA$, $RXJ$, and $RXU$ denote the ranks of these export quantities, where rank 1 denotes the highest rank and 3 denotes the lowest rank. (In the case of a tie, each would be accorded the higher rank.) Identify the realm matrix for the column vector $(RXA, RXJ, RXU)^{\mathrm{T}}$.

**c.** Define $E_1 \equiv (RXA < RXJ)$, $E_2 \equiv (RXA < RXU)$, and $E_3 \equiv (RXJ < RXU)$. Append the appropriate rows to your realm matrix in part b so that you now have identified the realm matrix for $(RXA, RXJ, RXU, E_1, E_2, E_3)$. Are the events $E_1$, $E_2$, and $E_3$ logically independent? Why or why not? Finally append another row to this realm matrix associated with the product event $E_1E_3$.

**d.** Graph the convex hull of $\mathscr{R}(\mathbf{E}_3)$.

**e.** Think for a bit and assert your $P(E_1)$ and your $P(E_1|E_3)$. If you cannot arrive at a precise number, then an interval bound would do.

**f.** Graph the convex hull of the event vector $(E_1, E_3, E_1E_3)$. Identify graphically the restrictions on your prevision vector $P(E_1, E_3, E_1E_3)$ that derive by coherency with your asserted number or interval for $P(E_1|E_3)$.

**g.** Identify the further restrictions that derive from your assertion of $P(E_1)$.

**h.** Using your graphical depiction, specify numerically (approximately) the bounds that coherency would place on your further assertion of $P(E_3|E_1)$.

**i.** Using the FTP II, set up problem h as a fractional programming problem and transform it into a linear programming problem for solution.

**4.** A Problem of Currency Exchange. Let $(A/B)_t$ denote the price of a unit of currency $B$ in terms of a unit of currency $A$ for the final trade that occurs in a currency market on day $t$. Further, define $(A \uparrow B) \equiv ((A/B)_{t+1} \geqslant (A/B)_t)$, the event that the final trading price of $B$ in terms of $A$ on day $t+1$ is at least as great as on day $t$. Consider the three currencies of the dollar, \$, the pound sterling, £, and the yen, $Y$. Thus, for example, you are considering quantities such as $(\$/Y)_t$ and $(£ \uparrow Y)$. Initially, suppose you assert the probabilities

$$P(\$ \uparrow £) = .4 \qquad \text{and} \qquad P((\$ \uparrow £)|(\$|£) \text{ or } (\$ \uparrow Y)) = .8.$$

**a.** Determine the cohering value of$P((\$ \uparrow £)$ or $(\$ \uparrow Y))$.

**b.** Identify and discuss the logical relations among $(\$ \uparrow £)$, $(\$ \uparrow Y)$, $(£ \uparrow Y)$, $((\$ \uparrow £)$ or $(\$ \uparrow Y))$, and $((\$ \uparrow £)|(\$ \uparrow £)$ or $(\$ \uparrow Y))$ in the context of this problem. *Hint*: Be careful. Pay attention to the operational definitions of these quantities. The pairs of parties involved in the commercial transactions defining the several events may be completely distinct from each other.

**c.** Define an event that is not a function of ($\$\uparrow£$), ($\$\uparrow Y$), and ($£\uparrow Y$) but is logically related to these three events. Denote this event by $B$. (You do not need to describe event $B$ in words. You may "define" $B$ merely by exhibiting the realm of the vector of all four events.)

**d.** Graph the convex hull of the realm of the vector of events $[B, (\$\uparrow£), ((\$\uparrow£)$ or $(\$\uparrow Y))]$.

**e.** Using the two probability assertions stated at the beginning of the problem, identify on your graph the region of cohering prevision vectors $[P(B)$, $P(\$\uparrow£)$, $P((\$\uparrow£)$ or $(\$\uparrow Y))]$.

**f.** Suppose you further assert the partial previsions $P(£\uparrow Y)\in[.25, .35]$ and $P(\$\uparrow Y)\in[.15, .25]$ and the conditional previsions $P((\$\uparrow£)|(\$\uparrow Y)) = .5$ and $P((£\uparrow Y)|(\$\uparrow Y))\in[.4, .6]$. Determine bounds on any further cohering assertion of $P((\$\uparrow£)|(£\uparrow Y))$.

**5.** A Problem on Product Quality. A timber products firm has sold seven container loads of chipboard to an overseas production firm under the stipulation that the shipment must pass specification tests at the port of arrival for it to be accepted. Three boards are to be selected for inspection, and two measurements are to be made on each board as follows: a 30-cm × 2-cm strip is to be cut from the center of each board; then the strip is to be weighed to the nearest gram and the tensile strength (breaking pressure) of the strip measured to the nearest kilogram per square centimeter. For each board selected ($i = 1, 2, 3$) the pair of measurements $(X_i, Y_i)$ is defined by

$$X_i = (\text{weight} < 27 \text{ grams}) - (\text{weight} > 29 \text{ grams}),$$

$$Y_i = (\text{tensile strength } < 15 \text{ kg/cm}^2).$$

The shipment will be accepted only if the vector $(\mathbf{X}_3, \mathbf{Y}_3)$ has at least five components that equal 0.

**a.** Using appropriate notation, specify the realm of the measurement $X_1$. What is the logical relation among $X_1$, $X_2$, and $X_3$? Without writing all its elements (i.e., using appropriate notation) specifying the realm of the vector $\mathbf{X}_3$.

**b.** Consider now the measurements for a single board $(X_1, Y_1)$. Write the realm matrix for this quantity pair.

**c.** Defining $Z_1 \equiv (X_1 \neq 0)$, append rows to the realm matrix of part b so that you construct the realm matrix for $(X_1, Y_1, Z_1, Y_1Z_1)$.

**d.** Plot the columns of the realm of $(Z_1, Y_1, Y_1Z_1)$ in three dimensions, and draw all the edges of their convex hull.

**e.** Suppose you assert $P(Z_1) = .10$. Show graphically on your figure from part d the restricted range that coherency would allow for your assertion of $P(Y_1, Y_1Z_1)$. State the geometric principle of coherent prevision that supports your argument.

**f.** To answer this part, begin by copying the convex hull of part d, and superscribe around it, lightly in pencil, the unit cube just to get your bearings. Suppose you assert only $P(Y_1|Z_1) = .50$. Show graphically the restriction that coherency would place on your assertion of $P(Z_1, Y_1, Y_1Z_1)$.

**g.** Combining the assertion suppositions of parts e and f, determine numerically the value of the cohering assertion of $P(Y_1Z_1)$, and show geometrically the bounds on any further cohering assertion of $P(Y_1)$.

**h.** Again supposing the assertions $P(Z_1) = .10$ and $P(Y_1|Z_1) = .50$, compute the cohering value for your probability that the weight of the first strip is between 27 and 29 grams (inclusive) and the tensile strength is strictly less than 15 kg/cm$^2$.

**i.** Formulate the linear programming problems that identify the upper and lower bounds on $P(Y_1)$ for it to cohere with $P(Z_1) = .10$ and $P(Y_1|Z_1) = .50$. How would the setup expand in order to include an additional assertion of $P(Y_1|X_1 = -1) = .90$?

**6.** A Problem on Product Reliability. Suppose that a newly produced dishwashing machine is put on test by running it through its standard prewash–wash–hot rinse cycle 50 times, repetitively. The machine can fail due to electrical problems, mechanical problems, and/or construction problems (leaking gaskets, etc.). The events that are of concern to the producer are whether

$E_1 \equiv$ (dishwasher fails for any reason before completing its first cycle);

$E_2 \equiv$ (dishwasher completes the first cycle without fault, but fails for any reason before completing 50 cycles);

$E_3 \equiv$ (dishwasher completes 50 cycles without failure);

$L \equiv$ (dishwasher fails the test for an electrical reason);

$M \equiv$ (dishwasher fails the test for a mechanical reason);

$C \equiv$ (dishwasher fails the test for a construction reason).

**a.** Discuss the logical relations among these six events, defining what it means for events to be logically dependent and identifying any logical relations among the six events that are evident from their definitions. Conclude by constructing the realm matrix for the vector of the six events.

**b.** Consider the event that the machine fails (any time before 50 complete cycles) for mechanical and electrical reasons. Append this event to the column vector of six events considered so far, and append the appropriate row to the realm matrix.

**c.** Suppose the engineers responsible assert the probabilities $P(E_3) = .90$ and $P(L|M) = .60$. First, define the conditional quantity $(L|M)$, append it to the vector of seven events considered so far, and append the appropriate row to the realm matrix. Then, using geometrical methods, identify the interval of possible values for a further assertion $P(M)$ that would cohere with these two probabilities. Explain briefly what you are doing, and state any properties of coherent prevision that your analysis relies upon. [*Hint*: The graphical analysis can be conducted in the realm space for the event vector $(E_3, LM, M)$.]

**d.** Suppose the event $E_2$ is refined further into

$F_1 \equiv$ (dishwasher completes the first cycle without fault but fails within the first 25 cycles);

$F_2 \equiv$ (dishwasher completes at least 25 cycles without fault but fails before completing 50 cycles).

Evidently, $E_2 = F_1 + F_2$. Now supposing that each $P(E_i) > 0$ and each $P(F_i) > 0$, which of the following conditional probabilities is larger in this problem: $P(F_2|\tilde{E}_1)$ or $P(F_2|\tilde{E}_1\tilde{F}_1)$? Why? (Just write out the equations that specify these two conditional probabilities and think about their component pieces.)

**e.** Failures on account of several factors would mean more severe penalty costs for the producer (via customers' warranty claims) than would failures based on a single factor. Based on all the probability assertions that have been specified so far, write out the form of a linear programming problem that would determine upper and lower bounds for the probability that the machine fails before 50 complete cycles for at least two reasons.

**f.** (Your answer to this question can be deferred until you have studied Section 3.7 on the judgment to regard events independently.) Suppose the electrical components of the machine are imported into the assembly area of the plant from another plant. Thus, the reliability of an electrical component used in a dishwasher is beyond this plant's control unless the electrical failure is precipitated by a mechanical failure or a construction error. Based on assurances from the quality control assessors from the supplying plant, suppose the engineers assert $P(L) = .05$, $P(L\tilde{C}) = .02$, and $P(C) = .04$. Are the events $L$ and $C$ regarded independently? Identify why or why not based on technical coherency reasons, and write a sentence or two motivating this opinion in the context of this specific story problem.

## 3.4 THE GENERAL PRODUCT RULE, THE THEOREM ON TOTAL PREVISION, AND BAYES' THEOREM

As we have learned, the fundamental theorem of prevision resolves virtually every probability problem that can be posed. For it exploits computationally the necessary and sufficient conditions for coherent prevision vectors, that they lie within the convex hull of the relevant realm.

Nonetheless, a few results that derive as special cases of this theorem are of great practical importance and worth studying explicitly. In this section we highlight three theorems: the general product rule for the probability of a product of events, the theorem of total probability, and Bayes' theorem. Discussion is minimal, and the applications may appear trite, for we use these results throughout the remainder of the text, developing more substantive applications.

**Theorem 3.3.** The General Product Rule for Probabilities of Product Events. If $E_1, \ldots, E_N$ are any events whatsoever, then coherency requires that

$$P(E_1 E_2 \cdots E_N) = P(E_1)P(E_2|E_1)P(E_3|E_1E_2)P(E_4|E_1E_2E_3)\cdots P(E_N|E_1E_2\cdots E_{N-1}).$$

Using product notation, we can write

$$P\left[\prod_i^N E_i\right] = \prod_{i=1}^N P\left[E_i \,\middle|\, \prod_{j=1}^{i-1} E_j\right], \qquad \text{where } \prod_{j=1}^{0} E_j \equiv 1.$$

You can prove this result easily yourself using induction on the product rule we have written simply in Theorem 3.1: $P(E_1E_2) = P(E_2|E_1)P(E_1)$.

***Example 1.*** One part of the post office problem we studied as Example 4 in Section 3.2 can be solved directly by using the product rule. Under the suppositions of the problem, what is your probability $P(R)$ that your friend receives the letter? Examining the realm matrix for the vector of events $(M, D, R)$, it is evident that $R = MDR$. This makes sense, since your friend can receive the letter only if it is both mailed and delivered. So according to the product rule,

$$P(R) = P(MDR) = P(M)P(D|M)P(R|DM).$$

Now $P(M) = .9$ and $P(D|M) = .9$, for the problem specifies that $P(\tilde{M}) = P(\tilde{D}|M) = .1$. Finally, the realm matrix again shows that $DM = D$. (The letter can be delivered only if it is mailed.) Thus, $P(R|DM) = P(R|D) = 1 - P(\tilde{R}|D) = .9$, and thus $P(R) = (.9)(.9)(.9) = .729$, as we found by solving the problem using the FTP.

The minor distractions involved in seeing that $R = MDR$ and $D = DM$ highlight an important point. It is not generally true that $P(E_3) = P(E_1E_2E_3) = P(E_1)P(E_2|E_1)P(E_3|E_2)$. In using the product rule, one must be very careful that the appropriate conditioning events be identified in every multiplicand. □

The most important and regular uses of the product rule occur in sequential forecasting. This procedure involves specifying probabilities for $(X_1 = x_1)$ for several possible values of $x_1$; and specifying probabilities for $(X_2 = x_2)|(X_1 = x_1)$ for several possible pairs $(x_1, x_2)$; and specifying probabilities for $(X_3 = x_3)|(X_2 = x_2)(X_1 = x_1)$ for several possible triples $(x_1, x_2, x_3)$; and so on. The product rule allows then the computation of probabilities for any entire sequence of events identified by the product $(X_1 = x_1)(X_2 = x_2)(X_3 = x_3)\cdots(X_T = x_T)$. A discursive example of such a problem appears in Section 5.3.5.2.

The product rule for coherent prevision, along with our understanding of the meaning of a partition, yield the following theorem on total prevision.

**Theorem 3.4.** Theorem of Total Prevision. Suppose the events $H_1, H_2, \ldots, H_N$ constitute a partition. Then for any quantity $X$, coherency requires that

$$P(X) = \sum_{i=1}^{N} P(X \mid H_i)P(H_i).$$

This result derives simply from the facts that:

**(1)** The events $\{H_1, \ldots, H_N\}$ constitute a partition, so $X = X\ [\sum_{i=1}^{N} H_i] = XH_1 + \cdots + XH_N$.
**(2)** Coherent prevision is a linear operator, so $P(X) = P(XH_1) + \cdots + P(XH_N)$.
**(3)** The product rule requires that for each $i$, $P(XH_i) = P(X \mid H_i)P(H_i)$.

In the special case when $X$ is an event, this theorem has long been called the theorem on total probability. O'Hagan (1988) alludes to this theorem, using subjectivist imagery as the rule for "extending the conversation" regarding $X$ to considerations of $X$ under the various scenarios specified by the partition. The stimulating articles of de Finetti (1977) and Scozzafava (1991) are both enjoyable and relevant.

***Example 2.*** Recall the simplest Example 1 of Section 3.2 of conditional prevision assertion, when you asserted your $P(H \mid M) = 177$ and $P(H \mid F) = 169$. Now the events $M$ and $F$ constitute a partition. So if you also assert $P(M) = .55$, the theorem on total probability requires that your

$$P(H) = P(H \mid M)P(M) + P(H \mid F)P(F) = 177(.55) + 169(.45) = 173.4.$$

If you don't assert your $P(M)$, then coherency would merely bound your $P(H)$ within the interval $[169, 177]$. This comment is generally relevant to theorems on coherent probabilities. A theorem cannot require you to make any particular assertions that appear in its formal statement. The results are merely cautionary, that if you do assert prevision values that appear in the statement of the theorem they had better satisfy the stated relationship. □

The theorem on total prevision has its greatest use in assessing your knowledge of quantities whose values continge on the values of other unknown quantities. How many tons of wheat will be harvested in the state of Kansas this summer? You might well say that the answer depends on how much it rains and how hot the temperature will be. If you would make a partition of events describing the range of rainfall and temperature possibilities, you may find it easier to assert your prevision for the wheat yield conditional on each partitioning event. Then your prevision for the wheat yield could be computed from these and from your probabilities for the constituents of this partition.

The product rule for coherent probability and the theorem of total probability together imply the celebrated Bayes' theorem, which formalizes the process of coherent inference that we shall call "learning in the small." Let us begin with a formal presentation and proceed to a discussion.

***Theorem 3.5.*** Bayes' Theorem. Suppose the events $H_1, H_2, \ldots, H_N$ constitute a partition. Then coherency requires that for any event $D$ assessed with $P(D) > 0$,

$$P(H_i|D) = \frac{P(D|H_i)P(H_i)}{\sum_{j=1}^{N} P(D|H_j)P(H_j)} \qquad \text{for each } i = 1, 2, \ldots, N.$$

Technically, the theorem follows simply from a dual application of the coherency requirement for conditional probabilities:

$$P(H_i|D)P(D) = P(DH_i) = P(D|H_i)P(H_i).$$

The stated result follows from dividing the equation by $P(D) > 0$, canceling $P(D)$ appropriately on the left side, and rewriting $P(D)$ in the right-side denominator as specified by the theorem on total probability.

Bayes' theorem has a simple interpretation which is described easily in terms of the letters chosen to denote the relevant events. The events $H_1, \ldots, H_N$ are supposed to represent exclusive and exhaustive measurable *hypotheses*; the event $D$ is supposed to represent some observable *data* which may or may not eventuate. The conditional probabilities denoted by $P(D|H_i)$ for $i = 1, 2, \ldots, N$ represent your probability assessment for this data eventuating, conditional upon each of the hypotheses listed. The probability assertions $P(H_1)$, $P(H_2), \ldots, P(H_N)$ represent your *prior probability distribution* for the hypotheses. Bayes' theorem, then, is a computational formula for determining your *posterior probability distribution* for the hypotheses conditional upon observing the data: $P(H_1|D), P(H_2|D), \ldots, P(H_N|D)$. The transformation of your prior distribution to your posterior distribution amounts to a specification of the coherent inference you assert regarding these hypotheses on the basis of evidence provided by the observation of the data.

To be more specific about this "data" event, $D$, notice that if $X$ is a quantity with realm $\mathscr{R}(X)$, then for any $x \in \mathscr{R}(X)$ the theorem allows

$$P[H_i|(X = x)] = \frac{P[(X = x)|H_i]P(H_i)}{\sum_{j=1}^{N} P[(X = x)|H_j]P(H_j)}, \qquad \text{for } i = 1, 2, \ldots, N.$$

We are ready for a simple computational example for you to get your bearings on the use of Bayes' theorem. We shall develop more important realistic examples of its use throughout the remainder of this text, but I'll leave you with a small-scale realistic example as a problem at the end of this section.

***Example 3.*** No text on probability theory would be complete without some example within the context of drawing colored balls from urns. The following one requires Bayes' theorem for its resolution.

Suppose you fill four indistinguishable urns with different compositions of colored balls:

Urn 1 contains 90 white balls and 10 black balls.
Urn 2 contains 80 white balls and 20 black balls.
Urn 3 contains 70 white balls and 30 black balls.
Urn 4 contains 60 white balls and 40 black balls.

One urn is selected from these four. Define the events $H_1$, $H_2$, $H_3$, and $H_4$ by

$$H_i \equiv (\text{Urn } i \text{ is selected}).$$

Thus, the events $\{H_1, H_2, H_3, H_4\}$ constitute a partition. Suppose you assert the probabilities

$$P(H_1) = P(H_2) = P(H_3) = P(H_4) = 1/4.$$

After the contents of the selected urn are scrambled, we begin to draw balls sequentially from the urn (without replacing them) and to record their color. Let $W_j$ denote the event that the ball drawn on the $j$th draw from the selected urn is white. Suppose you also assert the conditional probabilities $P(W_1|H_i) =$ proportion of white balls initially placed in Urn $i$. To begin, the theorem on total probability requires that your

$$P(W_1) = \sum_{i=1}^{4} P(W_1|H_i)P(H_i) = .9/4 + .8/4 + .7/4 + .6/4 = .75.$$

Now suppose you assert further your conditional prevision for the event that a second ball drawn is white given Urn $i$ is selected and given that a ball of known color is drawn from Urn $i$ and not replaced again equals the proportion of white balls among the balls *remaining* in the selected urn. Using similar notation, then, for example, your

$$P(W_2|H_1B_1) = 90/99) \qquad P(W_2|H_2W_1) = 79/99,$$
$$P(W_2|H_3B_1) = 70/99, \qquad P(W_2|H_4W_1) = 59/99.$$

In the spirit of these assertions, you might well imagine asserting similarly appropriate conditional probabilities when an even longer string of observed colored balls are drawn from an urn without replacement, for example,

$$P(W_3|H_2W_1B_2) = 79/98,\ P(B_4|H_4W_1W_2B_3) = 39/97, \text{ and}$$
$$P(B_9|H_3B_1B_2B_3W_4W_5B_6B_7B_8) = 24/92.$$

Having asserted all the probabilities noted thus far, Bayes' theorem specifies the coherency requirement that you assert the conditional previsions

$$P(H_i|B_1) = \frac{P(B_1|H_i)P(H_i)}{\sum_{i=1}^{4} P(B_1|H_i)P(H_i)}$$

and

$$P(H_i|W_1) = \frac{P(W_1|H_i)P(H_i)}{\sum_{i=1}^{4} P(W_1|H_i)P(H_i)}$$

for each value of $i$, 1, 2, 3, and 4. In particular,

$$P(H_1|B_1) = \frac{.1/4}{(.1 + .2 + .3 + .4)/4} = .1,$$

and

$$P(H_1|W_1) = \frac{.9/4}{(.9 + .8 + .7 + .6)/4} = .3.$$

Note that $P(H_1|B_1)$ and $P(H_1|W_1)$ do not sum to 1. It is the pair of conditional probabilities, $P(H_1|B_1)$ and $P(\tilde{H}_1|B_1)$, that must sum to 1.

## PROBLEMS

1. Continuing Example 3, compute the conditional probabilities $P(H_2|B_1)$, $P(H_3|B_1)$, and $P(H_4|B_1)$. Notice how and why $\sum_{i=1}^{4} P(H_i|B_1) = 1$. Now extending this example, can you determine your cohering assertion of $P(W_3|B_1W_2)$? [Use a corollary of the theorem on total probability which allows $P(W_3|B_1W_2) = \sum_{i=1}^{4} P(W_3|B_1W_2H_i)P(H_i)$, since $H_1, \ldots, H_4$ constitute a partition.]

2. Six cardboard cartons containing computer screens have been salvaged from the wreckage of a freight train boxcar. You do not know how many of the six screens have been damaged unless and until you would open each box. Let $D$ denote the unknown number of screens that are damaged. Assessing the site

of the wreckage, you decide to assert $P(D=d)=(7-d)/28$ for $d=0,1,2,\ldots,6$. Since the packages appear indistinguishable (you have not shaken them individually, and none of them appears unusual relative to the others, all of them looking somehat scuffed and scarred), you also decide to assert $P(D_1|(D=d))=d/6$, where $D_1$ is the event that the first carton you open is found to contain a damaged computer screen. Determine your cohering prevision $P(D_1)$ and your cohering conditional probabilities $P((D=d)|D_1)$ and $P(D=d)|\tilde{D}_1)$ for each $d=0,1,2,\ldots,6$.

**3.** A commercial plantation forest of *Pinus radiata* in Fiji is composed of plots of trees of different ages. Excluding the sections planted with seedlings up to 10-year-old trees, 35 percent of the trees are aged 10–20 years, 30 percent are aged 20–30 years, 25 percent are aged 30–40 years, and 10 percent are aged more than 40 years. After a devastating hurricane, you take a surveying flight around the plantations and assess your previsions for the proportion of trees felled within each of these age groups as 20 percent, 40 percent, 80 percent, and 90 percent, respectively. What is the cohering prevision for the proportion of all the trees that have been felled?

***Historical and Philosophical Note on Bayes' Theorem***

A moderating comment on our interpretation of Bayes' theorem is in order. The theorem does not necessarily describe coherent learning in a temporal sense. All the probability assertions that appear in the statement of Bayes' theorem are assertions made at some point in time as part of a coherent assessment of unknown quantities. Notice that in Example 3 we computed both of the conditional probabilities $P(H_i|B_1)$ and $P(H_i|W_1)$ that are required by Bayes' theorem. Both are meaningful specifications even if you do not ever observe the color of the first ball drawn. Even then, you could only possibly observe one of the colors. If you do observe that the selection of the first ball yields a black, say, you may well decide to change your probability assertion regarding $H_1$ to the appropriate conditional probability computed in the example. Doing so would involve (in Isaac Levi's terminology) your "confirmational commitment" to your previously asserted conditional probability values, a process described in Section 3.6 as "learning in the small" from the observation. But, strictly speaking, the assertion of your conditional probability $P(H_1|B_1)=.1$ has nothing to do with "changing your probability" for $H_1$ on the basis of evidence. This conditional probability is not a new evaluation or a revised evaluation of $H_1$ but a *different* evaluation of a completely different quantity. In sum, $P(H_1)$, $P(H_1|B_1)$, and $P(H_1|W_1)$ are three distinct assertions, which may all be meaningfully asserted at the same time. Of course, there are restrictions on their values if they are to cohere with one another.

Historically, Bayes' theorem has not always been so transparent, neither in its mathematical statement nor in its interpretation. The articles of Stigler (1982, 1983), Geisser (1985), Barnard (1958), and Deming (1963) are well worth reading.

Stigler (1983) presents an amusing account of the uncertain origins of the theorem, asserting that a complicated array of historical evidence points to the blind mathematician and professor of optics, Nicholas Saunderson, as the probable originator of the theorem. There are other named and unnamed contenders, including Bayes. Stigler computes the inferential support he asserts for Saunderson, using Bayes' theorem itself!

## 3.5 GENERAL CONDITIONAL PREVISION, $P(X|Y)$

In constructing the technical apparatus of our operational subjective theory through Chapter 2 we made explicit use of the fact that events and general quantities are not categorically different in any way. Events are merely quantities whose realms are the set $\{0, 1\}$. Then suddenly in Chapter 3 we have made an important definition of conditional prevision via $P(X|E)$ where $X$ is any general quantity (which may be an event) but $E$ was restricted to being an event. You may well reasonably wonder what's going on. Why do events appear to be taking on a special role in the important theory of statistical inference?

Actually, the construction of conditional prevision applies directly to conditional previsions of the form $P(X|Y)$ where $Y$ is a general quantity, too. Nonetheless, the predominant usefulness of the concept in practice pertains to situations when $Y$ is an event.

The plan of this brief section is, first, to specify the general theory of conditional prevision in a rather formalistic manner, although we shall see precisely what it means operationally. Then we identify how the general conditional prevision operator works in a coherent way, and we conclude with a real problem that was once considered to provide a serious challenge to the consistency of the operational subjective theory of uncertain knowledge as we have developed it. I leave to the interested reader the problem of resolving the apparent difficulty in terms of the general theory of conditional prevision. The solution to the problem and a discussion of relevant literature can be found in the self-contained article of Lad and Dickey (1990).

A complete understanding of conditional prevision can be achieved by extending the notion of conditional prevision beyond its restriction to assertions of $P(X|E)$, where $E$ is an event, to a characterization of conditional prevision in the more general form $P(X|Y)$. The important aspect of this construction is to identify the operational meaning of such a conditional assertion. Notice that we have already defined operationally the meaning of an assertion in the form $P(X|Y=3)$, say, since $(Y=3)$ is an event. We need now to identify a meaning for asserting a conditional prevision $P(X|Y)$ that does not require a specification of a single value for $Y$, but treats $Y$ as a quantity, just as it does $X$. The reader who is deeply immersed in the conventional characterization of formalist probability will need to be careful in thinking about this project. Usually the symbol $E(X|Y)$ would be understood to represent a functional defined by taking different values $E(X|Y=y)$ for various values of $y \in \mathcal{R}(Y)$. We do not mean something similar by

$P(X|Y)$. Be careful. Let us begin with a formal definition and then turn to a discussion.

***Definition 3.3.*** Let $X$ and $Y$ be any quantities, paired as a vector with realm $\mathscr{R}(X, Y)$. Your *conditional prevision for X given Y*, denoted $P(X|Y)$, is a number you specify with the understanding that you are thereby asserting your willingness to engage any transaction that would yield you a net gain of the amount $s[XY - YP(X|Y)]$, as long as $|s[xy - yP(X|Y)]| \leqslant S$ for every pair of numbers $(y, xy) \in \mathscr{R}(Y, XY)$. (Again we denote by $S$ the scale of your maximum stake.) □

The two quantities we are considering may be logically related in any way whatsoever: $\mathscr{R}(X, Y) \subseteq \mathscr{R}(X) \otimes \mathscr{R}(Y)$. The product $XY$ is also a quantity, and it is necessarily logically related to $X$ and $Y$, since

$$\mathscr{R}(X, Y, XY) = \{(x, y, z) | (x, y) \in \mathscr{R}(X, Y), \text{ and } z = xy\} \subset \mathscr{R}(X, Y) \otimes \mathscr{R}(XY).$$

***Definition 3.4.*** Having asserted your conditional prevision, $P(X|Y)$, the *conditional quantity X given Y*, denoted $(X|Y)$, is defined by

$$(X|Y) \equiv XY + (1 - Y)P(X|Y). \qquad \square$$

This definition of a conditional prevision is based upon a natural extension of the concept of a contingent transaction, which we used in defining $P(X|E)$. When you assert the number $P(X|Y)$, you agree to engaging transactions yielding a net gain of $s[XY - YP(X|Y)] = s[Y(X - P(X|Y))]$, once the measurements $X$ and $Y$ are recorded. That is, the *direction* of your net gain (or loss) depends on the arithmetic difference $(X - P(X|Y))$, but *the scale of your net gain depends on the numerical value of Y*. This is nothing more and nothing less than a glorified contingent transaction of $P(X|Y)$ for $X$, wherein both the scale of the transaction and the direction of the transaction are determined by the value of $Y$. Notice that if $Y = 0$, your net gain equals 0, whereas if $Y = 1$ your net gain equals $s[X - P(X|Y)]$. This is exactly the same as the possible net gains you achieve in the case that the quantity $Y$ is an event. But as $Y$ is here allowed to be a generic quantity, not merely an event, it may have possible values other than 0 and 1. For example, if $Y = 2$, your net gain would be $s(2[X - P(X|Y)])$. Moreover, $Y$ may be a quantity that does not include either 0 or 1 in its realm! Furthermore, $Y$ may even be a negative number, which would reverse the algebraic sign of your net gain from that of $[X - P(X|Y)]$.

Now assessing your $P(X|Y)$ in a practical problem may prove to be a real difficulty, as you might see by trying to assess your $P(S|\text{pH})$ in the salamander problem—not $P(S|(\text{pH} = 3.7))$, mind you, but $P(S|\text{pH})$. But theoretically, the definition of $P(X|Y)$ can be seen to complete the theory of prevision in a uniform way as it applies to any quantities whatsoever. The following requirement for the coherency of any assertion follows simply from the linearity of coherent prevision assertions.

**Theorem 3.6.** Let $X$ and $Y$ be any quantities. Then coherency requires that your prevision assertions must satisfy the equation

$$P(XY) = P(X \mid Y)P(Y).$$

This result follows directly from applying a linear prevision operator to the conditional quantity $(X \mid Y) \equiv XY + (1 - Y)P(X \mid Y)$ and simplifying the result by easy algebraic manipulations.

Understanding $P(X \mid Y)$ and its coherency condition in this way, you can easily prove the following theorem which characterizes the properties of a coherent conditional prevision operator as an abstract operator.

**Theorem 3.7.** In the following statements, $X$, $Y$, $X_1$, $X_2$, $Y_1$, and $Y_2$ represent generic quantities, and $K$ represents a constant. Coherency requires that prevision assertions satisfy the following:

**(i)** Supposing that $P(Y) \neq 0$, $P(X_1 + X_2 \mid Y) = P(X_1 \mid Y) + P(X_2 \mid Y)$.
**(ii)** $P(X \mid K) = P(X)$ for $K \neq 0$.
**(iii)** Supposing that $P(KY) \neq 0$, $P(X \mid KY) = P(X \mid Y)$.
**(iv)** Supposing that $P(Y_1 + Y_2) \neq 0$,

$$P(X \mid Y_1 + Y_2) = \frac{P(X \mid Y_1)P(Y_1) + P(X \mid Y_2)P(Y_2)}{P(Y_1 + Y_2)}.$$

**(v)** Supposing that $P(Y) \neq 0$ and that $\mathscr{R}(Y) = \{y_1, \ldots, y_K\}$,

$$P(X \mid Y) = \frac{\sum_{i=1}^{K} P[X \mid (Y = y_i)] y_i P(Y = y_i)}{\sum_{i=1}^{K} y_i P(Y = y_i)}.$$

As mentioned, you can prove these statements easily yourself. They merit here only the notice that while statement (i) identifies coherent conditional prevision as a linear operator on the argument quantities that appear to the left of the conditioning "bar," statement (iv) specifies that it is not a linear operator on the arguments appearing to the right of the conditioning bar. For $P(X \mid Y_1 + Y_2)$ must equal a *weighted* combination of $P(X \mid Y_1)$ and $P(X \mid Y_2)$, with weights proportional to the prevision assertions for the conditioning summand quantities. Notice that these are not necessarily convex weights, since there is no restriction that $P(Y_1)$ and $P(Y_2)$ be positive.

Statement (v) bears a special comment. It expresses the relationship between a coherent conditional prevision $P(X|Y)$ and the cohering prevision assertions for the individual conditional quantities $(X|Y = y_i)$. The conditional prevision $P(X|Y)$ must equal a *weighted average* of the prevision assertions $P(X|Y = y_i)$, where the weights are proportional to the associated values of the product $y_i P(Y = y_i)$ for the same identifying subscript $i$. It is worthwhile in this context to compare this representation for the conditional $P(X|Y)$ with the parallel representation for the unconditional $P(X)$. The theorem on total prevision, remember, requires that

$$P(X) = \sum_{i=1}^{K} P[X|(Y = y_i)]P(Y = y_i).$$

Thus, $P(X)$ is also a weighted average of the assertions $P[X|(Y = y_i)]$, but the weights are proportional merely to the corresponding probabilities, $P(Y = y_i)$.

This completes our technical treatment of general conditional prevision in this text. To conclude this section, let us pose a provocative problem that can be resolved using the concepts and apparatus of $P(X|Y)$, and move on.

***Example 1.*** Posing a Provocative Problem of Currency Exchange. Consider the following caricature of a currency exchange market in which dollars can be exchanged for pounds sterling. To keep the problem simple, suppose that there are three exclusive and exhaustive possibilities, denoted by the partition of events

$$\{E_1 \equiv (\$1 = £1), E_2 \equiv (\$1.5 = £1), \quad \text{and } E_3 \equiv (\$2 = £1)\};$$

where the parenthetic expression ($\$1 = £1$), for example, represents the number 1 if the public exchange rate offered by a specified bank at a specified time in the future is \$1 for £1, and represents the number 0 if not. Similar numerical meanings are ascribed to the events $E_2$ and $E_3$. This setup could be embellished to greater realism, allowing all possible rates of exchange and even buy-sell spreads. But these would only distract us from the point to be considered here.

As defined, the events $E_1$, $E_2$, and $E_3$ are unitless numbers, each equal to 0 or 1, whose sum equals 1. Suppose you assert your probabilities as $P(E_1) = P(E_2) = P(E_3) = 1/3$ on the basis of your asserted indifference to the prizes $\$E_1$, $\$E_2$, and $\$E_3$; this indifference is operationally identified by your willingness to exchange freely claims to $\$E_1$, $\$E_2$, $\$E_3$, and $\$(1/3)$.

An apparent problem arises if you consider in addition your willingness to exchange freely claims to fixed monetary prizes contingent on the values of $E_i$, but denominated in units of pounds sterling rather than dollars. Could you also assert your probabilities for each of the three events as equal to 1/3 by asserting your indifference among the contingent monetary prizes if they were denominated in pounds, via the indifference assertion $£E_1 \sim £E_2 \sim £E_3$? Evidently, if you have settled on your indifference among the prizes denominated in dollars, you would not want to assert your indifference among the "same prizes" denominated in pounds. For while you would clearly regard as equivalent the values £$E_1 \sim \$E_1$, you would only assess £$E_2 \sim \$(1.5)E_2$. For $E_2$ eventuates to equal 1 only if 1 pound sterling will exchange at the bank for \$1.5; and $\$(1.5)E_2$ is clearly preferable to $\$E_2$, which you avowedly regard as equivalent to $\$E_1$. Similarly, you assess £$E_3 \sim \$2E_3$, for $E_3$ eventuates to equal 1 only if 1 pound would yield \$2 at the bank, and this is more valuable still. Thus, if you regard the dollar-denominated prizes $\$E_1$, $\$E_2$, and $\$E_3$ equivalently, you do not regard the pound-denominated prizes £$E_1$, £$E_2$, and £$E_3$ equivalently. Your valuation of a pound-denominated prize, determined by the values of $E_1$, $E_2$, and $E_3$, apparently depends on the outcome values of these events. The prize of £1 associated with each contingency is equivalent to a different dollar-denominated prize, depending on the numerical values of the three continging events!

This type of situation generates a problem in utility theory that has been coined the problem of "state-dependent preferences." You are indifferent between \$1 and £1 if $E_1 = 1$, but you prefer £1 to \$1 if $(E_2 + E_3) = 1$. The question it poses for the operational subjective theory of probability is, "why should you conclude in such a state of uncertainty that your probabilities for each of the three events equal 1/3 on account of your equivaluation of the three events denominated in dollars, when you would clearly not equivalue these very same events when denominated in pounds sterling?" The operational meaning of de Finetti's definition of "coherent prevision" would be completely undermined if the numerical value of a coherent prevision assertion depends on the arbitrary units in which continging net gains would be denominated.

I leave this problem for you to ponder, hinting only that it is resolved completely by recognizing that net gains from the various transactions being considered can be specified in terms of transactions with contingent scales, the type of transaction that defines a general conditional prevision, $P(X|Y)$. A complete solution and discussion are given in Lad and Dickey (1990).

□

## 3.6 LEARNING IN THE LARGE AND LEARNING IN THE SMALL – STATISTICAL INFERENCE AND CONFIRMATIONAL COMMITMENT

The popular image of statistical inference that has emerged from the contemporary developments of "Bayesian statistical methodology" suggests that the difference between $P(X)$ and $P(X|E)$ is that the first expression represents an assertion about $X$ made "prior" to the observation of $E$, and that the second expression represents the assertion that must be made about $X$ "posterior" to the observation of $E$. The numerical difference $P(X|E) - P(X)$ is supposed to be a measure of what is learned about $X$ from the observation of $E$. On the basis of this imagery, the procedures of Bayesian statistical inference are often presumed to be completely routine once the relevant judgments are made. According to this interpretation, the learning process represents a substantive method of inferential logic. While there is much to be said for this imagery, it is only a crude manner of speaking. If taken literally, it is not only logically erroneous, but it also misses some very important aspects of your deciding questions of whether to *automate* specific decisions on the basis of your current state of knowledge, responding to recorded measurements that you will receive in the future. The goal of this section is to sort out these issues. Let us begin with two examples.

***Example 1.*** Suppose you are interested in the values of quantities that you will observe sequentially in time. It is January, and the quantities $X_1, X_2, \ldots, X_t, \ldots, X_{30}$ are the daily measurements of rainfall (in millimeters) to be recorded at a specific meteorological station in Kansas during June in the coming year. Based on your current state of knowledge you might assess your prevision for various quantities, such as $P(\mathbf{X}_{30})$, or $P(\sum_{i=1}^{30} X_i)$, $P(\sum_{i=1}^{10} X_i)$, $P(\sum_{i=11}^{20} X_i)$, and $P(\sum_{i=21}^{30} X_i)$, and/or your conditional previsions for still other quantities, such as $P[X_3|(X_1=0)]$, $P[X_3|(X_1>20)]$, or $P[\sum_{i=1}^{30} X_i|(\sum_{i=1}^{10} X_i<45)]$. Notice that you are making each of these assertions today, say in January. If our notation were completely explicit, the $P$ symbol denoting your prevision would be subscripted by symbols denoting "you, today." Your state of knowledge about the several quantities is presumably based upon your knowledge of past weather patterns in Kansas, in its neighboring states, and even in continental and global regions. Alternatively your knowledge may be based upon your evaluation of some professional weather forecaster's summary analysis of this information.

The discussion of the present section concerns the relationship between your current assessment of a conditional quantity, such as $P[X_3|(X_1=0)]$, and the prevision you will assert on a future day for $X_3$, such as the morning of June 2, after you have learned the precise value of the event $(X_1=0)$. We shall find that the mere requirement of coherency allows great freedom in this assessment. Nonetheless, there are various things you may (or may not) want to *assert* about this relationship based on your current state of information in January. □

Before defining notation to make the problem precise, let us lay some groundwork with a second example, which will suggest a different possible resolution to the same type of question.

***Example 2.*** It is January. You are a university teacher in your fifth year of teaching, examining your grading form for a class of 157 students to whom you have presented lectures during the recent term. On your desk you see the list denoting the names and marks of students. It includes

| Student Number | Student Name | Letter Grade | Numerical Value |
|---|---|---|---|
| . | . | C+ | 2.3 |
| . | . | C | 2.0 |
| 92 | . | B− | 2.7 |
| 93 | Nethercutt, Nancy | A | 4.0 |
| 94 | Nguyen, Li | A | 4.0 |
| 95 | Nimetz, Barbara | B+ | 3.3 |
| 96 | Norris, Kenneth | C | 2.0 |
| 97 | Norris, Thomas | A− | 3.7 |
| 98 | Nottingham, Gail | C+ | 2.3 |
| 99 | . | D | 1.0 |
| . | . | B+ | 3.3 |
| . | . | B− | 2.7 |

The letter grade for each student is translated into a numerical grade on the basis of the schedule

| A | A− | B+ | B | B− | C+ | C | C− | D | E |
|---|---|---|---|---|---|---|---|---|---|
| 4.0 | 3.7 | 3.3 | 3.0 | 2.7 | 2.3 | 2.0 | 1.7 | 1.0 | 0 |

Denoting by $X_i$ the numerical value of the grade evaluated for the work of the student associated with student number $i$, you obviously assess right now your $P(X_{93}) = 4.0$, $P(X_{94}) = 4.0$, $P(X_{95}) = 3.3$, $P(X_{96}) = 2.0$, $P(X_{97}) = 3.7$, and $P(X_{98}) = 2.3$, for you see the numerical values of $X_{93}, \ldots, X_{98}$ right in front of you. Although you have not summed the numerical values of your students' individual marks, suppose you are willing to assert your $P[\sum_{i=1}^{157} X_i] = 380$, and you assert your $P[360.0 \leqslant \sum_{i=1}^{157} X_i \leqslant 390.0] = .85$. As to conditional prevision, you are also willing right now to make the conditional prevision assertions

$$P\left[X_{94} \middle| \sum_{i=1}^{157} X_i = 361.1\right] = 4.0, \qquad P\left[X_{94} \middle| \sum_{i=1}^{157} X_i = 381.1\right] = 4.0,$$

$$P\left[X_{94} \middle| \sum_{i=1}^{157} X_i = 391.1\right] = 4.0,$$

as well as

$$P\left[X_{95} \middle| \sum_{i=1}^{157} X_i = 361.1\right] = 3.3, \qquad P\left[X_{95} \middle| \sum_{i=1}^{157} X_i = 381.1\right] = 3.3,$$

$$P\left[X_{95} \middle| \sum_{i=1}^{157} X_i = 391.1\right] = 3.3.$$

Again, you are quite willing to make all of these conditional prevision assertions because you are looking at the completed marking form for these individuals' marks right in front of you. You know for certain the values of the quantities that you now see.

You are scheduled over the next seven months to teach two more large classes of students during each of the next two quarter terms. Glancing at the grade list in front of you, you are aware that Mr. Li Nguyen is the only one of these six students whom you recognize personally. You have actually had a few interesting conversations with him during the past quarter term during your office consultation hours. With none of the other five students listed have you had any significant individual contact. You are fairly sure, and you hope even, that you will forget the individual grade information for the other five students by the time the end of next June arrives. For you feel that remembering such details can only detract from your attention to your other many responsibilities in the interim. However, Mr. Nguyen has made such an unusual impression on you that you feel sure you will remember then that his course grade is A. It would be difficult for you to forget that!

Now, what is the relation between your current prevision for the numerical grade of Ms. Barabara Nimetz as $P(X_{95}) = 3.3$, your current conditional prevision $P[X_{95}|\sum_{i=1}^{157} X_i = 381.1] = 3.3$, and the prevision you will assert next June for $X_{95}$ when you will be officially informed by the Office of Academic Affairs regarding the average of the numerical grades you have evaluated for these 157 students? Would you be incoherent in asserting right now that your prevision *for the prevision you would assert on 30 June for* $X_{95}$ after learning, say, that $\sum_{i=1}^{157} X_i = 381.1$ equals $(381.1 - 4.0)/156$? □

What is the difference between the situation of this example and the weather-forecasting Example 1?

The coherency of your several prevision assertions is a property of your asserted state of knowledge "today," or at whatever point in time you are making the assertions. In the way we have characterized *conditional* prevision and *conditional* quantities, they are not essentially different than your assertion of prevision and the definitions of any generic quantities. They are numbers representing some specific judgments you make today about the values of operationally defined measurements. Your assertion of conditional prevision has no necessary relationship with judgments you will make tomorrow or next month or with judgments you made last month about these same quantities. The prevision you will assert next month for any quantity is unknown to you now. For you do not know now precisely what state of mind you will be in at that future time or what judgments you will then make. As such, your prevision tomorrow is a quantity about which you may express your prevision today, if you wish. Similarly, you may have forgotten what prevision you asserted one month ago for any quantities. You can also assert your current prevision for your past prevision, if you like. If you recorded your prevision assertion one month ago, you can look up the recording in your record book to determine its value. Thus, your prevision tomorrow and your prevision yesterday are both quantities about which you can express your prevision today.

In fuller notation, then, we can distinguish the prevision you assert today (at time $t$) for a quantity $Y$ as $P_t(Y)$, from the prevision you will assert tomorrow (at time $t+1$) for $Y$ as $P_{t+1}(Y)$, from the prevision you asserted yesterday for $Y$ as $P_{t-1}(Y)$. Moreover, since $P_{t+1}(Y)$ and $P_{t-1}(Y)$ are both operationally defined quantities, it could be meaningful to assert your prevision today for the prevision you will assert tomorrow for $Y$, denoted by $P_t[P_{t+1}(Y)]$, and your prevision today for the prevision you asserted yesterday for $Y$, $P_t[P_{t-1}(Y)]$. Using this more complete notation, we can now formulate the critical question of this section.

Suppose $X_1, \ldots, X_{t-1}, X_t, X_{t+1}, \ldots, X_T$ denote quantities (or vectors of quantities) for which the observation $X_t$ is made during time period $t$. Let $P_j(X_1, \ldots, X_t, \ldots, X_T)$ denote your prevision for these quantities asserted in time period $j$. Thus, each component of the prevision vector $P_j(X_1, \ldots, X_t, \ldots, X_T)$ can be considered to be an element of a generic vector $X_j$. Moreover, even more complicated operationally defined quantities, such as $P_j[P_{j+3}(Y_{j-3})]$, can be considered to be elements of the vector $X_j$. [The quantity $P_j[P_{j+3}(X_{j-3})]$ would be your prevision in period $j$ for your prevision to be assessed in period $j+3$ for the value of the quantity $X_{j-3}$.] In principle, the adjunction of more and more prevision operators could be made, so an operationally defined quantity such as $P_j\{P_{j+1}[P_{j-1}(X_{j+2})]\}$ could be meaningful. However, our mental capacities are such that it becomes more and more difficult to bear down on exactly what the quantity is we are talking about as more prevision operations are adjoined.

The principles behind the point to be made in this section can be exposed by considering five sequentially observed quantities that do not include any previsions among them, denoted $X_{t-2}, X_{t-1}, X_t, X_{t+1}, X_{t+2}$. We identify two distinct states which you might recognize concerning relations among your prevision assessments $P_t(X_{t+2})$, $P_t(X_{t+2} | X_{t+1} = x_{t+1})$ and $P_t[P_{t+1}(X_{t+2})]$.

*State 1*

You think that tomorrow (period $t+1$) you will be more knowledgeable about $X_{t+2}$ than you are today (period $t$). That is, you previse that between now and tomorrow you will learn something (more and better) about $X_{t+2}$ than you know now. But you do not know now what it is that you will learn, otherwise you would already know it. This state of mind would be characterized by your current assertion that for every operationally distinguishable value of $\pi \in \min \mathscr{R}(X_{t+2})$. $\max(\mathscr{R}(X_{t+2})]$, your current conditional prevision $P_t[X_{t+2} | P_{t+1}(X_{t+2}) = \pi] = \pi$. In this state of mind concerning what you will learn about $X_{t+2}$, coherency of your prevision requires you to assert your $P_t[P_{t+1}(X_{t+2})] = P_t(X_{t+2})$, according to the theorem on total probability. For $X_{t+2}$ can be written as

$$X_{t+2} = \textstyle\sum_\pi X_{t+2}(P_{t+1}(X_{t+2}) = \pi),$$

where this summation runs over all operationally distinguishable values of $\pi$ lying within the interval $[\min \mathscr{R}(X_{t+2}), \max \mathscr{R}(X_{t+2})]$. For the events of the form $(P_{t+1}(X_{t+2}) = \pi)$ so defined would constitute a partition. Thus,

$$\begin{aligned} P_t(X_{t+2}) &= \textstyle\sum_\pi P_t[X_{t+2}(P_{t+1}(X_{t+2}) = \pi)] \\ &= \textstyle\sum_\pi P_t[X_{t+2} | (P_{t+1}(X_{t+2}) = \pi)] P_t(P_{t+1}(X_{t+2}) = \pi) \\ &= \textstyle\sum_\pi \pi P_t[P_{t+1}(X_{t+2}) = \pi] \\ &= P_t[P_{t+1}(X_{t+2})], \end{aligned}$$

according to standard rules of coherent prevision.

In summary, what does the state of mind called State 1 represent? You think now that at some future point in time you will have better knowledge about a quantity than you do now. But you do not know what that better knowledge will be. (Otherwise you would already be in that state of knowledge.) Thus, your current prevision for the specified quantity is the same as your current prevision for the prevision you will assert when you are in that improved state of knowledge!

Two comments are in order. First, there is nothing really different about the future and the past relative to states of mind in this regard. You might know now that you are in a less informed state of mind regarding $X$ than you used to be. You recognize now that you have forgotten what you once knew! In this state it would be reasonable for you to assert that your $P_t(X_{t+2}) = P_t[P_{t-1}(X_{t+2})]$. Moreover, just as your consideration of your knowledge in the past is no different from your consideration of your knowledge in the future with respect to this logic, your consideration of someone else's state of knowledge can be characterized in this same way. "I think Gail really knows a lot about $X$, much more than I do. But she hasn't told me what she knows. Whatever be the value of her prevision, that would be my prevision if she would tell me. But I don't know the value of her prevision. Thus, my prevision for $X$ is the same as my prevision for the value of Gail's prevision of $X$." The structure of this argument is the same as the structure of the argument based on State 1 about your current knowledge of your own different state of knowledge in either the future or the past.

Second, this extensive discussion of State 1 and its implications must surely excite the realization that it is not the only possible state of mind you may entertain about your own knowledge in the future, your own knowledge in the past, or even about someone else's state of knowledge. The alternative positions you might entertain in these three contexts are that you expect to forget what you now know, that you think you now know more than you used to know, and that you think you know more and better about $X$ than someone else does. Let us characterize these attitudes as representative of State 2.

*State 2*

You think that tomorrow (period $t + 1$) you will be less knowledgeable about $X_{t+2}$ than you are today (period $t$). That is, you previse that between now and tomorrow you will forget something related to $X_{t+2}$ that you know now. Two possibilities now arise. You might not have a good idea exactly what you will think at that future point in time, only that you now imagine it will be inferior in content value to what you know now. On the other hand, you might be able to say now how you expect your prevision in the future to be different from your prevision right now.

Example 2 described a context appropriate to this latter alternative. You happen to previse right now $P_t(X_{95}) = 3.3$, but only because you are looking at the grade list. You imagine, and you may even hope, that you will forget the specific marks of these faceless students, in order to free your mind for other responsibilities. So you might well now previse for your prevision 10 years from now for this $X_{95}$ as $P_t[P_{t+10}(X_{95})] = 2.3$, say, which is not equal to your $P_t(X_{95})$. Now since you are looking at the section of the completed marking form, notice that you also currently assert your $P_t(X_{97}) = 2.3$. Even though you do expect to forget this knowledge over the course of 10 years, you could well also currently previse your $P_t[P_{t+10}(X_{97})] = 2.3$, which happens to equal your $P_t(X_{97})$. Thus, it is not the condition $P_t(X) \neq P_t[P_{t+j}(X)]$ that characterizes State 2, but something different.

This state of mind we have labeled State 2 is characterized by your current assertion that for every operationally distinguishable value of $\pi$ within $[\min \mathscr{R}(X_{t+2}), \max \mathscr{R}(X_{t+2})]$, your current conditional prevision $P_t[X_{t+2} | P_{t+1}(X_{t+2}) = \pi] = P_t(X_{t+2})$. In this state of mind concerning what you will learn (that is, forget) about $X_{t+2}$, coherency of your prevision allows you to assert your $P_t[P_{t+1}(X_{t+2} | X_{t+1})]$ as any number at all within the boundaries of the realm $\mathscr{R}(X_{t+2})$. (You may at least presume that your prevision at time $t+1$ will be coherent.)

### *Historical Note*

The first really substantive and interesting development in statistical literature of the issues discussed above was presented by Michael Goldstein in a series of papers entitled "Revising Previsions: A Geometric Interpretation" (1981), "The Prevision of a Prevision" (1983), and "Temporal Coherence" (1985). In philosophical literature the issues had been discussed extensively Levi (1978, 1981) and Kyburg (1980). Goldstein's work identifies an intertemporal preference condition which, when honored, ensures that your attitudes about changes in your knowledge over time place you in State 1. The philosophical discussions between Kyburg and Levi centered mainly on the fact that your future prevision $P_{t+1}(X_{t+2})$, asserted after you have learned that $X_{t+1} = x$, say, need not actually equal your current conditional prevision $P_t[X_{t+2} | (X_{t+1} = x)]$. To assert your $P_{t+1}(X_{t+2})$ as equal to your $P_t[X_{t+2} | (X_{t+1} = x)]$ after observing $X_{t+1} = x$ is to *reaffirm* your assertion made at time $t$ that the observation $X_{t+1} = x$ would provide evidence for you to change your knowledge from $P_t(X_{t+2})$ to $P_{t+1}(X_{t+2}) = P_t[X_{t+2} | (X_{t+1} = x)]$. Levi labeled such a reaffirmed assertion as your *confirmational commitment* to the assertion. He recognized this commitment as an explicit requirement for the conduct of scientific inference in the way it is commonly presumed to proceed by Bayesian statisticians.

There are important practical benefits to recognizing the distinction between a *conditional* prevision asserted at some point in time and the *change* in your prevision over time as you make more observations and reflect on them. The benefits are that we can formalize criteria for deciding at time $t$ whether to automate a forecasting system by providing sequential forecasts of quantities that merely reports at each time $t + j$ the conditional prevision

$$P_t[X_{t+j+1} | (X_{t+j} = x_{t+j})(X_{t+j-1} = x_{t+j-1})(X_{t+j-2} = x_{t+j-2}) \cdots (X_{t+1} = x_{t+1})]$$

appropriate to the string of quantity values that are actually observed, or whether to make the sequential effort of evaluating the ever-current human forecast $P_{t+j}(X_{t+j+1})$ based on live judgment along with the string of observation values. It should be apparent that you would prefer to automate your current opinions into a forecasting system if you were in State 2 regarding your current opinions about your future opinions. That is, if you think you will forget soon what you now know, you would prefer to be making future decisions on the basis of your current conditional prevision assertions "remembered" through some automaton rather than be making "live" decisions on the basis of your expected deteriorating states of knowledge. In State 1 it is conceivable that you would prefer to be making live decisions or that you would just as soon automate your current assessments, depending on whether you expect to be learning "something" or not.

Intermediate between our State 1 and State 2 are other states which can be characterized formally. But a presentation of the detail is beyond the scope of this book.

Over the course of time, your knowledge develops in ways that are far richer than the mere accretion of known quantities. Just as our measurements of what's happening can never exhaustively describe what's happening, so also the mere listing of recorded measurements can never exhaustively describe anyone's knowledge of it. Operationally defined measurements are summary descriptions of both the state of nature and the state of our knowledge of it. Thus, it may well be that progressing from a state of knowledge about $X_{t+2}$ summarized by $P_t(X_{t+2} | X_{t+1} = x)$ to a state summarized by $P_{t+1}(X_{t+2})$ would involve a change for which $P_t(X_{t+2} | X_{t+1} = x)$ does not equal $P_{t+1}(X_{t+2})$, even if at time $t + 1$ you actually observe that $X_{t+1}$ equals this specified value of $x$!

Such a transition may be termed "learning in the large" as opposed to "learning in the small." Most of formal statistical theory pertains to the process of learning in the small. Informal procedures of learning in the large are most well known as "exploratory data analysis" in the coinage of Tukey (1977) or as "Bayesian data analysis" as described by Hill (1990).

The qualifying precaution of this section is that "learning in the small" about $X$ from the observation of $E$—that is, learning solely through the observation of $E = 1$—is completely specified by your conditional prevision $P(X | E)$, even before you have observed $E$. For $P(X \mid \tilde{E})$ is also a well-defined assertion. But only one of the two events $E$ and $\tilde{E}$ can be observed to equal 1.

## 3.7 ASSESSING QUANTITIES INDEPENDENTLY

In the remainder of this chapter we return to studying directly the properties of coherent inferential assertions that are relevant to research in empirical sciences. We begin with a characterization of the judgment to regard events independently, not because it is important for scientific applications—it is not—but because its common misunderstanding is the source of serious confusion in statistical scientific literature.

We have learned that the difference between your assertion of $P(E_1)$ and your assertion of $P(E_1 \mid E_2)$ amounts to the inferential evidence you consider the event $E_2$ to convey regarding $E_1$. Numerically, your coherent assertion of $P(E_1 \mid E_2)$ may be greater than or less than your $P(E_1)$, depending on how the conditioning upon $E_2$ changes your valuation of $E_1$. If you consider the value of $E_2$ to be completely irrelevant to your assessment of $E_1$, you would assert $P(E_1 \mid E_2) = P(E_1)$. In such a case we say you regard these two events independently. Otherwise you regard them dependently, since you consider the condition of $E_2$ to convey specific information regarding $E_1$. This concept of independence is often referred to as "stochastic" independence, to distinguish it from logical, functional, and linear independence.

We understand coherency to require that for any events $E_1$ and $E_2$, prevision assertions must respect the equalities

$$P(E_1E_2) = P(E_1 \mid E_2)P(E_2) = P(E_2 \mid E_1)P(E_1).$$

Dividing each equivalent by the product $P(E_1)P(E_2)$ yields the result that, as long as neither $P(E_1)$ nor $P(E_2)$ equal 0, the following ratios must be equal:

$$\frac{P(E_1E_2)}{P(E_1)P(E_2)} = \frac{P(E_1 \mid E_2)}{P(E_1)} = \frac{P(E_2 \mid E_1)}{P(E_2)}.$$

For each value of $i$ and $j$, these ratios identify the relative sizes of your conditional probability, $P(E_i \mid E_j)$, and your unconditional probability, $P(E_i)$. It is interesting that this ratio is necessarily symmetric with respect to $i$ and $j$. If you assert $P(E_1 \mid E_2) = P(E_1)$, and thus $P(E_2 \mid E_1) = P(E_2)$, you are asserting that neither event provides you any *evidential support* for the other. In such a case, the common ratio equals 1.

It is important to recognize that in specifying your probabilities, *you assert* your judgment to regard the events independently or not. The quantities are not independent or dependent in themselves. Someone else in another state of knowledge may regard them differently than you do. It would be a mistaken endeavor to try to "test" by observation whether two quantities are independent.

Either you regard them as informative about one another, or you don't. Do you?

***Example 1.*** Recall the expedition to search for red-backed salamanders, which we discussed in Example 4 of Section 3.2. Let $E_1$ be the event that you find at least one salamander on the meter-square plot under examination. Let $E_2$ be the event indicating that the prime minister of Italy on 14 May 1957 was a member of the Christian Democrat Party. Operationally, your assertion of the stochastic independence of $E_1$ and $E_2$ amounts to your asserting the equality of your $P(E_1)$ and your $P(E_1|E_2)$. You would only be willing to make such an assertion if you considered information about the value of $E_2$ to be completely uninformative to you regarding the value of $E_1$. I, Frank Lad, regard these events $E_1$ and $E_2$ independently. Do you? My $P(E_1) = .8$ and my $P(E_1|E_2) = .8$. □

Statistical procedures of scientific inference typically concern our studies of quantities we do consider to be informative about one another. We regard these dependently. The procedures codify the coherent process whereby we change our opinions about some quantities on the basis of information we gain concerning other quantities. We learn from our experience. Indeed, the logic of such learning procedures is the subject matter of this text.

Delaying further discussion of inference for the moment, in the remainder of this section we state some formal definitions and results concerning independent considerations of quantities.

***Definition 3.5.*** *You* are said to *regard two events independently* if you assert $P(E_1|E_2) = P(E_1)$, and $P(E_2|E_1) = P(E_2)$, which is then required by coherency. You regard three events independently if you regard all three pairs of them independently and you also assert $P(E_3|E_1E_2) = P(E_3)$. You *regard N events independently* if you regard all $N$ collections of $N-1$ of them independently and assert $P(E_N|E_1E_2\cdots E_{N-1}) = P(E_N)$. □

In a word, you are said to regard $N$ events independently if you regard none of the events $E_1$ through $E_N$ to be in any way informative about the others. The sequential definition of your regarding several events independently is necessary, since your assertion of *pairwise independence* of three events would not necessitate your assertion of complete independence among all three. The following problem, ascribed to S. N. Bernstein, provides a simple example.

***Example 2.*** A symmetric tetrahedonal die has one face colored only yellow, one face only red, one face only blue, and one face shows all three colors. Let symbols $Y$, $R$, $B$, and $A$ represent the events that each of these faces lies face down as a result of the throw, respectively. These four events constitute a partition. On the basis of your assessment of the symmetry of the die, suppose that you assert the probabilities $P(Y) = P(R) = P(B) = P(A) = .25$. In this context, do you regard independently the events that the downward face contains the colors yellow, red, and blue?

Denote by $E_1$, $E_2$, and $E_3$ the events that the downward face contains the colors yellow, red, and blue, respectively. Thus,

$$E_1 = Y + A, \qquad E_2 = R + A, \qquad \text{and } E_3 = B + A.$$

Coherence obviously requires that $P(E_1) = P(E_2) = P(E_3) = .5$.

Now for each value of $i$ and $j$, the product event $E_iE_j = A$. For example, $E_1E_2 = (Y + A)(R + A) = YR + YA + AR + A^2 = A$, since $Y$, $R$, and $A$ are incompatible, and $A^2 = A$. Thus, for each value of $i$ and $j$, your $P(E_i E_j) = P(A) = .25$, and, thus, $P(E_i | E_j) = P(E_iE_j)/P(E_j) = .25/.5 = .5 = P(E_i)$. So you do, evidently, regard each *pair* of the events $E_1$, $E_2$, and $E_3$ independently.

But consider $P(E_3 | E_1E_2)$, which must equal $P(E_1E_2E_3)/P(E_1E_2)$. In this context again, the product event $E_1E_2E_3$ also must equal $A$ as well. Thus, $P(E_3 | E_1E_2) = P(A)/P(A) = 1$, which does not equal $P(E_3)$. Thus, in light of your specified assertions, you do not regard *all three* of the events independently, even though you do regard each pair of them independently. □

The following trivial theorem is often taken to define independence, with our Definition 3.5 then understood as a theorem. Indeed, Theorem 3.8 is often quite useful for proving that events are regarded independently, as required by their cohering with other specific assertions.

***Theorem 3.8.*** If you regard $E_1$ and $E_2$ independently, then coherency requires that you assert $P(E_1E_2) = P(E_1)P(E_2)$. Conversely, if your $P(E_1E_2) = P(E_1)P(E_2)$ and both $P(E_1) > 0$ and $P(E_2) > 0$, then coherency requires you to regard $E_1$ and $E_2$ independently.

## PROBLEMS

1. In the context of Example 2, consider the events $F_1 = Y + R$, $F_2 = R + B$, and $F_3 = B + A$. Do you regard $F_1$ independently of $F_2$? $F_2$ independently of $F_3$? $F_1$ independently of $F_3$?

2. Show that if you regard $E_1$ and $E_2$ independently and your probability for each is strictly within $(0, 1)$, then coherency requires you also to regard the pairs of events $\tilde{E}_1$ and $\tilde{E}_2$, $\tilde{E}_1$ and $E_2$, and $E_2$ and $\tilde{E}_1$ independently.

3. Distinguish for yourself the concepts of logical dependence, linear dependence, functional dependence, and stochastic dependence among events. Which of these properties are characteristics of events themselves, and which are characteristics of someone's opinions about events?

**4.** Specify any third event that you regard independently of the events $E_1$ and $E_2$ described in Example 1.

For completeness' sake at this point, let us note how the concept of regarding events independently in the stochastic sense is extended to pertain to judgments regarding general quantities. The conditions of this extension are more severe than merely requiring that your $P(X_1X_2) = P(X_1)P(X_2)$. For now, let us merely state a formal definition. We shall study the matter in detail later.

***Definition 3.6.*** You *regard the quantities $X_1$ and $X_2$ independently* if for every pair $(x_1, x_2) \in \mathscr{R}(X_1, X_2)$ you assert $P[(X_2 = x_2)|(X_1 = x_1)] = P(X_2 = x_2)$ and $P[(X_1 = x_1)|(X_2 = x_2)] = P(X_1 = x_1)$. Equivalently, you regard every pair of events of the form $(X_1 = x_1)$ and $(X_2 = x_2)$ independently. □

***Definition 3.7.*** You *regard the N quantities $X_1, \ldots, X_N$ independently* if you regard all subcollections of them independently, and for each vector $\mathbf{x}_N \in \mathscr{R}(\mathbf{X}_N)$ you assert

$$P\left[(X_N = x_N)\Big|\prod_{i=1}^{N-1}(X_i = x_i)\right] = P(X_N = x_N).$$ □

This definition of asserting the stochastic independence of $N$ general quantities will lie fallow for the rest of this chapter. For the remaining developments concern a sequence of events rather than a sequence of general quantities. The expansion of these developments to the context of general quantities will occur in Chapters 5 and 7. At this point, let me merely pose a practical question to you in the context of two general quantities, which will serve to engage your thinking on a topic we study forthwith.

Let $X_1$ and $X_2$ equal the numbers of bushels of wheat yielded during the past year on two specific, but otherwise nondescript, acres of Kansas farmland. What are the numerical values of your $P(X_1)$ and $P(X_2)$? Do you regard these two quantities independently? (You may work in units of kilograms if you prefer.)

We understand that your judgment to regard several events independently precludes your willingness to use the observation values of some of them to make inferences about the yet unknown values of the others. Nonetheless, contemporary statistical literature, which is largely based upon ideas quite different from the operational subjective statistical viewpoint, abounds with references to statistical procedures that use observations from part of a sequence of "stochastically independent quantities" to estimate probabilities for the remaining quantities in the sequence. Surely there must be a confusion somewhere. The confusion is resolved by recognizing the structure of a different judgment which is appropriate to these problems, the judgment to regard a sequence of quantities exchangeably. In various extended forms, this topic will engage us through the remainder of this chapter and through most of this book.

## 3.8 ASSESSING QUANTITIES EXCHANGEABLY

Many collections of quantities are erroneously spoken of as "being stochastically independent." Typical examples are sequences of observations from controlled experiments, such as observations of plant yields from "identically treated" fields or quality control observations of output from a production system. Confused and confusing statements arise because stochastic independence is associated somehow with notions of causation. Often it is said that such observations are "known to be independent" because, for example, "the amount of bacterial growth on this Petri dish in this chamber couldn't have had any effect on the amount of growth in that dish (prepared in the same way) in that chamber." We have characterized the notion of stochastic independence quite differently. We would typically regard the amounts of bacterial growth on the two Petri dishes *dependently* because we are uncertain about both, and we would regard the measurement from one of them to convey some information to us regarding the measurement from the other. Indeed, this is precisely why we conduct sequences of experiments of the same type.

Our construction of statistical theory amounts to a formulation of mathematical representations for the various ways in which we deem the observation of some quantities to be informative to us about the values of others. A basic building block in this construction is the characterization of opinions that regard the several quantities exchangeably. You are said to regard a sequence of quantities exchangeably if there is a specific type of symmetry to your opinions about them. Formally, we say that you regard several quantities *exchangeably* if your probabilities for observing any two sequences of observation values are equal whenever the components of one sequence of observations is a mere *permutation* of the components of the other. You are said to regard a sequence of $N$ quantities exchangeably if your opinion about any one of them is the same as your opinion about any other one ; if your opinion about any two of them is the same as your opinion about any other two of them; and if your opinion about any $N-1$ them is the same as your opinion about any other $N-1$ of them. Let us begin our study with some examples.

***Example 1.*** A convoy of three trucks carrying frozen orange juice from Florida to Chicago approaches a weigh station on U.S. Interstate 75. Let $W_i$ denote the event that equals 1 if the $i$th truck weighs in at no more than its legal maximum weight per axle. Looking at the three trucks, you do not notice any remarkable feature that distinguishes one from the other. For example, none of them has especially distorted tires; the three have been traveling together in a convoy at roughly the same rate of speed; and none of them seems to shimmy unusually with its load. You doubt that any of the three trucks is overweight. But supposing one were, you have no reason to suspect it to be any one of them more than any other one. Similarly, supposing two are overweight, you have no reason to suspect any two of them as culprits over any other two. Suppose you assert the probabilities

$$P(W_1 W_2 W_3) = .984,$$
$$P(\tilde{W}_1 W_2 W_3) = P(W_1 \tilde{W}_2 W_3) = P(W_1 W_2 \tilde{W}_3) = .002,$$
$$P(W_1 \tilde{W}_2 \tilde{W}_3) = P(\tilde{W}_1 W_2 \tilde{W}_3) = P(\tilde{W}_1 \tilde{W}_2 W_3) = .001,$$
$$P(\tilde{W}_1 \tilde{W}_2 \tilde{W}_3) = .007.$$

You are said to regard the three events exchangeably, because you assess the same probability for any sequence involving a specific number of successes and failures (product event) irrespective of the order in which the successes and failures occur. For example, your probability that only the first is overweight, $P(\tilde{W}_1 W_2 W_3)$, is the same as your probability that only the second is overweight, $P(W_1 \tilde{W}_2 W_3)$, which is also the same as your probability that only the third is overweight, $P(W_1 W_2 \tilde{W}_3)$. You assess each of these probabilities as .002. □

***Example 2.*** A batch run of glass ceramic baking dishes has yielded 1000 dishes between the second and fourth hours of the morning workshift. The production engineering team is pleased with the conditions of the run. Ten dishes are selected from the batch for quality control testing. Each dish will be heated to 280°C and plunged into a cold water/ice slush (0°C). Let $S_i$ denote the event that equals 1 if the $i$th dish survives the shocking plunge without cracking. Generically, glass has an amorphous molecular structure which is extremely regular if it has been cooled into solid phase slowly enough to allow proper annealing. Cracking of dishes would identify voids, inclusions, or surface cracks in the structure, processing defects in manufacture which are matters of concern.

As a production engineer, you feel rather confident that each of the events in the vector $\mathbf{S}_{10}$ equals 1. Knowing, nonetheless, that it is possible that 1, 2, or even all 10 of them might equal 0, you consider the situation further. Suppose you realize that if exactly one of the tested dishes fails the test (one of the $S_i = 0$), you think it just as likely that it would be any one of the 10. Moreover, if 3, say, of the tested dishes fail, you regard it just as likely that the failures would be any 3 of the 10. This would be to assert that your probability for any particular product event involving 7 successes and 3 failures, such as $P(S_1 S_2 \tilde{S}_3 \tilde{S}_4 S_5 S_6 S_7 \tilde{S}_8 S_9 S_{10})$, is the same as your probability for any distinct product event that identifies three other specific dishes out of the 10 as those that fail, say $P(S_1 \tilde{S}_2 S_3 \tilde{S}_4 S_5 S_6 \tilde{S}_7 S_8 S_9 S_{10})$. When reading this product event notation, remember that $S_4 = 0$ implies that $\tilde{S}_4 = 1$.

In general, you are said to regard the quality of the 10 selected dishes exchangeably if for each value of $A \in \{1, 2, \ldots, 9\}$ your prevision is the same for any product event of the form $S_1 S_2 \cdots S_A \tilde{S}_{A+1} \tilde{S}_{A+2} \cdots \tilde{S}_{10}$, identifying $A$ specific successful events (no cracking) and $10 - A$ specific failure events, irrespective of the order in which the dishes exhibit cracking and noncracking. In this example, your $P(S_1 S_2 \tilde{S}_3 \tilde{S}_4 S_5 S_6 S_7 \tilde{S}_8 S_9 S_{10})$ is the same as your $P(S_1 S_2 S_3 S_4 S_5 S_6 S_7 \tilde{S}_8 \tilde{S}_9 \tilde{S}_{10})$, since the first of these product events derives from the second merely by rearranging the subscripts 3 and 4 with 9 and 10. These two product events each identifies a specific ordered way that you can observe 7 successes (and 3 failures) out of 10. You may already know that there are, thus, ${}^{10}C_7$ permutation events that are accorded this same probability. We review briefly in Section 3.8.1 the main principles of counting that are important to the study of exchangeable judgments. □

***Example 3.*** Peter's family through his father's line is largely male. He has four brothers and one sister. He has six uncles on his father's side and one aunt. His father had eight paternal uncles and two aunts. Amy's family through her mother's line has been more balanced in gender. She has two brothers and one sister, three maternal aunts, and three maternal uncles, and her mother had four maternal uncles and three aunts. Peter is assessing his opinions about the possible gender composition of a two-child family that he and Amy are planning together. Let $B_i$ denote the event indicating that their $i$th child is a boy, and $G_i$ denote the event indicating that their $i$th child is a girl: $G_i = \tilde{B}_i = 1 - B_i$.

After some consideration, Peter asserts the following probabilities regarding the gender composition of their desired family:

$$P(B_1B_2) = .7, \qquad P(B_1G_2) = P(G_1B_2) = .1, \qquad \text{and} \qquad P(G_1G_2) = .1.$$

We say that Peter regards the events $B_1$ and $B_2$ *exchangeably*, since he regards the gender *order* of the children as irrelevant to his assessed prevision for any particular family composition. That is, his assertions $P(B_1G_2)$ and $P(G_1B_2)$ are equal. Coherency then implies that $P(B_1) = P(B_2) = .8$, since coherency requires that

$$P(B_1) = P(B_1B_2) + P(B_1G_2) \qquad \text{and} \qquad P(B_2) = P(B_1B_2) + P(G_1B_2).$$

Furthermore, coherency also requires that Peter assert

$$P(B_2|B_1) = P(B_1B_2)/P(B_1) = .7/.8 = .8475 \neq P(B_2).$$

So Peter regards the genders of the two children exchangeably, but he does not regard these events independently, for he asserts $P(B_2|B_1) \neq P(B_2)$.

Peter's friend, Ron, is a student of genetics. Upon hearing these assertions, he rails against Peter's "foolishness." "The gender of one conception has no effect on the gender of a second conception," he argues. "I regard the gender outcome of your progeny as stochastically independent events. To be precise, my $P_{\text{Ron}}(B_1) = P_{\text{Ron}}(B_2) = P_{\text{Ron}}(B_2|B_1) = .51$. Hearing your family history, I understand your feelings about the high likelihood that your progeny will be boys. However, I think you should consider the situation in the following way. Generally, humans do not control the gender outcome of any particular conception. It is only natural to find that some large families will be predominantly male, while others are predominantly female. But that is just a matter of chance, not of different genetic or behavioral propensities in different human specimens. I think you would convince yourself of my point of view if we would study the family records of large families of the past century. Let us identify from geneological records many large families who had either four or five boys among their first five children, and find out how many of them also had a boy as their sixth child. If the proportion of boys among the sixth child born of each family is unduly large, I myself will be willing to reconsider my current opinion. But if the proportion of boys among these selected births is near to .51, I hope you will come to agree with me."

Hopefully, this discussion between Peter and Ron will stimulate some discussion and perhaps investigations among your own classmates. Let us dodge, here, the substantive and complex issues raised regarding gender and birth order in human populations. For our present purposes of introducing the concept of judging a sequence of events exchangeably, as opposed to judging them independently, let us conclude this example with the following technical point.

Ron avowedly regards the gender of Peter's children independently,with $P_{\text{Ron}}(B_i) = .51$. Thus, notice that he also regards them as exchangeably, since

$$P_{\text{Ron}}(B_1G_2) = P_{\text{Ron}}(B_1)P_{\text{Ron}}(G_2) = (.51)(.49) = P_{\text{Ron}}(B_2)P_{\text{Ron}}(G_1) = P_{\text{Ron}}(G_1B_2).$$

This turns out to be a general result: coherent opinions that regard events independently and with the same probability also regard them exchangeably, not vice versa. We study this matter more extensively in Section 3.9.

For a stimulating question, suppose that Peter and Amy plan to have two children, as mentioned. But they have aired the idea that they may even try to have three or four children, irrespective of the genders of the first two. Knowing his probability assertions as stated so far, is it possible that Peter can coherently regard the gender of all three or four children exchangeably? Think about it. But we are getting ahead of ourselves... as, perhaps, Peter and Amy may be getting ahead of themselves. □

***Example 4.*** Our first three examples concerned events that are regarded exchangeably. The concept of exchangeable judgments applies similarly to general quantities. In concluding Section 3.2 we considered the quantities $X_1$ and $X_2$ of bushels of wheat harvested on two specific, but otherwise nondescript, acres of Kansas farmland. For practical purposes, each $\mathcal{R}(X_i) = \{0, 1, \ldots, 100\}$, and the quantities are logically independent. Thus, for any choice of numbers $a_1$ and $a_2$ in $\mathcal{R}(X_i)$, I would assert my $P[(X_1 = a_1)(X_2 = a_2)] = P[(X_1 = a_2)(X_2 = a_1)]$. Now I don't have any reason to believe that $X_1$ will be larger or smaller than $X_2$. In vector notation, we would write that for any vector $\mathbf{a}_2 \in \{0, 1, \ldots, 100\}^2$ my $P(\mathbf{X}_2 = \mathbf{a}_2)$ is constant over all permutations of the components of $\mathbf{a}_2$. This manner of speaking also corresponds to our description of events regarded exchangeably. In the example of the glass baking dishes, each $\mathcal{R}(S_i) = \{0, 1\}$. To say, for example, that your probabilities of the form $P(S_1S_2S_3S_4S_5S_6S_7\tilde{S}_8\tilde{S}_9\tilde{S}_{10})$

are equal, no matter which of the numbered events are identified with the tildes, is equivalent to saying that your $P(\mathbf{S}_{10} = (1,1,1,1,1,1,1,0,0,0))$ is constant over all distinct permutations of the components of $(1,1,1,1,1,1,1,0,0,0)$. It is merely a matter of understanding notation to recognize that $P(S_1 S_2 \tilde{S}_3 \tilde{S}_4 S_5 S_6 S_7 \tilde{S}_8 S_9 S_{10}) = P(\mathbf{S}_{10} = (1,1,0,0,1,1,1,0,1,1))$. Think about it.

□

Our final example may appear contrived, since its content concerns a procedure of drawing balls from an urn, much as in the games of Bingo and Housie. However, the example is actually *paradigmatic* of any applied problem in which you regard a sequence of quantities exchangeably. It will also serve as a prelude to some easy problems that will help you fix some important ideas. Section 3.8.1 then provides a brief interlude in which we review the important principles of counting that will be used throughout this text. In Section 3.8.2 we present formal definitions of the judgment of exchangeability, and begin our analysis of its implications.

***Example 5.*** Suppose you fill an urn with $R$ red balls and $B$ black balls. Then the balls are thoroughly scrambled, and three balls are drawn successively from the urn *without replacing* the ones already drawn. Let $R_i$ denote the event equal to 1 if the $i$th ball drawn is red, and equal to 0 if the $i$th ball drawn is black. Since the balls are scrambled in the urn, suppose you assert your probabilities that any particular ball will be selected from the urn on any draw as 1 divided by the total number of balls in the urn at that time. Coherency would then require you also to assert

$$P(R_1 R_2 \tilde{R}_3) = P(R_1)P(R_2 | R_1)P(\tilde{R}_3 | R_1 R_2) = \frac{R(R-1)B}{(R+B)(R+B-1)(R+B-2)},$$

$$P(R_1 \tilde{R}_2 R_3) = P(R_1)P(\tilde{R}_2 | R_1)P(R_3 | R_1 \tilde{R}_2) = \frac{RB(R-1)}{(R+B)(R+B-1)(R+B-2)},$$

and

$$P(\tilde{R}_1 R_2 R_3) = P(\tilde{R}_1)P(R_2 | \tilde{R}_1)P(R_3 | \tilde{R}_1 R_2) = \frac{BR(R-1)}{(R+B)(R+B-1)(R+B-2)}.$$

(Notice the use of the product rule in the first step of these derivations.) These results state that you assert the *same probability* for any sequence of three draws that yields two reds and one black, *irrespective of the order* in which the reds and the black appear. You should be able to follow this same logic to show that coherency requires you to assert

$$P(R_1 \tilde{R}_2 \tilde{R}_3) = P(\tilde{R}_1 R_2 \tilde{R}_3) = P(\tilde{R}_1 \tilde{R}_2 R_3) = \frac{RB(B-1)}{(R+B)(R+B-1)(R+B-2)}.$$

For any sequence of three draws, you regard the order in which the colors appear to be irrelevant to your opinion about the sequence. We say you regard the colors of the three selected balls *exchangeably*. □

## PROBLEMS

1. Suppose the sequential draws from the urn are conducted by *replacing* the ball that is drawn once its color is noted and then remixing the balls before the next draw. Show that again your

$$P(R_1R_2\tilde{R}_3) = P(R_1\tilde{R}_2R_3) = P(\tilde{R}_1R_2R_3),$$
$$P(R_1\tilde{R}_2\tilde{R}_3) = P(\tilde{R}_1R_2\tilde{R}_3) = P(\tilde{R}_i\tilde{R}_2R_3).$$

Show that in this case, you also regard the events $R_1$, $R_2$, and $R_3$ independently.

2. Show that in the context of Example 5, you do not regard $R_1$ and $R_2$ independently, even though you do regard them exchangeably. [That is, show that your $P(R_2|R_1) \neq P(R_2)$, even though your $P(R_1\tilde{R}_2) = P(\tilde{R}_1R_2)$. *Hint*: Use the theorem on total probability to derive $P(R_2)$ from $P(R_2|R_1)P(R_1) + P(R_2|\tilde{R}_1)P(\tilde{R}_1)$ since $R_1$ and $\tilde{R}_1$ constitute a partition.]

3. Suppose the sequential draws are conducted by replacing the ball after its color is noted, but, in addition, another ball of the same color is added to the urn. Show that, again, the same two threesomes of probabilities are equal. Such a drawing process is called a "Polya urn process," since it was proposed and analyzed extensively by George Polya in the 1920s and 1930s.

4. Suppose the Polya urn process is amended so that $K$ additional balls of the same color are added to the urn after a ball is drawn. Again show that the same exchangeability property holds for your opinions about the sequence. (Note that our original example of drawing without replacement is a special case of this process in which $K = -1$; problem 1 specifies $K = 0$; and problem 3 specifies $K = 1$.)

5. Suppose the Polya urn process is amended so that $K$ additional balls of the contrasting color are added to the urn after a ball is drawn. Show that the property of exchangeability is no longer characteristic of your opinions.

### 3.8.1 Permutations and Combinations: Basic Principles of Counting

I presume you have already been introduced to problems of counting, particularly the enumeration of permutations and combinations. The following brief sequence of statements, questions, and answers reviews the logic of these principles as they apply to the concept of exchangeable judgments. If you are already comfortable with these principles, just skip this section. But if you can only remember the "formulas" for counting and cannot remember clearly how to derive them, or what they mean, you would do well to read this section.

An $N$-tuple is an *ordered* array of $N$ symbols, denoted by an expression such as $\mathbf{a}_N = (a_1, a_2, \ldots, a_{N-1}, a_N)$. For one example, any $N$-dimensional vector is an $N$-tuple. The *order* in which the components of a vector are listed is crucial to distinguishing the vector that is being represented. The vector $(2, 1)$ in two-dimensional space is distinct from the vector $(1, 2)$. Any English word constitutes another example of an $N$-tuple. For example, the word "read" is an ordered array of four letters. It is distinct from the words "dare" and "dear" because the order in which the letters appear in each of these 4-tuples is different; and each of these three words is distinct from the 4-tuple "rdae," which, although it is not an English word, is a 4-tuple of letters.

A problem with many applications in statistical inference is to determine how many distinguishable $N$-tuples can be constructed by merely rearranging the components of a specified $N$-tuple. The problem is so pervasive that we create a special word for a *rearrangement* of an $N$-tuple, a *permutation*. The problem can be solved by using the following definitions and principles.

0. Definition. The $N$-tuple $\mathbf{b}_N$ is said to be a *permutation* of the $N$-tuple $\mathbf{a}_N$ if the components of $\mathbf{b}_N$ are merely the components of $\mathbf{a}_N$ arranged in a different order. [Note: If a column vector $\mathbf{b}_N$ is a permutation of $\mathbf{a}_N$, then there is a matrix $\mathbf{\Pi}_{N,N}$ for which $\mathbf{b}_N = \mathbf{\Pi}_{N,N}\mathbf{a}_N$, where $\mathbf{\Pi}_{N,N}$ is constructed by reordering the columns of the $N \times N$ identity matrix, $\mathbf{I}_N$. Such a matrix $\mathbf{\Pi}_{N,N}$ is called a *permutation matrix*.]
1. A Basic Principle of Counting. If the construction of a $K$-tuple can be completed by a procedure involving $K$ sequential steps, in which there are $N_i$ different ways the $i$th step can be completed, then the total number of $K$-tuples that can be constructed by this procedure equals the product $N_1 N_2 \cdots N_K \equiv \prod_{i=1}^{K} N_i$.
2. Question. How many $K$-tuples can be constructed by picking their components sequentially from a collection of $N$ distinct objects, allowing repetitions among the components of the $K$-tuples?

   Answer. Since there are $N$ different objects that can be picked for each component, the total number of $K$-tuples that can be constructed equals $N^K$.
3. Question. How many permutations can be formed from the $N$-tuple $\mathbf{a}_N = (a_1, a_2, \ldots, a_N)$, where each component of $\mathbf{a}_N$ is distinct?

   Answer. Since there are $N$ different objects that can be picked for the first position, and $N-1$ different objects for the second position, and generally $N-i+1$ different objects for the $i$th position ($i = 1, 2, \ldots, N$), the total number of permutations that can be constructed equals the product $N(N-1)(N-2)\cdots(3)(2)(1)$. Since the product of a decreasing sequence of integers appears often in solutions to problems of counting, we give such a product a special notation: the symbol $N!$, read as "$N$ factorial," represents the product $N(N-1)(N-2)\cdots(3)(2)(1)$. That is, $N! \equiv \prod_{i=1}^{N} i$. By convention, we define $0! \equiv 1$.

**4.** QUESTION. How many $K$-tuples can be constructed by picking their components sequentially from a collection of $N$ distinct objects, but picking in such a way that each object in the $K$-tuple is distinct; that is, no pick is repeated?

ANSWER. Since there are $N-i+1$ different objects that can be picked for the $i$th component ($i=1,2,\ldots,K$), the total number of $K$-tuples that can be constructed equals $N(N-1)(N-2)(N-K+1)$, which can be written $N!/(N-K)!$. The answer to this question, too, we denote by a special symbol: ${}^{N}P_{K} \equiv \prod_{i=0}^{K-1}(N-i)$, which is equal to $N!/(N-K)!$ when $K$ is an integer between 0 and $N$. The symbol ${}^{N}P_{K}$ represents the number of ordered $K$-tuples that can be constructed from $N$ distinct objects, using no object more than once. In words, we read the symbol ${}^{N}P_{K}$ as "the number of permutations of $N$ things taken $K$ at a time."

**5.** QUESTION. How many recognizably distinct permutations can be formed from the letters of the word "Aotearoa"?

ANSWER. Since "aotearoa" represents an 8-tuple, we know that there are 8! permutations of these letters. However, in reaching this figure of 8!, for each *recognizable* ordering of the letters, there are $3! = 6$ permutations of the three a's and $2! = 2$ permutations of the o's that yield only one recognizably distinct ordering of the eight letters. Thus, we need to divide 8! by 6 and by 2 to count the number of recognizable permutations: $8!/(3!2!)$. This result can be easily generalized as follows.

**6.** ANOTHER COUNTING PRINCIPLE. If a collection of $N$ objects is made up of $N_1$ objects of one type, $N_2$ objects of a second type, ..., and $N_K$ objects of a $K$th type (that is, $N = \sum_{i=1}^{K} N_i$), then the total number of distinguishable $N$-tuples that can be constructed using the $N$ objects equals $N!/\prod_{i=1}^{K}(N_i!)$. We denote this number by ${}^{N}C_{(N_1,N_2,\ldots,N_K)}$ or, in vector notation, by ${}^{N}C_{\mathbf{N}_K}$. Following is a very important application of this general result, which we use in our development of properties of coherent prevision for sequences of events that you regard as exchangeably.

**7.** QUESTION. How many distinguishable permutations can be formed from an $N$-tuple composed of $A$ 1's and $(N-A)$ 0's?

ANSWER. $N!/[A!(N-A)!]$.

Another way in which this same question can be asked appeared in Example 2 of Section 3.8 concerning the testing of the glass baking dishes. How many different events can be formed by merely permuting the subscripts of the multiplicands of the product event $S_1S_2\tilde{S}_3\tilde{S}_4S_5S_6S_7\tilde{S}_8S_9S_{10}$? Notice that any permutation of the subscripts in this product expression that would result in the product of the same three inverted events, denoted with tildes, and the same seven events without tildes would not yield a different product event. For example, merely switching the subscript 4 with the subscript 8 would yield the expression $S_1S_2\tilde{S}_3\tilde{S}_8S_5S_6S_7\tilde{S}_4S_9S_{10}$, which is identical to the producing product event. For the product of 10 events does not depend on the order in which they are multiplied. The only permutations of subscripts in the product expression $S_1S_2\tilde{S}_3\tilde{S}_4S_5S_6S_7\tilde{S}_8S_9S_{10}$ that yield distinct events are those that result in a distinct identification of three events with tildes and seven events without tildes. Thus, our question regarding the glass baking dishes is equivalent to asking: How many distinct permutations can be formed from an $N$-tuple composed of three tildes and seven blanks? [ANSWER. 10!/(3!7!).

**8.** We need to consider one more question and introduce one more notation.

QUESTION. How many distinct *collections of* K objects can be selected from a collection of $N$ distinct objects?

ANSWER. Consider that a collection of objects, defined by a list of its members, is the same regardless of the order in which its members are listed. We already know that there are ${}^NP_K$ *distinguishable permutations* of $K$ objects selected from $N$ distinct objects. But in counting this number ${}^NP_K$, we count each *collection* of $K$ objects $K!$ times, since there are $K!$ distinct permutations of $K$ objects. Thus, the number of *distinct collections* of $K$ objects that can be selected from among $N$ distinct objects equals ${}^NP_K/K! = N!/[K!(N-K)!]$. This number, too, has so many applications that we denote it by its own symbol: ${}^NC_K \equiv N!/[K!(N-K)!]$ whenever $K$ is an integer between 0 and $N$. In words, we read ${}^NC_K$ as "the number of combinations of $N$ things taken $K$ at a time," or, in brief, "$N$ choose $K$."

**9.** Reexamining the question and answer posed in consideration 7 in light of the definition appearing in consideration 8, notice that the following odd- sounding statement is true: The number of *distinguishable permutations* that can be formed from an $N$-tuple composed of $A$ 1's and $(N-A)$ 0's is the number of *combinations* of $N$ things taken A at a time, ${}^NC_A$. Though this sentence may sound odd, its logic should be apparent. Each distinguishable permutation of the $N$-tuple is constructed merely by choosing the $A$ positions in which to place the 1's, leaving the $N-A$ positions for the 0's. Thus, we can ask, How many distinct events can be formed by permuting the subscripts in the product event $S_1S_2\tilde{S}_3\tilde{S}_4S_5S_6S_7\tilde{S}_8S_9S_{10}$? The answer is ${}^{10}C_7$, for each distinct product event is identified by a choice of 7 of the 10 multiplicand events to be distinguished without a tilde.

**10.** You should be able to show the following four useful results:

(i) ${}^NC_K = {}^NC_{(N-K)}$.

(ii) ${}^NC_K + {}^NC_{(K+1)} = {}^{(N+1)}C_{(K+1)}$. This result is the basis for Pascal's famous triangle. Each row of this triangle, generically call it the $N$th row, contains the sequence of numbers ${}^NC_0, {}^NC_1, {}^NC_2, \ldots, {}^NC_N$. Write out the first few rows of this triangle, beginning with row 0, and you will see this result in use. The value of a combination in any row equals the sum of two adjacent values in the previous row.

(iii) $(a+b)^N = \sum_{K=0}^{N} {}^N C_K\, a^{(N-K)} b^K$. This is called the "binomial expansion theorem" and can be proved by induction on $N$. First, notice that the result is true when $N = 1$. Then show that if it is true for $N$ it must be true for $N+1$. [In showing this, you will need to use the result stated as result (ii).] More generally, the "multinomial expansion theorem" states that $(\sum_{i=1}^{K} a_i)^N = \sum^* {}^N C_{(N_1, N_2, \ldots, N_K)} a_1^{N_1} a_2^{N_2} \cdots a_K^{N_K}$, where the summation $\sum^*$ runs over all vectors of nonnegative integers $(N_1, \ldots, N_K)$ whose components sum to $N$.

(iv) $2^N = \sum_{K=0}^{N} {}^N C_K$. This is a special case of the binomial expansion stated as result (iii), found by letting $a = b = 1$. If you sum the numbers in the first few rows of Pascal's triangle, you will see this result instantiated. More generally, $K^N = \sum^* {}^N C_{(N_1, N_2, \ldots, N_K)}$, a special case of the multinomial expansion theorem, found by setting $a_1 = a_2 = \cdots = a_K = 1$.

There are many good elementary textbook presentations of the basic principles of counting. Edwards (1987) presents an interesting historical study of Pascal's triangle and its philosophical and practical import in its own day.

### 3.8.2 Judging Events Exchangeably: Two Equivalent Definitions

In this section we study two different formal definitions of the judgment to regard $N$ events exchangeably. We shall discuss some of their implications and show them to be equivalent. The first definition formalizes the judgment that the order of the events is irrelevant to your probability assessment of the sequence.

***Definition 3.8a.*** *You are said to regard the events $E_1, E_2, \ldots, E_N$ exchangeably* if for each number $A \in \{0, 1, \ldots, N\}$ your prevision assertions of the form $P(E_1 E_2 \cdots E_A \tilde{E}_{A+1} \tilde{E}_{A+2} \cdots \tilde{E}_N)$ are identical for every permutation of the subscripts. □

By permuting the subscripts we mean respecifying which of the multiplicand events are inverted from $E$ to $(1-E)$ by the $\sim$ symbol and which of the events are not inverted. Any product of $N$ events, of which $N - A$ are inverted events, can be generated from the product event specified in the definition by rearranging the subscripts on the multiplicand events of the product. This does not mean rearranging the temporal order in which the events are to be observed. It merely means identifying a different product event among the ${}^N C_A$ distinct possibilities that can be generated by permuting the subscripts.

This definition generalizes easily. Suppose $\mathbf{X}_N$ is a vector of quantities with a common realm. You regard them exchangeably if, for any $\mathbf{a}_N \in \mathscr{R}(\mathbf{X}_N)$, your $P(\mathbf{X}_N = \mathbf{a}_N) = P(\mathbf{X}_N = \mathbf{b}_N)$ whenever $\mathbf{b}_N$ is a permutation of $\mathbf{a}_N$. Notice that if each $\mathscr{R}(X_i) = \{0, 1\}$ this definition reduces to the statement in Definition 3.8a concerning the permutation of the subscripts. (Refer to Example 4 of Section 3.8.) Before progressing further, you should try a simple problem or two.

## PROBLEMS

1. Show that if you regard the events $E_1, E_2, \ldots,$ and $E_N$ exchangeably, then you must regard the events $\tilde{E}_1, \tilde{E}_2, \ldots,$ and $\tilde{E}_N$ exchangeably.

2. Show that even if you regard $E_1$ and $E_2$ exchangeably, you need not regard $E_1$ and $\tilde{E}_2$ exchangeably. (You usually don't. When do you?)

It is worth mentioning that the judgment of exchangeability can be asserted meaningfully without your specifying your probability value for any particular sequence of the considered events. The exchangeability condition can be expressed merely by the stipulation that

$$P[E_1 E_2 \cdots E_A \tilde{E}_{A+1} \tilde{E}_{A+2} \cdots \tilde{E}_N - \text{perm}(E_1 E_2 \cdots E_A \tilde{E}_{A+1} \tilde{E}_{A+2} \cdots \tilde{E}_N)] = 0,$$

where $\text{perm}(E_1 E_2 \cdots E_A \tilde{E}_{A+1} \tilde{E}_{A+2} \cdots \tilde{E}_N)$ is any product event generated by permuting the subscripts of the multiplicands.

The following useful result establishes the relationship among cohering probability assertions for sequences of events that are regarded exchangeably and probability assertions concerning the sum of the events. In this theorem and throughout the remainder of this chapter, we use the notation $S_N \equiv \sum_{i=1}^{N} E_i$ to represent the sum of $N$ events.

**Theorem 3.9.** You regard the events $E_1, E_2, \ldots, E_N$ exchangeably if and only if for each number $A \in \{0, 1, \ldots, N\}$ your prevision assertions satisfy

$$P(E_1 E_2 \cdots E_A \tilde{E}_{A+1} \tilde{E}_{A+2} \cdots \tilde{E}_N) = P(S_N = A)/{}^N C_A,$$

for every permutation of subscripts in the product event on the left-hand side.

*Proof.* The event $(S_N = A) = \sum^* E_1 E_2 \cdots E_A \tilde{E}_{A+1} \tilde{E}_{A+2} \cdots \tilde{E}_N$, where the summation $\sum^*$ runs over all distinct product events generated by merely permuting the subscripts among the summands; for the sum $S_N = A$ if and only if $A$ of the events $E_1, \ldots, E_N$ in fact equal 1, and the remaining $N - A$ of the events equal 0. Each of the ${}^N C_A$ possible ways that this might eventuate specifies a product event of the form $E_1 E_2 \cdots E_A \tilde{E}_{A+1} \tilde{E}_{A+2} \cdots \tilde{E}_N$, for some permutation of the subscripts. Thus, by the additivity property of coherent prevision,

$$P(S_N = A) = \sum^* P(E_1 E_2 \cdots E_A \tilde{E}_{A+1} \tilde{E}_{A+2} \cdots \tilde{E}_N).$$

Since the judgment of exchangeability specifies that each of these ${}^N C_A$ summand probabilities are equal, the stated result must hold.

Considered in reverse, the theorem is true immediately, since it specifies the constancy of $P(E_1E_2\cdots E_A\tilde{E}_{A+1}\tilde{E}_{A+2}\cdots\tilde{E}_N)$ over permutations of the subscripts. □

Theorem 3.9 is very useful, both in applied statistical practice and in mathematical analysis. It is often easier to think about the sum of successes among events regarded exchangeably than it is to think about a particular ordered sequence. In our example of tests on baking dishes, an assessment of $P(S_{10}=7)$ appears much easier than, say, $P(S_1S_2\tilde{S}_3\tilde{S}_4S_5S_6S_7\tilde{S}_8S_9S_{10})$.

Probabilities for the constituents of the partition defined by the sum of $N$ events regarded exchangeably will be found so useful that we ought define a special notation to represent them. Let us denote your probabilities for the constituents of the partition $(S_N=0)$, $(S_N=1),\ldots,(S_N=N)$ by the vector $\mathbf{p}_N\equiv(p_{0,0},\ldots,p_{N,N})^{\mathrm{T}}$, where $p_{a,N}\equiv P(S_N=a)$. Notice that, although $\mathbf{p}_N$ is an $(N+1)$-dimensional vector, coherency requires that its components be nonnegative and sum to 1, since the events $(S_N=0)$, $(S_N=1),\ldots,(S_N=N)$ constitute a partition. Geometrically, Theorem 3.9 says that any coherent opinion that regards the events $E_1,\ldots,E_N$ exchangeably is representable by a member of the $N$-dimensional unit simplex of $(N+1)$-dimensional coordinate vectors,

$$\mathbf{S}^N\equiv\left\{(p_{0,N},p_{1,N},\ldots,p_{N,N})\,\middle|\,\sum_{i=0}^{N}p_{i,N}=1\text{ and every }p_{i,N}\geqslant 0\right\}.$$

We study geometrical concepts more extensively in Section 3.10. For now, let us apply Theorem 3.9 to the following matrix result, which is useful in reducing the amount of computations involved in applying the fundamental theorem of prevision to events you regard exchangeably.

Suppose the components of $\mathbf{E}_N$ are logically independent events that you regard exchangeably. Let $\mathbf{Q}_{2^N}$ represent the column vector of the $2^N$ constituents of the partition generated by $\mathbf{E}_N$. Then any coherent prevision assertion for the components of $\mathbf{Q}_{2^N}$ can be represented as

$$P(\mathbf{Q}_{2^N})=\mathbf{M}_{2^N,(N+1)}\mathbf{p}_N,$$

where $\mathbf{M}_{2^N,(N+1)}$ is a matrix whose row $i$ contains all 0's except for column $j+1$, where $j$ is the value of $S_N$ that would be implied if the $i$th component of $\mathbf{Q}_{2^N}$ equals 1. This variable specification of the form of $\mathbf{M}$ is necessary because of the arbitrariness allowed in ordering of the constituents of the partition vector.

For a small example, suppose you regard four logically independent events exchangeably. These events generate a partition of size 16. Suppose we order the components of $\mathbf{Q}_{16}$ according to an increasing size they would imply for the sum of the four events, $S_4$. The previous paragraph means that we can write

$$P\begin{pmatrix} \tilde{E}_1\tilde{E}_2\tilde{E}_3\tilde{E}_4 \\ \tilde{E}_1\tilde{E}_2\tilde{E}_3E_4 \\ \tilde{E}_1\tilde{E}_2E_3\tilde{E}_4 \\ \tilde{E}_1E_2\tilde{E}_3\tilde{E}_4 \\ E_1\tilde{E}_2\tilde{E}_3\tilde{E}_4 \\ \tilde{E}_1\tilde{E}_2E_3E_4 \\ \tilde{E}_1E_2\tilde{E}_3E_4 \\ E_1\tilde{E}_2\tilde{E}_3E_4 \\ \tilde{E}_1E_2E_3\tilde{E}_4 \\ E_1\tilde{E}_2E_3\tilde{E}_4 \\ E_1E_2\tilde{E}_3\tilde{E}_4 \\ \tilde{E}_1E_2E_3E_4 \\ E_1\tilde{E}_2E_3E_4 \\ E_1E_2\tilde{E}_3E_4 \\ E_1E_2E_3\tilde{E}_4 \\ E_1E_2E_3E_4 \end{pmatrix} = \begin{pmatrix} 1 & 0 & 0 & 0 & 0 \\ 0 & 1/4 & 0 & 0 & 0 \\ 0 & 1/4 & 0 & 0 & 0 \\ 0 & 1/4 & 0 & 0 & 0 \\ 0 & 1/4 & 0 & 0 & 0 \\ 0 & 0 & 1/6 & 0 & 0 \\ 0 & 0 & 1/6 & 0 & 0 \\ 0 & 0 & 1/6 & 0 & 0 \\ 0 & 0 & 1/6 & 0 & 0 \\ 0 & 0 & 1/6 & 0 & 0 \\ 0 & 0 & 1/6 & 0 & 0 \\ 0 & 0 & 0 & 1/4 & 0 \\ 0 & 0 & 0 & 1/4 & 0 \\ 0 & 0 & 0 & 1/4 & 0 \\ 0 & 0 & 0 & 1/4 & 0 \\ 0 & 0 & 0 & 0 & 1 \end{pmatrix} \begin{pmatrix} p_{0,4} \\ p_{1,4} \\ p_{2,4} \\ p_{3,4} \\ p_{4,4} \end{pmatrix},$$

where, remember, the vector $\mathbf{p}_4$ is constrained so that $\mathbf{1}_5^T\mathbf{p}_4 = 1$ and each $p_{i,4} \geqslant 0$. In the dense notation of the previous paragraph, we write $P(\mathbf{Q}_{16}) = \mathbf{M}_{16,5}\mathbf{p}_4$.

## PROBLEMS

**3.** Matrices of the form $\mathbf{M}_{2^N,(N+1)}$ will be important throughout our study of exchangeability. To firm your conception of such a matrix, construct the matrix $\mathbf{M}_{32,6}$, and think out the construction of $\mathbf{M}_{64,7}$ as a "thought construction."

**4.** You may like to write a computer program to generate the matrix $\mathbf{M}_{2^N,(N+1)}$ appropriate for your program realm.m that generates a realm matrix for $N$ logically independent events. (See problem 6 of Section 2.1.) The realm matrix that program generates does not order its columns according to the size of $S_N$, as we specified in our earlier example of $\mathbf{M}_{16,5}$. Think out why the following program, called MexchN.m, produces the $\mathbf{M}$ matrix appropriate to the realm $\mathbf{R}(\mathbf{E}_N)$ your realm.m program generates:

```
% MexchN.m A MATLAB program to generate the exchangeable M matrix
% appropriate to the "realm.m" program written in problem 6, of Section 2.1.
% Begin by setting K = 1, and N = some number.
% Then run the subroutine realm. Then
```

```
s = sum(RXN); M = zeros(size(s, 2), N + 1); NCJ = [1];
for j = 1:N, NCJ = [NCJ NCJ(1,j)*(N - j + 1)/j]; end
NCJINV = ones(size(NCJ))./NCJ;
for j = 1:size(s,2), M(j, s(j) + 1) = NCJINV(s(j) + 1); end
```

Run this program for $N = 4$ and examine the matrix $\mathbf{M}$ that it produces.

**5.** Our developments thus far seem to presume that you have asserted probabilities for each constituent of the partition $(S_N = 0), (S_N = 1), \ldots, (S_N = N)$ to yield the vector $\mathbf{p}_N$. Working the following problem should convince you that Theorem 3.9 is useful even if you assert much less. We discuss the following problem in detail in Section 3.11.2. Suppose you regard the events $E_1, E_2, E_3$, and $E_4$ exchangeably, and you assert the inequalities $.5 \leqslant P(S_4 \geqslant 3) \leqslant .75$ and $P(S_4 = 0) \leqslant .1$. Use the fundamental theorem of prevision and the matrix result $P(\mathbf{Q}_{2^N}) = \mathbf{M}_{2^N,(N+1)}\mathbf{p}_N$ to determine the coherent bounds on your prevision $P(S_4)$, on your $P[(S_4 = 0) + (S_4 = 4)]$, on your $P(E_4|(S_3 = 1))$, and on your $P(E_4|(S_3 \leqslant 1))$. Use a computer to determine these bounds by means of an LP algorithm. *Hint*: Initially, set up the problem in the form suggested by the FTP, ignoring the asserted judgment to regard $\mathbf{E}_4$ exchangeably. The dimension of the vector $\mathbf{q}$ in the LP setup is $16 \times 1$. Then realize that exchangeability requires that $\mathbf{q}_{16} = \mathbf{M}_{16,5}\mathbf{p}_5$, and reduce the dimension of the problem appropriately. Then run the appropriate linear programming routine.

Our second equivalent definition of exchangeability will be useful not so much in practice as in theoretical considerations. It hinges on your consideration of product events directly. Beginners and intermediate readers should move on to Section 3.9, perhaps reading the historical comments that conclude this section if you wish.

***Definition 3.8b.*** *You* are said to *regard the events* $E_1, E_2, \ldots, E_N$ *exchangeably* if for each positive integer $A \leqslant N$ your prevision for the product of any $A$ events chosen from among them is identical, no matter which $A$ events $E_1^*, E_2^*, \ldots, E_A^*$ are chosen from among $E_1, \ldots, E_N$ to be included in the product event $(E_1^* E_2^* \cdots E_A^*)$. □

**Theorem 3.10.** Definitions 3.8a and 3.8b are equivalent.

*Proof* ($\Rightarrow$). Assume Definition 3.8a. Choose any $A$ events $(1 \leqslant A \leqslant N)$ from among $E_1, \ldots, E_N$. Relabel them in any order as $E_1, \ldots, E_A$, and relabel the remaining events in any order as $E_{A+1}, \ldots, E_N$. If the product event $E_1 E_2 \cdots E_A$ is expanded as the product of itself with the number 1, expressed as $1 = (E_{A+1} + \tilde{E}_{A+1}) \cdots (E_N + \tilde{E}_N)$, the expansion looks like $E_1 E_2 \cdots E_A(E_{A+1} +$

$\tilde{E}_{A+1})\cdots(E_N + \tilde{E}_N)$, which equals

$(E_1 \cdots E_A E_{A+1} \cdots E_N)$

$+\,{}^{(N-A)}C_1$ events involving the product of $N-1$ events and 1 inverted event

$+\,{}^{(N-A)}C_2$ events involving the product of $N-2$ events and 2 inverted events

$+\cdots$

$+\,{}^{(N-A)}C_{(N-A)}$ events involving the product of $A$ events and $N-A$ inverted events.

Now according to Definition 3.8a, to regard the events $E_1, \ldots, E_N$ exchangeably means that your previsions are equal for all events involving the product of $N-j$ events and $j$ inverted events. Thus, exchangeability and coherency require

$$\begin{aligned} P(E_1E_2\cdots E_A) = P(E_1E_2\cdots E_N) &+ {}^{(N-A)}C_1 P(E_1\cdots E_{N-1}\tilde{E}_N) \\ &+ {}^{(N-A)}C_2 P(E_1\cdots E_{N-2}\tilde{E}_{N-1}\tilde{E}_N) \\ &+ \cdots + {}^{(N-A)}C_{(N-A-1)} P(E_1\cdots E_{A+1}\tilde{E}_{A+2}\cdots\tilde{E}_N) \\ &+ {}^{(N-A)}C_{(N-A)} P(E_1\cdots E_A\tilde{E}_{A+1}\cdots\tilde{E}_N). \end{aligned}$$

So for any positive integer $A \leqslant N$, for any choice of $A$ events $E_1^*, \ldots, E_A^*$ selected from among $E_1, \ldots, E_N$, this expansion would yield this same sum, $P(E_1E_2\cdots E_A)$. But this is the stipulation of Definition 3.8b. □

*Notational Interlude*

Since exchangeability requires that the probability $P(E_1E_2\cdots E_A)$ must be constant for any choice of $A$ multiplicand events from among $E_N$, we henceforth use $q_1, q_2, \ldots, q_N$ to denote the relevant probability values: $q_A \equiv P(E_1E_2\cdots E_A)$. This notation along with the earlier $p_{A,N} \equiv P(S_N = A)$ allow the final equality in the proof to be written as $q_A = \sum_{i=0}^{N} {}^{(N-A)}C_i\, p_{N-i,N}/{}^{N}C_{(N-i)}$. In fact, this same equation holds even when $A = 0$, for in this case it reduces to the statement of the coherency restriction, that $\sum_{i=0}^{N} p_{N-i,N} = 1$. So agreeing to call $q_0 \equiv 1$, we define the vector notation $\mathbf{q}_N \equiv (q_0, q_1, \ldots, q_N)$.

*Proof* ($\Leftarrow$). Assume Definition 3.8b. Choose any $A$ events from among $E_1, \ldots, E_N$. Relabel them in any order as $E_1, \ldots, E_A$, and relabel the remaining events in any order as $E_{A+1}, \ldots, E_N$. Identifying $\tilde{E}_i$ with $(1 - E_i)$, we have

$$\begin{aligned} E_1E_2\cdots E_A\tilde{E}_{A+1}\tilde{E}_{A+2}\cdots\tilde{E}_N &= E_1E_2\cdots E_A(1-E_{A+1})(1-E_{A+2})\cdots(1-E_N) \\ &= (E_1E_2\cdots E_A)\Bigg[1 - \sum_{i=A+1}^{N} E_i + \sum_{j>i=A+1}^{N} E_iE_j \end{aligned}$$

$$- \sum_{k>j>i=1}^{N} E_i E_j E_k$$

$$+ \cdots + (-1)^K \sum_{i_K > \cdots > i_2 > i_1 = A+1}^{N} E_{i_1} E_{i_2} \cdots E_{i_K}$$

$$\left. + \cdots + (-1)^{N-A} E_{A+1} E_{A+2} \cdots E_N \right],$$

where the sum notation $\sum_{k>j>i=A+1}^{N} E_i E_j E_k$, for example, denotes the sum of all products of three events $E_i E_j E_k$ for which $A+1 \leqslant i < j < k \leqslant N$. Since there are ${}^{(N-A)}C_3$ three-product events in such a sum, for example, and generally ${}^{(N-A)}C_K$ $K$-product events in such a sum $\sum_{i_k > \cdots > i_j > i_i = A+1}^{N} E_{i_1} E_{i_2} \cdots E_{i_k}$, we have

$$P(E_1 E_2 \cdots E_A \tilde{E}_{A+1} \tilde{E}_{A+2} \cdots \tilde{E}_N) = q_A - {}^{(N-A)}C_1 q_{A+1} + {}^{(N-A)}C_2 q_{A+2}$$
$$- \cdots + (-1)^{K} {}^{(N-A)}C_K q_{A+K} + \cdots + (-1)^{N-A} q_N$$

implied by the fact that prevision is constant over the product of any $A$ events chosen from $N$. Thus, $P(E_1 E_2 \cdots E_A \tilde{E}_{A+1} \tilde{E}_{A+2} \cdots \tilde{E}_N)$ is the same for any permutation of the subscripts, since any reordered product event can be decomposed into the same expression with the same values $q_A, q_{A+1}, \ldots, q_N$. This is the stipulation of Definition 3.8a. □

In the course of this proof we have derived the following two sets of linear equations relating the vectors $\mathbf{p}_N$ and $\mathbf{q}_N$: for each $A = 0, \ldots, N$

$$q_A = \sum_{i=0}^{N-A} \frac{{}^{(N-A)}C_i \, p_{N-i,N}}{{}^{N}C_{(N-i)}},$$

and

$$p_{A,N} = {}^{N}C_A \sum_{i=0}^{N-A} (-1)^i \, {}^{(N-A)}C_i \, q_{A+i}.$$

In matrix notation, these equations say $\mathbf{q}_N = \mathbf{T}(\mathbf{p} \to \mathbf{q})\mathbf{p}_N$, where $\mathbf{T}(\mathbf{p} \to \mathbf{q})$ is the $(N+1) \times (N+1)$-dimensional matrix whose component $t_{i,j}$ is

$$t_{i,j} = {}^{(N-i+1)}C_{(N-j+1)} / {}^{N}C_{(j-1)} \qquad (j \geqslant i).$$

and $\mathbf{p}_N = \mathbf{T}(\mathbf{q} \to \mathbf{p})\mathbf{q}_N$, where $\mathbf{T}(\mathbf{q} \to \mathbf{p})$ is the $(N+1) \times (N+1)$-dimensional matrix whose component $t_{i,j}$ is

$$t_{i,j} = (-1)^{(i+j)} \, {}^{N}C_{(i-1)} \, {}^{(N-i+1)}C_{(j-1)} \qquad (j \geqslant i).$$

Both of these matrices are upper triangular, as symbolized by the multiplicative factor $(j \geqslant i)$ in the expression for $t_{i,j}$.

Evidently, the matrix $\mathbf{T}(\mathbf{q} \to \mathbf{p})$ is the inverse of the matrix $\mathbf{T}(\mathbf{q} \to \mathbf{p})$. So if you regard $E_1, \ldots, E_N$ exchangeably, your prevision for any constituent of the partition they generate can be characterized by either $\mathbf{p}_N$ or $\mathbf{q}_N$. Although coherency requires that the components of $\mathbf{p}_N$ sum to 1, it places no such linear constraint on the components of $\mathbf{q}_N$. The components of $\mathbf{q}_N$ are constrained only to be nonnegative and weakly decreasing, this because for any value of $K$ the product event $E_1 E_2 \cdots E_K E_{K+1}$ cannot exceed the product event $E_1 E_2 \cdots E_K$. For $E_{K+1}$ can equal only either 1 or 0. Thus, a second geometrical representation of any coherent opinion that regards the events $E_1, \ldots, E_N$ exchangeably can be achieved via the set of $(N+1)$-dimensional coordinate vectors,

$$\mathbf{S}^{N^*} \equiv \{(1, q_1, \ldots, q_N) | 1 \geqslant q_1 \geqslant q_2 \geqslant \cdots \geqslant q_i \geqslant q_{i+1} \geqslant \cdots \geqslant q_N \geqslant 0\}.$$

We denote this set by $\mathbf{S}^{N^*}$ since it is a linear transformation of the unit simplex $\mathbf{S}^N$ defined earlier, through the transformation matrix $\mathbf{T}(\mathbf{p} \to \mathbf{q})$. Again, we return to geometrical concepts in Section 3.10.

Another interesting idea emerges as a corollary to the equivalence of our two definitions of exchangeability.

**Corollary 3.10.1.** If you regard the events $E_1, \ldots, E_N$ exchangeably, then for any $K < N$ and for any choice of $K$ events from $N$ you must regard $E_1^*, E_2^*, \ldots, E_K^*$ exchangeably.

This corollary follows immediately from Definition 3.8b, since this definition stipulates that for any value of $K \leqslant N$ your $P(E_1^* E_2^* \cdots E_K^*)$ is constant over all possible selections of $K$ events from among $E_1, \ldots, E_N$. The interest of this corollary stems from the fact that the *converse* of the corollary is not true. Even if for any choice of $N-1$ events from among $E_1, \ldots, E_N$ you regard the events $E_1, \ldots, E_{N-1}$ exchangeably, you need not regard the collection of all of them exchangeably! In fact, sometimes when you specify probabilities for events that you regard exchangeably, coherency will not allow you to regard them exchangeably with any larger collection of events that contain them! This situation will be studied in greater detail shortly. For now, the following examples and problems should get you thinking. We begin with a very simple example and proceed to a more involved one.

***Example 1.*** Suppose you regard $E_1$ and $E_2$ exchangeably, and you assert

$$P(\tilde{E}_1 \tilde{E}_2) = P(E_1 E_2) = 0, \qquad \text{and} \qquad P(E_1 \tilde{E}_2) = P(\tilde{E}_1 E_2) = 1/2.$$

If there were a third event $E_3$ that you regarded exchangeably with $E_1$ and $E_2$, coherency would require you to assert $P(\tilde{E}_1\tilde{E}_2\tilde{E}_3) = 0$ as well, for

$$P(\tilde{E}_1\tilde{E}_2) = P(\tilde{E}_1\tilde{E}_2\tilde{E}_3) + P(\tilde{E}_1\tilde{E}_2E_3) = 0.$$

Furthermore, your regarding $E_1$, $E_2$, and $E_3$ exchangeably would then also require

$$P(\tilde{E}_1E_2\tilde{E}_3) = P(E_1\tilde{E}_2\tilde{E}_3) = 0,$$

since the product events $(\tilde{E}_1E_2\tilde{E}_3)$ and $(E_1\tilde{E}_2\tilde{E}_3)$ can be obtained merely by permuting the subscripts of $(\tilde{E}_1\tilde{E}_2E_3)$, for which $P(\tilde{E}_1\tilde{E}_2E_3) = 0$.

In the same way, your asserting $P(E_1E_2) = 0$ would require both your $P(E_1E_2\tilde{E}_3) = 0$ and your $P(E_1E_2E_3) = 0$, since $E_1E_2\tilde{E}_3 + E_1E_2E_3 = E_1E_2$; and exchangeability would then require $P(E_1\tilde{E}_2E_3) = P(\tilde{E}_1E_2E_3) = 0$, again for the same reason of permutation symmetry.

Thus, given your original assertions $P(E_1\tilde{E}_2) = P(\tilde{E}_1E_2) = 1/2$, there cannot be any third event that you coherently regard exchangeably with $E_1$ and $E_2$. For if there were, coherency would require that you accord every constituent of the partition generated by $\mathbf{E}_3$ with probability 0, which would be incoherent.

This example corresponds to a situation in which there are two balls in a box, one of them red and one black, and you are to draw them out successively without replacement. $E_1$ represents the event that the first ball drawn is red, and $E_2$ represents the event that the second ball drawn is red. □

***Example 2.*** Consider three events for which you assert the probabilities

$$P\begin{pmatrix} \tilde{E}_1\tilde{E}_2\tilde{E}_3 \\ E_1\tilde{E}_2\tilde{E}_3 \\ \tilde{E}_1E_2\tilde{E}_3 \\ \tilde{E}_1\tilde{E}_2E_3 \\ E_1E_2\tilde{E}_3 \\ E_1\tilde{E}_2E_3 \\ \tilde{E}_1E_2E_3 \\ E_1E_2E_3 \end{pmatrix} = \begin{pmatrix} .1 \\ .1 \\ .05 \\ .05 \\ .15 \\ .15 \\ .2 \\ .2 \end{pmatrix}.$$

Coherency then requires that your

$$P(E_1\tilde{E}_2) = P(\tilde{E}_1E_2) = .25,$$
$$P(E_1\tilde{E}_3) = P(\tilde{E}_1E_3) = .25,$$
$$P(E_2\tilde{E}_3) = P(\tilde{E}_2E_3) = .2.$$

Even though you regard each pair of events exchangeably, you do not regard all three events exchangeably.

## PROBLEMS

6. Show that if you regard $E_1$ and $E_2$ exchangeably and you assert $P(E_1E_2) = P(\tilde{E}_1E_2) = P(E_1\tilde{E}_2) = 1/3$ and $P(\tilde{E}_1\tilde{E}_2) = 0$, then there could be a third event $E_3$ with which you regard them exchangeably. What must your probabilities for the relevant product events equal? Show further that there cannot be a fourth event that you can coherently regard exchangeably with these three.

7. Perhaps you can now solve the problems we posed concerning the sex of Peter and Amy's third and fourth possible children at the end of Example 3 of Section 3.8.

### *Historical References and Terminology*

The concept of the judgment to regard several events exchangeably is due to de Finetti, although neither the word "exchangeable" nor even the important mathematical results that he derived and which usually bear his name were first proposed by him. As to terminology, de Finetti originally used the word "equivalent" in place of exchangeable, as had Khinchin (1932), whereas Haag (1924) had used "symmetric," and Johnson (1924) had referred to the same structure as his "permutation postulate." The word "exchangeable" was proposed by George Polya in the discussion following de Finetti's paper "Sur la Condition de l'Equivalence Partielle," presented at the 1937 Geneva Colloquium on the Theory of Probability. De Finetti (1939) published extensive critical notes on all the papers and discussions that occurred there. His reference to Polya's suggestion appears on page 31. Recently, formalist literature such as Chow and Teicher (1988) has begun to use "interchangeable" to represent this condition.

The famous result now commonly called "de Finetti's representation theorem," on the algebraic representation of opinions that regard an infinite sequence of events exchangeably (we study it in the next section), had already been proved in a mildly different form by Haag (1924) in a paper read to the International Congress of Mathematicians in Toronto. Haag's paper documents the relevance of exchangeability to a whole series of problems in probability that had been discussed in the eighteenth and nineteenth centuries, particularly to contributions (including errors) by Laplace and Bertrand. Savage's cryptic note on Haag's paper in the 1972 Dover edition of *The Foundations of Statistics* suggested that Haag's work was somehow meager. But the representation of probabilities that are "symmetric" over both finite and infinite sequences is quite explicit in Haag. A historical article by Dale (1985) summarizes details of contributions in the works of Haag (1924), Johnson (1924), and de Finetti (1931). Nonetheless, de Finetti's conception of what exchangeability means as a personal judgment about quantities was truly novel, and is not widely understood even today.

An intriguing photograph of Bruno de Finetti, along with other famous participants at the Geneva Colloquium on Probability of 1937, has been published in *The Polya Picture Album* (Polya, 1987, p. 116). De Finetti appears in the very first row of the participants, but he is standing in a patch of glaring light in front of a bush, so that he is virtually invisible. Even the editor of the photo album, who identified the participants in the photo by means of a photo map, apparently did not see him there. De Finetti's frame is not traced on the map, though the frame of every other participant is traced, and the figures of Borel, Fréchet, Polya, Fisher, Levy, and other notables are identified. You almost have to know that de Finetti is there to see him. He would have been 31 years old. The photo is ironic, for de Finetti's imagination and clarity of thought shine brilliantly to those who have taken his ideas seriously. Despite his voluminous writings and his many presentations of ideas at international meetings, he seems to have laboured through the century virtually unseen.

## 3.9 CONDITIONAL PROBABILITIES FOR EVENTS REGARDED EXCHANGEABLY: THE SUM OF EVENTS AS A SUFFICIENT STATISTIC

The judgment to regard a sequence of events exchangeably has embedded within it a precise assertion of the inference to be made about the values of some events in the sequence on the basis of observed values of others. In Section 3.8.2 we learned two important features of prevision when you regard $N$ events exchangeably: your prevision for any particular sequence of the events and inverted events, which we represent by a specific product event, is determined by your probability distribution for the sum of the events; and if you regard $N$ events exchangeably, then you must also regard any subcollection of them exchangeably. The first of the two theorems we study in this section identifies an algebraic relationship that connects these two results.

With a view to the fact that our results have important application to the theory of survey sampling from finite populations, we denote the events under consideration by $E_1, E_2, \ldots, E_n, E_{n+1}, \ldots, E_N$. That is, $N$ denotes a number of events under consideration, and $n$ denotes a smaller number of them. Thus, $S_N$ denotes the sum of all $N$ events, and $S_n$ denotes the sum of $n$ of them. Similarly, $p_{A,N}$ denotes $P(S_N = A)$ and $p_{a,n}$ denotes $P(S_n = a)$.

**Theorem 3.11.** If you regard the events $E_1, \ldots, E_N$ exchangeably, then for any choice of $n$ events from among $E_1, \ldots, E_N$ coherency requires that your

$$P(S_n = a) = \sum_{A=a}^{N-(n-a)} \frac{{}^{A}C_{a}\,{}^{(N-A)}C_{(n-a)}}{{}^{N}C_{n}} P(S_N = A),$$

where $a$ and $n$ are any integers satisfying $0 \leqslant a \leqslant n \leqslant N$.

*Proof*. The product quantity $E_1E_2\cdots E_a\tilde{E}_{a+1}\tilde{E}_{a+2}\cdots\tilde{E}_n$, when multiplied by 1 in the form $1=(E_{n+1}+\tilde{E}_{n+1})\cdots(E_N+\tilde{E}_N)$, yields the representation

$$
\begin{aligned}
&E_1E_2\cdots E_a\tilde{E}_{a+1}\tilde{E}_{a+2}\cdots\tilde{E}_n(E_{n+1}+\tilde{E}_{n+1})\cdots(E_N+\tilde{E}_N)\\
&\quad=(E_1\cdots E_a\tilde{E}_{a+1}\cdots\tilde{E}_N)\\
&\qquad+{}^{(N-n)}C_1 \text{ events involving the product of } a+1 \text{ events and } N-a-1 \text{ inverted events}\\
&\qquad+{}^{(N-n)}C_2 \text{ events involving the product of } a+2 \text{ events and } N-a-2 \text{ inverted events}\\
&\qquad+\cdots\\
&\qquad+{}^{(N-n)}C_{(N-n)} \text{ events involving the product of } a+N-n \text{ events and } n-a \text{ inverted events.}
\end{aligned}
$$

Thus, since exchangeability requires that your prevision for any $N$-product event involving $j$ events and $N-j$ inverted events must equal $P(S_N=j)/{}^NC_j$, the additivity property of coherent prevision requires that

$$
\begin{aligned}
P(E_1E_2\cdots E_a\tilde{E}_{a+1}\tilde{E}_{a+2}\cdots\tilde{E}_n)&=\frac{P(S_N=a)}{{}^NC_a}+\frac{{}^{(N-n)}C_1P(S_N=a+1)}{{}^NC_{(a+1)}}\\
&\quad+\frac{{}^{(N-n)}C_2P(S_N=a+2)}{{}^NC_{(a+2)}}\\
&\quad+\cdots\\
&\quad+\frac{{}^{(N-n)}C_{(N-n)}P(S_N=a+N-n)}{{}^NC_{(a+N-n)}}\\
&=\sum_{A=a}^{N-(n-a)}\frac{{}^{(N-n)}C_{(A-a)}}{{}^NC_A}P(S_N=A).
\end{aligned}
$$

Now the implied judgment of exchangeability regarding the events $E_1,\ldots,E_n$ requires that $P(S_n=a)={}^nC_aP(E_1E_2\cdots E_a\tilde{E}_{a+1}\tilde{E}_{a+2}\cdots\tilde{E}_n)$. Multiplying the summation by ${}^nC_a$ then yields the representation for $P(S_n=a)$ stated in the theorem, due to the fact that

$$
{}^nC_a\,{}^{(N-n)}C_{(A-a)}/{}^NC_A={}^AC_a\,{}^{(N-A)}C_{(n-a)}/{}^NC_n,
$$

an algebraic identity which can be shown easily by expanding both sides.

This theorem yields an interesting and important corollary. □

**Corollary 3.11.1.** If you regard the events $E_1,\ldots,E_N$ exchangeably, then coherency requires that for any integers $a, A$, and $n$ satisfying $0\leqslant a\leqslant n\leqslant N$,

$a \leqslant A \leqslant N$, and $n - a \leqslant N - A$, you must assert the conditional probabilities

$$P(S_n = a | S_N = A) = {}^{A}C_a \, {}^{(N-A)}C_{(n-a)} / {}^{N}C_n.$$

*Proof*. Since the events $(S_N = 0), \ldots, (S_N = N)$ constitute a partition, the theorem on total probability requires that $P(S_n = a) = \sum_{A=0}^{N} P(S_n = a | S_N = A) P(S_N = A)$. But if $S_N < a$, it is impossible that $S_n = a$, since it is necessary that $S_n \leqslant S_N$. Thus, $P(S_n = a | S_N = A) = 0$ if $A < a$. Similarly, if $S_N > N - (n - a)$, it is also impossible that $S_n = a$, since it is necessary that $n - S_n \leqslant N - S_N$. Thus, also, $P(S_n = a | S_N = A) = 0$ if $A > N - (n - a)$. Thus, the summation reduces to

$$P(S_n = a) = \sum_{A=a}^{N-(n-a)} P(S_n = a | S_N = A) P(S_N = A);$$

but Theorem 3.11 specifies that

$$P(S_n = a) = \sum_{A=a}^{N-(n-a)} \left[ \frac{{}^{A}C_a \, {}^{(N-A)}C_{(n-a)}}{{}^{N}C_n} \right] P(S_N = A).$$

Since both of these equations must hold for any vector of probabilities $(P(S_N = 0), P(S_N = 1), \ldots, P(S_N = N))$ within the $N$-dimensional simplex, it must be true that for any integers $a$ and $A$ satisfying $0 \leqslant a \leqslant A \leqslant N$ and $a \geqslant n - (N - A)$,

$$P(S_n = a | S_N = A) = {}^{A}C_a \, {}^{(N-A)}C_{(n-a)} / {}^{N}C_n. \quad \square$$

This corollary makes sense intuitively. If you knew that the sum of the $N$ events were $A$, and you regard in the same way all permutations of any $N$-product event involving $A$ events and $N - A$ inverted events, your probability for the event $(S_n = a)$ should equal the proportion of all possible selections of $n$ events from among $E_1, \ldots, E_N$ that would include among the selected events $a$ of the $A$ events that equal 1 and $n - a$ of the $N - A$ events that equal 0. Principles of counting show this proportion to equal ${}^{A}C_a \, {}^{(N-A)}C_{(n-a)} / {}^{N}C_n$. If you are already familiar with the hypergeometric probability distribution, you will understand Corollary 3.11.1 to say that if you regard the components of $\mathbf{E}_N$ exchangeably, then your conditional distribution for $S_n$ given $(S_N = A)$ is hypergeometric, $H(n, A, N)$, a terminology we develop in Chapter 5.

Our next theorem identifies an important feature of the coherent inferences that must be made concerning events $E_{n+1}, E_{n+2}, \ldots, E_N$ on the basis of specific values of the conditioning events $E_1, \ldots, E_n$, when all $N$ events are regarded exchangeably. These inferences can be expressed generically in the form of the conditional probabilities

$$P(E_{n+1} E_{n+2} \cdots E_{n+A-a} \tilde{E}_{n+A-a+1} \tilde{E}_{n+A-a+2} \cdots \tilde{E}_N | E_1 E_2 \cdots E_a \tilde{E}_{a+1} \tilde{E}_{a+2} \cdots \tilde{E}_n)$$

for any permutation of all subscripts. The idea here is that we are conditioning on some particular ordered observation of $a$ successes among $n$ sampled events, and we are making inferences regarding some particular ordered observation of $A-a$ successes among the remaining $N-n$ events. Let us state and prove a representation theorem for this problem of finite inference, and then discuss the meaning and implications of each of its lines.

**Theorem 3.12.** Suppose you regard the events $E_1, \ldots, E_N$ exchangeably, and let $a$, $A$, and $n$ be any integers satisfying $0 \leqslant a \leqslant n \leqslant N$, $a \leqslant A \leqslant N$, and $n-a \leqslant N-A$, with $P(S_n = a) > 0$. Then coherency requires that for any permutation of the subscripts, your inferential conditional probabilities satisfy

$$
\begin{aligned}
P(E_{n+1}E_{n+2}\cdots E_{n+A-a}\tilde{E}_{n+A-a+1}\tilde{E}_{n+A-a+2}\cdots \tilde{E}_N | E_1E_2\cdots E_a\tilde{E}_{a+1}\tilde{E}_{a+2}\cdots\tilde{E}_n)\\
= P[E_{n+1}E_{n+2}\cdots E_{n+A-a}\tilde{E}_{n+A-a+1}\tilde{E}_{n+A-a+2}\cdots\tilde{E}_N|(S_n=a)]\\
= P(S_N = A|S_n = a)/{}^{(N-n)}C_{(A-a)}\\
= \frac{{}^{A}C_a\,{}^{(N-A)}C_{(n-a)}P(S_N=A)/{}^{(N-n)}C_{(A-a)}}{\sum_A^{N-(n-a)}{}^{A}C_a\,{}^{(N-A)}C_{(n-a)}\,P(S_N=A)}.
\end{aligned}
$$

*Proof.* The first equality follows immediately from the coherency requirement that

$$
\begin{aligned}
P(E_{n+1}E_{n+2}\cdots E_{n+A-a}\tilde{E}_{n+A-a+1}\tilde{E}_{n+A-a+2}\cdots\tilde{E}_N|E_1E_2\cdots E_a\tilde{E}_{a+1}\tilde{E}_{a+2}\cdots\tilde{E}_n)\\
= \frac{P(E_1E_2\cdots E_a\tilde{E}_{a+1}\tilde{E}_{a+2}\cdots\tilde{E}_nE_{n+1}E_{n+2}\cdots E_{n+A-a}\tilde{E}_{n+A-a+1}\tilde{E}_{n+A-a+2}\cdots\tilde{E}_N)}{P(E_1E_2\cdots E_a\tilde{E}_{a+1}\tilde{E}_{a+2}\cdots\tilde{E}_n)}\\
= \frac{P((S_n=a)E_{n+1}E_{n+2}\cdots E_{n+A-a}\tilde{E}_{n+A-a+1}\tilde{E}_{n+A-a+2}\cdots\tilde{E}_N)/{}^{n}C_a}{P(S_n=a)/{}^{n}C_a},
\end{aligned}
$$

this final line deriving from the previous one by counting (in both the numerator and denominator) the number of permutations of subscripts among the first $n$ multiplicands that yield events supporting $S_n = a$.

The second equality in the theorem follows from the first, along with the fact that the product event $[(S_N = A)(S_n = a)]$ is identical to the sum of events, $\sum^*[(S_n = a)\ E_{n+1}E_{n+2}\cdots E_{n+A-a}\tilde{E}_{n+A-a+1}\tilde{E}_{n+A-a+2}\cdots\tilde{E}_N]$, where the summation $\sum^*$ runs over all distinct permutations of the subscripts of the multiplicand events $E_{n+i}$. There are ${}^{(N-n)}C_{(A-a)}$ such permutations.

The final equality in the theorem derives from applying Bayes' theorem to the conditional probability $P(S_N = A|S_n = a)$, using Corollary 3.11.1. It provides a computational formula for the inferential probability in question, depending only on the probabilities $P(S_N = a), P(S_N = a+1), \ldots, P(S_N = a+N-n)$. □

The first equality in Theorem 3.12 states that the inference regarding $E_{n+1}, \ldots, E_N$ on the basis of $E_1, \ldots, E_n$ depends on the conditioning events only through their sum. This interesting result introduces a very important concept in statistics which we study in detail in the next subsection.

The second equality of Theorem 3.12 points out that if you regard $N$ events exchangeably, then conditional on the values of any $n$ of them, you must regard the remaining $N - n$ of them exchangeably; for your conditional probability for any sequence of the remaining events is constant over permutations of its subscripts. The third equality merely provides a computational formula.

### 3.9.1 Sufficient Statistics of Fixed Dimension

If you regard $N$ events exchangeably, then the inference you assert about any $N - n$ of these events on the basis of observing the values of $n$ of them does not depend at all on the order in which the successes were observed among the $n$ events. Your conditional probability is a function of these $n$ observations only through their sum. As far as your inference is concerned, all the information about events $E_{n+1}, \ldots, E_N$ that is contained in the entire collection of individual observations $E_1, \ldots, E_n$ is also contained in the mere observation of their sum, $S_n$.

This concept is so important and has such an important generalization in statistical theory that we make a special terminology to describe it. To preview this terminology, we say that if you regard $N$ events exchangeably then you must regard the sum of any $n$ of these events as a sufficient statistic for your inference about the remaining $N - n$ events. The number of conditioning observed events and their sum constitute sufficient statistics of fixed dimension. Let us proceed with some formal definitions.

***Definition 3.9.*** Let $\mathbf{X}_N$ be any vector of $N$ quantities. Any quantity defined as a function of $\mathbf{X}_N = (X_1, \ldots, X_N)^{\mathrm{T}}$ is called a *statistic*: $S \equiv S(\mathbf{X}_N)$. A vector of $K$ statistics, $\mathbf{S}_K(\mathbf{X}_N) = (S_1(\mathbf{X}_N), \ldots, S_K(\mathbf{X}_N))$, is called a *statistic of dimension K*. □

For example, the sum of $N$ quantities, which we have been denoting by $S_N$, is a statistic. Similarly, the average of $N$ quantities, which we denote by $\bar{X}_N = N^{-1}S_N$, is a statistic, as are:

$\max_{i=1,\ldots,N} X_i$, the maximum of the quantities;
$\sum_{i=1}^{N}(X_i - \bar{X}_N)^2$, the sum of their squared deviations about their average;
$\max_{i=1,\ldots,N}|X_i - \bar{X}_N|$, their maximum absolute deviation from their average.

Notice, too, that the quantities $X_1, \ldots, X_N$ themselves are each statistics, since formally they are describable via the functions $S_i(\mathbf{X}_N) \equiv X_i$. Thus, the quantity vector $\mathbf{X}_N$ itself is a statistic of dimension $N$: $\mathbf{S}_N(\mathbf{X}_N) \equiv \mathbf{X}_N$. In general, you should be recognizing that a statistic is a functional summary of some data.

How good is the summary? The concept of a *sufficient statistic* represents what we consider to be an ideal property of a summary. Suppose there are $N$ quantities that interest you, and you can observe $n$ of them. If you judge that all the information contained in the $n$ individual observations, $\mathbf{X}_n$, relevant to your inference about the remaining $N-n$ quantities, is also contained in the summary statistic, $S(\mathbf{X}_n)$, we say you regard the summary statistic as *sufficient* for your inference about $\mathbf{X}_N$. Let us record this definition formally.

***Definition 3.10.*** Let $\mathbf{X}_N$ be a vector of any $N$ quantities, with realm $\mathscr{R}(\mathbf{X}_N)$, and let $\mathbf{X}_n$ be a vector of any $n$ components of $\mathbf{X}_N$. You are said to *regard $S(\mathbf{X}_n)$ as a sufficient statistic* for inference about $\mathbf{X}_N$ if for any $n<N$ and any vector $\mathbf{x}_N=(x_1,\ldots,x_n,x_{n+1},\ldots,x_N)\in\mathscr{R}(\mathbf{X}_N)$, you assert

$$P[(X_{n+1}=x_{n+1})(X_{n+2}=x_{n+2})\cdots(X_N=x_N)|(\mathbf{X}_n=\mathbf{x}_n)]$$
$$=P[(X_{n+1}=x_{n+1})(X_{n+2}=x_{n+2})\cdots(X_N=x_N)|(S(\mathbf{X}_n)=S(\mathbf{x}_n))].$$

You are said to regard the vector of statistics $\mathbf{S}_K(\mathbf{X}_n)=(S_1(\mathbf{X}_n),\ldots,S_K(\mathbf{X}_n))$ as a *sufficient statistic of (fixed) dimension* $K$ for inference about $\mathbf{X}_N$ if for any value of $n$ satisfying $1\leqslant n<N$, and any vector $\mathbf{x}_N\in\mathscr{R}(\mathbf{X}_N)$, you assert

$$P[(X_{n+1}=x_{n+1})(X_{n+2}=x_{n+2})\cdots(X_N=x_N)|(\mathbf{X}_n=\mathbf{x}_n)]$$
$$=P[(X_{n+1}=x_{n+1})(X_{n+2}=x_{n+2})\cdots(X_N=x_N)|(\mathbf{S}_K(\mathbf{X}_n)=\mathbf{S}_K(\mathbf{x}_n))]. \quad \square$$

Two important aspects of the concept of sufficient statistics are worth stressing. First is that the sufficiency of a statistic (or of a vector of $K$ statistics) for inference about a vector of quantities is defined relative to your knowledge (opinion, judgment) about the quantities. Other people, who may not regard the quantities in the same way you do, need not regard "your" sufficient summary of the data as sufficient for their inferences. Fair enough.

Second to note is that the $n$ observed quantities, $\mathbf{X}_n$, always constitute a sufficient statistic of dimension $n$ that is deemed sufficient for inference about $\mathbf{X}_N$ by everyone. The drawback of the entire observed quantity vector as a sufficient statistic is that its dimension is just as large as the data itself. It amounts to no summarizing of the data at all. A major concern of the theory of inferential statistics is to identify small-sized, yet sufficient summaries of large numbers of observations. Thus, the second admirable quality of a vector of sufficient statistics is that it have fixed dimension . Then the summary provided by the sufficient statistics reduces the size of the information contained in the $n$ quantities to the size of merely $K$ quantities, no matter how large $n$ might be.

For now, let us conclude this discussion by restating as a corollary the feature of conditional probabilities we learned in Theorem 3.12, using the language of sufficiency. We return to the concept of sufficient statistics of fixed dimension in our discussion of partial exchangeability (Section 3.12) and in Chapter 5.

**Corollary 3.12.1.** If you regard the events $E_1,\ldots,E_N$ exchangeably, then, for any $n<N$, you regard the number of conditioning events, $n$, and their sum, $S_n$, as sufficient for your inference about the remaining $N-n$ events.

This corollary is merely an application of the definitions of a statistic and of a sufficient statistic to the first two equalities stated in Theorem 3.12. The converse of the corollary is also true: if you regard the sum of any $n < N$ events $\mathbf{E}_N$ as sufficient for your inference about the remaining $N - n$ events, then you must regard the events $\mathbf{E}_N$ exchangeably. (You may prove this yourself.)

The remainder of this section and the next one present some technical analysis that is not appropriate for a beginning student. A beginner should move to Section 3.11.

### 3.9.2 The Algebraic Transformation of the Unit Simplex $S^N$ to $S^n$

Let us return to examine more closely a technical feature of Theorem 3.11 which we neglected temporarily. The theorem identified the algebraic relationship between your probability distribution for the sum of $n$ events chosen from among $N$ events that you regard exchangeably and your distribution for the sum of the $N$ events themselves. Using our now familiar notation $\mathbf{p}_N = (p_{0,N}, p_{1,N}, \ldots, p_{N,N})^{\mathrm{T}} \equiv [P(S_N = 0), \ldots, P(S_N = N)]^{\mathrm{T}}$ and $\mathbf{p}_n = (p_{0,n}, p_{1,n}, \ldots, p_{n,n})^{\mathrm{T}} \equiv P[(S_n = 0), \ldots, P(S_n = n)]^{\mathrm{T}}$, the theorem stated that your opinion about the sum of any $n$ of the $N$ quantities must satisfy the linear equations

$$p_{a,n} = \sum_{A=a}^{N-(n-a)} [{}^{A}C_{a}\,{}^{(N-A)}C_{(n-a)}/{}^{N}C_{n}]\,p_{A,N} \qquad \text{for } a = 0, 1, 2, \ldots, n,$$

these equations being linear in the components of $\mathbf{p}_N$.

In matrix form these equations specify a *linear* transformation matrix $\mathbf{T}_{(n+1),(N+1)}$: $\mathbf{S}^N \rightarrow \mathbf{S}^n$ by the equation $\mathbf{p}_n = \mathbf{T}_{(n+1),(N+1)}\mathbf{p}_N$, where the components

$$\begin{aligned} T_{a,A} &= [{}^{A}C_{a}\,{}^{(N-A)}C_{(n-a)}/{}^{N}C_{n}] && \text{for } a = 0, 1, \ldots, n \quad \text{and } A = a, \ldots, N-(n-a), \\ &= 0 && \text{otherwise.} \end{aligned}$$

It is worthwhile to write out a few components of the matrix $\mathbf{T}_{(n+1),(N+1)}$ in a small-sized case and to become familiar with some of its properties. The following matrix displays the components of $\mathbf{T}_{5,9}$ without their common divisor, which is ${}^{8}C_{4}$. In general, the lower left corner of the matrix [below the main diagonal vector $(T_{0,0}, T_{1,1}, \ldots, T_{n,n})$] and the upper right corner of the matrix [above the top diagonal vector $(T_{0,n}, T_{1,(n+1)}, \ldots, T_{n,N})$] are filled with 0's. The columns of the matrix sum to 1, for they specify the conditional probability distributions for $S_n$ conditioned on $(S_N = A)$ for the various possible values of $A$. (Remember this displayed matrix requires division by ${}^{8}C_{4}$.)

$$\begin{pmatrix} {}^{0}C_{0}{}^{8}C_{4} & {}^{1}C_{0}{}^{7}C_{4} & {}^{2}C_{0}{}^{6}C_{4} & {}^{3}C_{0}{}^{5}C_{4} & {}^{4}C_{0}{}^{4}C_{4} & 0 & 0 & 0 & 0 \\ 0 & {}^{1}C_{1}{}^{7}C_{3} & {}^{2}C_{1}{}^{6}C_{3} & {}^{3}C_{1}{}^{5}C_{3} & {}^{4}C_{1}{}^{4}C_{3} & {}^{5}C_{1}{}^{3}C_{3} & 0 & 0 & 0 \\ 0 & 0 & {}^{2}C_{2}{}^{6}C_{2} & {}^{3}C_{2}{}^{5}C_{2} & {}^{4}C_{2}{}^{4}C_{2} & {}^{5}C_{2}{}^{3}C_{2} & {}^{6}C_{2}{}^{2}C_{2} & 0 & 0 \\ 0 & 0 & 0 & {}^{3}C_{3}{}^{5}C_{1} & {}^{4}C_{3}{}^{4}C_{1} & {}^{5}C_{3}{}^{3}C_{1} & {}^{6}C_{3}{}^{2}C_{1} & {}^{7}C_{3}{}^{1}C_{1} & 0 \\ 0 & 0 & 0 & 0 & {}^{4}C_{4}{}^{4}C_{0} & {}^{5}C_{4}{}^{3}C_{0} & {}^{6}C_{4}{}^{2}C_{0} & {}^{7}C_{4}{}^{1}C_{0} & {}^{8}C_{4}{}^{0}C_{0} \end{pmatrix}$$

## PROBLEMS

1. Construct the matrices $\mathbf{T}_{3,4}$, $\mathbf{T}_{3,5}$, $\mathbf{T}_{3,6}$, and $\mathbf{T}_{3,7}$. Once you have understood the idea, look ahead to Table 3.1 in Section 3.10.2 to notice that the columns printed there are the columns of these matrices, except for the first and last columns, which are $(1, 0, 0)^{\mathrm{T}}$ and $(0, 0, 1)^{\mathrm{T}}$ in every matrix.

2. Show that $\mathbf{T}_{(n+1),(N+1)} = \prod_{j=1}^{N-n} \mathbf{T}_{(n+j),(n+j+1)}$. *Hint*: Don't try to grind this out algebraically. Merely think about what the transformation matrix does in any instance.

## 3.10 (PRELUDE) BARYCENTRIC COORDINATES FOR THE UNIT SIMPLEX

We have discovered algebraically the importance of the unit simplex

$$\mathbf{S}^N = \left\{ \mathbf{p}_N = (p_{0N}, p_{1N}, \ldots, p_{NN})^{\mathrm{T}} \,\middle|\, \sum_{i=1}^{N} p_{iN} = 1 \qquad p_{iN} \geqslant 0 \right\},$$

for characterizing any coherent opinion that regards the events $\mathbf{E}_N$ exchangeably. To deepen our understanding of the judgment of exchangeability, it is helpful to become familiar with the geometrical representation of such a simplex, using what is called a "barycentric coordinate system." First, notice that any member of the unit simplex can be represented as a linear combination of the vertices of the simplex, $\mathbf{e}_{0,N}, \mathbf{e}_{1,N}, \ldots, \mathbf{e}_{N,N}$, where $\mathbf{e}_{i,N}$ denotes column $i+1$ of the identity matrix $\mathbf{I}_{N+1}$. A generic simplex member, $\mathbf{p}_N$, can be expressed as $\mathbf{p}_N = p_{0,N}\mathbf{e}_{0,N} + p_{1,N}\mathbf{e}_{1,N} + \cdots + p_{N,N}\mathbf{e}_{N,N}$. The coefficients $p_{i,n}$ are called the *barycentric coordinates* of $\mathbf{p}_N \in \mathbf{S}^N$.

A graphical depiction of an $N$-dimensional unit simplex is achieved through the convex hull of the unit vectors $\mathbf{e}_{0,N}, \mathbf{e}_{1,N}, \ldots, \mathbf{e}_{N,N}$. For example, when $N = 2$, these vectors are $\mathbf{e}_{0,2} = (1, 0, 0)$, $\mathbf{e}_{1,2} = (0, 1, 0)$, and $\mathbf{e}_{2,2} = (0, 0, 1)$. Figure 3.4 displays these three vectors in three-dimensional space, and you can see that the three of them lie on a plane in two dimensions. The convex hull of these three points is the equilateral triangle in Figure 3.5. This is the same triangle that is situated in three-dimensional space in Figure 3.4, merely laid flat on the page for easy viewing. The vertices of this triangle are the points denoted by the vectors (1, 0, 0), (0, 1, 0), and (0, 0, 1).

The coordinates of any point within the two-dimensional simplex (any point within the equilateral triangle shown in Figure 3.5) can be determined by the following three-step procedure, portrayed in Figure 3.6.

1. Draw the lines that bisect each vertex angle and continue through the triangle as the perpendicular bisector of the opposite edge.

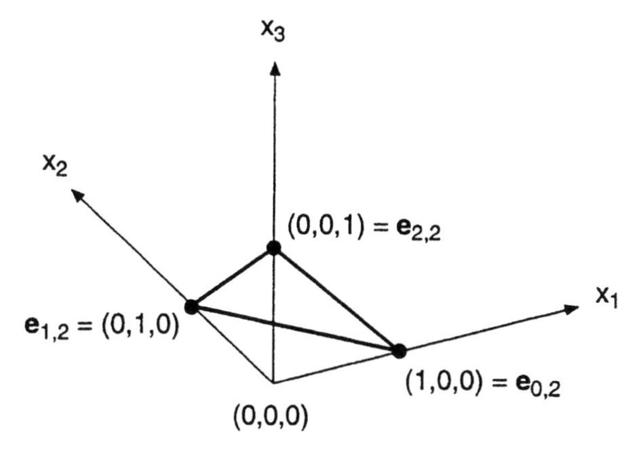


**Figure 3.4.** The basis vectors for $\mathbf{S}^2$, $\mathbf{e}_{02} = (1, 0, 0)$, $\mathbf{e}_{12} = (0, 1, 0)$, and $\mathbf{e}_{22} = (0, 0, 1)$, shown as points in three-dimensional space.

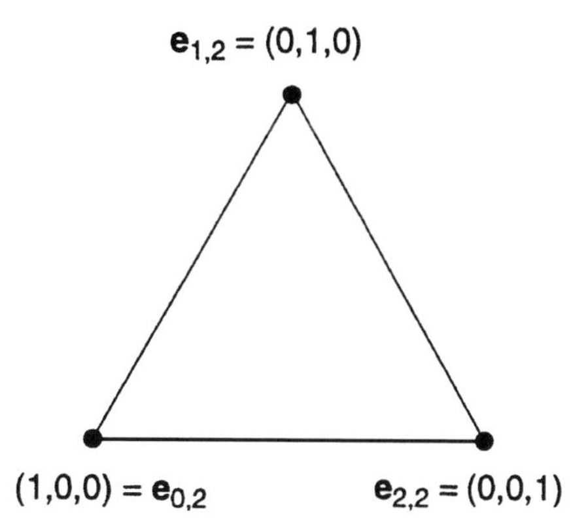


**Figure 3.5.** The equilateral triangle that constitutes the convex hull of these three vectors.

2. Project the point in question, orthogonally, to each of these perpendicular bisectors.
3. For each such projection, determine the proportion of the bisector that is covered by the line segment connecting the edge opposite the vertex to the projected point. This proportion is the numerical value of the coordinate of $\mathbf{p}_N$ that multiplies the associated vertex in the equation

$$\mathbf{p}_N = p_{0,N}\, \mathbf{e}_{0,N} + p_{1,N}\, \mathbf{e}_{1,N} + \cdots + p_{N,N}\, \mathbf{e}_{N,N}.$$

An example depicting the point (.12, .58, .30) appears in Figure 3.6. In studying this figure, you should see that the center point of the equilateral triangle represents the vector $\mathbf{p}_2 = (1/3, 1/3, 1/3)$. Review your understanding as well by studying the coordinates for the midpoints of each edge of the triangle.

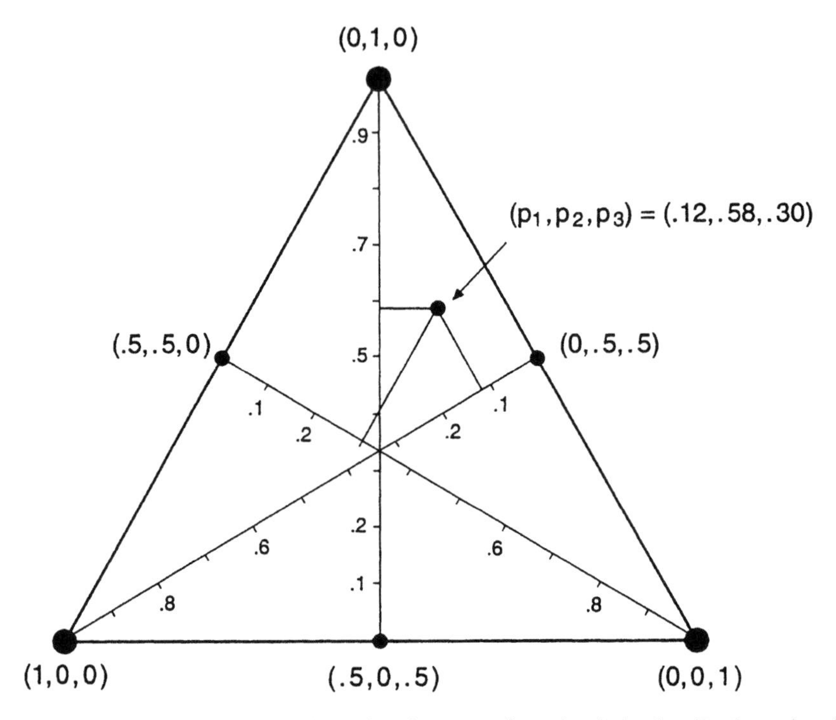


**Figure 3.6.** Barycentric coordinates for points in a two-dimensional simplex. To determine the coordinates of a point, project the point orthogonally to each line that bisects a vertex angle and bisects its opposite edge. The numerical value of each coordinate is the proportion of this line covered from the edge to the projection point.

## PROBLEMS

1. Pick some other point in Figure 6 and determine its coordinates.

2. Use this procedure in reverse to determine where in the Figure lies the point associated with $\mathbf{p}_2 = (.61, .13, .26)$. Try $(.03, .41, .56)$ and $(.03, .11, .86)$.

A three-dimensional unit simplex includes all convex combinations of the unit vectors $\mathbf{e}_{0,3} = (1, 0, 0, 0)$, $\mathbf{e}_{1,3} = (0, 1, 0, 0)$, $\mathbf{e}_{2,3} = (0, 0, 1, 0)$, and $\mathbf{e}_{3,3} = (0, 0, 0, 1)$. We display one in the next section. Depicted in space, this simplex constitutes a regular tetrahedron whose every face is an equilateral triangle. The coordinates of any point $\mathbf{p}_3 = (p_{0,3}, p_{1,3}, p_{2,3}, p_{3,3})$ in this simplex are determined by the proportions of the lines connecting the vertices to the centers of their opposing faces that are covered by the line connecting the center of this face with the projection of the point onto this line. The center point of the three-dimensional simplex is the vector $(1/4, 1/4, 1/4, 1/4)$. If $p_{33} = 0$, then the vector $(p_{03}, p_{13}, p_{23}, 0)$ lies within the triangular face opposing the basis vector $\mathbf{e}_{33} = (0, 0, 0, 1)$. Thus, it lies appropriately in a two-dimensional simplex.

## 3.10 A GEOMETRICAL REPRESENTATION OF ALL COHERENT OPINIONS THAT REGARD TWO EVENTS EXCHANGEABLY

In this section we take a geometrical excursion to see how the judgment to regard two events exchangeably fits into the class of all coherent assessments of the partition generated by two events. Then within the class of all judgments that regard these two events exchangeably, we make an exhaustive study of conditions under which such a judgment can be extended to regard three or more events exchangeably. By the conclusion of this section, we shall understand de Finetti's representation theorem for exchangeable assertions that are infinitely exchangeably extendible.

Consider the partition generated by two logically independent events, $E_1$ and $E_2$. Its constituents are the product events $\tilde{E}_1\tilde{E}_2$, $\tilde{E}_1E_2$, $E_1\tilde{E}_2$, and $E_1E_2$. The collection of all coherent prevision assertions for these four events is the simplex $\mathbf{S}^3 = \{\mathbf{p}_3 = (p_0, p_1, p_2, p_3) | \sum_{i=0}^{3} p_i = 1, \text{and every } p_i \geqslant 0\}$, where $p_i$ is a prevision for the $i$th constituent in the order listed above: $p_0 = P(\tilde{E}_1\tilde{E}_2)$, $p_1 = P(\tilde{E}_1E_2)$, $p_2 = P(E_1\tilde{E}_2)$, and $p_3 = P(E_1E_2)$. Any coherent prevision assertion for these four constituents is representable by a unique vector $\mathbf{p}_3 \in \mathbf{S}^3$, and each vector $\mathbf{p}_3 \in \mathbf{S}^3$ represents a coherent prevision for the four constituents.

The unit simplex $\mathbf{S}^3$ is representable by the regular tetrahedron shown in Figure 3.7. Any coherent prevision vector for the constituents can be written as a convex combination of the vertices (unit vectors) of the simplex:

$$(p_0, p_1, p_2, p_3) = p_0(1, 0, 0, 0) + p_1(0, 1, 0, 0) + p_2(0, 0, 1, 0) + p_3(0, 0, 0, 1).$$

If the events $E_1$ and $E_2$ are regarded exchangeably, then the vector of probabilities for the constituents of this partition must have its middle two components identical: $p_1 = P(\tilde{E}_1E_2) = P(E_1\tilde{E}_2) = p_2$. Let us denote the set of all coherent prevision assertions that assess $E_1$ and $E_2$ exchangeably by

$$\mathscr{E} = \left\{ (p_0, p_1, p_2, p_3) \,\middle|\, \sum_{i=0}^{3} p_i = 1, p_1 = p_2, \quad \text{and each } p_i \geqslant 0 \right\} \subset \mathbf{S}^3.$$

One vector that meets this criterion is $(0, 1/2, 1/2, 0) \in \mathscr{E}$. It is identified explicitly in Figure 3.7 as the midpoint of one edge of the pyramid. Two other vectors that have this property are the vertices of the pyramid, $(1, 0, 0, 0)$ and $(0, 0, 0, 1)$. Furthermore, any vector in the pyramid that has this property of identical middle components can be written as a convex combination of these three vectors. For

$$(a, b, b, c) = a(1, 0, 0, 0) + 2b(0, 1/2, 1/2, 0) + c(0, 0, 0, 1) \qquad \text{with } a + 2b + c = 1.$$

Thus, the set of all coherent exchangeable assessments of $E_1$ and $E_2$ is described geometrically by the plane of points within the simplex that connects these three vectors.

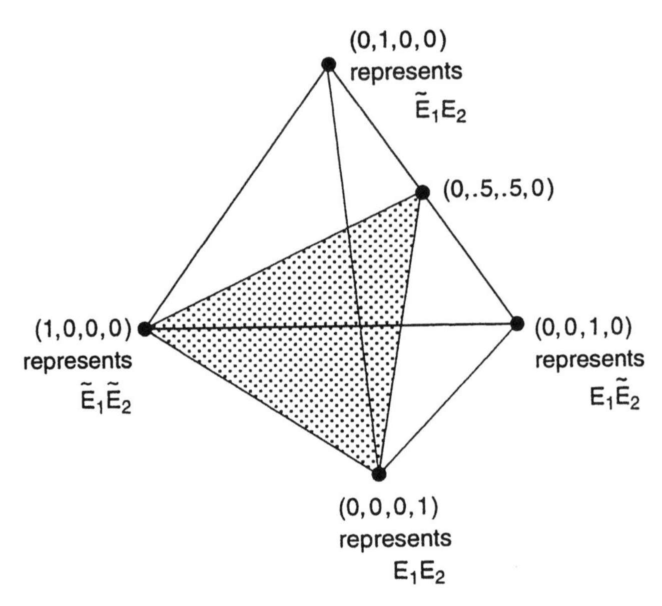


**Figure 3.7.** The three-dimensional simplex $\mathbf{S}^3$ of all coherent prevision vectors for the partition generated by two logically independent events. The dotted plane is the set $\mathscr{E}$ of coherent previsions that assess the two events exchangeably.

If you regard $E_1$ and $E_2$ exchangeably, your opinions about $E_1$ and $E_2$ are representable by a point that lies somewhere on the dotted plane in Figure 3.7 that slices the simplex in half (or in the spirit of prevision bounds, by some convex body within this plane).

Let us pull out this plane representing exchangeable assessments of $\mathbf{E}_2$ and examine it. The plane appears in Figure 3.8, embedded into an equilateral triangle for comparative purposes. The smaller triangle in the Figure is composed of 4-tuples, but they are constrained twice: each 4-tuple has its middle two components equal; and the four components sum to 1. Thus, this triangle can be interpreted as a two-dimensional simplex. The two equal components represent the common prevision for two distinct product events, $\tilde{E}_1 E_2$ and $E_1 \tilde{E}_2$. Thus, this number cannot exceed 1/2. If we now agree to use the notation $(p_0, p_1/2 = p_1/2, p_2)$ to denote 4-tuples in this triangle, we use Theorem 3.9 Section 3.8.2, which characterized exchangeable assessments of $N$ events by previsions for the partition defined by their sum. For using this notation, we have

$$p_0 = P(\tilde{E}_1 \tilde{E}_2) = P(S_2 = 0),$$

$$p_1 = P(\tilde{E}_1 E_2) + P(E_1 \tilde{E}_2) = P(S_2 = 1),$$

$$p_2 = P(E_1 E_2) = P(S_2 = 2).$$

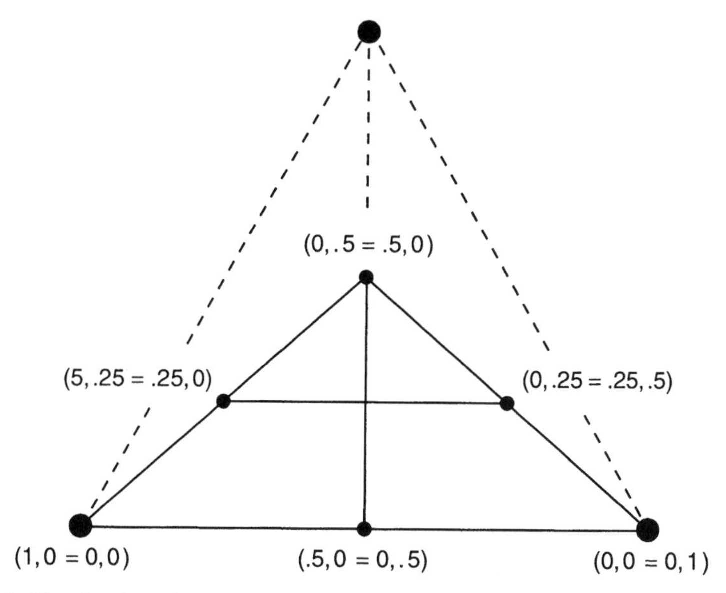


**Figure 3.8.** The triangle $\mathscr{E}$ of points that represent exchangeable assessments of $\mathbf{E}_2$. (This triangle has been removed from the simplex $\mathbf{S}^3$ displayed in Figure 3.7.) Each point within this triangle specifies a triple, denoted by $(P(\tilde{E}_1\tilde{E}_2), P(\tilde{E}_1E_2) = P(E_1\tilde{E}_2), P(E_1E_2))$, or $(p_0, p_1/2 = p_1/2, p_2)$, where $p_i$ denotes $P(S_2 = i)$. The midpoint of the triangle is (.25, .25 = .25, .25). The external equilateral triangle is depicted merely for comparative purposes.

Each horizontal line within the triangle of exchangeable assessments represents all the points that have some particular value for their interior two components. The points from the left edge to the right edge of any horizontal line within the triangle exhaust the coherent possibilities of the form $(p_0, p_1/2 = p_1/2, 1 - p_0 - p_1)$ as $p_0$ ranges downward from $1 - p_1$ to 0. For example, the horizontal line that bisects the altitude of the exchangeable triangle has endpoints (.5, .25 = .25 ,0) and (0, .25 = .25, .5), and the midpoint (.25, .25 = .25, .25).

### 3.10.1 Points below the Parabola: Mixtures of Stochastically Independent Assessments

The next observation we can make is that the triangle $\mathscr{E}$ contains all points of the form $[(1-\pi)^2, (1-\pi)\pi = \pi(1-\pi), \pi^2]$, as long as $\pi \in [0, 1]$. These points trace out a parabola within $\mathscr{E}$ that has its apex at (.25, .25 = .25, .25) and passes through the two base vertices of the triangle (1, 0 = 0, 0) and (0, 0 = 0, 1), as depicted in Figure 3.9. To see that the points on this arc represent those specified algebraically, note that if $x$ is any number in the interval [0, .25], for example $x = 3/16$, and if $x$ is identified with $\pi(1-\pi) \equiv x$ the solutions to this quadratic equation in $\pi$ are $\pi^+ = 1/2 + \sqrt{1-4x}/2$ and $\pi^- = 1/2 - \sqrt{1-4x}/2$. These two values for $\pi$ then specify two points on the parabola of the form $[(1-\pi)^2, x = x, \pi^2]$. In the case of

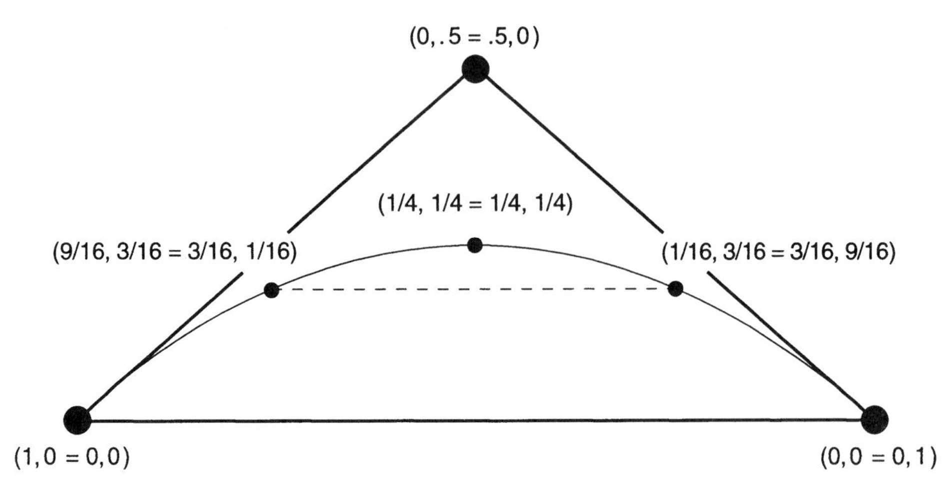


**Figure 3.9.** The parabola of points $[(1-\pi)^2, (1-\pi)\pi = \pi(1-\pi), \pi^2]$ shown within the triangle $\mathscr{E}$ of previsions representing exchangeable assessments of $\mathbf{E}_2$.

$x = 3/16$, the solutions are $\pi^+ = 1/4$ and $\pi^- = 3/4$. These yield the two 4-tuples $(9/16, 3/16 = 3/16, 1/16)$ and $(1/16, 3/16 = 3/16, 9/16)$, which are identified explicitly as two of the points on the parabola in Figure 3.9.

Further examination of the vectors $[(1-\pi)^2, (1-\pi)\pi = \pi(1-\pi), \pi^2]$ yields the awareness that they correspond to *stochastically independent* assessments of $E_1$ and $E_2$ with identical probabilities. For if your $P(E_1) = P(E_2) = \pi$ and your $P(E_1E_2) = P(E_1)P(E_2)$, then your $P(E_1E_2) = \pi^2$, while $P(\tilde{E}_1E_2) = P(E_1\tilde{E}_2) = \pi(1-\pi)$ and $P(\tilde{E}_1\tilde{E}_2) = (1-\pi)^2$. Thus, your $P(E_i|E_j) = P(E_i)$. This parabola's lying within $\mathscr{E}$ represents geometrically the fact that independent and identical assessment of two events is a special case of the judgment to regard them exchangeably.

Now the theory of convexity implies that any point in the triangle below the inverted parabola can be represented as a convex mixture of points on the parabola itself. That is, any point within the triangle lying below the parabola can be represented in the form

$$\left(\int_0^1 (1-\pi)^2\, dM(\pi), \int_0^1 \pi(1-\pi)\, dM(\pi) = \int_0^1 \pi(1-\pi)\, dM(\pi), \int_0^1 \pi^2\, dM(\pi)\right),$$

where the mixing function $M(\cdot)$ is a nondecreasing function satisfying $M(a) = 0$ when $a < 0$, and $M(a) = 1$ when $a \geqslant 1$. The integrals in this representation are Stieltjes integrals, so $M$ may be continuous or a step function, or a combination of the two: when $M(\cdot)$ is a step function, these integrals reduce to the sums

$$\left(\sum_{i=1}^{N} (1-\pi_i)^2 m(\pi_i), \sum_{i=1}^{N} (1-\pi_i)\pi_i\, m(\pi_i) = \sum_{i=1}^{N} \pi_i(1-\pi_i) m(\pi_i), \sum_{i=1}^{N} \pi_i^2\, m(\pi_i)\right),$$

where $m(\pi_i)$ is the associated mass function; when $M(\cdot)$ is differentiable, with $dM(\pi) = m(\pi)\,d\pi$, the integrals reduce to the Riemann integrals

$$\left(\int_0^1 (1-\pi)^2 m(\pi)\,d\pi, \int_0^1 \pi(1-\pi)m(\pi)\,d\pi = \int_0^1 \pi(1-\pi)\,m(\pi)\,d\pi, \int_0^1 \pi^2 M(\pi)\,d\pi\right),$$

We study distribution functions such as $M(\cdot)$ in greater detail in Chapter 5.

The representation of any point below the parabola in this form is not unique. For example, the vector $(5/16, 3/16 = 3/16, 5/16)$ that lies on the midpoint of the dotted line through the parabola in Figure 3.9 can be represented either as the simple average of the endpoints of that line, viz.,

$$(1/2)(9/16, 3/16 = 3/16, 1/16) + (1/2)(1/16, 3/16 = 3/16, 9/16),$$

corresponding to the mixture function $M(\pi) = 1/2(1/4 \leqslant \pi < 3/4) + (\pi \geqslant 3/4)$, or as a weighted average of the points at the two base vertices of the triangle and the point at the apex of the parabola, viz.,

$$(1/8)(1, 0 = 0, 0) + (3/4)(1/4, 1/4 = 1/4, 1/4) + (1/8)(0, 0 = 0, 1),$$

corresponding to the mixture function $M(\pi) = (1/8)(\pi \geqslant 0) + (7/8)(1/2 \leqslant \pi < 1) + (\pi \geqslant 1)$. In fact, there are many such representations for any point beneath the parabola.

In summary, prevision assertions corresponding to points beneath the parabola can be represented as convex "mixtures" of previsions that regard the events $E_1$ and $E_2$ identically and independently with $P(E_1) = P(E_2) = \pi$, for various values of $\pi \in [0, 1]$.

### 3.10.2 Points above the Parabola: The Extendability of Exchangeable Assessments

What about the points above the parabola? We already know how we can represent them algebraically in terms of vectors from larger-dimensioned simplices, on account of Theorem 3.11, which holds for every point in the triangle: for each value of $0 \leqslant a \leqslant n \leqslant N$,

$$p_{a,n} = \sum_{A=a}^{N-(n-a)} [{}^{A}C_a \, {}^{(N-A)}C_{(n-a)} / {}^{N}C_n] p_{A,N},$$

or, as we had denoted this result in matrix form, $\mathbf{p}_n = \mathbf{T}_{(n+1),(N+1)} \mathbf{p}_N$, described in Section 3.9.2. But it will also be fruitful to identify the relationship between this representation and the form we now know to characterize points beneath the parabola:

$$p_{an} = {}^{n}C_a \int_0^1 \pi^a (1-\pi)^{n-a}\, dM(\pi).$$

Identifying this relationship requires that we define what is meant by opinions that regard $E_1$ and $E_2$ *exchangeably in such a way that the judgment of exchangeability can be coherently extended* from the class of events containing only themselves to a class containing $2 + K$ members. We have already seen in Example 1 of Section 3.8.2 that there are opinions regarding $E_1$ and $E_2$ exchangeably that cannot possibly regard these two exchangeably with any other quantities. Thus, the following definition contains a recognizable stringency.

***Definition 3.11.*** If you regard the events $\mathbf{E}_N$ exchangeably, and your prevision is characterized by the vector $\mathbf{p}_N = (p_{0,N}, \ldots, p_{N,N})^{\mathrm{T}} \in \mathbf{S}^N$, then *you regard them as exchangeably extendable to dimension* $N + K$ if there is some vector $\mathbf{p}_{N+K} = (p_{0,(N+K)}, \ldots, p_{(N+K),(N+K)}) \in \mathbf{S}^{N+K}$ that reduces to $\mathbf{p}_N \in \mathbf{S}^N$ via the linear transformation $\mathbf{T}_{(N+1),(N+K+1)}$ specified in Section 3.9.2. □

The extendability of your exchangeable assessment of $\mathbf{E}_N$ is a formal property of your assertion vector, $\mathbf{p}_N$. You need not identify any events that you do regard exchangeably with $\mathbf{E}_N$ in order for your opinions to be recognized as exchangeably extend*able* to dimension $N + K$. You may even avow that you regard $E_1, \ldots, E_N$ exchangeably among themselves but with no others, even though your opinions are exchangeably extendable. The extendability of your exchangeable assessment is defined by the character of your previsions *for the events* $\mathbf{E}_N$ *themselves*. The following examples will obviate this point.

***Example 1.*** Suppose you regard $E_1$ and $E_2$ exchangeably, and you assert $P(\tilde{E}_1\tilde{E}_2) = .1$, $P(\tilde{E}_1 E_2) = P(E_1\tilde{E}_2) = .1$, and $P(E_1 E_2) = .7$. These previsions would be implied by coherency if your previsions for these two and a third event, $E_3$, were $P(\tilde{E}_1\tilde{E}_2\tilde{E}_3) = .1$, $P(\tilde{E}_1\tilde{E}_2 E_3) = P(\tilde{E}_1 E_2\tilde{E}_3) = P(E_1\tilde{E}_2\tilde{E}_3) = 0$, $P(E_1 E_2\tilde{E}_3) = P(\tilde{E}_1 E_2 E_3) = P(E_1\tilde{E}_2 E_3) = .1$, and $P(E_1 E_2 E_3) = .6$. Thus, we would say you regard $E_1$ and $E_2$ as exchangeably extendable to dimension 3. But these specific exchangeable judgments regarding the three events are not exchangeably extendable to four events. Try it! Actually, your assessments of $E_1$ and $E_2$ are exchangeably extendable to any dimension. We shall learn why shortly. But if they are extended to these specific assertions shown for three events, then they cannot be coherently extended to four events. □

***Example 2.*** Reconsider Peter and Amy's prospective family which introduced the notion of exchangeability in Example 3 of Section 3.8. Peter's previsions for the gender order of a two-child family were exactly those specified in the preceding example. Thus, Peter regards the gender of the first two children as exchangeably extendable to dimension 3. Yet he and Amy are not even planning to have a third child with which he would regard the gender of the first two exchangeably! (Peter may even suspect that they would try for a third child only if they do not have a child of each sex among their first two, and if those two are both boys they may look into the possibility of "doing something about it" on the next try. Such an attitude would preclude his regarding the sexes of three children exchangeably, even though his asserted opinions about two children are exchangeably extendable.) □

## PROBLEM

1. Are the probabilities for the weights of the trucks in the orange juice convoy (Example 1 Section 3.8) exchangeably extendable to degree 4?

Let us now use the concept of the extendability of exchangeable assessments to complete our understanding of those points in the triangle of exchangeable previsions that lie above the inverted parabola.

Reviewing our definition in the context of two events, if you regard $E_1$ and $E_2$ as exchangeably extendable to dimension $N$ and your prevision for the partition they generate is characterized by $\mathbf{p}_2 = (p_{0,2}, p_{1,2}, p_{2,2})^{\mathrm{T}} \in \mathbf{S}^2$, then there must be some vector $\mathbf{p}_N = (p_{0,N}, \ldots, p_{N,N})^{\mathrm{T}}$ in $\mathbf{S}^N$ for which $\mathbf{p}_2 = \mathbf{T}_{3,(N+1)}\mathbf{p}_N$, where $\mathbf{T}_{3,(N+1)}$ is the matrix defined in Section 3.9.2.

One important property of a unit simplex is that it is spanned by its vertices, $\mathbf{e}_{0,N}, \mathbf{e}_{1,N}, \ldots, \mathbf{e}_{N,N}$, which are the column vectors of the identity matrix, $\mathbf{I}_{N+1}$. Since the transformation $\mathbf{S}^N$ to $\mathbf{S}^2$ via $\mathbf{T}_{3,(N+1)}$ is linear, the subset of vectors in $\mathbf{S}^2$ representing assessments that are exchangeably extendable to dimension $N$ must be spanned by the transformed vertices of $\mathbf{S}^N$, which are merely the column vectors of $\mathbf{T}_{3,(N+1)}$. To see this, notice that if $\mathbf{p}_2$ represents an assessment of $E_1$ and $E_2$ as exchangeably extendable to dimension $N$, then it must be expressible in the form

$$\mathbf{p}_2 = \mathbf{T}_{3,(N+1)}\mathbf{p}_N = \mathbf{T}_{3,(N+1)} \sum_{j=0}^{N} p_{j,N}\, \mathbf{e}_{jN} = \sum_{j=0}^{N} p_{j,N} \mathbf{T}_{3,(N+1)} \mathbf{e}_{j,N},$$

where $p_{0,N}, p_{1,N}, \ldots, p_{N,N}$ are nonnegative coefficients that sum to 1, and each vector $\mathbf{T}_{3,(N+1)}\mathbf{e}_{j,N}$ is a vertex of $\mathbf{S}^N$, transformed into $\mathbf{S}^2$. (The final equality in this line derives from the defining property of linear transformations: the transformation of any linear combination of vectors equals the same linear combination of the transformations of the vectors.) These transformed vertices are identical to the columns of the matrix $\mathbf{T}_{3,(N+1)}$.

Table 3.1 lists the columns of the $\mathbf{T}_{3,(N+1)}$ matrices associated with the assessments of $E_1$ and $E_2$ as exchangeably extendable to degree $N$ for values of $N = 3$, 4, 5, and 6. For each value of $N$ there would be $N+1$ transformed unit vectors. Neither the first nor the last column of $\mathbf{T}_{3,(N+1)}$ appear in the table for any $N$, since $\mathbf{T}_{3,(N+1)}\mathbf{e}_{0,N} = (1,0,0)^{\mathrm{T}}$ and $\mathbf{T}_{3,(N+1)}\mathbf{e}_{N,N} = (0,0,0)^{\mathrm{T}}$ for every value of $N$. (Refer to Section 3.9.2 to examine the transformation matrix there.) Thus, Table 3.1 lists only the remaining $N-1$ column vectors of $\mathbf{T}_{3,(N+1)}$ for each value of $N$. The column heading $\mathbf{Te}_{jN}$ denotes that column is the transformation of the basis vector $\mathbf{e}_{j,N}$ as it appears in $\mathbf{S}^2$.

Figure 3.10 identifies within $\mathbf{S}^2$ the vectors that appear in the columns of Table 3.1. For $N = 3$, 4, 5, and 6, the prevision vectors $(p_0, p_1/2 = p_1/2, p_2)$ that specify

**Table 3.1. Transformations of Unit Vectors from $\mathbf{S}^N$ into $\mathbf{S}^2$: $N = 3, 4, 5, 6$**

| | $\mathbf{Te}_{13}$ | $\mathbf{Te}_{23}$ | $\mathbf{Te}_{14}$ | $\mathbf{Te}_{24}$ | $\mathbf{Te}_{34}$ | $\mathbf{Te}_{15}$ | $\mathbf{Te}_{25}$ | $\mathbf{Te}_{35}$ | $\mathbf{Te}_{45}$ | $\mathbf{Te}_{16}$ | $\mathbf{Te}_{26}$ | $\mathbf{Te}_{36}$ | $\mathbf{Te}_{46}$ | $\mathbf{Te}_{56}$ |
|---|---|---|---|---|---|---|---|---|---|---|---|---|---|---|
| $p_{02}$ | 1/3 | 0 | 1/2 | 1/6 | 0 | 3/5 | 3/10 | 1/10 | 0 | 2/3 | 2/5 | 1/5 | 1/15 | 0 |
| $p_{12}$ | 2/3 | 2/3 | 1/2 | 2/3 | 1/2 | 2/5 | 3/5 | 3/5 | 2/5 | 1/3 | 8/15 | 3/5 | 8/15 | 1/3 |
| $p_{22}$ | 0 | 1/3 | 0 | 1/6 | 1/2 | 0 | 1/10 | 3/10 | 3/5 | 0 | 1/15 | 1/5 | 2/5 | 2/3 |

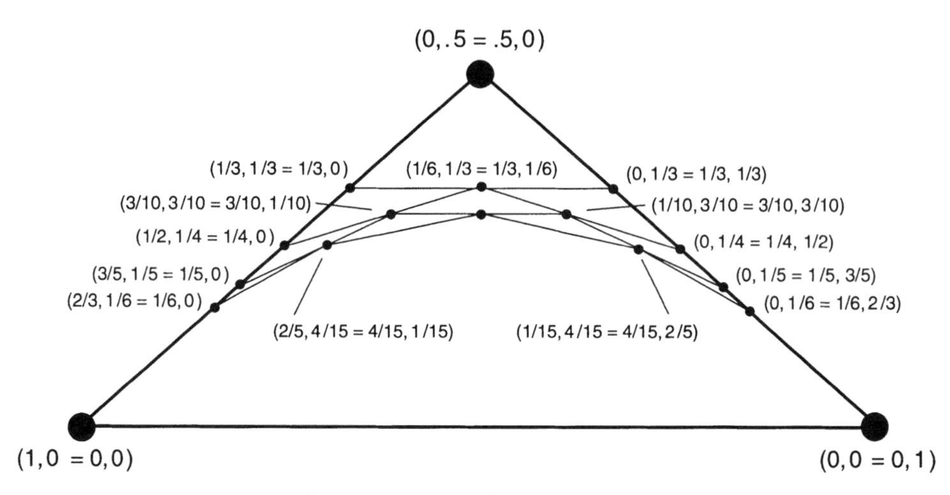


**Figure 3.10.** Basis vectors of $\mathbf{S}^N$ transformed into $\mathbf{S}^2$ via the transformation $\mathbf{T}_{3,(N+1)}$ for $N = 3, 4, 5$, and 6, yielding $[P(\tilde{E}_1\tilde{E}_2), P(\tilde{E}_1E_2) = P(E_1\tilde{E}_2), P(E_1E_2)]$.

$E_1$ and $E_2$ as exchangeably extendable to dimension $N$ are those within the convex hull of vectors $\mathbf{Te}_{1N}, \mathbf{Te}_{2N}, \ldots, \mathbf{Te}_{(N-1)N}$ that are shown in the table, along with the base vertices of the triangle, $(1, 0 = 0, 0)$ and $(0, 0 = 0, 1)$. Remember when associating these vectors with the points designated in the figure that the middle component, $p_{12}$, should be separated into two equal components. For example, we have been writing a vector like $\mathbf{Te}_{26}$ as $(2/5, 4/15 = 4/15, 1/15)$.

Figure 3.11 illustrates the shrinking of the convex sets of previsions that specify $E_1$ and $E_2$ as exchangeably extendable to dimension $N$ for increasing values of $N$. The convex hulls of transformed vertices are shaded in darker colors as $N$ increases. The upper tip of the triangle contains all the points representing assessments of exchangeability that are not extendable even to dimension 3. Thus, this tip of the triangle is not shaded at all. No prevision vectors with middle components exceeding $1/3$ are exchangeably extendable at all. At the opposite extreme, all points lying beneath the parabola of vectors $\{((1-\pi)^2, (1-\pi)\pi = \pi(1-\pi), \pi^2) \mid \pi \in [0, 1]\}$ represent assessments of $E_1$ and $E_2$ as infinitely exchangeably extendable. For any finite $N$ they are exchangeably extendable to degree $N$.

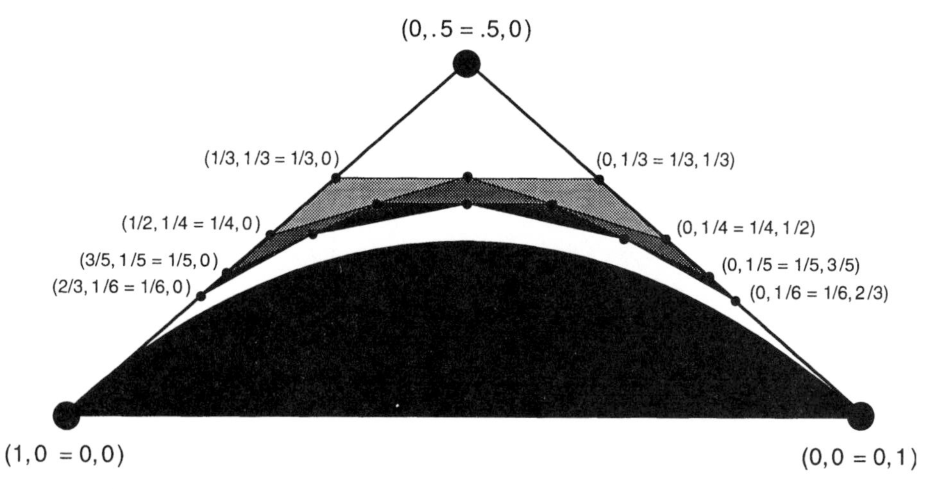


**Figure 3.11.** The triangle of previsions regarding $E_1$ and $E_2$ exchangeably. Regions shaded in darker color depict subregions of assessments that are exchangeably extendable to higher dimensions.

### 3.10.3 De Finetti's Representation Theorem for Infinitely Extendable Exchangeable Distributions

It is apparent from this graphical exposition that the family of assessments regarding $E_1$ and $E_2$ as exchangeably extendable to dimension $N$ converges onto the set of all the points bounded by the parabola of points $((1-\pi)^2, (1-\pi)\pi = \pi(1-\pi), \pi^2)$ as $N$, the dimension of extendability, increases. All points on or below the parabola are said to be *infinitely exchangeably extendable* since they are exchangeably extendable to any finite degree. This is the burden of de Finetti's representation theorem for infinitely extendable exchangeable judgments, which follows.

**Theorem 3.13.** If you regard the events $\mathbf{E}_N$ as infinitely exchangeably extendable, then your $P(S_N = a)$ can be represented for each $a \in \{0, 1, \ldots, N\}$ as

$$P(S_N = a) = {}^N C_a \int_0^1 \theta^a (1-\theta)^{(N-a)}\, dM(\theta)$$

for some distribution function $M(\cdot)$.

*Proof.* Suppose you regard $\mathbf{E}_N$ as exchangeably extendable to dimension $N+K$. Then for some vector $\mathbf{p}_{N+K} = (p_{0,(N+K)}, \ldots, p_{(N+K),(N+K)})^{\mathrm{T}} \in \mathbf{S}^{N+K}$, your probability $P(S_N = a)$ can be written as

$$\sum_{A=0}^{N+K} T_{(a+1),A}\, p_{A,(N+K)} = \sum_{A=0}^{N+K} [{}^A C_a\, {}^{(N+K-A)}C_{(N-a)} / {}^{(N+K)}C_N]\, p_{A,(N+K)}.$$

Multiplying and dividing this expression by ${}^N C_a$ yields the equivalent form

$$ {}^N C_a \sum_{A=0}^{N+K} \frac{A(A-1)\cdots(A-a+1)(N+K-A)(N+K-A-1)\cdots[N+K-A-(N-a-1)]}{(N+K)(N+K-1)\cdots(N+K-a+1)(N+K-a)(N+K-a-1)\cdots[N+K-(N-1)]} \times p_{A,(N+K)} $$

Now using the notation $\theta = A/(N+K)$ and $M_{N+K}(\theta) = \sum_{i=0}^{N+K} p_{i,(N+K)}(i \leqslant \theta(N+K))$, this summation can be written as the Stieltjes integral

$$ {}^N C_a \int_0^1 \frac{\theta[\theta - 1/(N+K)]\cdots[\theta-(a-1)/(N+K)](1-\theta)[1-\theta-1/(N+K)]\cdots[1-\theta-(N-a-1)/(N+K)]}{1[1-1/(N+K)]\cdots[1-(a-1)/(N+K)][1-a/(N+K)]\cdots[1-(N-1)/(N+K)]} \times dM_{N+K}(\theta) $$

$$ \to {}^N C_a \int_0^1 \theta^a (1-\theta)^{(N-a)}\, dM(\theta) $$

for some distribution function $M(\cdot)$, when $K \to \infty$.

The limiting conclusion is a consequence of the analytic result that the product of a finite number of multipliers converges uniformly to the limit of their products. (Notice there are only $N$ multipliers in the numerator and in the denominator of the integrand.) □

Diaconis and Freedman (1980c) present details of this result even more explicitly,identifying a precise bound on the closeness of the limiting result for any specific $N$ and $K$, as follows.

**Theorem 3.14.** If you regard $\mathbf{E}_N$ as exchangeably extendable to degree $N+K$, then for some monotonic function $M(\theta)$ $(0 \leqslant \theta \leqslant 1)$ with $M(0) \geqslant 0$ and $M(1) = 1$,

$$ \left| P(E_1 E_2 \cdots E_a \tilde{E}_{a+1} \cdots \tilde{E}_N) - \int_0^1 \theta^a (1-\theta)^{(N-a)}\, dM(\theta) \right| \leqslant c(N,k) \qquad (0 \leqslant a \leqslant N), $$

where $c(N,K) = 4N/(N+K)$.

That is, the maximum difference between your $P(E_1 E_2 \cdots E_a \tilde{E}_{a+1} \cdots \tilde{E}_N)$ and the integral representation $\int_0^1 \theta^a (1-\theta)^{(N-a)}\, dM(\theta)$ is bounded precisely by

$4N/(N+K)$:

$$\max_{0\leqslant a\leqslant N}\left|P(E_1E_2\cdots E_a\tilde{E}_{a+1}\cdots\tilde{E}_N)-\int_0^1\theta^a(1-\theta)^{(N-a)}\,dM(\theta)\right|\leqslant\frac{4N}{N+K}.$$

It is worthwhile dwelling on the meaning of the mixing distribution function $M(\theta)$. Examine the function $M_{N+K}(\theta)=\sum_{i=0}^{N+K}p_{i(N+K)}(i\leqslant\theta(N+K))$, which appears in the exact Stieltjes integral representation of $P(E_1E_2\cdots E_a\tilde{E}_{a+1}\cdots\tilde{E}_N)$. Notice that if there were $K$ more events that you regard exchangeably with $\mathbf{E}_N$, then $M_{N+K}(\theta)$ would represent your probability for the event that the sum of all $N+K$ quantities is less than or equal to $\theta(N+K)$ or, equivalently, for the event that the proportion of all the events that you regard exchangeably does not exceed $\theta$. The limiting distribution function $M(\theta)$, then, would represent your opinion about the proportion of 1's that would appear in an infinite sequence of these events.

In referring to "distribution functions" at this stage, we are stepping ahead of ourselves once again. For our systematic study of distribution functions will not take place until Chapter 5. But life is awkward. I presume that if you have been following this section completely either you were familiar with the mathematical aspects of distribution functions before you began or you have already studied Chapter 5. Otherwise, you have been reading more casually, just to get the main ideas.

In the next few sections we describe the main statistical applications of the judgment to regard events exchangeably—problems involving sampling from a finite population and, more generally, observations from designed experiments.

### *Historical Notes*

The presentation in this section of judgments regarding $E_1$ and $E_2$ as exchangeably extendable to dimension $N+K$ comes from an article by Diaconis (1977) entitled "Finite Forms of de Finetti's Theorem on Exchangeability." The reference to de Finetti's theorem alludes to de Finetti's (1937) proof that if you regard an infinite sequence of events exchangeably, then there is a unique distribution function $F(\cdot)$ for which your $P(E_1\cdots E_x\tilde{E}_{x+1}\cdots\tilde{E}_N)$ is representable as $\int_0^1\theta^x(1-\theta)^{(N-x)}\,dF(\theta)$ for every $x$ and $N$. De Finetti's original proof was rather intricate. Heath and Sudderth (1976) presented an expository article proving this result by means of the limit of representations of $P(S_N=x)$ when $N$ is finite. The ideas in Diaconis' paper had been developed in a different way by Crisma (1971), who worked with the characterization of exchangeability in terms of asserted probabilities for products of events rather than for sums; this involves the vector $\mathbf{q}_N=(1,q_1,q_2,\ldots,q_N)^{\mathrm{T}}\equiv(1,P(E_1),P(E_1E_2),\ldots,P(E_1E_2\cdots E_N))^{\mathrm{T}}$ that we studied in Section 3.8. Crisma's geometrical scheme and further developments appear in English in Crisma (1982). (You may derive them for yourself as a problem shortly.) The article by Daboni (1982) in the same volume in also relevant. Further quantitative results have been developed by Wood (1992).

## PROBLEMS (Crisma's Analysis in Two and Three Dimensions)

The following problems are completed most easily by writing appropriate computer programs.

1. Recall the matrix $\mathbf{T}(\mathbf{p} \to \mathbf{q})$ from Section 3.8.2, which transforms the simplex $\mathbf{S}^N$ to the space $\mathbf{S}^{N*} = \{\mathbf{q}_N = (1, \mathbf{q}_1, \ldots, \mathbf{q}_N)^T | 1 \geqslant q_1 \geqslant \cdots \geqslant q_{N-1} \geqslant q_N \geqslant 0\}$. Applying this transformation to both sides of the equation $\mathbf{p}_2 = \sum_{j=0}^{N} p_{jN} \mathbf{T}_{3,(N+1)} \mathbf{e}_{jN}$ yields $\mathbf{q}_2 = \sum_{j=0}^{N} p_{jN} \mathbf{T}(\mathbf{p} \to \mathbf{q}) \mathbf{T}_{3,(N+1)} \mathbf{e}_{jN}$. Compute this transformation of the columns of $\mathbf{T}_{3,(N+1)}$ for $N = 3$, 4, 5, and 6 by using the appropriate matrix

$$\mathbf{T}(\mathbf{p}_2 \to \mathbf{q}_2) = \begin{pmatrix} 1 & 1 & 1 \\ 0 & .5 & 1 \\ 0 & 0 & 1 \end{pmatrix}$$

to transform the column vectors shown in Table 3.1 into corresponding vectors $\mathbf{q}_2$. Plot the resultant component pairs $(q_1, q_2)$ in two-dimensional space (ignoring $q_0 = 1$), along with the points (0, 0) and (1, 1), which correspond to transformations of the vectors (1, 0, 0) and (0, 0, 1). For each value of $N$ the convex hull of the plotted points identifies the prevision vectors $\mathbf{q}_2$ that are exchangeably extendable to an assertion $\mathbf{q}_N$. Use the same matrix, $\mathbf{T}(\mathbf{p}_2 \to \mathbf{q}_2)$, to identify the transformed points on the parabola $((1-\pi)^2, 2\pi(1-\pi), \pi^2)$ for $\pi \in [0, 1]$. Draw the transformed curve on the same graph. The convex hull of this curve contains all vectors $\mathbf{q}_2$ that represent infinitely exchangeably extendable assertions regarding $\mathbf{E}_2$. Why is the assertion $\mathbf{q}_2 = (1, .5, .1)$ not exchangeably extendable to any $\mathbf{q}_3$ of the form $(1, .5, .1, q_4)$ for $q_4 \in [0, .1]$? (If you cannot see why using "off the cuff" checking, try using the fundamental theorem of prevision to show that the constraints yield no feasible solution.)

2. Repeat problem 1 for three dimensions. That is, suppose you regard $\mathbf{E}_3$ exchangeably. Compute the matrices $\mathbf{T}_{4,(N+1)}$ whose columns are the extreme vectors in the three-dimensional simplex representing opinions that are exchangeably extendable to degree $N$ for $N = 5, 6, 7, 8, 9, \ldots$. Then transform these columns from $\mathbf{S}^4$ to $\mathbf{S}^{4*}$ by the appropriate matrix $\mathbf{T}(\mathbf{p}_3 \to \mathbf{q}_3)$. Similarly, transform the curve of vectors $\{((1-\pi)^3, 3\pi(1-\pi)^2, 3\pi^2(1-\pi), \pi^3) | \pi \in [0, 1]\}$ using $\mathbf{T}(\mathbf{p}_3 \to \mathbf{q}_3)$, and trace the curve in three-dimensional space. The convex hull of this curve identifies within $\mathbf{S}^{3*}$ the set of infinitely exchangeably extendable previsions regarding $\mathbf{E}_3$.

3. Identify geometrically the space of all *independent* assessments of $E_1$ and $E_2$ within the simplex of coherent probabilities for the partition they generate: $(P(\tilde{E}_1\tilde{E}_2), P(\tilde{E}_1E_2), P(E_1\tilde{E}_2), P(E_1E_2))$. You should be able to identify the curve of independent exchangeable distributions lying within this "sheet" of independent distributions. Which side of the sheet contains distributions for which

$P(E_2|E_1) > P(E_2)$? Use your answer to conclude that this is a necessary condition for the infinite extendability of exchangeable assessments of $E_1$ and $E_2$. [*Hint*: Visualize the image of the two-dimensional set of pairs $(p_1, p_2) \in [0, 1]^2$ as mapped onto $((1-p_1)(1-p_2), (1-p_1)p_2, p_1(1-p_2), p_1p_2) \in S^3$.]

## 3.11 SAMPLING FROM A FINITE POPULATION

Of all the trout in this mountain lake, how many have a dragonfly larva in their stomachs? Of these mice that I inoculate today, how many will catch a virus to which I will expose them next week? Of all the girls with recorded age 16 in the 1980 Census of Manhattan, how many had completed a university degree before reaching their twenty-first birthday? Of all men recorded as black males of age 21 in Washington, DC by the 1990 census, how many will be unemployed on 17 November 1998? Of all the sales of General Motors common stock on the New York Stock Exchange today, how many are "short sales"? Of all the floppy disks that have been produced on this production line during the past hour, how many will be recognized as acceptable for initialization by a computer? Of all the radioactive particles that I shoot at this lead barrier, how many will register on a film located behind it?

GENERIC QUESTION. Of all the units of a specific finite population of units, how many have some specified characteristic?

Let us denote by $E_i$ the event that the $i$th unit has the specified characteristic. Denoting the number of units in the population by $N$, the answer to this generic question is $S_N = \sum_{i=1}^{N} E_i$. The events $E_i$ are called the "constituents of $S_N$." What is the numerical value of $S_N$?

Your prevision for $S_N$ represents a specific aspect of your uncertain knowledge of the answer to the generic question. You could change the state of your knowledge by observationally identifying the values of some of the events that constitute $S_N$. Do these six trout that I've caught with a rod have dragonfly larvae in their stomachs? If you would like to change your state of uncertain knowledge about any one of the questions listed by a process of partial observation, how many and precisely which of the constituent events would you like to observe? Does the order in which you would observe them matter to you? You could conceivably observe them all. Then you would have identified observationally the exact value of $S_N$. Do you actually want to observe them all?

To answer these questions, it is helpful to consider still another question: do you regard the events that constitute $S_N$ exchangeably?

For several of the questions listed, I must say that I do not. My prevision for the constituents of the partition generated by $\mathbf{E}_N$, such as $E_1\tilde{E}_2\tilde{E}_3E_4E_5\tilde{E}_6E_7 \cdots \tilde{E}_{N-1}E_N$ would be greater for those product events involving

**More Events Corresponding To**

Trout lolling near the shore
Mice with lower weight
Girls recorded living on the block of 73rd Street and York
Men recorded living in Shaw
Sales transacted at a price below that of the previous transaction

and

**More Inverted Events Corresponding To**

Trout swimming the deeps
Mice with higher weight
Girls recorded living on the block of 126th Street and Amsterdam
Men recorded living in Georgetown
Sales transacted at a price above that of the previous transaction

than it would be for product events that involve the same number of multiplicand events and inverted events, but with the opposite relative balance of multiplicand events and inverted events within the two subgroups.

Of course, to apply this "knowledge of mine" to a particular group of trout lying on the grass, I would have to know which of them had been caught in the middle of the lake and which had been caught lolling near the shoreline. Even having expressed this opinion about the trout, I also aver that if I did not know which of the trout had been caught deep and which had been caught shallow, I would regard exchangeably the events identifying whether each of them has a dragonfly larva in its stomach. On one hand, this remark only serves to accentuate our recognition that the judgment to regard a sequence of events exchangeably depends on one's state of knowledge about the events. On the other hand, this remark also suggests an important result concerning the procedure of "scrambling" events that you do not regard exchangeably.

### 3.11.1 Scrambling the Units in a Population

Expanding on my avowed opinion about the fish in the lake, I would say that if the fish in the lake were numbered in sequence according to the rule that the unit number of a fish increases with the distance of the fish from the shoreline, and $E_i$ denotes the event that the $i$th numbered fish has a dragonfly larva in its stomach, then, if $i < j$, my $P(E_i) > P(E_j)$. I do not regard the events $E_1, \ldots, E_N$ exchangeably when I know their identifying units are determined in this way. However, suppose that the identifying unit numbers for the fish in the lake were not ordered by their distance from the shore, but were determined merely by sequentially removing balls from an urn of balls numbered 1 through $N$ that had been thoroughly scrambled. The fish identified as unit 1 in the population according to the distance ordering would be identified as unit $s_1$ in the scrambled ordering, where $s_1$ is the number of the draw on which ball 1 is removed from the urn. Similarly, the fish identified as unit 2 according to the distance ordering would be identified as unit $s_2$ according to the scrambled ordering, and so on. Now I do aver that I regard the events $E_{s_1}, E_{s_2}, \ldots, E_{s_N}$ exchangeably, for in this unit identification scheme I no longer know which fish is which. When I now compare event $E_{17}$ and $E_{32}$, I no longer know which event corresponds to a fish swimming closer to shore.

It turns out that my regarding the scrambled events exchangeably is not just a peculiar decision on my part. Such an avowal is required by coherency. Let us develop some terminology and state this result formally.

***Definition 3.12.*** If we identify the distinct units of a population by the integers 1 through $N$, then the unit designated by the integer $i$ is called the $i$th *unit of the population.* If the identifying unit number is chosen on the basis of some operationally defined characteristic of the unit, then the units are said to be *ordered units.* □

Operationally defined characteristics of the units can be used to define events, one for each unit. Thus, the events $E_1, E_2, \ldots, E_N$ correspond to numerical indicators of whether the units numbered $1, 2, \ldots, N$, respectively, exhibit a characteristic: $E_i \equiv$ (unit $i$ exhibits the characteristic denoted by $E$).

***Definition 3.13.*** Suppose the unit identifying numbers for the $N$ units in a population are designated according to some selection procedure. Denoting the $N$ unit numbers by $(U_1, U_2, \ldots, U_N)$, *you* are said to *regard the designation as a scrambling procedure with respect to the events* $\mathbf{E}_N$ if, for each vector of integers $(u_1, u_2, \ldots, u_N)$ that is a permutation of the $N$-tuple $(1, 2, \ldots, N)$, you assert

$$P(\mathbf{U}_N = \mathbf{u}_N) = \frac{1}{N!} = P((\mathbf{U}_N = \mathbf{u}_N) | E_1^* E_2^* \cdots E_K^*)$$

for every choice of $K \leqslant N$ events, $E_K^*$, from among $\mathbf{E}_N$. If the unit identifications are selected by such a procedure, you are said to regard the population units as *scrambled units.* □

In a word, the scrambling criteria amount to your regarding the unit identifying numbers for the events independently of the events themselves. From our study of independent judgments, we know these criteria imply that for each possible unit labeling, $\mathbf{u}_N$, and for each choice of $K \leqslant N$ events from among $\mathbf{E}_N$, your

$$P(E_1^* E_2^* \cdots E_K^*) = P(E_1^* E_2^* \cdots E_K^* | (\mathbf{U}_N = \mathbf{u}_N)).$$

We are ready to state and prove the important theorem that you must regard exchangeably any events whose unit designations are scrambled.

**Theorem 3.15.** Let $\mathbf{E}_N$ denote events that you do not regard exchangeably. Suppose you assert your probabilities for the constituents of the partition they generate as $P(\mathbf{Q}(\mathbf{E}_N)) = (q_1, q_2, \ldots, q_{2^N})$, where these assertions do not respect the exchangeability conditions. Now let the component events of $\mathbf{E}_N$ be relabeled as $F_{u_1}, F_{u_2}, \ldots, F_{u_N}$, according to a scrambling procedure. (Each event is reidentified by its new label but not by its old label.) Then coherency requires you to regard the events $F_{u_1}, F_{u_2}, \ldots, F_{u_N}$ exchangeably.

*Proof.* Your asserted probabilities for the constituents of the partition vector $\mathbf{Q}(\mathbf{E}_N)$ specify your probabilities for the products of each choice of $K$ events from among the components of $\mathbf{E}_N$. Each such probability can be computed as a sum of your asserted probabilities for specific constituents of the partition generated by $\mathbf{E}_N$, denoted in the theorem by $\mathbf{q}_{2^N}$. Now, if you consider the product of any $K$ events, $\mathbf{F}_K$, chosen from among $F_{u_1}, F_{u_2}, \ldots, F_{u_N}$, you have no way of knowing which of the events originally denoted as $E_1, \ldots, E_N$ appear among these $K$ multiplicands. Of course, conditional upon their identification, you assert the appropriate conditional probability which can be computed from $\mathbf{q}_{2^N}$, as asserted unconditionally with the knowledge of which unit is which.

Now as there are ${}^N C_K$ distinct choices of $K$ events from $N$ that you regard as equilikely prospects for $\mathbf{F}_K$, coherency requires that for any distinct integers $u_i \neq u_j \neq u_k \neq \cdots \neq u_N$, each within the set $\{1, 2, \ldots, N\}$, your

$$P(F_{u_i}) = \sum_{i=1}^{N} P(E_i) \Big/ {}^N C_1,$$

$$P(F_{u_i} F_{u_j}) = \sum_{j>i=1}^{N} P(E_i E_j) \Big/ {}^N C_2,$$

$$P(F_{u_i} F_{u_j} F_{u_k}) = \sum_{k>j>i=1}^{N} P(E_i E_j E_k) \Big/ {}^N C_3,$$

$$\vdots$$

$$P(F_{u_1} F_{u_2} \cdots F_{u_N}) = P(E_1 E_2 \cdots E_N) / {}^N C_N,$$

where the probabilities $P(E_i E_j E_k \cdots E_r)$ are computed from appropriate components of your asserted $P(\mathbf{Q}_{2^N})$. Thus, according to Definition 3.8b of exchangeability, you regard the events $F_{u_1}, F_{u_2}, \ldots, F_{u_N}$ exchangeably, since your prevision for the product of any $K$ of them is identical, no matter which $K$ of the $N$ events appear in the product as multiplicands. □

The practice of observing a sample of $n$ items from a population of size $N$, wherein the units to be sampled are chosen according to a procedure that you regard as a scrambling procedure, has long been called the practice of *simple random sampling.* By and large, this methodology, along with embellished variants, has been considered essential to the practice of statistical inference by objectivist statistical theorists of the twentieth century. Observations of units in a population that are not selected by random sampling have even been considered by many as unfit for statistical analysis.

In a historically important and controversial public seminar, Savage (1961) challenged the notion that "randomization" played any crucial role in the theory of statistical inference. He noted the result we stated as Corollary 3.12.1, that if you regard a sequence of events exchangeably then you regard the sum of any $n$ of the unknown events as sufficient for your inference about the remaining $N - n$ events, no matter how the $n$ events were selected. In this light, he criticized certain objectivist procedures of "statistical inference" that allow the inference made from a selection of observations to depend in a crucial way on how the units were selected for observation. At the time, Savage's argument was considered wild talk indeed. "A theory of statistics that is not based on random samples?!!!"

Developments since 1961 have followed two lines. In the first place, detailed analysis of what is called "the likelihood principle" has identified the incongruities of certain procedures of "objectivist inference" that depend on the way the observations are chosen. These developments began with the pathbreaking papers of Birnbaum (1962, 1972), continued through a controversial two decades, and largely concluded in the extensive summary monograph by Berger and Wolpert (1984) with modification by Hill (1987). Much of this discussion has been conducted within the context of objectivist formulations of the problem, even several contributions with subjectivist content and conclusions. This is noted particularly in Lane's comments at the end of the Berger and Wolpert monograph.

In the second place, Theorem 3.15 does qualify any suggestion that randomized selection of units in a population should play no role in the practice of statistical inference. Although I may regard the events $\mathbf{E}_N$ exchangeably, and thus regard the sum of any $n$ events selected from among them as sufficient for inference about the rest, perhaps you do not regard them exchangeably. If the units to be observed are selected by a procedure we both regard as scrambled, then you and I could both use the exchangeability formulas for making inferences about $S_N$ on the basis of the observations $E_{u_1}$, $E_{u_2}, \ldots, E_{u_n}$. Moreover, even for myself, if I do not regard the events exchangeably, but my qualifications to exchangeability conditions are so meager that it is not worth my effort to keep track of all the characteristics of the units that affect my judgment of nonexchangeability, then selecting the units by a scrambling procedure may be worthwhile—even if I don't care what you are going to infer from my observations. Of course, these general statements depend on the specific costs of sampling according to various procedures. These must be considered explicitly in a complete study of sampling strategy.

### 3.11.2 Inference Based on Simple Scrambled Sampling

Suppose you have 200 white mice of the same age to the nearest week (32 weeks), and you are interested in the question "How many of these mice that I inoculated today will catch a virus to which I will expose them next month?" The mice have been fed similar diets, and their measured weights when inoculated differed by no more than 3 grams. Distinctions you might be able to make among the mice are so meager that it would be virtually worthless to attempt to take advantage of them. Thus, you do not tag them individually. When left to mingle in a cage, you consider them to scramble themselves. You would not really care in which order you observe them when you determine whether each has caught a virus. You regard exchangeably the events that the individual inoculated mice catch the virus. To the extent that your avowed judgment of exchangeability is merely an approximation of your opinion, determining which mice to examine by using a scrambling procedure would ensure a precise judgment of exchangeability.

In fact, if you do observe the mice, you will observe them in some order. Suppose we denote the events in the order in which you observe them, by $E_1, E_2, \ldots, E_n, \ldots, E_{200}$, where $E_i$ identifies whether the $i$th mouse tested is infected with the virus. Conceivably, you could test all 200 mice, or you might stop testing once you have examined only some $n < 200$ of them. The quantity that answers your question of interest, "how many mice that I inoculate will catch the virus?", is $S_N \equiv \sum_{i=1}^{N} E_i$. Your prevision, $P(S_N)$, represents a specific aspect of your uncertain knowledge about the efficacy of the inoculation syrum you have prepared for conducting this experiment. You may express your uncertainty in other forms too, for example, by asserting probabilities such as $P(a \leqslant S_N \leqslant b)$ for various values of $a$ and $b$. You may even express your knowledge about $S_N$ in complete detail by asserting your probability distribution for $S_N$ via the assertions $P(S_N = 0), P(S_N = 1), \ldots, P(S_N = N)$; or you may express less specific knowledge by partial assertions such as $P(S_N) \in [c, d]$ or $P(e \leqslant S_N \leqslant f) \in [g, h]$.

Suppose you could observe the values of $E_1$ through $E_n$, and identify some product event, say, $E_1 \tilde{E}_2 E_2 E_4 \tilde{E}_5 \cdots E_{n-1} \tilde{E}_n$ as 1, and all other constituents of the partition generated by $E_1, \ldots, E_n$ as 0. What would you infer from this observation? That is, "What is your $P(S_N | E_1 \tilde{E}_2 E_2 E_4 \tilde{E}_5 \cdots E_{n-1} \tilde{E}_n)$?" That is, what inference would you make about $S_N$ if only you could know precisely the values of $n$ particular constituents of $S_N$? Since you expressly regard the events $E_1, \ldots, E_N$ exchangeably, your coherent answer to this question depends only on what you are able to assert about $S_N$. In the next two subsections we examine the requirements of coherent inference on the basis of information gained through simple scrambled sampling under two conditions: (1) you assert a complete distribution to represent your uncertainty about $S_N$; (2) you assert something less detailed than a full distribution.

#### *3.11.2.1 Inference from Sample Information, Supposing That a Distribution for $S_N$ Has Been Asserted*

At the extreme end of uncertainty specification, represented by an exhaustive assertion of a probability distribution, $P(S_N = 0), P(S_N = 1), \ldots, P(S_N = N)$, the theorems of Section 3.9 state the coherent determination of the conditional previsions $P(S_N = A | E_1 E_2 \cdots E_a \tilde{E}_{a+1} \cdots \tilde{E}_n)$ for any selection of $n$ events from among $E_1, \ldots, E_N$ and for any value of $a = 0, 1, \ldots, n$ and for any permutation of the subscripts in conditioning product event:

$$\begin{aligned} P(S_N = A | E_1 E_2 \cdots E_a \tilde{E}_{a+1} \cdots \tilde{E}_n) &= P(S_N = A | S_n = a) \\ &= P(S_n = a | S_N = A) P(S_N = A) / P(S_n = a) \\ &= \frac{{}^{A}C_a \, {}^{(N-A)}C_{(n-a)} P(S_N = A)}{\sum_{A=a}^{N-(n-a)} {}^{A}C_a \, {}^{(N-A)}C_{(n-a)} P(S_N = A)}. \end{aligned}$$

The first equality is due to the fact that the sum of $n$ events is a sufficient statistic for inference about the remaining events with which they are judged exchangeably (Theorem 3.12 and its corollary); the second line derives from a routine application of Bayes' theorem (Theorem 3.5); and the final computational formula is due to exchangeability again (Theorem 3.11).

***Example 1.*** For expositional simplicity, let us reduce the size of the problem regarding inoculated mice to a population of size 10. Suppose you think that the inoculation will be quite conducive to developing antibodies against the virus. Defining $S_{10}$, as the number of mice who do catch the virus despite inoculation, suppose you assert the probabilities

$$P(S_{10} = A) \approx 3/4^{(A+1)} \qquad \text{for } A = 0, 1, \ldots, 10.$$

Suppose you check four of these mice and find that one has the virus. Your conditional probabilities $P(S_{10} = A | S_4 = 1)$ can be computed from Bayes' theorem on the basis of the exchangeability condition that

$$P(S_4 = 1 | S_{10} = A) = {}^{A}C_1 \, {}^{(10-A)}C_3 / {}^{10}C_4 \qquad \text{for } A = 0, 1, \ldots, 10.$$

Table 3.2 presents the arithmetical evaluations of the various components of Bayes' theorem based on these assertions.

Since the asserted probabilities, $P(S_{10} = A)$, represent your opinions before you do any experimenting, and the computed probabilities $P(S_{10} = A | S_4 = 1)$ incorporate what you assert should be learned from the observation $S_4 = 1$, these two sequences of probabilities are called your *prior probability mass function* for $S_{10}$ and your *posterior probability mass function* for $S_{10}$ given $S_4 = 1$, respectively. For reasons to be specified in due time, the sequence of probabilities $P(S_4 = 1 | S_{10} = A)$ for various values of $A$ is called the likelihood function for $S_{10}$. For now, the important thing for you to realize computationally is that the third column of Table 3.2 is derived from the component-by-component products of the first two columns. The final column is the third column divided by its sum.

□

**Table 3.2. Numerical Components of Bayes' Theorem for Inference about the Inoculated Mice**

| $A$ | $P(S_{10}=A)$ | $P(S_4=1 \mid S_{10}=A)$ | Product | $P(S_{10}=A \mid S_4=1)$ |
|---|---|---|---|---|
| 0 | .75 | 0 | 0 | 0 |
| 1 | .1875 | 4/10 | .075 | .6990 |
| 2 | .046875 | 8/15 | .025 | .2330 |
| 3 | .011719 | 1/2 | .005860 | .0546 |
| 4 | .002929 | 8/21 | .001116 | .0108 |
| 5 | .000732 | 5/21 | .000303 | .0028 |
| 6 | .000183 | 8/70 | .000021 | .0002 |
| 7 | .000046 | 1/30 | .000002 | .0000 |
| 8 | .000012 | 0 | 0 | 0 |
| 9 | $3 \times 10^{-7}$ | 0 | 0 | 0 |
| 10 | $1 \times 10^{-7}$ | 0 | 0 | 0 |

## PROBLEMS

The first two problems can be computed easily by hand or by using MATLAB or a spreadsheet. It would be worthwhile to write the program for such computations to clinch your ideas on the computational inference formula. More sizable problems virtually require a computer for their numerical resolution. If you write a program, you can solve the final two problems numerically. Otherwise, just work out the logic to set up the computations.

1. An ichthyologist is interested in whether five mollies (a species of tropical fish) still show signs of "ich," a virus that infects fish gills, after their tank was chemically treated three days ago . Let $K$ denote the number of fish in the tank that exhibit symptoms of ich. Suppose the ichthyologist regards these five events exchangeably and asserts the probabilities $P(K=k)=(6-k)/21$ for $k=0,1,2,3,4,5$. Suppose that a lab assistant examines two of the mollies, and finds that one of them shows signs of ich and that the other does not. Determine the inference about the remaining three fish to be made by the ichthyologist on the basis of this information.

2. Repeat problem 1, supposing that the ichthyologist asserts the probabilities

$$P(K=k) = {}^{5}C_{k}\,{}^{20}C_{(5-k)}/{}^{25}C_{5} \qquad \text{for } k=0,1,\ldots,5.$$

3. The reliable lifetimes of 100 dishwashers that are produced on the assembly line at an appliance plant are regarded exchangeably. Fifty of them have been sold locally, and 50 have been exported to another country. Let $\mathbf{E}_{100}$ denote the vector of events that each of the machines fails within one year of use in a customer's home to the extent that a warranty claim is made against the company by its purchaser. Let $S_{100}$ denote the sum of these 100 events. Based

on the results of in-plant testing on similar machines and on historical experience with previous models, the engineers assess the following probabilities regarding $S_{100}$:

$$P(S_{100} = s) \approx K4^S/s! \qquad \text{for } s = 0, 1, \ldots, 100,$$

where $K$ is a proportionality constant that ensures these probabilities sum to 1. Now suppose it is learned that 3 of the 50 machines sold locally have had warranty claims made on them within one year of service.

**a.** Identify the inference that this observation on the local 50 machines provides to the engineers regarding the experience of the 50 machines sold in the export market. That is, determine the probabilities $P(S_{100} = s | S_{50} = 3)$.

**b.** Thinking about the substance of this problem, which do you think is greater numerically, $P(S_{100} \geqslant 7)$ or $P(S_{100} \geqslant 7 | S_{50} = 3)$? How about $P(S_{100})$ compared to $P(S_{100} | S_{50} = 3)$? What is the difference between these two comparisons?

**c.** Sketch approximately on a graph the prior p.m.f. for $S_{100}$, the likelihood for $S_{100}$ based on the data, and the posterior conditional p.m.f. for $S_{100}$. Label your axes and label each function.

**4.** Twenty-five thousand live sheep arrive by boat at the port of Medina three days before the end of Ramadan. The sheep are numbered with ear tags and are penned in numerically identified locations on the ship. Let $E_i$ denote the event that sheep numbered $i$ is found, upon examination, to have scabies, an infectious disease. Twenty-five sheep are to be selected by drawing their tag numbers from a box of 25,000 scrambled numbers. Let $F_1, \ldots, F_{25}$ denote the events that the first, second,..., twenty-fifth selected sheep is found to have scabies.

**a.** Do you regard the events $\mathbf{E}_{25}$ exchangeably? Why or why not? How about the events $\mathbf{F}_{25}$? Do you regard the components of $\mathbf{E}_{25}$ exchangeably with the components of $\mathbf{F}_{25}$?

**b.** Let $S$ denote the total number of sheep in the shipment that have contracted scabies. Suppose you assert the probabilities $P(S = 0) = .99$ and $P(S = s) = K\exp(-10^{-9}s)$ for $s = 1, 2, \ldots, 25{,}000$, where $K$ is an appropriate proportionality constant to ensure that these probabilities sum to 1. Write the equation that would represent the conditional probability mass function for $S$ given that 1 out of 25 sheep selected by the scrambling procedure is found to have scabies. State how the judgment of exchangeability is relevant to this construction.

The following example introduces a family of applications based on "capture-recapture methods" which are widely used in field zoology for the estimation of population sizes. It applies the inferential formula for $P(S_N = A | E_1 E_2 \cdots E_a \tilde{E}_{a+1} \cdots \tilde{E}_n)$ under exchangeability in a context that inverts the situation of the problems you have just completed. In the problems already considered, the size of a population is known, and the unknown quantity is the number of these population units who exhibit a specified characteristic. In the following example, the number of units who exhibit a specified characteristic is known; but the size of the population is unknown, and is subject to inferential argument. Let us develop the general argument in the context of an example.

***Example 2. Capture-Recapture Methods in Field Zoology.*** Suppose you are a field zoologist interested in the number of trout, $N$, that live in a small, high-mountain lake. Upon observing the size of the lake, taking some measurements of its depth in various places, and knowing something about the thickness of the ice crust in some other lakes during the previous winter freeze, you assert probabilities for the partition of events $(N = 0), (N = 1), \ldots,$ and $(N = \mathbb{N})$, where $\mathbb{N}$ is the maximum number of trout that could be packed into a lake of this size.

A capture-recapture sampling method proceeds as follows. First, a large net is laid across the lake from bank to bank. After the net has been left to rest for a day or two so that the disturbance it has generated subsides, the net is drawn sometime during the period of an evening mosquito feed. As a result, $K$ fish are captured, tagged, and released. Then the net is laid once again and left a few days to rest. Again during an evening mosquito feed, the net is raised, yielding $M$ trout, of which $T$ are found to have tags and $M - T$ not to have them. What should the result of this experiment tell you about the population size $N$?

Based upon your own experience and on trout-fishing lore, you know that trout scramble themselves throughout the lake during time periods of the mosquito feed. Thus, you regard exchangeably the events $E_1, \ldots, E_M$ identifying whether each fish caught in the second netting is found to be tagged or not, and you regard these events exchangeably with the tagging condition of the fish positioned in other parts of the lake who have not been caught. In this situation, the information contained in the detailed observations of $\mathbf{E}_M$ that is relevant to the size of $N$ is summarized in the observation of the sufficient statistic $S^K(\mathbf{E}_M) = T$. (The superscript $K$ on the sum function, $S^K(\cdot)$, denotes the context that $K$ of the $N$ fish were tagged during the initial capture.) The coherent inference asserted from this new information is computable from Bayes' theorem, using the formula

$$P[(S^K(\mathbf{E}_M) = T) | (N = n)] = {}^K C_T \, {}^{(n-K)} C_{(M-T)} / {}^n C_M \qquad \text{for } n = 0, 1, \ldots, \mathbb{N}.$$

This is the standard conditional probability formula for the sum of $M \leqslant n$ events regarded exchangeably, given that the sum of $n$ events equals $K$. Since exactly $K$ fish were caught and tagged on the first capture, the sum of all $N$ events is known to equal $K$. The variable component in these conditional probabilities is $n$, the number of events involved in the sum. A routine application of Bayes' theorem yields

$$P[N = n|(S^K(\mathbf{E}_M) = T)] = \frac{[{}^K C_T\, {}^{(n-K)}C_{(M-T)}/{}^n C_M]P(N = n)}{\sum_{j=0}^{\mathbb{N}}[{}^K C_T\, {}^{(j-K)}C_{(M-T)}/{}^j C_M]P(N = j)}$$

for each of these values of $n$. Expanding the combinatoric expressions and canceling common factors of the numerator and denominator that do not involve $j$ yield the computational formula, for each $n$,

$$P[N = n|(S^K(\mathbf{E}_M) = T)] = \frac{[{}^{(N-M)}P_{(K-T)}/{}^n P_K]P(N = n)}{\sum_{j=0}^{\mathbb{N}}[{}^{(j-M)}P_{(K-T)}/{}^j P_K]P(N = j)}.$$

Notice the use of the permutation symbol in this formula, ${}^N P_K \equiv N!/(N-K)!$, according to statement 4 of Section 3.8.1.

Notice how this example inverts the context of what is known and what is unknown in the sampling problem involving the inoculated mice. Whereas we knew the number of mice, but were unsure of how many catch the virus, in this fish-sampling problem we know how many fish were caught and tagged in the first drawing of the net, but we are uncertain of how many fish there are in all!

□

## PROBLEM

**5.** A nature conservation research team desires to know more precisely the number of possums who currently live on 50 contiguous hectares of Rata Forest in Westland, New Zealand. Let $N$ denote the unknown number of possums living there. Based on their understanding of possum behavior and of possum incidence in various regions of New Zealand, the research team recognizes the conceivable range (realm) for $N$ as the integers from 0 through 999. Furthermore, after a modicum of introspection and discussion, the team members agree on an approximate consensus assertion of the probabilities

$$P(N = n) = Kn^{38.0625}\exp(-.3125n) \qquad \text{for } n = 0, 1, \ldots, 999.$$

The proportionality constant $K$ is determined to ensure that the sum of these probabilities equals 1. (By Chapter 7 you will recognize that these probabilities constitute a discrete analogue of a gamma density function, if you do not already. Thus, your $P(N) \approx 125$ and your $V(N) \approx 400$. But this feature is irrelevant to our concerns just now. Merely treat this problem as one of logical setup and computation.)

A field study is conducted as follows. Nonlethal possum traps are located at grid points within the forest. The next day the traps are revisited, and 38 "captured" possums are tagged by marking the soles of their left rear feet with an indelible ink. The possums are released, and the traps are removed from their locations. One week later, the traps are set once again in new positions, each about 15 meters from its original location. On this occasion, 33 possums are found in the traps, and eight of them have a marked left rear foot.

Using Bayes' theorem, determine the conditional probabilities for $(N = n)$ based on this data. Explain how the judgment of exchangeability is used in this inferential specification. Would you have any misgivings about asserting exchangeability in this way? You may draw or compute graphs to aid your exposition.

The work of Seber (1973) on *The Estimation of Animal Abundance* presents a synthesis of embellished variations on this problem, including, for example, the possibility of mortality and fertility experienced by the population over a long period between capture and recapture. Whitehead (1990) develops an application that allows migration into and out of the study area between captures. Both of these authors analyze the problem from a "likelihood" point of view, to be discussed. Their results, nonetheless, provide useful and important pieces of the subjectivist analysis.

Our development of inference based on capture-recapture methods is of historical and philosophical interest. Objectivist statisticians, who would like to have a theory of statistical inference that is not based on anyone's subjective assertions, use capture-recapture data to "estimate" the value of $N$ on the basis of properties of a portion of the inference formula,

$$P[(S^K(\mathbf{E}_M) = T)|(N = n)] = {}^K C_T\, {}^{(n-K)}C_{(M-T)}/{}^n C_M \qquad \text{for } n = 0, 1, \ldots, \mathbb{N}.$$

For reasons that are discussed in detail in Section 7.4.2, this piece of the formula is commonly called the likelihood function for $N$, evaluated at $n$, based on the data that $S(\mathbf{E}_M) = T$. The likelihood function is denoted by $L_N(n; T, K, M)$. This function would represent your probability that $T$ of $M$ fish netted the second time would be found tagged, but only if you knew that (if this event were conditioned upon) $N = n$ to begin with. By analyzing algebraically the quotient function

$$Q(n) \equiv L_N(n + 1; T, K, M)/L_N(n; T, K, M),$$

it is easy to show that the function $L_N(n; T, K, M)$ achieves its maximum value at $\tilde{n} = [[MK/T]] - 1$, where $[[x]]$ denotes the greatest integer contained in $x$. (As an exercise, you should show that $Q(n) > 1$ if and only if $n < MK/T - 1$.)

Objectivist statisticians regard $\tilde{n}$ as a good "estimate" of $N$ on the basis of the capture-recapture data, choosing to ignore the zoologist's other information about the size of the lake, the depth of the ice crust, and so on. A presentation of this problem worth examining is in the classic text of Feller (1958), a book also worth perusing more generally. In a nondogmatic may it expounds a very different point of view than ours, and it has been extremely influential. Questions for you: Where has the objectivist statistician's opinions or knowledge entered the prescribed "estimation" procedure? What happens in science when such objective procedures yield "unreasonable" results? Let us return to the development of our text.

Of course one's assessments of $P(S_N = 0), P(S_N = 1), \ldots, P(S_N = N)$ may be whatever they may be, modulo the restriction of coherency that they sum to 1. But if your probabilities happen to be representable in a certain form, we can develop an instructive, amusing, and computationally useful result. The context of this result is the paradigmatic context for exchangeability, of drawing balls from an urn.

***Example 3. Distributions in Hypergeometric Form.*** Suppose an urn contains $N$ balls, of which $X$ are colored red and $N - X$ are colored black. If the balls are well scrambled in the urn, I would regard exchangeably the events $\mathbf{E}_N$ that each of the $N$ balls drawn sequentially from the urn without replacement is red. (Recall the problems you did at the end of Section 3.8.) This implies that I assert the conditional probabilities

$$P(S_n = x \mid S_N = X) = {}^{X}C_x\, {}^{(N-X)}C_{(n-x)} / {}^{N}C_n \qquad \text{for } x = 0, 1, \ldots, n,$$

as long as $x \leqslant X$ and $n - x \leqslant N - X$. Now suppose that I do not know precisely the values of $X$ and $N - X$, but I know that they were themselves determined by drawing $N$ balls without replacement from a larger scrambled urn that contained $\chi$ red balls and $\eta - \chi$ black balls, where $\chi$ and $\eta$ are known integers. Thus, I would also regard exchangeably the events $\mathbf{E}_N$ that each of the $N$ balls drawn is red; so, I would assert as my probabilities

$$P(S_N = X) = {}^{\chi}C_X\, {}^{(\eta-\chi)}C_{(N-X)} / {}^{\eta}C_N \qquad \text{for } x = 0, 1, \ldots, N,$$

as long as $X \leqslant \chi$ and $N - X \leqslant \eta - \chi$. For I regard the draws exchangeably, and I know that the sum of $\eta$ draws from the larger urn would yield $\chi$ reds. This quotient formula for $P(S_N = X)$ equals the proportion of all the distinct selections of $N$ balls from these $\eta$ that would include $X$ red balls and $N - X$ black balls. (In Chapter 5 we learn to say that my distribution for $S_N$ is in hypergeometric form.)

Having expressed these probabilities, $P(S_N = X)$, and these conditional probabilities, $P(S_n = x \mid S_N = X)$, coherency requires the assertion of the probabilities

$$P(S_n = x) = {}^{\chi}C_x\, {}^{(\eta-\chi)}C_{(n-x)} / {}^{\eta}C_n \qquad \text{for } x = 0, 1, \ldots, n,$$

and

$$P(S_N = X \mid S_n = x) = \frac{{}^{(\chi-x)}C_{(X-x)}\, {}^{[\eta-\chi-(n-x)]}C_{[N-X-(n-x)]}}{{}^{(\eta-n)}C_{(N-n)}}$$

$$\text{for } X = x, x+1, \ldots, N-(n-x).$$

The first equation follows (after some algebraic manipulation) from applying the theorem on total probability to $P(S_n = x)$, using the conditional probabilities

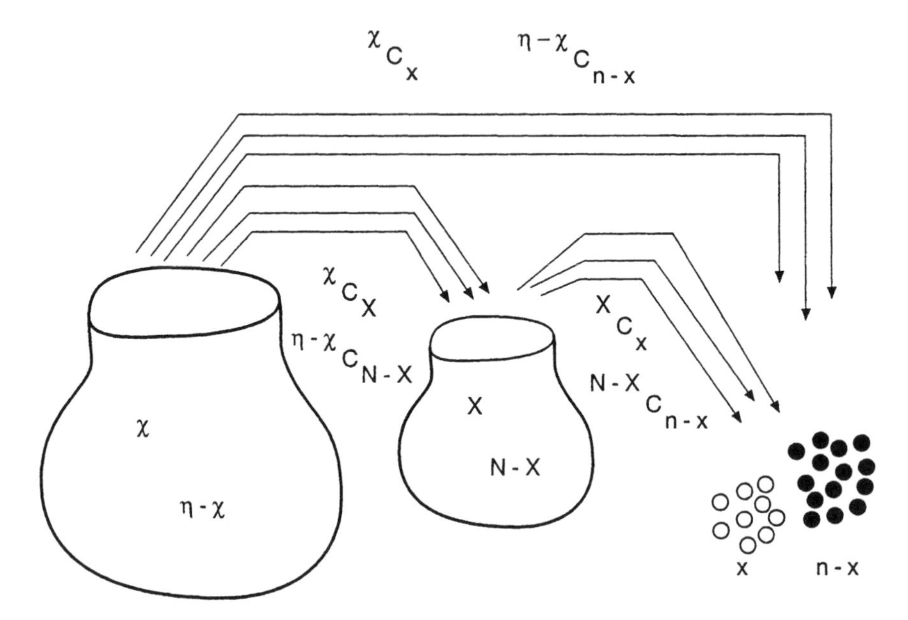


**Figure 3.12.** The unknown population of $N$ balls in the middle urn, of which some number $X$ are red and $N - X$ are black, has been selected from the large urn of known original composition: $\chi$ red and $\eta - \chi$ black balls, scrambled. The asserted probability of drawing $x$ red balls in $n$ draws from the middle urn (of unknown composition) is the same as the asserted probability of drawing $x$ red balls in $n$ draws directly from the larger source urn of known composition: $\chi$ red balls and $\eta - \chi$ black balls.

$P(S_n = x | S_N = X)$ and $P(S_N = X)$. Because the probability distribution for $S_n$ is in hypergeometric form, this first result says that the probability of drawing $x$ red balls in $n$ draws from the urn of unknown composition is the same as the probability of drawing $x$ reds in $n$ draws from the larger urn of known composition, $\chi$ reds and $\eta - \chi$ blacks. Figure 3.12 illustrates this result.

The second equation follows by applying Bayes' theorem to the asserted hypergeometric probabilities. The inferential probability $P(S_N = X | S_N = x)$ is also simple to interpret in the context of drawing balls from urns. Notice that the conditional probabilities $P(S_N = X | S_n = x)$ can be expressed equivalently as $P(S_N - x = X - x | S_n = x)$ for each $X - x \in \{0, 1, \ldots, (N - n)\}$. Thus, the second result says that upon observing the number of red balls in $n$ draws from the urn of size $N, S_n = x$, your opinion about the number of red balls remaining in the urn, $S_N - x$, would be the same as your opinion about the number of red balls among $\eta - n$ balls drawn from the larger urn, now containing only $\eta - n$ balls of which $\chi - x$ are red and $(\eta - \chi) - (n - x)$ are black, as Figure 3.13 illustrates.

Now recall problems 1 and 2 of this section, which concerned a laboratory ichthyologist's opinions about the fish with ich. In order to present some relevant numbers for your computation, I had first proposed the probabilities $P(K = k) = (6 - k)/21$ for each $k = 0, 1, 2, 3, 4, 5$. In the second problem I proposed instead the probabilities $P(K = k) = {}^5C_k\, {}^{20}C_{(5-k)}/{}^{25}C_5$ for the same values of $k$.

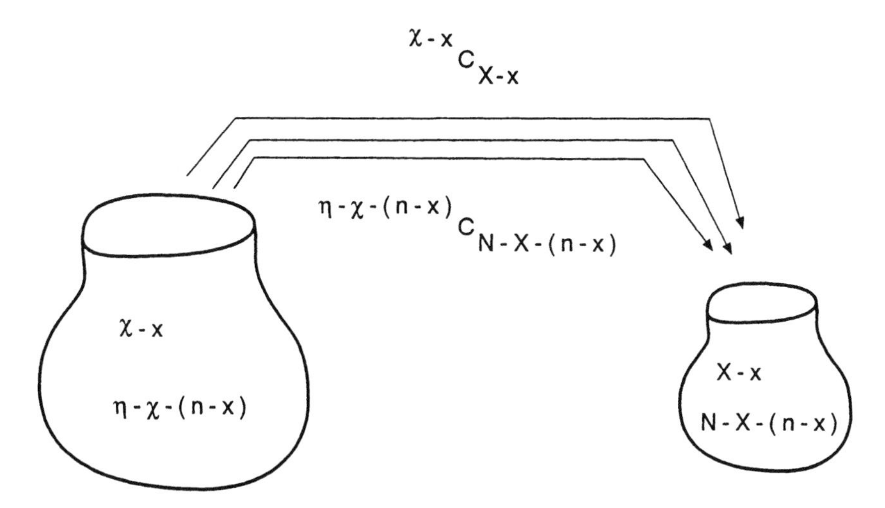


**Figure 3.13.** Conditioning upon the event of drawing $x$ red balls in $n$ draws without replacement from the scrambled middle urn of unknown composition shown in Figure 3.12, the inference asserted about the number of red balls remaining in the middle urn, $X - x$, is to regard this number as you regard the number of reds drawn in $N - n$ draws from a reduction of the original urn, achieved by removing $x$ reds and $n - x$ blacks. The resulting urn would contain $\chi - x$ red balls and $\eta - \chi - (n - x)$ black balls.

These latter probabilities are in hypergeometric form. Thus, you can use the algebraic formula for $P(S_N - x = X - x | S_n = x)$ in this urn context to determine the coherent inferences asserted from the conditioning information that one of two netted fish is found to have the ich. In fact, to solidify your understanding, why don't you do so now. Then we shall continue a discussion.

[Return to problem 2. Specify the prior distribution for $K$ in terms of the numbers $\chi$ and $\eta$ described in this example. Using the result we developed for $P(S_N = X | S_n = x)$, determine the inferential probabilities of problem 2 directly from this result.]

Now evidently, in problem 2 there is no tank of mollies from which the five fish in the treated tank have been "selected without replacement." Although the numbers 5, 20, and 25 that appear in the formula for $P(K = k)$ play the algebraic role of the numbers $\chi$, $\eta - \chi$, and $\eta$ in the urn problem, they do not characterize a larger population of fish from which the five in the tank have been drawn. These numbers just happen to characterize the ichthyologist's opinions about the number of sick fish in the tank. (Notice that in problem 1 they would not!) Nonetheless, a terminology has developed in statistical literature on sampling theory that refers to $\chi$ and $\eta$ and as "superpopulation parameters." This nomenclature is somewhat amusing, since there is usually no superpopulation from which the actual population has been selected. These parameters $\chi$ and $\eta$ characterize your knowledge of $X$. They do not characterize $X$ itself. Nor do they characterize a process that generates $X$. To someone else, in another state of knowledge about X that does not bear this similarity of uncertainty to the situation of drawing balls without replacement from urns, opinions about $X$ may not even be expressible in a functional form involving $\chi$ and $\eta$. □

We discuss hypergeometric probability distributions in more detail in Chapter 5. It might be well to read this portion in tandem with that chapter.

### *3.11.2.2 Inference from Sample Information, Supposing Less Detailed Assertions Are Made about $S_N$*

The wondering reader may wonder, "Hey! Why have we gone into such detail to study inference for events regarded exchangeably while supposing that I have asserted my entire probability distribution for their sum? What if I can't easily specify my probabilities for all these events. Is there still anything that can be said?"

If you regard $E_1, \ldots, E_N$ exchangeably, but are less specific in your assessment of the quantity $S_N$, you could use the fundamental theorem of prevision to compute your inference about $S_N$ on the basis of learning the value of $S_n$. Let us outline briefly how this is done. You have already thought about this procedure if you have done problem 5 of Section 3.8.2.

Consider $N$ logically independent events $E_1, \ldots, E_N$, which can be represented as $\mathbf{E}_N = \mathbf{R}_{N,2^N}\mathbf{Q}_{2^N}$. Recall from Section 3.8.2 that your regarding the events $E_1, \ldots, E_N$ exchangeably implies that your prevision for the constituent vector $\mathbf{Q}_{2^N}$ is representable as $P(\mathbf{Q}_{2^N}) = \mathbf{M}_{2^N,(N+1)}\mathbf{p}_N$, where $\mathbf{p}_N$ is the vector of probabilities $\mathbf{p}_N = P[(S_N = 0), (S_N = 1), \ldots, P(S_N = N)]$ and $\mathbf{M}_{2^N,(N+1)}$ is the appropriate reduction matrix.

Now suppose that you do not want to assert all the $N+1$ previsions $\mathbf{P}_N = P[(S_N = 0), P(S_N = 1), \ldots, P(S_N = N)]$, but you do want to make $K$ different assertions about $S_N$, representable in the form $\mathbf{A}_{K,(N+1)}\,\mathbf{p}_N \leqslant \mathbf{b}_K$. Included in the $K$ assertions represented in this way could be such details as $b_1 \leqslant P(S_N) \leqslant b_2$ or $P(S_N \leqslant x) \leqslant b_3$, where $x$ is an integer, and $b_i$ denote various numbers that you specify. For $P(S_N)$ can be expressed as the vector product $(0\ \ 1\ \ 2\ \ 3\ \ \cdots\ \ N)\mathbf{p}_N$, and $P(S_N \leqslant x)$ can be expressed as $(\mathbf{1}^{\mathrm{T}}_{(1+X)}\ \ \mathbf{0}^{\mathrm{T}}_{(n-x)})\mathbf{p}_N$, where $(\mathbf{1}^{\mathrm{T}}_{(q+x)}\ \ \mathbf{0}^{\mathrm{T}}_{(n-x)})$ is the row vector whose first $1+x$ components equal 1 and whose final $n-x$ components equal 0.

In such a scenario of your assertions regarding $\mathbf{E}_N$, what would be your coherent inference about $S_N$ conditional upon the information that the sum of the first $n$ events equals $a$; that is, $S_n = a$? The fundamental theorem of prevision part II allows us to compute cohering inferences of the form $P(S_N | S_n = a)$ and $P(S_N = A | S_n = a)$ for any value of $A$ from 0 through $N$. The logic of applying the fundamental theorem of prevision is direct, and the computational procedure can be stated simply, defining the following notation:

1. Construct the realm matrix $\mathbf{R}(\mathbf{E}_N) = \mathbf{R}_{N,2^N}$.
2. Append the quantities $(S_n = a)$, $S_N(S_n = a)$, and $(S_N = A)(S_N = a)$ to the vector $\mathbf{E}_N$, and compute the corresponding rows of the augmented realm matrix. Denote these rows by $\mathbf{r}_a$, $\mathbf{r}_{a,S_N}$, and $\mathbf{r}_{a,A}$.

Using this notation, we can now state that if you regard the events $\mathbf{E}_N$ exchangeably and you make $K$ assertions about $S_N$ that are representable in the form $\mathbf{A}_{K,(N+1)}\mathbf{p}_N \leqslant \mathbf{b}_K$, then coherency requires that you assert the conditional prevision interval

$$P(S_N|S_n = a) \in [p_l(S_N|S_n = a), p_u(S_N|S_n = a)],$$

and for any value of $A$ from 0 through $N$ that you assert

$$P(S_N = A|S_n = a) \in [p_l(S_N = A|S_n = a), p_u(S_N = A|S_n = a)],$$

where these interval boundaries are computable, respectively, from the following fractional programming problems:

Find the $(N+1)$-dimensional vectors $\mathbf{p}_N = (p_0, p_1, \ldots, p_N)$ for which

$$\begin{aligned}
p_l(S_N|S_n = a) &= \min(\mathbf{r}_{a,S_N}\mathbf{M}_{2^N,(N+1)}\mathbf{p}_N/\mathbf{r}_a\mathbf{M}_{2^N,(N+1)}\mathbf{p}_N),\\
p_u(S_N|S_n = a) &= \max(\mathbf{r}_{a,S_N}\mathbf{M}_{2^N,(N+1)}\mathbf{p}_N/\mathbf{r}_a\mathbf{M}_{2^N,(N+1)}\mathbf{p}_N),\\
p_l(S_N = A|S_n = a) &= \min(\mathbf{r}_{a,A}\mathbf{M}_{2^N,(N+1)}\mathbf{p}_N/\mathbf{r}_a\mathbf{M}_{2^N,(N+1)}\mathbf{p}_N),\\
p_u(S_N = A|S_n = a) &= \max(\mathbf{r}_{a,A}\mathbf{M}_{2^N,(N+1)}\mathbf{p}_N/\mathbf{r}_a\mathbf{M}_{2^N,(N+1)}\mathbf{p}_N),
\end{aligned}$$

each subject to the linear constraints

$$\begin{aligned}
\mathbf{A}_{K,(N+1)}\mathbf{p}_N &\leqslant \mathbf{b}_K,\\
\mathbf{1}^{\mathrm{T}}_{(N+1)}\mathbf{p}_N &= 1,\\
\mathbf{p}_N &\geqslant \mathbf{0}_{(N+1)}.
\end{aligned}$$

The bounding interval for each of these conditional prevision assertions has been stated marginally. The joint partially asserted prevision polytope for all of these conditional previsions would be the image of the common feasible set of solutions to these programming problems under the vector of transformations defined by the various objective functions.

In the scenario we have been discussing, all the previsions mentioned as inputs to the inference problem have involved assertions regarding $S_N$ in the form $\mathbf{A}_{K,(N+1)}\mathbf{p}_N \leqslant \mathbf{b}_K$. However, the FTP can deal with asserted conditional previsions as well. In fact, it allows you to specify some or all of your conditional previsions $P(S_N|S_n = a)$, for various values of $a$, and to determine the coherent consequences of these assertions for your $P(S_N = A)$, or $P(S_N)$, or whatever. Some interesting results in this context appear in the recent article of Lad, Deely, and Piesse (1995).

In practical applications, it is often very easy and useful to specify some of your knowledge in the form of bounds on conditional previsions. These can be assessed by thinking about how you would react to seeing data of various sorts. The FTP accepts this form of input into an inference problem quite readily. If you would like to try a simple problem, return to problem 5 in Section 3.8.2.

We are ready to turn our attention to a useful general extension of the concept of exchangeable judgments.

## 3.12 ASSERTIONS OF PARTIAL EXCHANGEABILITY

Having exhaustively studied the judgment to regard a sequence of events exchangeably, we would do well to consider a generalization of this judgment that has broader applications to the statistics of 2 × 2 tables, to the design of experiments, and to measurements ordered sequentially in time. In this concluding section we continue our study of exchangeability within the context of events rather than general quantities, but we shall extend the concept of exchangeability to regarding a sequence of events only *partially exchangeably*. Formalities of judging general quantities exchangeably are examined only in Chapters 5 and 7.

Our studies have led us to realize that the judgment to regard events exchangeably is equivalent to regarding the number of events, $N$, and their sum, $S(\mathbf{E}_N) = \sum_{i=1}^{N} E_i$, as sufficient for your inference regarding some of them on the basis of observing the values of others. For any vectors $\mathbf{e}_N$ and $\mathbf{f}_N \in \mathcal{R}(\mathbf{E}_N)$, your $P(\mathbf{E}_N = \mathbf{e}_N) = P(\mathbf{E}_N = \mathbf{f}_N)$ as long as $S(\mathbf{e}_N) = S(\mathbf{f}_N)$. Their common probability must equal $P[S(\mathbf{E}_N) = S(\mathbf{e}_N)]/{}^N C_{S(\mathbf{e}_N)}$.

The characterization of exchangeable judgments through the concept of sufficiency is the easiest interpretable way to generalize the concept. Let us begin with a general definition and then study its application in three different contexts.

***Definition 3.14.*** You *regard the events* $\mathbf{E}_N$ *partially exchangeably with respect to the vector of statistics* $\mathbf{S}_K(\mathbf{E}_N) \equiv [S_1(\mathbf{E}_N), S_2(\mathbf{E}_N), \ldots, S_K(\mathbf{E}_N)]$ if your $P(\mathbf{E}_N = \mathbf{e}_N) = P(\mathbf{E}_N = \mathbf{f}_N)$ for any choice of vectors $\mathbf{e}_N$ and $\mathbf{f}_N \in \mathcal{R}(\mathbf{E}_N)$ for which $\mathbf{S}_K(\mathbf{e}_N) = \mathbf{S}_K(\mathbf{f}_N)$, and the function defining the statistics vectors satisfies some recursion formula, $\mathbf{S}_K(\mathbf{E}_{J+1}) = \mathbf{S}_K^*(\mathbf{S}_K(\mathbf{E}_J), E_{J+1})$ for any $J < N$. □

Evidently, your common probability for these two events must equal $P[\mathbf{S}_K(\mathbf{E}_N) = \mathbf{S}_K(\mathbf{e}_N)]/N^*(\mathbf{S}_K(\mathbf{e}_N))$, where $N^*$ is the number of elements in $\mathcal{R}(\mathbf{E}_N)$ that yield the same sufficient statistics vector value, $\mathbf{S}_K(\mathbf{e}_N)$. Your asserting your probability distribution for the vector $\mathbf{S}_K(\mathbf{E}_N)$ would characterize your probability distribution for $\mathbf{E}_N$. The recursion formula restriction on $\mathbf{S}_K(\cdot)$ means that the sufficient statistics for any $J+1$ events can be computed from the sufficient statistics for any $J$ of them, together with the value of the $(J+1)$st event. Thus, the statistics vector is sufficient not merely for your inference about the next event, but also for your computing sufficient statistics based on more events.

In the following subsections we motivate the judgment of partial exchangeability in three situations of its application. Section 3.12.1 is relevant to the statistics of 2 × 2 tables. It supposes a judgment of exchangeability regarding events within identifiable subgroups of a collection of sampled events, but rejects exchangeability regarding events in different subgroups. Section 3.12.2 is relevant to considerations of experimental design. It supposes the judgment of exchangeability among events defined by a specific "treatment type," but rejects exchangeability among events defined by different treatment types. Finally, Section 3.12.3 is relevant to the analysis of events occurring in an ordered time series. It supposes a judgment of exchangeability among events selected from a temporal sequence, but only under a condition that the histories of the proximate event(s) preceding the selected events satisfy specifiable conditions.

While motivating these distinct types of judgment by examples, we construct the specific structure of partial exchangeability appropriate to each. Theorems will be stated merely as summaries of discussions in which they have been essentially proved. Relying on your understanding of extendable exchangeability, aspects of the analysis pertaining to the extendability of partially exchangeable assertions will merely be stated without proof.

### 3.12.1 Partial Exchangeability: Exchangeability within Subgroups

Suppose that $N$ taxpayers are selected from the 1990 U.S. tax returns by some scrambling procedure. Consider the events $E_{1j}$ and $E_{2j}$, defined by

$$E_{1j} \equiv (\text{The declared taxable income of a taxpayer } j \text{ exceeds } \$18{,}000),$$

$$E_{2j} \equiv (\text{Taxpayer } j \text{ has completed a high school diploma}).$$

Using $\mathbf{E}_{2j}$ to denote the column vector $(E_{1j}, E_{2j})^{\mathrm{T}}$, suppose we define

$$\mathbf{E}_{2,N} \equiv (\mathbf{E}_{21}, \mathbf{E}_{22}, \ldots, \mathbf{E}_{2N}) = \begin{pmatrix} E_{11}E_{12}\cdots E_{1N} \\ E_{21}E_{22}\cdots E_{2N} \end{pmatrix},$$

which is the matrix of these column pairs of events for $N$ persons selected by scrambling the tax returns.

Note carefully the distinction between $\mathbf{E}_{2N}$ and $\mathbf{E}_{2,N}$. The former vector represents the final column of the latter matrix. The events $E_{11}, \ldots, E_{1N}$ will be referred to as "first-component" events, and the events $E_{21}, \ldots, E_{2N}$ as "second-component" events of the matrix of event vectors, $\mathbf{E}_{2,N}$.

Now in what sense do you regard the component events of $\mathbf{E}_{2,N}$ exchangeably? On the one hand, our scrambling Theorem 3.15 ensures that if you consider the unit identifications to be scrambled, then you must regard the vector of first component events exchangeably, and you must also regard the vector of second component events exchangeably. Furthermore, it would ensure that you must regard the vector *pairs* $\mathbf{E}_{21}, \ldots, \mathbf{E}_{2N}$ exchangeably. What, precisely, would this mean?

Notice that there are four distinct elements in the realm of any vector pair: $\mathscr{R}(\mathbf{E}_{2j}) = \{(0,0)^{\mathrm{T}}, (0,1)^{\mathrm{T}}, (1,0)^{\mathrm{T}}, (1,1)^{\mathrm{T}}\}$. Furthermore, $\mathscr{R}(\mathbf{E}_{2,N}) = [\mathscr{R}(\mathbf{E}_{2j})]^N$, for the component events of $\mathbf{E}_{2,N}$ are *logically* independent. Now the observation of $\mathbf{E}_{2,N}=\mathbf{e}_{2,N}\in\mathscr{R}(\mathbf{E}_{2,N})$ would identify that some number $S_{00}(\mathbf{e}_{2,N})$ of the $\mathbf{E}_{2j}$ pairs equal $(0, 0)^{\mathrm{T}}$, some $S_{01}(\mathbf{e}_{2,N})$ of the $\mathbf{E}_{2j}$ pairs equal $(0, 1)^{\mathrm{T}}$, some $S_{10}(\mathbf{e}_{2,N})$ of the $\mathbf{E}_{2j}$ pairs equal $(1, 0)^{\mathrm{T}}$, and some $S_{11}(\mathbf{e}_{2,N})$ of the $\mathbf{E}_{2j}$ pairs equal $(1, 1)^{\mathrm{T}}$. Since the units of the $\mathbf{E}_{2j}$ vectors are selected by scrambling, you must assign the same probability to any event defined by a matrix of event outcomes $(\mathbf{E}_{2,N}=\mathbf{e}_{2,N})$ for which the vector of statistics $(S_{00}(\mathbf{e}_{2,N}), S_{01}(\mathbf{e}_{2,N}), S_{10}(\mathbf{e}_{2,N}), S_{11}(\mathbf{e}_{2,N}))$ is equal. For you avow that the unit indicators give you no information about which of the vectors $\mathbf{E}_{2j}$ are the ones that yield each of the possibilities in $\mathscr{R}(\mathbf{E}_{2j})$.

Before proceeding, review your understanding of the following questions and answers involving combinatorics. First, notice that the size of $\mathscr{R}(\mathbf{E}_{2,N})$ is $2^{2N}$. (For each component of the matrix may be 0 or 1.) Now define the vector of four functions, $\mathbf{S}_4:\mathscr{R}(\mathbf{E}_{2,N})\rightarrow\mathscr{R}(\mathbf{S}_4)$, by

$$\mathbf{S}_4(\mathbf{e}_{2,N})\equiv\left(\sum_{j=1}^{N}\tilde{e}_{1j}\tilde{e}_{2j},\ \sum_{j=1}^{N}\tilde{e}_{1j}e_{2j},\ \sum_{j=1}^{N}e_{1j}\tilde{e}_{2j},\ \sum_{j=1}^{N}e_{1j}e_{2j}\right).$$

These four functions identify how many of the column vectors $\mathbf{e}_{2j}$ of $\mathbf{e}_{2,N}$ equal $(0, 0)^{\mathrm{T}}$, how many equal $(0, 1)^{\mathrm{T}}$, how many equal $(1, 0)^{\mathrm{T}}$, and how many equal $(1, 1)^{\mathrm{T}}$, respectively. It will be convenient in what follows to denote these statistics by the vector $\mathbf{S}_4=(S_{00}, S_{01}, S_{10}, S_{11})$, whose generic realm element will be denoted by $\mathbf{s}_4=(s_{00}, s_{01}, s_{10}, s_{11})$.

**Q.** What is the realm of the vector $\mathbf{S}_4$?

**A.** $\mathscr{R}(\mathbf{S}_4)=\{\mathbf{s}_4|$ the components are nonnegative integers, summing to $N\}$.

**Q.** For how many elements $\mathbf{e}_{2,N}\in\mathscr{R}(\mathbf{E}_{2,N})$ is $\mathbf{S}_4(\mathbf{e}_{2,N})=(s_{00}, s_{01}, s_{10}, s_{11})$?

**A.** ${}^{N}C_{(s_{00},s_{01},s_{10},s_{11})}=N!/[s_{00}!\ \ s_{01}!\ \ s_{10}!\ \ s_{11}!]$. [Refresh yourself on the multiple combinatoric counting principle 6 of Section 3.8.1. Also recognize that the sum $\sum_{\mathbf{s}_4\in\mathscr{R}(\mathbf{S}_4)}{}^{N}C_{\mathbf{s}_4}=4^N$, on account of the multinomial expansion theorem, as explained in statements 10 (iii) and (iv) of Section 3.8.1. This is consoling, for the size of $\mathscr{R}(\mathbf{E}_{2,N})$ is $4^N$.]

Your recognition of the scrambling procedure means that you assert equal probability to any matrix of event outcomes that yields the same sufficient statistics: if $\mathbf{S}_4(\mathbf{e}_{2,N})=\mathbf{S}_4(\mathbf{f}_{2,N})$, then your $P(\mathbf{E}_{2,N}=\mathbf{e}_{2,N})=P(\mathbf{E}_{2,N}=\mathbf{f}_{2,N})$. It should also be evident that, for any matrix realm element $\mathbf{e}_{2,N}\in\mathscr{R}(\mathbf{E}_{2,N})$, your

$$P(\mathbf{E}_{2,N}=\mathbf{e}_{2,N})=P[\mathbf{S}_4(\mathbf{E}_{2,N})=\mathbf{S}_4(\mathbf{e}_{2,N})]/{}^{N}C_{\mathbf{S}_4(\mathbf{e}_{2,N})}.$$

We can conclude these considerations, applying the general Definition 3.14, with the realization that you regard the $2N$ events composing $\mathbf{E}_{2,N}$ partially exchangeably with respect to the four-dimensional vector of statistics $\mathbf{S}_4(\mathbf{e}_{2,N})$. The qualifying computational requirement for sufficient statistics vectors is satisfied by $\mathbf{S}_4(\mathbf{e}_{2,J+1})=\mathbf{S}_4(\mathbf{e}_{2,J})+\mathbf{S}_4(\mathbf{e}_{2(J+1)})$. The sufficient statistics vector $\mathbf{S}_4$ augments from $\mathbf{S}_4(\mathbf{e}_{2,J})$ by adding 1 to the component appropriate to the value of $\mathbf{e}_{2(J+1)}$.

Our conclusion to this point derives by coherency merely from your regarding the vector *pairs* exchangeably. You regard the first component events exchangeably and the second component events exchangeably. But you regard the *entire* collection of events only partially exchangeably, with respect to the vector of statistics $\mathbf{S}_4$. This vector, by the way, summarizes the observations $\mathbf{E}_{2,N} = \mathbf{e}_{2,N}$ by means of a $2 \times 2$ table.

Let us now address the structure of coherent inference that must be asserted when events are regarded partially exchangeably in this way. You should begin your part in these considerations by assessing values for your assertions of $P(E_{1j})$, $P(E_{2j})$, $P(E_{1j}|E_{2j})$, $P(E_{1j}|\tilde{E}_{2j})$, $P(E_{2j}|E_{1j})$, and $P(E_{2j}|\tilde{E}_{1j})$ for the taxpayer example. (Yes, review the operational definitions of these events and assert your probabilities. Notice that the definitions are based on taxpayers' taxable incomes, not on household incomes.) So that we have some numbers to work with, I'll tell you that as I write this, sitting at my desk in Christchurch, New Zealand, April 1993, I assert my

$$P(E_{1j}) = .4, \qquad P(E_{1j}|E_{2j}) = .6, \qquad P(E_{1j}|\tilde{E}_{2j}) = .1,$$

and

$$P(E_{2j}) = .6, \qquad P(E_{2j}|E_{1j}) = .9, \qquad P(E_{2j}|\tilde{E}_{1j}) = .4.$$

You may be surprised by (disagree with) my probabilities, and maybe not. (Check for yourself that at least my previsions are coherent. Are yours?) My prevision values are based on my foggy knowledge about incomes and educational attainment in the United States in 1990, about the age distribution of taxpayers, and the economic knowledge that in 1990 it was very difficult, though not impossible, to contract a high-wage or high-salary job without having completed a high school education. You should recognize, by the way, that on the basis of the probabilities I've asserted, coherency requires me to assert the prevision $P(\mathbf{S}_4(\mathbf{E}_{2,N})) = N(.36, .24, .04, .36)$. But I have not yet asserted my complete probability distribution for $\mathbf{S}_4(\mathbf{E}_{2,N})$.

Before proceeding, I will highlight one feature of my opinions. In my foregoing statement I asserted several conditional probabilities, such as $P(E_{1j}|E_{2j}) = .6$. The second component event for unit $j$ describes the educational status of the same person whose income status is described by $E_{1j}$. Do notice that if $k \neq j$ there is no reason that my $P(E_{1j}|E_{2k})$ need equal my $P(E_{1j}|E_{2j})$. I mention this fact here merely as a prelude to a different type of inference problem that we study in the next subsection.

Also notice that I explicitly do not regard the first component events $E_{1j}$ exchangeably with the second component events $E_{2j}$. For I do not even assess them with the same probabilities.

Now suppose I would like to inform myself better about the subject of education and income in the United States in 1990 by observing the values of $\mathbf{E}_{2,N}$ corresponding to $N$ selected taxpayers, and making inferences from this sample regarding the values of $\mathbf{E}_{2(N+1)}$ for another taxpayer selected by the scrambling procedure. Inferential assertions of special interest would be the values of

$$P(E_{1(N+1)}|(\mathbf{E}_{2,N}=\mathbf{e}_{2,N})) \quad \text{and} \quad P(E_{2(N+1)}|(\mathbf{E}_{2,N}=\mathbf{e}_{2,N})),$$
$$P(E_{1(N+1)}|E_{2(N+1)}(\mathbf{E}_{2,N}=\mathbf{e}_{2,N})) \quad \text{and} \quad P(E_{2(N+1)}|E_{1(N+1)}(\mathbf{E}_{2,N}=\mathbf{e}_{2,N})),$$
$$P(E_{1(N+1)}|\tilde{E}_{2(N+1)}(\mathbf{E}_{2,N}=\mathbf{e}_{2,N})) \quad \text{and} \quad P(E_{2(N+1)}|\tilde{E}_{1(N+1)}(\mathbf{E}_{2,N}=\mathbf{e}_{2,N})).$$

Do you agree that these inferential probabilities would be interesting to compute and to compare with my assertions of the same six probabilities prior to the conditioning observation $(\mathbf{E}_{2,N}=\mathbf{e}_{2,N})$? Please think about them for a moment, because I shall continue now with merely computational technicalities. The differences in the arrays of assertions exhibit what is to be learned about the relationship between income and education from the sample observations.

On the basis of your regarding the vectors $\mathbf{E}_{21},\ldots,\mathbf{E}_{2(N+1)}$ exchangeably, each of these inferential probabilities can be computed from the general result that for any $\mathbf{e}_2\in\mathscr{R}(\mathbf{E}_{2(N+1)})=\{(0,\ 0)^{\mathrm{T}},\ (0,\ 1)^{\mathrm{T}},\ (1,\ 0)^{\mathrm{T}},\ (1,\ 1)^{\mathrm{T}}\}$ and $\mathbf{e}_{2,N}\in\mathscr{R}(\mathbf{E}_{2,N})$, if $P(\mathbf{E}_{2,N}=\mathbf{e}_{2,N})>0$ then

$$P[(\mathbf{E}_{2(N+1)}=\mathbf{e}_2)|(\mathbf{E}_{2,N}=\mathbf{e}_{2,N})]=P[(\mathbf{E}_{2(N+1)}=\mathbf{e}_2)(\mathbf{E}_{2,N}=\mathbf{e}_{2,N})]/P(\mathbf{E}_{2,N}=\mathbf{e}_{2,N})$$
$$=\frac{P[\mathbf{S}_4(\mathbf{E}_{2,N+1})=\mathbf{S}_4(\mathbf{e}_{2,N})+\mathbf{S}_4(\mathbf{e}_2)]/{}^{(N+1)}C_{\mathbf{S}_4(\mathbf{e}_{2,N})+\mathbf{S}_4(\mathbf{e}_2)}}{P(\mathbf{S}_4(\mathbf{E}_{2,N})=\mathbf{S}_4(\mathbf{e}_{2,N})]/{}^{N}C_{\mathbf{S}_4(\mathbf{e}_{2,N})}}.$$

This result is a statement of finite coherent inference in the context of $2(N+1)$ events regarded partially exchangeably in the way appropriate to this example. It is a multiple-combinatoric extension of Theorem 3.12 in Section 3.9 which pertains to events that are regarded completely exchangeably. Algebraically, it follows from the feature of conditional probability that if $P(B)>0$ then $P(A|B)=P(AB)/P(B)$.

You can perhaps imagine a companion result that if your partially exchangeable opinions about $\mathbf{E}_{2,(N+1)}$ are infinitely exchangeably extendable, then for some distribution function $M(\boldsymbol{\theta}_4)$ on the simplex, $\mathbf{S}^3$,

$$P[(\mathbf{E}_{2,(N+1)}=\mathbf{e}_2)|(\mathbf{E}_{2,N}=\mathbf{e}_{2,N})]=\int_{\mathbf{S}^3}\prod_{i=1}^{4}\theta_i^{S_i(\mathbf{e}_2)}\,dM(\boldsymbol{\theta}_4|\mathbf{S}_4(\mathbf{E}_{2,N})=\mathbf{S}_4(\mathbf{e}_{2,N}))$$

and

$$dM(\theta_4|\mathbf{S}_4(\mathbf{E}_{2,N})=\mathbf{S}_4(\mathbf{e}_{2,N}))=\frac{\prod_{i=1}^{4}\theta_i^{S_i(\mathbf{e}_{2,N})}\,dM(\boldsymbol{\theta}_4)}{\int_{\mathbf{S}^3}\prod_{i=1}^{4}\theta_i^{S_i(\mathbf{e}_{2,N})}\,dM(\boldsymbol{\theta}_4)}.$$

These statements presume the notation $S_1(\mathbf{e}_{2,N})\equiv S_{00}(\mathbf{e}_{2,N})$, $S_2(\mathbf{e}_{2,N})\equiv S_{01}(\mathbf{e}_{2,N})$, $S_3(\mathbf{e}_{2,N})\equiv S_{10}(\mathbf{e}_{2,N})$, and $S_4(\mathbf{e}_{2,N})\equiv S_{11}(\mathbf{e}_{2,N})$.

A discussion of details regarding the mixing distribution function $M(\cdot)$ is best deferred until Chapter 5. Once you learn about the Dirichlet distribution, you will realize it yields a very simple computational resolution of predictive probabilities of the form $P[(\mathbf{E}_{2,(N+1)} = \mathbf{e}_2) | (\mathbf{E}_{2,N} = \mathbf{e}_{2,N})]$.

As to the wider applicability of partially exchangeable judgments, the substantive context of the measurements we are considering could be transferred to different subject matters quite easily. For example, the event measurements considered might identify various aspects of the plots where we were searching for salamanders at the beginning of this chapter, $E_{1j}$ denoting whether the $j$th plot contains at least one salamander, and $E_{2j}$ denoting whether the pH measurement on the $j$th plot exceeds 3.8. More expansively, the concept of partial exchangeability can obviously apply to more than two measurements made on a particular unit. For example, if we made further measurements on the person whose tax form was selected, we might define longer vectors of events denoted by $\mathbf{E}_{Kj} \equiv (E_{1j}, E_{2j}, \ldots, E_{Kj})^{\mathrm{T}}$ whose 3rd, 4th, ..., and $K$th components identify aspects of the person's age, sex, marital status, race, geographical location, employment history, and so on. In the context of the salamander study, the 3rd, 4th, ..., and $K$th components could identify aspects of the humidity, temperature, or salinity of the soil on the $j$th plot.

The inferential results we have generated in the context of two measurements made on each unit of a population generalize to the situation of $K$ measurements in a straightforward way. The dimension of the sufficient statistics vector becomes $2^K$ rather than 4. You need merely keep track of how many of the sampled population units exhibit each of the $2^K$ possible values of $\mathbf{E}_K \in \mathscr{R}(\mathbf{E}_K)$. This is a powerful result, since we need only track $2^K$ sufficient statistics no matter how large $N$ might be.

The following three problems should help you firm your conceptions of the issues involved in computing inferences under this form of partially exchangeable judgments. All require a computer for their completion. I suggest you complete the first one to yield some results that will afford an interesting comparison for a related problem that will be proposed to illuminate the ideas of the next section. The second and third problems here can then be treated as thought problems.

## PROBLEMS

**1.** Suppose that you regard the three event pairs $\mathbf{E}_{21}, \mathbf{E}_{22}, \mathbf{E}_{23}$ exchangeably. The realm of the sufficient statistic vector $\mathbf{S}_4$ contains 20 elements. Suppose you assert the following probabilities for the 20 events of the partition generated by $\mathbf{S}_4$, each of the form $(\mathbf{S}_4 = \mathbf{s}_{4j})$:

| $s_{1i}$ | 3 | 0 | 0 | 0 | 2 | 2 | 2 | 1 | 0 | 0 | 1 | 0 | 0 | 1 | 0 | 0 | 1 | 1 | 1 | 0 |
|---|---|---|---|---|---|---|---|---|---|---|---|---|---|---|---|---|---|---|---|---|
| $s_{2i}$ | 0 | 3 | 0 | 0 | 1 | 0 | 0 | 2 | 2 | 2 | 0 | 1 | 0 | 0 | 1 | 0 | 1 | 1 | 0 | 1 |
| $s_{3i}$ | 0 | 0 | 3 | 0 | 0 | 1 | 0 | 0 | 1 | 0 | 2 | 2 | 2 | 0 | 0 | 1 | 1 | 0 | 1 | 1 |
| $s_{4i}$ | 0 | 0 | 0 | 3 | 0 | 0 | 1 | 0 | 0 | 1 | 0 | 0 | 1 | 2 | 2 | 2 | 0 | 1 | 1 | 1 . |
| $74\,P(\mathbf{S}_4 = \mathbf{s}_{4i})$ | 4 | 3 | 2 | 1 | 8 | 6 | 4 | 7 | 5 | 3 | 6 | 4 | 2 | 5 | 3 | 1 | 4 | 3 | 2 | 1 |
| $N^*(\mathbf{s}_{4i})$ | 1 | 1 | 1 | 1 | 3 | 3 | 3 | 3 | 3 | 3 | 3 | 3 | 3 | 3 | 3 | 3 | 6 | 6 | 6 | 6. |

Notice that the row of this information showing the probabilities $P(\mathbf{S}_4 = \mathbf{s}_{4i})$ needs to be divided by 74. This is both for visual ease in your reading the table, and so that you can generate intuitions regarding the assessments being made. Study the numerical structure of these assessments to gain some feeling for the opinions they represent. The final row of information displays $N^*(\mathbf{s}_{4i})$, the number of events $(\mathbf{E}_{2,3} = \mathbf{e}_{2,3})$ that yield the same sufficient statistic value as $\mathbf{s}_{4i}$. Can you identify some of the individual elements $\mathbf{e}_{2,3} \in \mathcal{R}(\mathbf{E}_{2,3})$ that would yield these sufficient statistics vectors? Using the results presented in this section, compute the following probabilities:

**a.** $P(E_{11}), P(E_{21}), P(E_{21}|E_{11}), P(E_{21}|\tilde{E}_{11}), P(E_{11}|E_{21}), P(E_{11}|\tilde{E}_{21})$;

**b.** $P(E_{21}|E_{12}), P(E_{21}|\tilde{E}_{12}), P(E_{11}|E_{22}), P(E_{11}|\tilde{E}_{22})$.

**c.** Compute each of the probabilities listed in parts a and b again, conditioned additionally upon the event $(\mathbf{E}_{23} = \mathbf{e}_2)$ for each $\mathbf{e}_2 \in \{(0, 0)^T, (0, 1)^T, (1, 0)^T, (1, 1)^T\}$. [*Hint*: The probabilities of part a should be easy to compute directly from the probabilities for the sufficient statistics vector possibilities. Just think about which possible values for the sufficient statistics vector could allow the relevant events associated with the problems of part a to occur. But for parts b and c, you need to start the problem from the beginning, creating a realm matrix for $\mathbf{E}_6$, the event vector composed of the stacked columns of $\mathbf{E}_{2,3}$. Append the sufficient statistic vector $\mathbf{S}_4$ to $\mathbf{E}_6$, using their functional definitions to append their appropriate values to the realm matrix. Construct a vector, $\mathbf{q}_{64}$, of the associated probabilities for each column of this realm matrix on the basis of the distribution of $\mathbf{S}_4$ given in the problem. To determine any probabilities needed for parts b or c, merely construct the appropriate product event as a function of $\mathbf{E}_6$, and append a row for it to the realm matrix. The probability for that event will then equal that row of the realm matrix times $\mathbf{q}_{64}$.]

**d.** Think about the numerical values of your results, comparing appropriate ones, such as $P(E_{11})$, $P(E_{11}|P(E_{21})$, $P(E_{11}|E_{22})$, $P(E_{11}|E_{21}(E_{23} = (0, 0)^T))$ and $P(E_{11}|E_{22}(E_{23} = (0, 0)^T))$.

**e.** Once you have learned about the Dirichlet distribution, compute the array of probabilities based on the initial mixing function of $\boldsymbol{\theta}_4 \sim D(4, 3, 2, 1)$.

**2.** Suppose the column vectors of the matrix $\mathbf{E}_{K,N}$ are stacked one atop the other as a $KN \times 1$ column vector, denoted by $\mathbf{E}_{KN,1}$. Identify the matrix $\mathbf{M}$ that specifies the coherency restriction that the partial exchangeability assertion regarding $\mathbf{E}_{K,N}$ places on the partition it generates: $P(\mathbf{Q}(\mathbf{E}_{KN,1})) = \mathbf{M}P(\mathbf{S}_{2^K}(\mathbf{E}_{KN,1}))$. Write a computer program to generate the matrix $\mathbf{M}$ appropriate to the realm matrix for events that is produced by the program realm.m of problem 6 of Section 2.1. [*Hint*: Append the vector $\mathbf{S}_{2^K}(\mathbf{E}_{KN,1})$ to the column vector $\mathbf{E}_{KN,1}$ in its stacked form. Then imagine the realm matrix for the expanded vector $(\mathbf{E}_{KN,1}^T \mathbf{S}_{2^K}^T)^T$. Realize that any columns of this realm matrix whose final $2^K$ row elements are equal identify constituents of the partition

$\mathbf{Q}(\mathbf{E}_{KN,1})$ that must be accorded equal probabilities, each with the value $P(\mathbf{S}_{2^K}(\mathbf{E}_{KN,1}) = \mathbf{s}_{2^K}(\mathbf{e}_{KN,1}))/{}^N C_{\mathbf{s}_{2^K}}$.] The developments in the present subsection have been presented in the context that you assert your probability distribution for the sufficient statistic $\mathbf{S}_{2^K}(\mathbf{E}_{2,N})$. Think about how your matrix $\mathbf{M}$ would be used with the fundamental theorem of prevision if you assert partial exchangeability in the form we have developed, but you asserted merely a few 'ballpark" previsions for some of the events defined by $\mathbf{S}_{2^K}(\mathbf{E}_{2,N})$.

**3.** Reconsider problem 1. Suppose you assert merely the probabilities you determined for part a along with partial exchangeability. Use the program you wrote for problem 2 along with a computer to determine coherent bounds on the probabilities proposed in parts b and c of problem 1.

**4.** Show that if you regard the vectors $\mathbf{E}_{K1}, \mathbf{E}_{K2}, \ldots, \mathbf{E}_{KN}$ exchangeably, then within any subgroup of these $N$ units that exhibit the same values for $K-1$ of their event components, you must regard the $K$th component events exchangeably.

### 3.12.2 Partial Exchangeability: Exchangeability within Treatments

Consider the general subject matter of controlled experiments. For examples, consider event vectors $\mathbf{E}_{21}, \ldots, \mathbf{E}_{2N}$ for which the component events

$E_{1i}$ identify some yield characteristic of an hectare of corn plants sown with a particular type of seed and fertilized using one procedure, and

$E_{2i}$ identify the same characteristic of an hectare of corn plants sown with the same type of seed but fertilized using another procedure

or for which the component events

$E_{1i}$ identify some health characteristic of an AIDS patient receiving one type of treatment, and

$E_{2i}$ identify the same characteristic of an AIDS patient receiving another type of treatment.

Undoubtedly, you can imagine a plethora of such examples. In a well-designed experiment of this type, one typically would select subjects (patients, fields for sowing) that are indistinguishable in terms of other characteristics that the experimenter/theorist regards as relevant information concerning this characteristic. For example, supposing that the measurement on AIDS patients pertain to weight loss over a specified time interval of treatment, one would want to select as subjects patients who are in a similar stage of infection at the time the treatment begins, who have a similar bone structure and weight history prior to infection, who have a similar level of emotional support from friends and family, and so on.

For the fertilizer experiment, one would select fields for sowing that have similar sun exposure, soil composition, drainage pattern, and so on. We shall return to this idea in more detail.

For such experiments defining a sequence of event vectors $\mathbf{E}_{21}, \ldots, \mathbf{E}_{2N}$, the judgment to regard these $2N$ events partially exchangeably would be merited. But the vector of sufficient statistics would have even lower dimension than in the type of problems considered in Section 3.12.1. For here, your probability assertions $P(\mathbf{E}_{2,N} = \mathbf{e}_{2,N})$ would be constant over any matrix for which merely the row sums, $S_1 \equiv \sum_{j=1}^{N} e_{1j}$ and $S_2 \equiv \sum_{j=1}^{N} e_{2j}$, are constant. (Of course it is required that the number of experiments, $N$, be specified as well.) The four sufficient statistics of our preceding problem, $S_{00}$, $S_{01}$, $S_{10}$, and $S_{11}$, can generate these two statistics, since $S_1 = S_{10} + S_{11}$ and $S_2 = S_{01} + S_{11}$, but the four statistics cannot be generated merely from the knowledge of $S_1$ and $S_2$. The reason that the number of required statistics reduces in the context of an experimental setup is that the unit $j$ of the first row vector identifies a different person or a different hectare of land than the unit $j$ of the second row vector. It does not matter to you, for example, which of the 1's in the first row also correspond to 1's in the second row, because the positioning of the units in the first and the second rows is arbitrary.

Notice that the number of vectors $\mathbf{e}_{2,N} \in \mathscr{R}(\mathbf{E}_{2,N})$ that would yield the same sufficient statistics $S_1(\mathbf{e}_{2,N})$ and $S_2(\mathbf{e}_{2,N})$ equals the product ${}^N C_{S_1} {}^N C_{S_2}$. If the number of subjects receiving the first treatment were different from the number receiving the second treatment, the two distinct sizes of $N$ also need to be stated. In the first case the experimental design is said to be balanced, and in the second, unbalanced. We attend in what follows merely to the setup of balanced designs, both to simplify matters and to allow easy comparisons to be made with our results of the previous section.

A well-designed experiment is performed in such a way that the judgment of partial exchangeability with respect to the sufficient statistics $N$, $S_1$, and $S_2$ would be made. These concepts easily generalize to the situation in which you experiment on $K$ different treatments.

Let us summarize the inferential conclusions that can be made regarding controlled experiments, stating a theorem that parallels the conclusions of Section 3.12.1 regarding data sampled from a finite population.

**Theorem 3.16.** If $\mathbf{E}_{K1}, \ldots, \mathbf{E}_{KN}$ are vectors of events identifying characteristics of $KN$ units selected from an experiment involving $K$ different treatments, then coherency requires that if you regard each row vector of $\mathbf{E}_{K,N}$ exchangeably, then you regard all $KN$ events partially exchangeably with respect to the statistics identifying the total number of experiments and the $K$ row sums:

$$\mathbf{S}(\mathbf{E}_{K,N}) = (N, S_1(\mathbf{E}_{K,N}), S_2(\mathbf{E}_{K,N}), \ldots, S_K(\mathbf{E}_{K,N})), \qquad \text{where } S_i(\mathbf{E}_{K,N}) \equiv \sum_{j=1}^{N} E_{ij}.$$

(For simplicity, the remaining statements of this theorem are made in the context of $K = 2$. They can be easily generalized.)

Coherent inference regarding the values of the next experimental event vector, $\mathbf{E}_{2(N+1)}$, on the basis of the observation $\mathbf{E}_{2,N} = \mathbf{e}_{2,N}$ must respect the equation

$$P[(\mathbf{E}_{2(N+1)} = \mathbf{e}_2) | (\mathbf{E}_{2,N} = \mathbf{e}_{2,N})]$$
$$= \frac{P[\mathbf{S}(\mathbf{E}_{2,N+1}) = \mathbf{S}(\mathbf{e}_{2,N}) + \mathbf{S}(\mathbf{e}_2)]/[{}^{(N+1)}C_{S_1(\mathbf{e}_{2,N}) + S_1(\mathbf{e}_2)}\, {}^{(N+1)}C_{S_2(\mathbf{e}_{2,N}) + S_2(\mathbf{e}_2)}]}{P[\mathbf{S}(\mathbf{E}_{2,N}) = \mathbf{S}(\mathbf{e}_{2,N})]/[{}^{N}C_{S_1(\mathbf{e}_{2,N})}\, {}^{N}C_{S_2(\mathbf{e}_{2,N})}]}$$

for any $(\mathbf{e}_{2,N}\mathbf{e}_2) \in \mathscr{R}(\mathbf{E}_{2,N+1}) = \{0, 1\}^{2(N+1)}$.

Asserting your probability distribution for $\mathbf{S}(\mathbf{E}_{2,N+1})$ would characterize completely any coherent inferences regarding components of $\mathbf{E}_{2,N+1}$.

If your partially exchangeable opinions about $\mathbf{E}_{2,N}$ are infinitely exchangeably extendable, then for any $N$, for some distribution function on the unit square, $M(\boldsymbol{\theta}_2)$, your inferences are representable as

$$P[(\mathbf{E}_{2(N+1)} = \mathbf{e}_2) | (\mathbf{E}_{2,N} = \mathbf{e}_{2,N})]$$
$$= \int \prod_{i=1}^{2} \theta_i^{S_i(\mathbf{e}_2)} (1 - \theta_i)^{1 - S_i(\mathbf{e}_2)}\, dM(\boldsymbol{\theta}_2 | \mathbf{S}(\mathbf{E}_{2,N}) = \mathbf{S}_2(\mathbf{e}_{2,N})),$$

where

$$dM(\boldsymbol{\theta}_2 | \mathbf{S}(\mathbf{E}_{2,N}) = \mathbf{S}(\mathbf{e}_{2,N})) = \frac{\prod_{i=1}^{2} \theta_i^{S_i(\mathbf{e}_{2,N})} (1 - \theta_i)^{N - S_i(\mathbf{e}_{2,N})}\, dM(\boldsymbol{\theta}_2)}{\int \prod_{i=1}^{2} \theta_i^{S_i(\mathbf{e}_{2,N})} (1 - \theta_i)^{N - S_i(\mathbf{e}_{2,N})}\, dM(\boldsymbol{\theta}_2)}.$$

The final statement of the theorem specifying coherent inference in the context of infinitely extendable partially exchangeable judgments is almost always applicable to problems involving controlled experiments. For we usually imagine about our experiments that if only we had enough time, patience, and patients (experimental units), continuing our experimentation would not induce us to relinquish our judgments of exchangeability.

Note that if the distribution function $M(\boldsymbol{\theta}_2)$ is separable so that $dM(\boldsymbol{\theta}_2) = dM(\theta_1)\, dM(\theta_2)$, then treatment-type-1 events are regarded independently of treatment-type-2 events. This would be to affirm that experimental outcomes from one treatment tell you nothing about experiments on the other treatment. It is also worthy of interest that in the context of Section 3.12.1, if the mixing distribution mentioned there, $M(\boldsymbol{\theta}_4)$, represents independent judgments regarding the component proportions in the two rows of $\mathbf{E}_{2,N}$, then the inferential equation listed in that section also reduces to the inferential equation of Theorem 3.16 in the special case of a separable $M(\boldsymbol{\theta}_2)$.

A nonseparable mixing distribution function $M(\boldsymbol{\theta}_2)$ represents an uncertain opinion regarding the proportion of successes within the two treatment types for which you are willing to learn about one of the proportions from observations on units treated by the other type. Lad and Deely (1994) provide a simple example of this type of partial exchangeability in the context of using urine sugar data to learn about blood sugar levels, although the main focus of the article is on still another question.

## PROBLEM

**1.** Suppose that you regard the events in each row of the event pairs $\mathbf{E}_{21}$, $\mathbf{E}_{22}$, and $\mathbf{E}_{23}$ exchangeably, where each row of events represents outcomes of experiments based on a specific treatment type. The realm of the sufficient statistics vector $\mathbf{S}_2$ contains 16 elements, $\mathscr{R}(\mathbf{S}_2) = \{0, 1, 2, 3\}^2$. Suppose you assert the following probabilities for the 16 events defined by $\mathbf{S}_2$:

| | | | | | | | | | | | | | | | | |
|---|---|---|---|---|---|---|---|---|---|---|---|---|---|---|---|---|
| $s_{1i}$ | 0 | 0 | 0 | 0 | 1 | 1 | 1 | 1 | 2 | 2 | 2 | 2 | 3 | 3 | 3 | 3 |
| $s_{2i}$ | 0 | 1 | 2 | 3 | 0 | 1 | 2 | 3 | 0 | 1 | 2 | 3 | 0 | 1 | 2 | 3 |
| $74\ P(\mathbf{S}_2 = \mathbf{s}_{2i})$ | 4 | 8 | 7 | 3 | 6 | 8 | 8 | 3 | 6 | 6 | 6 | 3 | 2 | 2 | 1 | 1 |
| $N^*(\mathbf{s}_{2i})$ | 1 | 3 | 3 | 1 | 3 | 9 | 9 | 3 | 3 | 9 | 9 | 3 | 1 | 3 | 3 | 1. |

In order to make some interesting comparisons possible, these probabilities were derived from those specified in problem 1 of Section 3.12.1, by computing the joint probabilities for the possible values of $S_1 \equiv S_{10} + S_{11}$ and $S_2 \equiv S_{01} + S_{11}$ that are implied by those probabilities for the vector $\mathbf{S}_4 = (S_{00}, S_{01}, S_{10}, S_{11})$. Notice that the implied marginal probabilities for values of $S_1$ and $S_2$ are $P(S_1 = 0, 1, 2, 3) = (22, 25, 21, 6)/74$ and $P(S_2 = 0, 1, 2, 3) = (18, 24, 22, 10)/74$, so the joint probabilities for the values of the vector $\mathbf{S}_2$ do not equal the products of these marginal probabilities. The components of $\mathbf{S}_2$ are not regarded independently.

**a.** Using the results presented in this section, compute the following probabilities: $P(E_{11})$, $P(E_{21})$, $P(E_{21} | E_{11})$, $P(E_{21} | \tilde{E}_{11})$, $P(E_{11} | E_{21})$, $P(E_{11} | \tilde{E}_{21})$.

**b.** Notice that these latter four conditional probabilities are identical with your $P(E_{21} | E_{12})$, $P(E_{21} | \tilde{E}_{12})$, $P(E_{11} | E_{22})$, $P(E_{11} | \tilde{E}_{22})$. (For the unit $j$ corresponds to a different person (or field) engaged in a first type of treatment event than the unit $j$ does for a second treatment event.)

**c.** Compute each of the listed probabilities again, conditioned additionally upon the event $(\mathbf{E}_{23} = \mathbf{e}_2)$ for each $\mathbf{e}_2 \in \{(0, 0)^T, (0, 1)^T, (1, 0)^T, (1, 1)^T\}$.

**d.** Compare your answers with those from problem 1 of Section 3.12.1.

### 3.12.3 Partial Exchangeability: Exchangeability within Subsequences

In this section we study a form of reasonable opinions about a sequence of experiments described by Diaconis and Freedman (1980a) which, if seemingly contrived, is paradigmatic of a general type of experiments that have long been studied under the title of "Markov chains." Just as we have found that probability distributions for events regarded exchangeably are often representable as mixtures of independent distributions, we find here that a certain type of judgment of partial exchangeability for sequential experiments yields to analysis in the form of mixtures of Markov chains. This terminology will be clear after an example and some analysis.

Consider an experiment in which a thumbtack is thrown into the air and allowed to land on the floor. We define an event, $E_0$, to equal 1 if the tack lands point upward and to equal 0 if it lands point to the floor. When the tack comes to rest again after the initial fall, its orientation is recorded (as 1 or 0), and it is then given an energetic flick with the fingers from the very same position in which it has come to rest. Again its orientation is recorded, and it is flicked again from whatever position in which it rests. The process of experimentation continues in this way. Denoting the sequence of results of such experiments by $E_0, E_1, E_2, \ldots, E_N$, let us study what might be reasonable opinions about these events.

To begin, regarding these events exchangeably hardly seems reasonable. The physical activity of applying an energetic flick of the fingers to a tack lying point up after the previous flick seems different from the activity of applying a flick to a tack lying point to the floor after the previous flick. Thus, I (and you?) would assess differently my probability that a string of 10 flips, say, yields five successive 1's followed by five successive 0's than that a string of 10 flips yields alternating 1's and 0's. However, I would regard exchangeably all events $E_i$ for which the preceding event, $E_{i-1}$, were found to equal 0, and I would regard exchangeably all events $E_j$ for which $E_{j-1}$ were found to equal 1. For the conditions of a flicking experiment appear symmetric within each of these groups.

Notational Caution Since the generation of $E_0$, occurs by a procedure that is different from the procedure generating the rest of the events, we are denoting it specially as $E_0$. Nonetheless, when we want to denote all the events under consideration, we use the vector symbol $\mathbf{E}_N \equiv (E_0, E_1, \ldots, E_N)$. Thus, be especially aware that in this subsection vectors denoted by $\mathbf{E}_N$ or $\mathbf{e}_N$ contain $N+1$ components.

Further consideration suggests the following unifying feature to most people's opinions about the results of such a sequence of experiments. I would assert the same probability to any string $\mathbf{E}_N = \mathbf{e}_N$ that involves the same starting point and the same number of *transitions* of $E_t$ to $E_{t+1}$ showing 0 to 0, 0 to 1, 1 to 0, and 1 to 1. The statistics $E_0$, $T_{00}(\mathbf{e}_N)$, $T_{01}(\mathbf{e}_N)$, $T_{10}(\mathbf{e}_N)$, and $T_{11}(\mathbf{e}_N)$ are sufficient for my inference about an event $E_{N+1}$. Let us make a formal definition, display a simple example, and state the relevant theorem.

***Definition 3.15.*** Let $\mathbf{E}_N = (E_0, E_1, \ldots, E_N)$ denote logically independent events. *You regard the events partially exchangeably as a mixture Markov chain* if you assert $P(\mathbf{E}_N = \mathbf{e}_N) = P(\mathbf{E}_N = \mathbf{f}_N)$ whenever $\mathbf{T}(\mathbf{e}_N) = \mathbf{T}(\mathbf{f}_N)$, where $\mathbf{T}(\mathbf{e}_N)$ denotes the vector of statistics

$$\Big[ T_0(\mathbf{e}_N) = e_0,\ T_{00}(\mathbf{e}_N) = \sum_{j=1}^{N} \tilde{e}_j \tilde{e}_{j-1},\ T_{01}(\mathbf{e}_N) = \sum_{j=1}^{N} e_j \tilde{e}_{j-1},$$
$$T_{10}(\mathbf{e}_N) = \sum_{j=1}^{N} \tilde{e}_j e_{j-1},\ T_{11}(\mathbf{e}_N) = \sum_{j=1}^{N} e_j e_{j-1} \Big]. \qquad \square$$

The statistics vector $\mathbf{T}$ is defined by the function $\mathbf{T}: \mathscr{R}(\mathbf{E}_N) \to \mathscr{R}(\mathbf{T})$, where

$$\begin{aligned} \mathscr{R}(\mathbf{T}) = \{\mathbf{t}_5 = (t_0, t_{00}, t_{01}, t_{10}, t_{11}) | &\text{each component } t \text{ is an integer, } t_0 \in \{0,1\}, \\ &\text{if } t_0 = 0 \text{ then } t_{00} + t_{01} \geqslant 1 \text{ and } t_{01} - t_{10} \in \{0,1\}, \\ &\text{if } t_0 = 1 \text{ then } t_{10} + t_{11} \geqslant 1 \text{ and } t_{10} - t_{01} \in \{0,1\}, \\ &\text{and } t_{00} + t_{01} + t_{10} + t_{11} = N\}. \end{aligned}$$

These conditions on vectors in $\mathscr{R}(\mathbf{T})$ derive from the fact that if $E_0 = 0$ (making $t_0 = 0$) then at least one $E_i = 0$ is followed by an $E_{i+1} = 0$ or $E_{i+1} = 1$ (making $t_{00} + t_{01} \geqslant 1$). Furthermore, in this case the number of switches from an $E_i = 0$ to an $E_{i+1} = 1$ can differ from the number of switches from an $E_i = 1$ to an $E_{i+1} = 0$ by only 0 or 1. For to switch from an $E_i = 1$ to an $E_{i+1} = 0$, there must have been a switch from an event $E_j = 0$ to an $E_{j+1} = 1$ prior to the occurrence of $E_i$. Thus, the difference in the statistic values $t_{01} - t_{10}$ can only be either 0 or 1. The same logic governs the situation when $E_0 = 1$.

Notice that the value of the statistics vector $T(\mathbf{E}_N)$ identifies the value of the final component, $E_N$, depending on the value of $E_0$ and on the difference $T_{01}(\mathbf{E}_N) - T_{10}(\mathbf{E}_N)$ if $E_0 = 0$ or on the difference $T_{10}(\mathbf{E}_N) - T_{01}(\mathbf{E}_N)$ if $E_0 = 1$. This feature of the statistics vector allows you to compute the vector of $T(\mathbf{E}_{N+1})$ from $T(\mathbf{E}_N)$ and $E_{N+1}$, as required by Definition 3.14 of partial exchangeability with respect to a statistics vector.

Table 3.3 exhibits values of $\mathbf{T}(\mathbf{E}_4 = \mathbf{e}_4)$ for each $\mathbf{e}_4 \in \mathscr{R}(\mathbf{E}_4)$. Perusing it should help you develop your intuitions about this problem.

Determining the size of the realm $\mathscr{R}(\mathbf{T}(\mathbf{E}_N))$ for a general problem, and counting the number of elements $\mathbf{e}_N \in \mathscr{R}(\mathbf{E}_N)$ that get mapped onto each possible value of $\mathbf{t} \in \mathscr{R}(\mathbf{T})$, are complicated but solvable problems. Without deriving them here, let us merely state the following solutions:

**(i)** The size of $\mathscr{R}(\mathbf{T}(\mathbf{E}_N))$ is $N^*(\mathscr{R}(\mathbf{T}(\mathbf{E}_N))) \equiv N^2 + N + 2$.

**(iia)** If $K$ is a positive integer, and $K - i$ is an even number, then the size of the set $\{\mathbf{e}_N \in \mathscr{R}(\mathbf{E}_N) | \mathbf{T}(\mathbf{e}_N) = (0, N-K, (K-i)/2, (K-i)/2, i)\}$ equals

$$N^*(0, N-K, (K-i)/2, (K-i)/2, i) = {}^{[N-K+(K-i)/2]}C_{(K-i)/2}\ {}^{[i+(K-i)/2-1]}C_{[(K-i)/2-1]}.$$

**Table 3.3. Mappings of $\mathbf{e}_4 \in \mathscr{R}(\mathbf{E}_4)$ into $\mathbf{t}(\mathbf{e}_4) \in \mathscr{R}(\mathbf{T}(\mathbf{E}_4))$**[a]

| $e_0, e_1, e_2, e_3, e_4$ | $t_0$ | $t_{00}$ | $t_{01}$ | $t_{10}$ | $t_{11}$ | Type | $N^*(\mathbf{t}(\mathbf{e}_4))$ |
|---|---|---|---|---|---|---|---|
| 0, 0, 0, 0, 0 | 0 | 4 | 0 | 0 | 0 | 1 | 1 |
| 0, 1, 0, 0, 0 | 0 | 2 | 1 | 1 | 0 | 2 | 3 |
| 0, 0, 1, 0, 0 | 0 | 2 | 1 | 1 | 0 | 2 | 3 |
| 0, 0, 0, 1, 0 | 0 | 2 | 1 | 1 | 0 | 2 | 3 |
| 0, 0, 0, 0, 1 | 0 | 3 | 1 | 0 | 0 | 3 | 1 |
| 0, 1, 1, 0, 0 | 0 | 1 | 1 | 1 | 1 | 4 | 2 |
| 0, 1, 0, 1, 0 | 0 | 0 | 2 | 2 | 0 | 5 | 1 |
| 0, 1, 0, 0, 1 | 0 | 1 | 2 | 1 | 0 | 6 | 2 |
| 0, 0, 1, 1, 0 | 0 | 1 | 1 | 1 | 1 | 4 | 2 |
| 0, 0, 1, 0, 1 | 0 | 1 | 2 | 1 | 0 | 6 | 2 |
| 0, 0, 0, 1, 1 | 0 | 2 | 1 | 0 | 1 | 7 | 1 |
| 0, 1, 1, 1, 0 | 0 | 0 | 1 | 1 | 2 | 8 | 1 |
| 0, 1, 1, 0, 1 | 0 | 0 | 2 | 1 | 1 | 9 | 2 |
| 0, 1, 0, 1, 1 | 0 | 0 | 2 | 1 | 1 | 9 | 2 |
| 0, 0, 1, 1, 1 | 0 | 1 | 1 | 0 | 2 | 10 | 1 |
| 0, 1, 1, 1, 1 | 0 | 0 | 1 | 0 | 3 | 11 | 1 |
| 1, 0, 0, 0, 0 | 1 | 3 | 0 | 1 | 0 | 12 | 1 |
| 1, 1, 0, 0, 0 | 1 | 2 | 0 | 1 | 1 | 13 | 1 |
| 1, 0, 1, 0, 0 | 1 | 1 | 1 | 2 | 0 | 14 | 2 |
| 1, 0, 0, 1, 0 | 1 | 1 | 1 | 2 | 0 | 14 | 2 |
| 1, 0, 0, 0, 1 | 1 | 2 | 1 | 1 | 0 | 15 | 1 |
| 1, 1, 1, 0, 0 | 1 | 1 | 0 | 1 | 2 | 16 | 1 |
| 1, 1, 0, 1, 0 | 1 | 0 | 1 | 2 | 1 | 17 | 2 |
| 1, 1, 0, 0, 1 | 1 | 1 | 1 | 1 | 1 | 18 | 2 |
| 1, 0, 1, 1, 0 | 1 | 0 | 1 | 2 | 1 | 17 | 2 |
| 1, 0, 1, 0, 1 | 1 | 0 | 2 | 2 | 0 | 19 | 1 |
| 1, 0, 0, 1, 1 | 1 | 1 | 1 | 1 | 1 | 18 | 2 |
| 1, 1, 1, 1, 0 | 1 | 0 | 0 | 1 | 3 | 20 | 1 |
| 1, 1, 1, 0, 1 | 1 | 0 | 1 | 1 | 2 | 21 | 3 |
| 1, 1, 0, 1, 1 | 1 | 0 | 1 | 1 | 2 | 21 | 3 |
| 1, 0, 1, 1, 1 | 1 | 0 | 1 | 1 | 2 | 21 | 3 |
| 1, 1, 1, 1, 1 | 1 | 0 | 0 | 0 | 4 | 22 | 1 |

[a]The column headed "Type" displays a number identifying the 22 distinct elements of $\mathscr{R}(\mathbf{T}(\mathbf{E}_4))$. The column $N^*(\mathbf{t}(\mathbf{e}_4))$ designates the number of elements $\mathbf{e}_4$ yielding this type of $\mathbf{t}(\mathbf{e}_4)$.

**(iib)** If $K$ is a positive integer, and $K - i$ is an odd number, then the size of the set $\{\mathbf{e}_N \in \mathscr{R}(\mathbf{E}_N) | \mathbf{T}(\mathbf{e}_N) = (0, N-K, (K-i+1)/2, (K-i-1)/2, i)\}$ equals

$$N^*(0, N-K, (K-i+1)/2, (K-i-1)/2, i)$$
$$= {}^{[N-K+(K-i-1)/2]}C_{(K-i-1)/2}\ {}^{[i+(K-i+1)/2-1]}C_{[(K-i+1)/2-1]}.$$

Before you try the following problems, you may wish to examine Table 3.3, and notice how both of these counting results apply.

## PROBLEMS

**1.** Show that if $\mathbf{T}(\mathbf{e}_N) = \mathbf{T}(\mathbf{f}_N)$, then $e_N = f_N$. (The final components are equal.)

**2.** Show that if your $P(T_{01}(\mathbf{e}_N) = 0) = 1$, then the only elements of the partition generated by $\mathbf{E}_N$ that have positive probability are the events $(\mathbf{E}_N = \mathbf{e}_N)$ for which $\mathbf{e}_N = (\mathbf{1}_K \quad \mathbf{0}_{N+1-K})$ for $K = 0, 1, 2, \ldots, N+1$.

**3.** For the general problem of $\mathbf{E}_N$ derive the formula for the size of $\mathscr{R}(\mathbf{T}(\mathbf{E}_N))$ as $N^2 + N + 2$, and derive the formula for the number of elements of $\mathscr{R}(\mathbf{E}_N)$ that are mapped into each $\mathbf{t} \in \mathscr{R}(\mathbf{T}(\mathbf{E}_N))$.

The analysis of this form of partially exchangeable judgments is rather complex, technically, but it succumbs to a logical analysis similar in structure to arguments we have made already. To provide both a flavor and some technical detail of the results, I directly state the following partial theorem (governing the condition when $E_0 = 0$ and $E_N = 0$), relying on your understanding of conditional probability to provide its proof. More extensive discussion appears in references that follow the theorem.

**Theorem 3.17.** If you regard the events $\mathbf{E}_{N+M} = (E_0, E_1, \ldots, E_{N+M})$ partially exchangeably as a mixture Markov chain, then under presumptions a and b, which follow, coherency requires the concomitant assertions

$$P[\mathbf{T}(\mathbf{E}_N) = \mathbf{t}(\mathbf{e}_N)] \,|\, \mathbf{T}(\mathbf{E}_{N+M}) = \mathbf{t}(\mathbf{e}_N) + \mathbf{t}(\mathbf{e}_M)] = \frac{N^*(\mathbf{t}(\mathbf{e}_N)) N^*(\mathbf{t}(\mathbf{e}_M))}{N^*(\mathbf{t}(\mathbf{e}_N) + \mathbf{t}(\mathbf{e}_M))}.$$

(a) $\mathrm{t}(\mathbf{e}_N) = (0, N-K, (K-i)/2, (K-i)/2, i)$ where $K - i$ is an even number;

(b) $\mathbf{t}(\mathbf{e}_M) = (0, M-S, ]](S-j)/2[[, [[(S-j)/2]], j)$, where $]]x[[$ denotes the smallest integer not less than $x$, and $[[x]]$ denotes the greatest integer contained in $x$. (Thus, notice that if $S - j$ is even, $]](S-j)/2[[ = [[(S-j)/2]]$. Otherwise, they differ by 1.)

If you assert additionally the probabilities $P[\mathbf{T}(\mathbf{E}_{N+M}) = \mathbf{t}(\mathbf{e}_{N+M})]$ for each $\mathbf{t}(\mathbf{e}_{N+M}) \in \mathscr{R}(\mathbf{T}(\mathbf{E}_{N+M}))$, then the inferential probabilities $P[\mathbf{T}(\mathbf{E}_{N+M}) = \mathbf{t}(\mathbf{e}_N) + \mathbf{t}(\mathbf{e}_M) \,|\, \mathbf{T}(\mathbf{E}_N) = \mathbf{t}(\mathbf{e}_N)]$ can be computed from these asserted probabilities and this specification of $P[\mathbf{T}(\mathbf{E}_N) = \mathbf{t}(\mathbf{e}_N) \,|\, \mathbf{T}(\mathbf{E}_{N+M}) = \mathbf{t}(\mathbf{e}_N) + \mathbf{t}(\mathbf{e}_M)]$ via Bayes' theorem. If your partially exchangeable assertions regarding $\mathbf{E}_N$ are infinitely extendible as a mixture Markov chain and you assert $P(T_{01}(\mathbf{E}_N) = 0)$ and $P(T_{10}(\mathbf{E}_N) = 0)$ both as numbers strictly within $(0, 1)$, then your probability $P(\mathbf{E}_N = \mathbf{e}_N)$ is representable as

$$\int_{\mathscr{R}(\boldsymbol{\theta}_5)} \theta_0^{e_0} (1 - \theta_0)^{1 - e_0} \theta_{00}^{T_{00}(\mathbf{e}_N)} \theta_{01}^{T_{01}(\mathbf{e}_N)} \theta_{10}^{T_{10}(\mathbf{e}_N)} \theta_{11}^{T_{11}(\mathbf{e}_N)} \, dM(\boldsymbol{\theta}_5)$$

for some distribution function $M(\cdot)$ on $\mathscr{R}(\boldsymbol{\theta}_5) = \{\boldsymbol{\theta}_5 | (\theta_0, \theta_{01}, \theta_{11}) \in [0,1]^3,$ $\theta_{00} + \theta_{01} = 1$, and $\theta_{10} + \theta_{11} = 1\}$. If your $P(T_{10}(\mathbf{e}_N) = 0) = 1$, then your partially exchangeable assertions regarding $\mathbf{E}_N$ are infinitely extendable as a mixture Markov chain, but the integral mixture representation of $P(\mathbf{E}_N = \mathbf{e}_N)$ does not hold. Rather, in this case your probability assertions $P(\mathbf{E}_N = \mathbf{e}_N)$ are nonzero only for the $N+2$ vectors of the form $\mathbf{e}_N = (\mathbf{0}_K \quad \mathbf{1}_{N+1-K})$ for $K = 0, 1, 2, \ldots, N+1$. Denoting your probabilities for these $N+2$ vectors by $\mathbf{p}_{N+2} \equiv (p_0, p_1, \ldots, p_{N+1})$, these are extendable by any sequences of the form $\mathbf{p}_\infty^* \equiv (p_0^*, p_1^*, \ldots, p_N^*, p_{N+1}^*, \ldots)$ for which $p_i^* = p_i$ when $i \leqslant N$, and $\sum_{i=N+1}^\infty p_i^* = p_{N+1}$. Similar representations hold if your $P(T_{01}(\mathbf{e}_N) = 0) = 1$.

When appropriate, asserting the condition $P(T_{10}(\mathbf{e}_N) = 0) = 1$ expresses the recognition that once the sequence of events exhibits a 1 in some component, all remaining components of $\mathbf{E}_N$ must also equal 1. In such a case, the condition of $E = 1$ is said to be asserted as an "absorbing state" for the sequence.

Theorem 3.17 was partially proved by Diaconis and Freedman (1980b), with some of the final detail provided by Zaman (1986). The concept of partial exchangeability was introduced by de Finetti (1938b) and subsequently. But the unifying concept of partial exchangeability with respect to specified sufficient statistics was developed by Martin-Lof and Lauritzen. The expository article of Diaconis and Freedman (1984) reviews many other applications of the sufficiency concept and contains a good bibliography. The text of Martin (1975) presents results concerning several forms of mixture distribution $M(\boldsymbol{\theta}_5)$ that allow easy computation of inferential conditional probabilities.

With this presentation we conclude our study of partial exchangeability here. Given the generality with which partially exchangeable judgments have been characterized in terms of sufficient statistics, you should understand that there are many forms in which its logic will be relevant to your concerns. We study some of them explicitly in Chapter 8.

CHAPTER 4

# Related Forms for Asserting Uncertain Knowledge

*Now the fields flew by, the hedges and the ditches, ghastly in the train's light, or appeared to do so, for in reality it was the train that moved, across a land forever still.* *Watt*, SAMUEL BECKETT

## 4.1 INTRODUCTION

We have motivated and formalized the language of coherent prevision for expressing your uncertain knowledge about quantities. Indeed, whenever a question has been posed as to what you know about some quantity, a response was formulated in terms of your prevision or partial prevision assertion. Even our brief allusions to your assertion of a probability distribution in Sections 3.8–3.11 concerned your prevision for the constituents of a partition. Some questions arise as to the sufficiency of the notion of prevision to account for different types of uncertain knowledge about a given quantity.

Consider two quantities with identical realms. Suppose that you happen to assert the same prevision for each of them. Isn't it possible to feel quite certain that the one quantity is very close to your prevision value while being quite uncertain about the magnitude of the other? How is the language of prevision able to account for such distinct types of uncertain knowledge about the two quantities when your previsions for the two quantities are identical? What if someone wants to assert a median for a quantity $X$, or a mode for $X$, or some quantiles of a personal probability distribution for $X$?

As we shall see in this chapter, the formalities of prevision provide a basis for representations of each of these and of many other forms of uncertain knowledge, each with their own distinct characteristics. Mathematically, this is due to the fact that you may assess your prevision not only for any quantity $X$ but also for any other quantity that can be defined by a function of $X$. We study some of the many forms for representing your uncertainty about $X$, each of which can be specified in terms of your prevision for a specific quantity, focusing on the characteristics of your assertions in these various forms that are necessitated by the condition that your previsions be coherent. Mathematically, we use the fact that coherent prevision assertions delineate a *linear functional*—that is, a linear operator defined on a linear space of functions of quantities. We have already alluded to the structure of prevision as a linear functional at the end of Section 2.5. It is time now for a formal definition.

**Definition 4.1.** Suppose $X$ is a quantity with realm $\mathscr{R}(X) = \{x_1, x_2, \ldots, x_K\}$. Let $\mathscr{G}_X$ denote a set of real-valued functions that are well defined over the domain specified by $\mathscr{R}(X)$. A set of functions, $\mathscr{G}_X$, is said to be a *linear space of functions* if whenever two functions $G_1(\cdot)$ and $G_2(\cdot)$ are members of $\mathscr{G}_X$, then for any real numbers $a_1$ and $a_2$ the function $G_3(\cdot) \equiv a_1 G_1(\cdot) + a_2 G_2(\cdot)$ is also a member of $\mathscr{G}_X$. Any real-valued function $P$: $\mathscr{G}_X \to \mathbb{R}$ is said to be a *functional on* $\mathscr{G}_X$. A functional $P$ is said to be a *linear functional on* $\mathscr{G}_X$ if

$$P[a_1 G_1(\cdot) + a_2 G_2(\cdot)] = a_1 P[G_1(\cdot)] + a_2 P[G_2(\cdot)]$$

whenever the functions $G_1(\cdot)$, $G_2(\cdot)$, and $G_3(\cdot)$ are in the space of functions $\mathscr{G}_X$. If $\mathscr{G}_X$ is a linear space of functions, a subset of functions $\mathscr{B}_X \subset \mathscr{G}_X$ is said to be a *basis* for $\mathscr{G}_X$ if every member of $\mathscr{G}_X$ can be generated from a linear combination of members of $\mathscr{B}_X$. Finally, a linear functional is said to be *bounded* if there exists a basis $\mathscr{B}_X$ and a number $b$ for which $P[|B(X)|] < b$ for every function $B(\cdot) \in \mathscr{B}_X$. □

Definition 4.1 can be generalized, merely by understanding "the quantity $X$" to represent "the vector of quantities $\mathbf{X}$" and understanding the members $x_i$ of $\mathscr{R}(X)$ to be vectors of the same dimension as $\mathbf{X}$.

Since a quantity $X$ is a number and equals some member of its realm, $\mathscr{R}(X) = \{x_1, x_2, \ldots, x_K\}$, any real-valued function $G$ defines another quantity, $Y \equiv G(X)$, as a number and as a member of its realm, $\mathscr{R}(Y) = \{G(x_1), G(x_2), \ldots, G(x_K)\}$. We say that your prevision assertions constitute a functional on such functions of $X$, since you may assert your prevision to represent your uncertain knowledge about any quantity, and thus about any function of $X$.

Using this language, we can restate Theorem 2.5 of Section 2.6 and its corollary as specifying that coherent prevision assertions, $P(\mathbf{X}_N)$, define a bounded linear functional on the space of linear functions of $\mathscr{R}(\mathbf{X}_N)$. For the theorem states that coherent previsions must satisfy the linearity restriction that $P(\mathbf{s}_N^{\mathrm{T}}\mathbf{X}_N) = \mathbf{s}_N^{\mathrm{T}}P(\mathbf{X}_N)$ for any vector of coefficients $\mathbf{s}_N$. Furthermore, we can restate the fundamental theorem of prevision as specifying the bounds within which a coherent prevision specified over the linear space of linear functions of a vector, $\mathbf{X}_N$, can always be extended to a coherent linear functional over linear functions of $\mathbf{X}_N$ and another quantity, $X_{N+1}$, even when this additional quantity is defined by a nonlinear function of $\mathbf{X}_N$. In the field of analysis, the general extension theorem of this type for linear functionals is called the Hahn-Banach theorem.

We have made extensive use of the practical consequences of these results. Here we study in greater detail the implications of coherency for prevision assertions within larger classes of functions of $G_1(X)$ and $G_2(X)$, not just linear functions. In an important sense, we have already exhausted these implications in the fundamental theorem of prevision. Our aim here is to achieve another level of understanding the structure of coherent prevision and its various related forms for specifying uncertain knowledge.

***Example 1.*** Suppose you can measure the quantities $L$, $W$, and $H$ which specify, respectively, the length, width, and height of the interior of a shipping container. Looking at the container, you may specify your previsions $P(L)$, $P(W)$, and $P(H)$. Now you may be interested in several aspects of the container for different reasons. For example, you may be interested in the area of the floor, defined by the product quantity $LW$; or the volume of the container, defined by the product $LWH$; or the length of the diagonal of width by height, defined by $(W^2 + H^2)^{1/2}$. These quantities are defined via functions of $L$, $W$, and $H$, though they are not linear functions. Thus, the six quantities are logically related, although they are not linearly related. But if you would assert your previsions for all of these quantities, coherency requires that your prevision specifies a linear operator on linear functions of the six quantities $L, W, H, LW, LWH$, and $(W^2 + H^2)^{1/2}$. □

***Example 2.*** Suppose the components of a vector $\mathbf{X}_N$ denote the accounting values of the capital assets of a company over $N$ consecutive years. An investor may be interested not only in the components of $\mathbf{X}_N$ but in other quantities too that can be derived from them, for example in the annual rates of increase in these components. These are also quantities, defined by $R_t \equiv (X_t - X_{t-1})/X_{t-1}$, for $t = 2, \ldots, N$. Or the investor may be interested in the *stability* of these annual rates of increase, defined, say, by the quantity $S_N \equiv \sum_{t=2}^{N}(R_t - \bar{R}_N)^2/(N-1)$, where $\bar{R}_N \equiv \sum_{t=2}^{N} R_t/(N-1)$ is the average annual rate of increase over these years. Again, coherent prevision assertions would specify a linear functional over the space of linear functions of the quantities $\mathbf{X}_N$, $\mathbf{R}_N$, $\bar{R}_N$, and $S_N$. The components of these vectors of quantities themselves are logically related, though they are not all linearly related. □

***Example 3.*** A Generic Review . Suppose you consider your prevision for the quantities $X$ and $Y$, where $Y$ is defined by some function of $X$, $G(X)$. Yet you have distinct reasons for being interested in each quantity. In such situtations, the realm of the quantity vector $(X, Y)^{\mathrm{T}}$ is $\mathscr{R}(X, Y)^{\mathrm{T}} = \{(x, y)^{\mathrm{T}} | x \in \mathscr{R}(X), y = G(x)\}$. This is the type of situation we are considering when we talk about a quantity that is defined operationally by a function of another. Evidently, such quantities are logically related because they are functionally related. □

***Example 4.*** (Be careful!) It is important that you know precisely what quantity you are considering when it is suggested to you casually by a name. In situations where one quantity is defined by a function of some others, such as the volume quantity $V \equiv LWH$ in Example 1, another quantity can often be defined operationally to represent the volume of the container, say, but need not be restricted to equal the product $LWH$. For example, another quantity $V^*$ may be defined to equal the number of liters of water that fit into the container. Suppose the measuring device used to determine the quantities $L$, $W$, and $H$ is distinguishable to the nearest centimeter. So the realms for each of these (logically unrelated) quantities is the set $\mathscr{R}(L) = \mathscr{R}(W) = \mathscr{R}(H) = \{0, 1, 2, \ldots, M\}$, where $M$ is the largest measurement possible with this "stick." Thus, the realm of the quantity $V$, defined by the product $LWH$, is the set $\mathscr{R}(V) = \{v \mid v = lwh$ for some $l \in \mathscr{R}(L)$, $w \in \mathscr{R}(W)$, $h \in \mathscr{R}(H)\} \subset \{1, 2, \ldots, M^3\}$. Now suppose that the measuring device used to determine the quantity $V^*$ is distinguishable to the nearest centiliter. (In the metric system of measurement, 1 liter is defined to equal 1 cubic decimeter = 1000 cubic centimeters. Hence, a centiliter is defined as 10 cubic centimeters.) Thus, in units comparable to the cubic centimeters which delineate members of $\mathscr{R}(V)$, the realm of $V^*$ would be $\mathscr{R}(V^*) = \{10, 20, \ldots, M^*\}$. Not only are the sets $\mathscr{R}(V)$ and $\mathscr{R}(V^*)$ distinct, allowing that the quantity $V$ may not be contained in $\mathscr{R}(V^*)$, but it is possible that $V < V^*$, that $V = V^*$, or that $V > V^*$. In fact, the quantities $V$ and $V^*$ are completely logically independent! Of course, the probability you assert for events $((V, V^*) = (v, v^*))$ may well be miniscule for many of the elements $(v, v^*) \in \mathscr{R}(V, V^*)$, but that is another matter.

Related issues of measurement definitions arise in many scientific and commercial fields. In economics the quantity theory of money specifies that the supply of money times the velocity at which it circulates equals the product of the average price level with the total output for which money is exchanged: $Mv = PT$. In electrical engineering Ohm's law specifies that voltage equals current times resistance: $V = IR$. The fine points of measurement may well be ignored when casually discussing theoretical scientific issues, but when it comes time to "put your money where your mouth is" by asserting your prevision for various measurements, it is best to pay close attention to the precise details of "measurement" procedure. □

We begin this chapter with an exposition of how the concept of prevision allows for the practical assertion of your uncertain knowledge in related forms of your mode, median, or general quantiles of your probability distribution. Each of these "alternative" forms of asserting your knowledge amounts to a specific procedure for reducing the convex hull of the realm of a quantity vector to a partial prevision polytope. In particular, we dwell on several properties of variance assertions that are required by the coherency of previsions that define them. Next, we extend our awareness of coherency restrictions beyond linearity by noticing a standard inequality regarding previsions for convex functions of a quantity, commonly called Jensen's inequality. The chapter continues with an exposition of the mathematical properties of coherent prevision as it is characterized as a "generalized mean" function. In particular, we identify the relation of prevision assertions to utility theory via the unique property of "associative mean" functions for representing conditional previsions. Finally, we conclude by portraying the economic theory of portfolio selection from the subjectivist perspective, noticing how the mean-variance portrayal of preference orderings fits within the context of coherent prevision as a linear operator.

## 4.2 PRACTICAL EXPRESSIONS OF PREVISION VIA YOUR MODE, MEDIAN, QUANTILES, VARIANCE, COVARIANCE, AND CORRELATION

The burden of this section is to establish that many forms of uncertain knowledge specification that are often touted as "alternatives" to prevision are in fact defined by your prevision or partial prevision assertion for specific functions of some quantity of interest.

Let us begin by defining your "mode" for $X$. Intuitively, your mode(X) is the member of $\mathscr{R}(X)$ that you regard as "the most likely" value of $X$.

***Definition 4.2.*** Let $X$ be a quantity with realm $\mathscr{R}(X) = \{x_1, \ldots, x_K\}$. You assert *your mode for* $X$, denoted by mode($X$), as some number within the realm $\mathscr{R}(X)$, when you assert the inequality $P(X = \text{mode}(X)) \geqslant P(X = x_i)$ for every $x_i \in \mathscr{R}(X)$. □

Your mode for $X$ need not necessarily be unique. As an extreme example, if you regard every constituent of the partition generated by $X$ as equilikely, then every member of $\mathscr{R}(X)$ would qualify as your mode. But such an observation is largely a red herring. If you are thinking about a quantity for which the assertion of your mode is a natural and helpful activity in getting your bearings on the problem, go ahead and assert it. If not, think about something worthwhile. To assert identical probabilities over every constituent of the partition generated by $X$, for example, is a momentous assertion of its own. It matters not at all that such assertions fail to specify a unique mode.

Asserting your mode for $X$ constitutes a specification of a partial prevision polytope for the constituents of the partition generated by $X$. For your asserting mode($X$) is equivalent to the asserting $K - 1$ linear inequalities in the form $P(X = \text{mode}(X)) \geqslant P(X = x_i)$ for every other value of $x_i \in \mathscr{R}(X)$. Using the full structure allowed by asserting prevision inequalities, the notion of a modal assertion can be weakened in a useful way. Suppose that the elements of $\mathscr{R}(X)$ are ordered according to size by $x_1 < x_2 < \cdots < x_K$. If you are willing to assert $P(X = x_i) \geqslant P(X = x)$ for any $x \leqslant x_i$, and $P(X = x_j) \geqslant P(X = x)$ for any $x \geqslant x_j > x_i$, but you have not yet determined even your ordering of previsions for events of the form $(X = x)$ for values of $x$ within the bounds $x_i \leqslant x \leqslant x_j$, then the interval $[x_i, x_j]$ could be called your asserted modal boundary for $X$.

Evidently, we could continue such extensions of the concept of a modal assertion indefinitely. The important thing to see is that any such assertions are subsumed within the framework of asserting a partial prevision polytope.

## PROBLEMS

1. Today is 28 October 1988. During the past summer months, the plains states of the United States have experienced a great heat wave and drought. Discussions of the theory of global warming abound. Consider a quantity $X$ defined by the measured rainfall recorded in Wichita, Kansas, for June 1989. Can you specify your mode for $X$? a modal boundary for $X$? Can you assert your prevision for $X$? a prevision boundary?

2. The Canterbury plains of New Zealand are also in the midst of their worst drought in 20 years. Observations recorded at the Meteorological Office of the Christchurch Airport for rainfall (in millimeters) are printed in today's *The Press* include

| | |
|---|---|
| Yesterday 9 A.M. to 11 P.M. | nil |
| To date this month | 8 |
| To date this year | 250 |
| To date last year | 500 |
| Average to date | 540 |
| Average for October | 47 |
| Wettest October (1947) | 137 |
| Driest October (1961) | 3 |

Can you specify your modes or modal boundaries for the corresponding quantities appearing in *The Press* on 28 October, 1990? how about your previsions or prevision intervals?

In operational subjective terms, virtually any number that you might consider to be a characteristic of your uncertain opinions can be specified in terms of a prevision or partial prevision assertions. Your median for a quantity or, more generally, any quantile assertion would provide another case in point. Intuitively, your median for a quantity is a number for which you assert equal probabilities that the quantity would exceed it and that the quantity would not exceed it. Similarly, your quantiles are specified numbers for which you assert some probability for an exceedence, and 1 minus that probability for nonexceedence. Formalizing precise details of this idea is sometimes awkward. Let's consider one minimal proposal before some concluding commentary. The important point here is that defining your median or any quantile assertions is achieved in terms of prevision assertions.

***Definition 4.3.*** Let $X$ be a quantity with realm $\mathscr{R}(X) = \{x_1, \ldots, x_K\}$, and suppose that these elements of $\mathscr{R}(X)$ are ordered by size according to $x_1 < x_2 < \cdots < x_K$. To assert the number $x^*$ as your *median for* $X$ is to assert your $P(X \leqslant x^*) = .5$. More generally, if you identify a number $q_J$ for which you assert $P(X \leqslant q_J) = J/100$, then $q_J$ is your *Jth percentile for* $X$. □

In some cases your opinions may be such that either no number satisfies the intuitive idea of a median, or that many numbers do. Consider the following examples. In the case that $\mathscr{R}(X) = \{1, 2, 3, 4\}$, and you would want to assert $P(X = 1) = .1$, $P(X = 2) = .5$, $P(X = 3) = .3$, and $P(X = 4) = .1$, no number $x^*$ exists for which $P(X \leqslant x^*) = .5$. Alternatively, in the case that you would want to assert $P(X = 1) = .1$, $P(X = 2) = .4$ $P(X = 3) = .3$, and $P(X = 4) = .2$, any number within the interval $[2, 3)$ would satisfy the definition of your median($X$). In either case, of course, you have exhausted whatever you might want to say by your assertions of the four probabilities.

We have subsumed any substantive usefulness to the notions of your asserting a mode or a median for $X$ in Section 2.11 when we gave precise meaning to your assertion of the inequality $\mathbf{a}_N^T P(\mathbf{X}_N) \leqslant b$ in Definition 2.19. The only purpose of our current discussion is to recognize explicitly how these "alternative" measures of uncertain knowledge achieve their operational specification in terms of the notion of prevision and partial prevision assertions. This same feature is true of the notions of variance, covariance, and correlation, which are also operationally definable in terms of your prevision assertions for functions of quantities. Let's review these briefly.

***Definition 4.4.*** Let $(X, Y)$ be quantities with realm $\mathscr{R}(X, Y) = \{(x_1, y_1), \ldots, (x_K, y_K)\}$. Then $X^2$, $Y^2$, and $XY$ are also logically related quantities, and the realm of the vector $(X, Y, XY, Y^2, X^2)$ is $\mathscr{R}(X, Y, XY, Y^2, X^2) = \{(x, y, xy, y^2, x^2) \mid (x, y) \in \mathscr{R}(X, Y)\}$. Supposing that you assert your previsions $P(X)$ and $P(Y)$, *you assert your variances for* $X$ *and* $Y$ *and your covariance* between $X$ and $Y$ when you assert previsions for the quantities $[X - P(X)]^2$, $[Y - P(Y)]^2$, and $[(X - P(X))(Y - P(Y))]$, respectively. We use the notation

$$V(X) \equiv P([X - P(X)]^2, \quad V(Y) \equiv P([Y - P(Y)]^2),$$

and

$$\mathrm{Cov}(X, Y) = P[(X - P(X))(Y - P(Y))].$$

Having asserted your variances and covariance, the matrix

$$\mathbf{\Sigma}_{\mathrm{X,Y}} \equiv \begin{pmatrix} V(X) & \mathrm{Cov}(X, Y) \\ \mathrm{Cov}(X, Y) & V(Y) \end{pmatrix}$$

is called *your variance-covariance matrix* for $X$ and $Y$. In terms of these assertions, your *correlation between* $X$ *and* $Y$ is defined as the quotient of your covariance and the square root of the product of your variances. We write

$$\mathrm{Cor}(X, Y) \equiv \mathrm{Cov}(X, Y)/[V(X)V(Y)]^{1/2}. \qquad \square$$

Historically, the importance of a variance as a measure of uncertainty has stemmed from its role in providing a bound on the probability of extreme values of a quantity via the Bienaymé–Chebyshev inequality, which we shall study shortly, and in characterizing members of the family of normal distributions, which we study in Chapter 7. On both of these counts, statistical theory has circumvented by now the importance of the variance as a measure per se, except for the purpose of making crudely approximate statements. Nonetheless, the properties of a coherent variance assertion are well worth knowing about, so we linger on them briefly now.

### 4.2.1 Coherency Restrictions on Variance and Covariance Assertions

Suppose you assert your prevision for a quantity $P(X)$. The inequality of Bienaymé–Chebyshev establishes an upper bound for a cohering probability that $X$ differs from your $P(X)$ by at least $K$ units. The simplest way to derive their result is as a corollary to another inequality, called Markov's inequality.

**Theorem 4.1.** Markov's Inequality. If $\min \mathscr{R}(Y) \geqslant 0$, then for any constant $K > 0$, coherency requires that $P(Y) \geqslant K\, P(Y \geqslant K)$.

*Proof.* Coherency requires that

$$P(Y) = \sum_{y \in \mathscr{R}(Y)} y\, P(Y = y) \geqslant \sum_{\substack{y \in \mathscr{R}(Y) \\ y \geqslant K}} K\, P(Y = y) = K\, P(Y \geqslant K). \qquad \square$$

**Corollary 4.1.1.** Bienaymé–Chebyshev (B–C) Inequality. Let $X$ be any quantity for which you assert $P(X)$. Then for any constant $t > 0$, coherency requires that

$$P(|X - P(X)| \geqslant t) \leqslant V(X)/t^2.$$

*Proof.* Let $Y = [X - P(X)]^2$, and let $K = t^2$. The stated inequality then follows as a special case of Theorem 4.1, remembering that $V(X) \equiv P([X - P(X)]^2)$.

$\square$.

Derived in this way, the B–C inequality can be understood to provide jointly a bound on each of the previsions $P(|X - P(X)| \geqslant t)$ and $V(X)$ in terms of one another. If you assert a value for one of these, coherency establishes the bound for the other assertion as stated in Corollary 4.1.1.

In 1853, I. J. Bienaymé derived this inequality with the purpose of establishing that the variance of a quantity is an optimal measure of the dispersion of its probability distribution. The text of Heyde and Seneta (1977) on the general attitudes and contributions of Bienaymé to statistical theory merits examination. Though the inequality was quite adequate and appropriate to Bienaymé's purpose in understanding what has been called the weak law of large numbers, two qualifications must be made to your understanding this result in an operational subjective context. First, the bound is not necessarily the tightest bound in any application to bounded discrete quantities, the realistic case. The tightest bound on $P(|X - P(X)| \geqslant t)$ induced by your assertion of $P(X)$ and $P([X - P(X)]^2)$ is computable via the fundamental theorem of prevision, which will provide a lower bound for your probability of the extreme event as well. Second, suppose that two quantities are assessed with the same prevision, $P(X) = P(Y)$, but their variances are assessed differently, $V(X) > V(Y)$. Neither is it true that $P(|X - P(X)| > K)$ must exceed $P(|Y - P(Y)| > K)$, nor even is it true that the tightest cohering assertion of $P(|X - P(X)| > K)$ must exceed $P(|Y - P(Y)| > K)$. You can work out an example for yourself in the following problems or read the solution with a discussion in Lad, Dickey, and Rahman (1992).

## PROBLEMS

**1.** Suppose $\mathscr{R}(X) = \{1, 2, 3, 4, 5\}$, and you assert $P(X) = 2.2$ and $V(X) = 7.0$. What is the Bienaymé–Chebyshev upper bound for $P(|X - 2.2| \geqslant 1.8)$? Determine the upper and lower bounds on a cohering assertion of $P(|X - 2.2| \geqslant 1.8)$, using the fundamental theorem of prevision. Repeat the computation, presuming instead that you assert $V(X) = 7.6$. Think about the size of your upper bounds for the two numerical computations in the context of the B–C inequality.

**2.** Under what conditions will the Bienaymé–Chebyshev upper bound agree with the upper bound provided by the FTP?

As a bottom line to these considerations, if you would like to assess your probabilities for extreme values of various quantities, go ahead and assess them. There is no need to work through the intermediary of an assessed variance. Moreover, if you would like to understand the coherent implications of whatever you assert as your probabilities of extreme values, go ahead and compute them.

Before we leave our direct consideration of a variance assertion, the following three useful results are a worthy matter of record. For brevity we discard the theorem-proof format, stating the results informally and proving them within their statement.

First, a variance and covariance are representable in terms of your previsions for quantities and for their products.

1. For any quantities $X$ and $Y$, the linearity of coherent prevision requires that

$$V(X) \equiv P([X - P(X)]^2) = P(X^2) - [P(X)]^2$$

$$\mathrm{Cov}(X, Y) \equiv P[(X - P(X))(Y - P(Y))] = P(XY) - P(X)P(Y).$$

Both these results and the following one derive from the multiplicative expansion of the product quantity and the linearity of coherent prevision.

Second, since variances and covariances are defined in terms of previsions, the principle of coherency regulates the array of assertions that are appropriate to them. Specifically, a variance-covariance matrix specifies a quadratic form that is nonnegative definite.

2. Concerning quantities defined by a linear combination of other quantities, $Z \equiv aX + bY$, for any real scalars $a$ and $b$,

$$\begin{aligned} V(Z) &\equiv P([Z - P(Z)]^2) = P([aX + bY - aP(X) - bP(Y)]^2) \\ &= a^2 V(X) + b^2 V(Y) + 2ab\,\mathrm{Cov}(X, Y) \\ &= (a \quad b)\boldsymbol{\Sigma}_{\mathbf{X},\mathbf{Y}}(a \;\; b)^{\mathrm{T}} \geqslant 0. \end{aligned}$$

Thus, recognizing $V(Z)$ as a quadratic form in $(a \quad b)$ about $\boldsymbol{\Sigma}_{\mathbf{X},\mathbf{Y}}$, coherency requires that any covariance matrix be nonnegative definite with a nonnegative determinant: $|\boldsymbol{\Sigma}_{\mathbf{X},\mathbf{Y}}| = V(X)V(Y) - [\mathrm{Cov}(X, Y)]^2 \geqslant 0$. This, in turn, implies the bounds on any correlation, $\mathrm{Cor}(X, Y) \in [-1, 1]$. The variance result extends simply to linear combinations of $N$ variables: $V(\mathbf{s}_N^{\mathrm{T}}\mathbf{X}_N) = \mathbf{s}_N^{\mathrm{T}}\,\boldsymbol{\Sigma}_{N,N}\,\mathbf{s}_N$, where $\boldsymbol{\Sigma}_{N,N}$ denotes an asserted variance-covariance matrix for $\mathbf{X}_N$.

The third coherence requirement for variances we shall examine requires some formalities of definition that may appear contrived and even far-fetched at this point. Nonetheless, it does have a regular application in the specification of forecast variances, as we shall see in Chapters 5 and 7. In what follows I present the definitions and the results fairly densely with minimal discussion. For we have not yet even addressed formally the notion of your joint distribution for two quantities nor your conditional distribution for one quantity given the value of another. Thus, the discussion here is somewhat out of place, but it would be distracting to position it later in the text where it is applied. You will be referred back to a consideration of these results and their meaning in the later chapters.

Your asserted variance for a quantity must relate in a specific way to your conditional previsions and conditional variances, when informed by the value of another quantity. The formulation of this relation requires the technical notion of a partitioned prevision and a partitioned variance.

***Definition 4.5.*** Let $\{\mathbf{Y} = \mathbf{y}\}$ denote the partition generated by a quantity $Y$, with realm $\mathscr{R}(Y) = \{y_1, y_2, \ldots, y_K\}$. If you have asserted conditional previsions for a quantity $X$ via $P(X\,|\,Y = y_i)$ for each $y_i \in \mathscr{R}(Y)$, then your *partitioned prevision for $X$ given this partition* is the quantity defined by

$$\begin{aligned} P(X\,|\,\{\mathbf{Y} = \mathbf{y}\}) \equiv (Y = y_1)P(X\,|\,Y = y_1) + (Y = y_2)P(X\,|\,Y = y_2) \\ + \cdots + (Y = y_K)P(X\,|\,Y = y_K). \end{aligned}$$

Similarly, if for each $y_i \equiv \mathscr{R}(Y)$ you also assert your conditional variances $V(X\,|\,Y = y_i) \equiv P[(X - P(X\,|\,Y = y_i))^2\,|\,Y = y_i]$, your *partitioned variance for $X$* is the quantity defined by

$$V(\mathrm{X}|\{\mathbf{Y}=\mathbf{y}\}) \equiv (Y=y_1)V(X|Y=y_1)+(Y=y_2)V(X|Y=y_2) \\ +\cdots+(Y=y_K)V(X|Y=y_K). \qquad \square$$

Notice those sentences with some care: a partitioned prevision and a partitioned variance are both quantities! The definitions of the quantities $P(X|\{\mathbf{Y}=\mathbf{y}\})$ and $V(X|\{\mathbf{Y}=\mathbf{y}\})$ presuppose your assertion of the conditional previsions and conditional variances whose numerical values are the linear coefficients of the partition events whose linear combination defines these "partitioned quantities." These unknown quantities, then, have the numerical values of your conditional prevision and conditional variance, conditioned on the specific value of $Y$ that obtains, which you are uncertain about. Otherwise, these quantities are nowise different from any other quantity, and you may assert your prevision and your variance for them.

With this understanding, the following lines derive the coherency requirements that your prevision for your partitioned prevision must equal your asserted $P(X)$ and that your $V(X)$ must equal your prevision for your partitioned variance plus your variance for your partitioned prevision.

3. Let $X$ and $Y$ denote any quantities, with $\mathscr{R}(Y)=\{y_1, y_2, \ldots, y_K\}$. Furthermore, let $\{\mathbf{Y}=\mathbf{y}\}$ denote the partition generated by $Y$: $\{(Y=y_1), (Y=y_2), \ldots, (Y=y_K)\}$. Then

**a.**

$$P(P(X|\{\mathbf{Y}=\mathbf{y}\})) = \sum_{i=1}^{K} P(Y=y_i)P(X|Y=y_i) = P(X),$$

according to the linearity of coherent prevision and the theorem on total prevision; and

**b.**

$$\begin{aligned} V(X) &= P([X-P(X)]^2) \\ &= P([X-P(X|\{\mathbf{Y}=\mathbf{y}\})+P(X|\{\mathbf{Y}=\mathbf{y}\})-P(X)]^2) \\ &= P([X-P(X|\{\mathbf{Y}=\mathbf{y}\})]^2) \\ &\quad + 2P([X-P(X|\{\mathbf{Y}=\mathbf{y}\})][P(X|\{\mathbf{Y}=\mathbf{y}\})-P(X)]) \\ &\quad + P([P(X|\{\mathbf{Y}=\mathbf{y}\})-P(X)]^2) \\ &= P(V(X|\{\mathbf{Y}-\mathbf{y}\}))+V(P(X|\{\mathbf{Y}=\mathbf{y}\})). \end{aligned}$$

How sure are you about the value of $X$? (In a word, how large is your $V(X)$?) Coherency requires your answer to relate in a specific way to how sure you expect to be if you learn the value of $Y$ and how sure you are regarding what your prevision for $X$ is when conditioned upon the value of $Y$ that obtains.

### 4.2.2 Asserting a Correlation: Do We Need an FTP Part III?

Definition 4.4 identified your correlation between $X$ and $Y$ as the ratio of your covariance to the square root of the product of your variances. There is nothing unusual about the notion of correlation derived from your previsions in a context that you have asserted values of your $P(X)$, $P(X^2)$, $P(Y)$, $P(Y^2)$, and $P(XY)$. Notice, however, that your correlation is not defined until you have asserted these previsions.

If your prevision for the vector of quantities $(X, X^2, Y, Y^2, XY)$ is merely bounded within some partial prevision polytope, you could undoubtedly specify the cohering bounds on an associated "correlation assertion" as a programming problem with the usual linear restrictions. The relevant objective function would then be an awkward-looking quotient of functions which you can delineate for yourself if you understand the FTP part II. Historical attentions to "the correlation between variables" notwithstanding, I wonder, however, what would be the purpose of identifying these bounds? What operational use can be made of the fact that your $\text{Cov}(X, Y)/[V(X)V(Y)]^{1/2}$ must lie within some interval, above and beyond the fact that the boundary-determining vector of previsions lies within its asserted polytope? I am somewhat puzzled about this.

The reverse side of the question is perhaps more relevant. Suppose, in assessing my uncertainty about the quantities $X$ and $Y$, I would want to begin by asserting that my correlation between them is within, say, [.4, .6]? Of course this could be understood as a shorthand way of asserting a corresponding partial prevision polytope for the vector $(X, X^2, Y, Y^2, XY)$. But I am stymied by the problem of identifying an operationally defined transaction whose prospective net gain I can consider in order to assert my interval. Directly, the asserted bounds on $\text{Cor}(X, Y)$ place bounds only on the products of prices for these vector components. In what context are the products of prices useful? Perhaps we should merely recognize our understanding of the mathematical meaning of a correlation as a convenient way to specify five-dimensional assertions, as formalized in Section 2.10.1.

Technically, moreover, we should notice that a resolution to the meaning of a correlation assertion per se would identify a truly nonlinear programming problem as a model for still another variety of the fundamental theorem of prevision. Asserting an interval bound or even an exact value for a correlation, without specifically identifying assertion values for the products and powers of $X$ and $Y$, would identify a nonlinear constraint on the components of the $\mathbf{q}$ vector that appears in the formulation of the FTP. Computational solutions to such problems have been identified for many forms of objective function, both linear and nonlinear, although they can be achieved only in much-smaller-scale problems.

A related problem, of even greater practical merit, should be mentioned. In a practical problem involving the quantities $X$, $Y$, $W$, and $Z$, it seems meaningful to assert a conditional prevision ordering of the form $P(X|Y) \geqslant P(W|Z)$, even without having asserted a numerical value for either of these. Of course, if I would specify at least one number that sits between these ordered assertions, the contents of the claims can be reduced to linear constraints on $\mathbf{q}$, as we have noted in the discussion of FTP part II. However, if only the inequality is asserted, the implied constraints on $\mathbf{q}$ are quadratic, as you can determine for yourself. This situation, too, would specify the context for stating formally a part III of the FTP, involving nonlinear constraints as inputs to the relevant programming problems. Some problems seem to motivate a weak ordering of a massive array of conditional previsions as a starting point to the assessment of uncertainties, before any particular numerical details have been specified. The exploration in Lad and Coope (1995) provides an example. Is it satisfactory to specify the meaning of conditional prevision orderings as the assertion of a many- dimensional polytope? An analysis of this question would be useful, but we turn our attention now to a consideration of coherency restrictions on previsions for convex functions.

## 4.3 PREVISION ORDERINGS FOR CONVEX FUNCTIONS OF QUANTITIES

We have learned that a coherent prevision assertion must equal some convex combination of the realm elements for the quantity under consideration. Coupled with the general theory of convex functions, this result directly implies a definite ordering to one's previsions for the powers of a positive quantity. The argument is easily stated, beginning with a reminder of the definition of a convex function.

***Definition 4.6.*** *A function $g(\cdot)$ is convex over the interval $[a, b]$ if for any vector $\mathbf{x}_N$ of ordered components satisfying $a \leqslant x_1 \leqslant x_2 \leqslant \cdots \leqslant x_N \leqslant b$, and for any nonnegative coefficient vector $\mathbf{q}_N$, whose components sum to 1,*

$$g\left(\sum_{i=1}^{N} q_i x_i\right) \leqslant \sum_{i=1}^{N} q_i\, g(x_i).$$

(The function is *concave* over the interval if this last inequality is reversed.) □

**Theorem 4.2.** Jensen's Inequality. Suppose the realm of a quantity, $\mathscr{R}(X) = \{x_1, \ldots, x_K\}$, is ordered by $x_i < x_{i+1}$ for each $i$. If a function $g(\cdot)$ is convex over the interval $[x_1, x_K]$ then coherency requires that $g[P(X)] \leqslant P[g(X)]$.

*Proof*. Since the quantity $g(X)$ is defined as $\sum_{i=1}^{K} g(x_i)(X = x_i)$, coherency requires that $P[g(X)] = \sum_{i=1}^{K} q_i g(x_i)$ for some nonnegative coefficient vector $\mathbf{q}_N$ whose components sum to 1. But the convexity of $g(\cdot)$ then requires that

$$g[P(x)] = g\left(\sum_{i=1}^{K} q_i x_i\right) \leqslant \sum_{i=1}^{K} q_i g(x_i) = P[g(X)]. \qquad \square$$

The direction of Jensen's inequality reverses for concave functions. An important application of the inequality provides an ordering to coherent previsions for powers of quantities.

**Corollary 4.2.1.** For any strictly positive quantity $X$, if $\alpha < \beta$ are any real numbers, then coherency requires that $[P(X^\alpha)]^{1/\alpha} \leqslant [P(X^\beta)]^{1/\beta}$. Included in this ordering is the limiting value, $\lim_{\alpha \to 0}[P(X^\alpha)]^{1/\alpha}$ which yields $\exp[P(\log(X))]$.

*Proof.* Notice that $X^\beta = (X^\alpha)^{\beta/\alpha}$ is defined by a function of $X^\alpha$: $g(y) = y^{\beta/\alpha}$. This function is convex over $\mathscr{R}(X^\alpha)$ as long as $\beta > 0$ and $\alpha \neq 0$. Thus, $P(X^\beta) \geqslant [P(X^\alpha)]^{\beta/\alpha}$, which establishes the result for this case by taking the $\beta$-root of both sides. When $\alpha < \beta < 0$, the same function $g(\cdot)$ is concave, so the prevision inequality reverses, but this inequality reverses again when the $\beta$-root is taken. The limiting result as $\alpha \to 0$ is achieved by applying L'Hôpital's rule to the representation of $[P(X^\alpha)]^{1/\alpha}$ as $(\sum_{i=1}^K x_i^\alpha q_i)^{1/\alpha} = \exp[\alpha^{-1}\log(\sum_{i=1}^K x_i^\alpha q_i)]$, for an appropriate vector $\mathbf{q}_K$. Applying the rule yields $\exp[\sum_{i=1}^K q_i \log(x_i)]$, since the exponential function is continuous. But this is a representation for $\exp[P(\log(X)]$, as specified in the theorem. □

## 4.4 COHERENT PREVISION AS AN ASSOCIATIVE GENERALIZED MEAN

In motivating our development of prevision, we remarked that prevision is a specific summary of your uncertain knowledge that is sufficient for the purpose of some particular interest you have in $X$. [In the standard procedure of prevision elicitation, indeed, we portrayed your interest in $X$ as the net gain you would achieve by engaging a transaction whose bottom line continges on the difference between $X$ and $P(X)$.] In one sense, the summarizing quality of prevision as a formulation of uncertain knowledge was a source of grief for us, for we admitted the feeling that our knowledge can be too complex to be represented in full by a number, just as our experience of what's happening can be too complex to be measured in full by one number. Mathematically, the summarizing quality of the prevision functional identifies it as a generalized mean function.

The technical specification of a generalized mean requires an expansive formulation of the concept of a summary of uncertain knowledge about $X$ that is sufficient for a purpose. A complete discussion is beyond the scope of this text, and a beautiful historical and substantive review at an advanced level has been provided recently by Muliere and Parmigiani (1993). Daboni (1984) presents an axiomatic discussion. However, a brief introduction here is worthwhile to provide another glimpse of the important connection between subjective probability theory and utility.

We shall learn that the form of the generalized mean function specified by prevision assertions is the only form of generalized mean that has an associative property, which characterizes the use of conditional previsions in making decisions.

A generalized mean function is a procedure for summarizing a vector of numbers. The sufficiency of the summary is evaluated relative to its capacity to achieve a specified purpose. Formally, we define a mean function $M(\cdot)$ relative to a function of interest, $I(\cdot)$.

***Definition 4.7.*** Suppose the mapping rule defining a function $M$: $\mathbb{R}^N \to \mathbb{R}$ is well defined for any positive integer $N$, and let $I$: $\mathbb{R}^N \to \mathbb{R}$ be any function, called an "interest function." The function $M(\cdot)$ specifies a *generalized mean function relative to* $I(\cdot)$ if $I(\mathbf{v}_N) = I(M(\mathbf{v}_N)\mathbf{1}_N)$ for any $\mathbf{v}_N \in \mathbb{R}^N$. □

The motivation behind this definition should not be lost in its formality. Suppose you are interested in the value of some function $I(\mathbf{v}_N)$, but you do not know the components of the argument vector, $\mathbf{v}_N$. In such a case it would be fortunate to know the generalized mean value of $\mathbf{v}_N$ relative to $I(\cdot)$, say $M(\mathbf{v}_N) = m_I$, for we could then easily compute the value of $I(\mathbf{v}_N)$ as $I(m_I\mathbf{1}_N)$, even without knowing the components of $\mathbf{v}_N$. For the generalized mean function summarizes the values of $\mathbf{v}_N$ for the purpose identified by $I(\cdot)$.

***Example 1.*** Suppose $X$ is a quantity that identifies a length measurement, with realm $\mathscr{R}(X) = \{x_1, \ldots, x_K)$. Are you interested in the measurement of length itself: $X = I_1(x_1, \ldots, x_K) \equiv \sum_{i=1}^K x_i(X = x_i)$? Or are you really interested in the area covered by a square whose sides equal this length: $X^2 = I_2(x_1, \ldots, x_K) \equiv \sum_{i=1}^K x_i^2(X = x_i)$? In either case the appropriate function $I(\cdot)$ represents your interest in $X$. The unknown components of the vector $\mathbf{v}_K \in \mathbb{R}^K$ are the values of the partition events $(X = x_1)$ through $(X = x_K)$. As another example, are you interested in the net monetary gain you will accrue in a commercial transaction: $G = I_1(g_1, \ldots, g_K) = \sum_{i=1}^K g_i(G = g_i)$? Or are you interested in how much enjoyment (utility) you will incur as a result of this income: $U = I_U(g_1, \ldots, g_K) = \sum_{i=1}^K U(g_i)(G = g_i)$? □

***Example 2.*** A purely mathematical example of a generalized mean function is that the arithmetic average of a vector $\mathbf{x}_N$ is a generalized mean relative to the function $I_S(\cdot) = \Sigma(\cdot)$ which maps any vector onto the sum of its components. For $\mathbf{\Sigma}(\mathbf{x}_N) = \mathbf{\Sigma}(\bar{x}_N\mathbf{1}_N)$. But the arithmetic average is not a generalized mean relative to the function $I_P(\cdot) = \Pi(\cdot)$, which maps any vector onto the product of its components. Relative to the product function $\Pi(\cdot)$, the geometric average, or $(\prod_{i=1}^N x_i^N)^{1/N}$, provides a generalized mean. □

How does coherent prevision identify a generalized mean function, and what salient properties does it exhibit? Consider a quantity $X$ with the realm $\mathscr{R}(X) = \{x_1, \ldots, x_K\}$, Suppose we define a function $M$: $\mathbb{R}^N \to \mathbb{R}$ for any $N \leqslant K$ by

$$M(\mathbf{y}_N) \equiv P\left( \sum_{i=1}^N y_i(X = x_i) \middle| \sum_{i=1}^N (X = x_i) \right).$$

To begin, it should be evident that, according to this definition, $M(\mathbf{x}_K) = P(X)$. Moreover, for any $N \leqslant K$, $M(\mathbf{x}_N) = P(X|\sum_{i=1}^{N}(X = x_i))$, your conditional prevision. In general, the value of $M(\mathbf{y}_N) = P(Y|\sum_{i=1}^{N}(X = x_i))$, where $Y$ is the quantity defined by the function $I{:}\mathscr{R}(X) \to \{y_1, y_2, \ldots, y_K\}$.

Second, notice the mathematical property that, at least in the case where each $P(X = x_i) = p_i > 0$, the value of $M(\mathbf{y}_N)$ must equal a convex combination of its components, due to relevant coherency conditions:

$$M(\mathbf{y}_N) = \sum_{i=1}^{N} y_i p_i \Big/ \sum_{i=1}^{N} p_i, \qquad \text{where each } p_i > 0.$$

Finally, if a vector $\mathbf{y}_K$ is modified so that any $N$ of its components are replaced by their mean value, yielding $\mathbf{y}^*_{N,K} \equiv (M(\mathbf{y}_N)\mathbf{1}_N, y_{N+1}, y_{N+2}, \ldots, y_K)$, the function value $M(\mathbf{y}^*_{N,K})$ is identical to $M(\mathbf{y}_K)$. This is the property that defines the feature of $M(\cdot)$ as an "associative mean" function, so it is worth dwelling briefly on why this is so.

Presuming $Y$ is a quantity defined by $I{:}\mathscr{R}(X) \to \{y_1, y_2, \ldots, y_K\}$, suppose we define another quantity, $Y^*_{N,K}$, in terms of your asserted conditional prevision, as

$$Y^*_{N,K} = I^*_{N,K}(X) \equiv \sum_{i=1}^{N} M(\mathbf{y}_N)(X = x_i) + \sum_{i=N+1}^{K} y_i(X = x_i).$$

Your cohering prevision for this derived quantity must then be

$$\begin{aligned} P[I^*_{N,K}(X)] &= M(\mathbf{y}_N) \sum_{i=1}^{N} P(X = x_i) + \sum_{i=N+1}^{K} y_i P(X = x_i) \\ &= P\left(Y \,\middle|\, \sum_{i=1}^{N} (X = x_i)\right) P\left(\sum_{i=1}^{K} (X = x_i)\right) \\ &\quad + P\left(Y \,\middle|\, \sum_{i=N+1}^{N} (X = x_i)\right) P\left(\sum_{i=N+1}^{K} (X = x_i)\right) \\ &= P(Y) = P(I(X)), \end{aligned}$$

as required by the theorem on total prevision. This theorem allows that one's complete assessment of a quantity can be achieved by considering it under an array of exclusive and exhaustive scenarios. Such a conceptual division of the various possibilities is characteristic of scientific theorizing, experimentation, and analysis.

A central result is that every associative generalized mean function can be represented in the form $M_I(\mathbf{x}_N) = I^{-1}(\sum_{i=1}^{N} I(x_i)q_i)$ for some monotonic function $I(\cdot)$. This was discovered independently by Kolmogorov (1930), Nagumo (1930), and de Finetti (1931), whose analysis is discussed in the classic work of Hardy, Littlewood, and Polya (1934). A comprehensive formulation of more recent extensions of the analysis has been achieved by Dickey (1986).

The importance of the characterization of associative mean functions for subjectivist methods is due to its relevance to the connection between prevision and the theory of utility. In practical decisions, when the unknown consequences of your activities are measured in monetary terms, you are typically concerned not with the monetary yield per se, but with the amount of pleasure or utility this yield will provide. Indeed, concern with this problem led us to stipulate our definition of prevision in Chapter 2 as the price at which you would engage in transactions whose volume is modulated by the feature that the scale of your maximum stake is not to be exceeded. Over small monetary gains, most anyone's utility increases linearly with money. When the very nature of the activities being contemplated precludes such a modulation (we examine one shortly) their yield must be assessed according to more general forms of prevision. Here enters the relevance of the associative mean theorem.

Presuming that your utility increases with money, it specifies an invertible "utility function," $U(\cdot)$. The Kolmogorov–Nagumo–de Finetti theorem requires that as long as you would like to follow a method of assessing the unknown satisfaction you might incur as a consequence of several available actions by considering their utilities separately under an array of exclusive and exhaustive scenarios, you must assess each action on the basis of a generalized prevision of the form

$$P_U(X(a)) = U^{-1}(P(U(X(a)))) = U^{-1}\left(\sum_{i=1}^{K} U(x_i(a))p_i(a)\right).$$

Thus, the characterization of associative means specifies a desirable property of Bayesian decision-making procedures based on expected utility.

We conclude this chapter with an example of the generalized utility assessment of the portfolio selection problem which had been resolved by de Finetti (1940) in the context of apportioning insurance risk.

## 4.5 AN EXAMPLE FROM THE THEORY OF PORTFOLIO SELECTION

Both for conceptual reasons germane to the overall outlook of this book and for economic theoretical reasons, we conclude with a discussion of the mean-variance theory of portfolio selection. The focus of the commentary will be on the subjectivist characterization of the problem and on the extent to which the nonlinear utility function it presumes still allows for a resolution in terms of the linear prevision functional.

Suppose $\mathbf{X}_t = (X_{1t}, X_{2t}, \ldots, X_{Nt})^{\mathrm{T}}$ is the vector of unknown rates of return to be achieved by $N$ different stocks over a specified time interval $[t, t+1]$. (Notation for the dimension of the vector $\mathbf{X}_t$ is suppressed. The subscript $t$ represents time period considered. Throughout this section, vector symbols printed in bold should be presumed to represent vectors of dimension $N$.) The components of $\mathbf{X}_t$

are defined by

$$X_{it} \equiv (P_{i(t+1)} - P_{it} + D_{it})/P_{it},$$

which is the rate of return achieved by buying one share of stock $i$ at time $t$ for price $P_{it}$, receiving the dividend payment $D_{it}$ during period $t$, and then selling the share one period later at time $t+1$ for price $P_{i(t+1)}$.

A portfolio selection amounts to your choice of a vector $\mathbf{s} = (s_1, \ldots, s_N)^T$ in the $(N-1)$-dimensional unit simplex $\mathbf{S}^{N-1}$, with the interpretation that your total investment, $I$, is to be composed of investments in several selected stocks, with $s_i$ being the proportion of $I$ invested in stock $i$. Making the portfolio choice $\mathbf{s}$ means that $s_i I/P_{it}$ shares of stock $i$ are to be bought at time $t$ for each $i = 1, \ldots, N$. Thus, the rate of return achieved by holding this portfolio during period $t$ through $t+1$, written as a function of your portfolio selection $\mathbf{s}$, is a convex combination of the rates of return achieved by the individual stocks:

$$R_t(\mathbf{s}) = \sum_{i=1}^{N} \frac{s_i I}{P_{it}} \frac{[P_{i(t+1)} - P_{it} + D_{it}]}{I} = \sum_{i=1}^{N} s_i X_{it} = \mathbf{s}^T \mathbf{X}_t.$$

Linearity of coherent prevision requires that your assertion $P[R_t(\mathbf{s})] = \mathbf{s}^T P(\mathbf{X}_t)$.

The simplest theory of portfolio selection is based on the supposition that your previsions for the rates of return from various stocks are constant with respect to the various portfolio selections $\mathbf{s}$ that you might make. This is appropriate when the scale of investment, $I$, is small relative to the total number of stock shares available. We are familiar with assessing our valuations for uncertain financial rewards in the context of very-small-scale transactions, for which the maximum net gain or loss does not exceed the scale of our maximum stake. In the context of such miniscule transactions, which govern the definition of prevision, the solution to the choice problem is trivial: choose the portfolio $\mathbf{s}$ for which your $P[R_t(\mathbf{s})]$ is a maximum. This would amount to placing your total investment in the particular stock for which your prevision for its return is the greatest.

However, such a small-scale problem and its solution are irrelevant to investment practice. Real portfolio decisions considered by investment market participants are never such small-scale problems. Indeed, for a professional portfolio manager, the scale of contemplated investment decisions is often far greater even than the manager's total personal annual income. Choosing a portfolio $\mathbf{s}$ that maximizes the manager's $P[R_t(\mathbf{s})]$ cannot resolve the decision problem. The mean-variance theory of portfolio selection is designed to address the more complicated problem of large-scale ventures. The theory still requires that the scale of transactions be small enough relative to the number of shares outstanding in the entire market that your prevision for the rates of return on individual stocks be constant relative to your choice of your portfolio-apportioning vector $\mathbf{s}$. Yet the scale is recognized to be large enough that your utility valuation for the uncertain rate of return deriving from a portfolio choice $\mathbf{s}$ is not merely an increasing function of your prevision for its rate of return but is also a decreasing function of your variance for the portfolio return.

The portfolio selection problem is to determine the portfolio choice, **s**, for which your utility valuation function, $U(\mathbf{s}) = U(P[R(\mathbf{s})], V[R(\mathbf{s})])$, is a maximum. It is presumed that the partial derivative of your utility with respect to $P[R(\mathbf{s})]$ is positive and that with respect to $V[R(\mathbf{s})]$ is negative.

A precise specification of your utility function would pose a forbidding task. Rather than specifying a detailed utility function $U[P(X), V(X)]$, the problem can be resolved by identifying the set of feasible $[P(R(\mathbf{s})), V(R(\mathbf{s}))]$ combinations that are achievable from various possible portfolios **s**. Among all possibilities, the *set of efficient portfolios* are those for which your associated $(P, V)$ assertions have the following two properties: among all portfolios that are assessed with the same prevision for their return, an efficient portfolio is one that is assessed with the smallest variance; similarly, among all portfolios that are assessed with the same variance for their return, an efficient portfolio is one that is assessed with the greatest prevision.

As long as your opinions about rates of return from individual stocks do not depend on your portfolio choice, your prevision and variance for the return achievable from the choice **s** are

$$P[R(\mathbf{s})] = \mathbf{s}^{\mathrm{T}} P(\mathbf{X}),$$
$$V(R(\mathbf{S})] = \mathbf{s}^{\mathrm{T}} V(\mathbf{X})\mathbf{s} = \mathbf{s}^{\mathrm{T}}[P(\mathbf{X}\mathbf{X}^{\mathrm{T}}) - P(\mathbf{X})P(\mathbf{X})^{\mathrm{T}}]\mathbf{s}.$$

Here $V(\mathbf{X})$ is your asserted variance-covariance matrix for the returns from the components of **X**. Thus, the points along the efficient boundary of feasible $(P, V)$ combinations can be computed easily using Lagrangian methods: determine vectors **s** that maximize $P[R(\mathbf{s})]$ subject to variance constraints, or determine vectors **s** that minimize $V[R(\mathbf{s})]$ subject to prevision constraints.

The general theory of portfolio selection is a specification of the logic of all sorts of investment strategies. It includes investment decisions on scales so great that the strategist's prevision and variance for the rate of return on each particular stock (or other investment instrument) depend on which portfolio the strategist chooses. "If I do this, what do I think will happen to **X**? If I do that, what do I think...?" If you would attempt to "corner" the silver market, for example, you would need to act on such a scale that this type of thinking is necessary. If you are acting on a small scale relative to the market size (even though it may be a large scale relative to your own fortune), your assessment problem for rates of return is simplified in that your opinions of the several possible rates of return would not vary with the portfolio choices which are feasible for you. The mean-variance theory is applicable to this scale of problem.

This rather specialized example has been discussed to highlight three important ideas, foundational and technical. In a general perspective it has relevance to foundational issues underlying the theories of efficient markets and rational expectations which have dominated economic theorizing over the past two decades. As a technical matter it exemplifies the scope of problems that can be resolved within the superficially limited domain of the linear prevision functional.

First, especially if you are familiar with portfolio theory, notice how this problem has been characterized differently than its standard formulation, proposed in objectivist statistical terms. It is typically presumed that "the expected rates of return" from individual stocks are unknown objective constants, properties of the stochastic structure of random market behavior, regardless of any and every market participant's portfolio choice. Participants' personal expectations of the objective, yet in-principle-unobservable "true" expected rates of return, are even categorized as "correct" or "incorrect" according to whether or not they equal the true expected rates. Quite to the contrary, we have formalized the problem of participant strategy in a market quite generally, without making use of the notion of any supposedly "correct" prevision or variance for currently unknown rates of return from an investment, nor has there been supposed any "true distribution" of presumed randomness in the market itself. The history of investment activity (sequential prices and volumes of trades, and sequential dividend payments by firms) is the result of the activities of many financial strategists, with their own personal uncertain knowledge, influence, and financial bases. At any time in history, the several strategists may be largely in agreement or largely in disagreement. Thus, we each behave some way, and some history of our trading prices and quantities is generated. Since we each act in terms of own personal opinion (or some "expert's" opinion if the portfolio choice is left to a manager) each person may well choose a different portfolio and a different time to adjust it, all in keeping with efficient choice theory. However, there are no market laws that require anyone to regard the opinions of others in any particular relation to one's own opinions. Specifically, the oft-heard assertion that the average of all market participants' expectations of price must equal the supposedly correct expectation cannot even be meaningfully entertained.

The second idea exposed by this extended example clarifies the extent to which the theory of prevision is a theory of a *linear* functional. If prevision is a linear functional, how and why should the theory allow that optimal portfolio choice takes consideration of your variance assertions into account? The answer is that the scale of transactions under consideration involve possible yields so great in size that your utility function is nonlinear in the prospective rates of return over their entire realm. While variance evaluations require your consideration of features of the uncertain return vector that are not linear in $\mathbf{X}$, nevertheless these return features *are linear* in the quantities composing $\mathbf{XX}^T$. In the context of a linear prevision functional, these are merely further quantities that are logically, even functionally, related to $\mathbf{X}$. There is no contradiction in the linearity of coherent prevision and the mean-variance theory of portfolio selection.

Finally, this example reiterates in still another way that the operational subjective theory of prevision is specified modulo the general theory of utility.

The mean-variance theory allows that the utility you achieve from a portfolio choice is a function of both your prevision and your variance of return. It is your utility valuation that ultimately governs your decisions. Prevision assertions are gauged in tandem with your utility valuations.

Historical discussion of de Finetti's (1940) contribution to this aspect of financial theory, in the context of insurance problems, appears in Daboni and Pressacco (1987), who compare it with the classic work of Markovitz (1952).

CHAPTER 5

# Distribution Functions

*Mr. Graves pronounced his th charmingly. Turd and fart, he said, for third and fourth. Watt liked these venerable saxon words.*

*Watt*, SAMUEL BECKETT

## 5.1 INTRODUCTION

Having characterized your coherent prevision as a linear functional over the space of real-valued functions defined on $\mathscr{R}(X)$, let us consider some questions:

> So prevision can be assesed for any function of $X$. Must prevision be assessed distinctly for each function of $X$? Is there any end to the introspection necessary to specify *completely* my uncertain knowledge of $X$? How much thinking about $X$ must I do in order to be able to detect *mechanically* whether any particular behavior with consequences dependent on $X$ is coherent?

As one disturbed statistician recently posed the concern: "the problem is that there is too much thinking required in subjective Bayesian statistics."

Happily, we can characterize the extent of introspection needed for you to have assessed (via the implications of coherency) your prevision for every function of a bounded discrete quantity $X$. As we shall see, this characterization is not unique. But perhaps the simplest characterization can be made in terms of your *probability distribution* for $X$. Let us state a definition and a theorem, and then discuss some alternative characterizations.

***Definition 5.1.*** Let $X$ be a quantity with realm $\mathscr{R}(X) = \{x_1, x_2, \ldots, x_K\}$, the elements presumed to be ordered from smallest to largest by their subscripts. If you assert your probabilities for every constituent of the partition generated by $X$ as $P(X = x_1)$, $P(X = x_2), \ldots, P(X = x_K)$, then the function $F(x) \equiv P(X \leqslant x) = \sum_{i=1}^{K} P(X = x_i)(x_i \leqslant x)$, defined for any real $x$ is called your *probability distribution function* for $X$. The function defined by $f(x) \equiv P(X = x)$ is called your *probability mass function* for $X$. □

In order to distinguish notationally your probability distribution for a quantity $X$ from your distribution for another quantity $Y$, a tagging subscript is commonly attached to the corresponding distribution function symbol. Thus, for example, we may write $F_X(\cdot)$ for the former and $F_Y(\cdot)$ for the latter. Similarly, the corresponding probability mass functions would be tagged $f_X(\cdot)$ and $f_Y(\cdot)$. The subscript denotation is suppressed in informal notation whenever the corresponding quantity is obvious from context.

We have defined your probability distribution function for a quantity, but the definition can be read as applying to a vector of quantities as well. You may read the symbol $X$ and each of the realm elements $x_i$ in Definition 5.1 as denoting vectors. In this context, to write $x_i \leqslant x$ would mean that every component of the vector $x_i$ does not exceed the corresponding component of the vector $x$. We defer a detailed presentation of probability distributions for vectors of quantities to Section 5.3.5.

Properties of Coherent Probability Distributions. Any coherent probability distribution function, $F_X(\cdot)$, asserted for a quantity $X$ with realm $\mathscr{R}(X) = \{x_1, x_2, \ldots, x_K\}$ has the following properties:

1. The function is nondecreasing: if $a < b$, then $F(a) \leqslant F(b)$.
2. The function is bounded by 0 and 1: for any real $x$, $0 \leqslant F(x) \leqslant 1$.
3. The function has a finite number of jump points, each of which is an element of the realm of $X$. The sizes of the jumps are indicated by the values of the associated p.m.f. $f(\cdot)$. Symbolically we write

$$\text{for any } x \in \mathscr{R}(X), \qquad [F(x) - \lim_{a \to x(-)} F(a)] = f(x) \geqslant 0,$$

$$\text{for any } x \notin \mathscr{R}(X), \qquad [F(x) - \lim_{a \to x(-)} F(a)] = f(x) = 0.$$

4. The distribution function can be expressed as the sum of a mass function. For any real number $a$, $F(a) = \sum_{i=1}^{K} f(x_i)(x_i \leqslant a)$.

As stated, these properties pertain to probability distributions for any quantity $X$ with a finite discrete realm. Extensions of the concept to infinite realms requires modifications that are best addressed in de Finetti (1974a, chap. 6).

It is now apparent precisely how extensive is the introspection needed to assess your probability distribution for a quantity, $X$. If $X$ is an event, for example, your assessment of only $P(X)$ implies your probability distribution for $X$, since $P(X = 1) = P(X)$, and $P(X = 0) = 1 - P(X)$; but if $\mathscr{R}(X)$ has $K$ elements, you need assess your prevision for $K - 1$ of the partition events. For the sum of all $K$ of the probabilities over a partition must equal 1. If $X$ is a vector of $N$ logically independent quantities, each with $K$ elements in its realm, you need to assess prevision for $K^N - 1$ events in order to specify your probability distribution for $X$.

For many applied problems, however, you need not assess your entire probability distribution for $X$ to determine how you would behave in order to achieve you desired ends. Often a "ballpark" assertion regarding the value of $X$ is sufficient. When you are asked, "how far is it from Cleveland to Chicago?" your inquisitor may wish to know merely whether traveling this distance by automobile would typically involve a trip of an afternoon, a full day, or several days. You need not assert your full probability distribution for the distance to provide a satisfactory answer. Asserting your $P(X)$ and your $P(|X - P(X)| > 100)$, say, may suffice (where $X$ is the highway distance measured in kilometers).

Nonetheless, the study of probability distribution functions is of great importance for problems of scientific inference, both conceptually and practically. Let us begin by examining the conceptual importance of the probability distribution function in characterizing uncertain knowledge.

## 5.2 THE CONCEPTUAL IMPORTANCE OF A PROBABILITY DISTRIBUTION

The conceptual importance of your probability distribution function derives from the fact that *your prevision for any quantity*, defined by a function of $X$, *is implied* via coherency by your specification of your probability distribution function. Let us write this as a theorem.

**Theorem 5.1.** If $F(x)$ is your probability distribution function for a quantity $X$ with realm $\mathscr{R}(X) = \{x_1, \ldots, x_K\}$, then coherency requires that your prevision for any quantity $Y$ defined by a function of $X$, $Y \equiv g(X)$, be computed as a sum over probabilities that specify your distribution function:

$$P[g(X)] = \sum_{i=1}^{K} g(x_i) P(X = x_i).$$

*Proof*. The result is an immediate consequence of the facts that $Y$ is a linear combination of the constituents of the partition generated by $X$, and that coherent prevision is a linear functional. Since $Y \equiv g(X) = \sum_{i=1}^{K} g(x_i)(X = x_i)$, linearity requires that

$$P(Y) = P[g(X)] = P\left[\sum_{i=1}^{K} g(x_i)(X = x_i)\right] = \sum_{i=1}^{K} g(x_i) P(X = x_i). \qquad \square$$

Thus, a specification of your probability distribution for $X$ would completely characterize your coherent prevision functional over the space of real-valued functions of $X$, as we had promised.

A disclaimer to the importance of this result should be mentioned immediately. Although the result is pleasant, it does not *uniquely* distinguish the importance of probability distribution function. Your probability distribution for $X$ is not the only possible characterization of your coherent prevision functional, $P$. We can easily develop some other recognizable and important characterizations. Suppose, for example, that you assert your previsions for several of the powers of the quantity $X$, say your $P(X)$, $P(X^2), \ldots,$ and $P(X^{K-1})$. Since for any value of $J$, the quantity $X^J$ can be written as $X^J = \sum_{i=1}^K x_i^J(X = x_i)$, we can write a linear relation between the powers of $X$ and the constituents of the partition generated by $X$—that is, $(X = x_1), (X = x_2), \ldots, (X = x_K)$—as

$$\begin{pmatrix} 1 \\ X \\ X^2 \\ \vdots \\ X^{K-1} \end{pmatrix} = \begin{pmatrix} 1 & 1 & \cdots & 1 \\ x_1 & x_2 & \cdots & x_K \\ x_1^2 & x_2^2 & \cdots & x_K^2 \\ \vdots & & & \\ x_1^{K-1} & x_2^{K-1} & \cdots & x_K^{K-1} \end{pmatrix} \begin{pmatrix} (X = x_1) \\ (X = x_2) \\ (X = x_3) \\ \vdots \\ (X = x_K) \end{pmatrix},$$

summarized by a matrix equation, using obvious notations, $\mathbf{X}^K = \mathbf{V}_{K,K}(X)\mathbf{Q}_K(X)$. A matrix with the form of $\mathbf{V}_{K,K}(X)$ is commonly called a *Vandermonde matrix*.

On account of the linear relation $\mathbf{X}^K = \mathbf{V}_{K,K}(X)\mathbf{Q}_K(X)$, we can also identify a linear relation between your assessed distribution function $F(x)$, which is defined by your prevision for the partition, $P(\mathbf{Q}_K(X))$, and your prevision for the powers of $X$, $P(\mathbf{X}^K)$. Linearity of coherent prevision requires that your $P(\mathbf{X}^K) = \mathbf{V}_{K,K}(X)P(\mathbf{Q}_K(X))$ or, equivalently, $P(\mathbf{Q}_K(X)) = \mathbf{V}_{K,K}^{-1}(X)P(\mathbf{X}^K)$. Thus, we conclude the following theorem, which is equivalent to Theorem 5.1.

**Theorem 5.2.** If $\mathscr{R}(X) = \{x_1, x_2, \ldots, x_K)$ and $P(\mathbf{X}^K)$ denotes the vector of your previsions $P(1, X, X^2, X^3, \ldots, X^{K-1})^T$, then coherency requires that your prevision for any quantity defined by a function of $X$, $Y \equiv g(X)$, can be computed as

$$P(Y) = P[g(X)] = [g(x_1), g(x_2), \ldots, g(x_K)]\mathbf{V}_{K,K}^{-1}(X)P(\mathbf{X}^K).$$

In fact, your prevision for *any* $K-1$ distinct powers of $X$ would similarly characterize your prevision for any function of $X$. In the preceding discussion each exponent $J$ which appears in the vector $\mathbf{X}^K$ and the matrix $\mathbf{V}_{K,K}(X)$ can be replaced by a generic power, $\alpha_J$. Even more generally, your prevision for any $K-1$ *functions* of $X$ that specify linearly independent vectors of possible observations would characterize your prevision for any other function of $X$.

Suppose, for example, that the function $g(X)$ in Theorem 5.2 is the exponential function $g(X) = \exp(tx)$ for some specified value of $t$. Then coherency requires that $P[g(X)]$ equal

$$\begin{aligned} P[\exp(tX)] &= (e^{tx_1} \quad e^{tx_2} \quad \cdots \quad e^{tx_K})\mathbf{V}_{K,K}^{-1}(X)P(\mathbf{X}^K) \\ &= (e^{tx_1} \quad e^{tx_2} \quad \cdots \quad e^{tx_K})P(\mathbf{Q}_K(X)). \end{aligned}$$

Thus, your asserting your $P[\exp(tx)]$ for $K-1$ distinct values of $t \neq 0$ is a third procedure sufficient for characterizing your opinions about any function of $X$. Such a characterization provides an intuitive motivation for the theory of *moment generating functions*, a topic we discuss in Section 5.3.3.

In conclusion, we can summarize the conceptual importance of the probability distribution function as due to its complete characterization of your uncertain knowledge about $X$. Although it is important to remember that there are other characterizations, too, your probability distribution function is often the easiest to assess within a recognizable degree of approximation. Other characterizations are mainly useful in proofs of theorems, such as the central limit theorem, which identify forms of uncertain knowledge about quantities when the knowledge assertions satisfy specified conditions.

Before studying some particular useful forms of *probability* distributions, we identify in Section 5.3 the features of a larger family of functions that are classified in mathematics as "distribution functions." We shall find that several members of this larger family have applicability to computational procedures of statistical inference.

To conclude this section, let us mention three mathematical asides on the Vandermonde matrix. First, it can be shown using mathematical induction that the matrix $\mathbf{V}_{K,K}(X)$ is nonsingular; moreover, using induction and elementary row operations on matrices, the determinant of this matrix can be shown to equal the product of the differences between its components:

$$|\mathbf{V}_{K,K}(X)| = \prod_{j>i=1}^{K} (x_j - x_i).$$

Proofs of these propositions are suggested as problems in most texts on linear algebra as applications of the principles of nonsingularity and of matrix row operations. A complete derivation of the determinant appears in Franklin (1968).

Second, the Vandermonde matrix appears in the algebraic expression for the $(K-1)$-degree polynomial that passes through $K$ specified points in two-dimensional space, $(x_1, y_1), (x_2, y_2), \ldots, (x_K, y_K)$:

$$\mathbf{a}^{\mathrm{T}}\mathbf{V}_{K,K}(X) = (a_0 \quad a_1 \quad a_2 \quad \cdots \quad a_{K-1})\mathbf{V}_{K,K}(X) = (y_1 \quad y_2 \quad \cdots \quad y_K) \equiv \mathbf{y}^{\mathrm{T}}.$$

Thus, its inverse is used to calculate the coefficients of the polynomial function that passes through those points:

$$\mathbf{a}^{\mathrm{T}} = \mathbf{y}^{\mathrm{T}}\mathbf{V}_{K,K}^{-1}(X).$$

This solution equation, which is linear in $\mathbf{y}$ and polynomial in $X$, was known to Lagrange and is commonly called the *Lagrange interpolating polynomial.* An algebraic formula appears in the text of Moursund and Duris (1967).

Finally, the powers of $X$ provide the simplest example of what is called a "Chebyshev system of functions." A classic discussion of the significance and application of the power functions appears in Polya and Szego (1976, Vol. II,

Part 5, Chap. 1). Karlin and Studden (1966) present a mathematical treatise on applications of the general formulation.

## 5.3 MATHEMATICAL CHARACTERIZATION OF A DISTRIBUTION FUNCTION

Mathematically, the concept of a *distribution function* is formalized even more broadly than the particular type of function we have identified as a *probability* distribution function. Consider the following definition.

***Definition 5.2.*** Any nondecreasing function $G: \mathbb{R} \to [0, 1]$ with the properties that $\lim_{a \to -\infty} G(a) = 0$ and $\lim_{a \to \infty} G(a) = 1$ is said to be a *distribution function.* The symbol $\mathbb{R}$ denotes the set of real numbers. □

We can distinguish three general types of functions that have these properties: (1) step, (2) absolutely continuous, (3) continuous but not absolutely continuous. The first type, called a *discrete* distribution function, is a *step function* that is nondecreasing everywhere except at a finite or countable number of points. The sizes of the jumps at these points define the associated discrete mass function. Metaphorically, a discrete distribution corresponds to the partition of a finite mass into a finite or countable number of clumps. The proportions of the total mass that are clumped at each designated unit specify the sizes of the jumps.

The second type of distribution function is called an *absolutely continuous* distribution function. It is a function that can be represented by a *Riemann integral of a density function,* $G(t) = \int_{-\infty}^{t} g(x)\,dx$ for some density function, $g(x) = D_x G(x)$. Since $G$ is nondecreasing, a density function must be everywhere nonnegative, $g(x) \geqslant 0$, and have a finite integral: $\int_{-\infty}^{\infty} g(x)\,dx = 1$.

The mathematical study of real analysis has yielded the knowledge that not every continuous function $G$ can be represented in this way. The third type of distribution function includes those that are *continuous but not absolutely continuous.* "To say that $G(x)$ is continuous means that for each $\varepsilon > 0$ however small, every interval whose length is less than some suitable number $\delta$ contains a mass smaller than $\varepsilon$. To say that $G$ is absolutely continuous means something more: that the same is true of the mass contained in any arbitrary number of intervals of total length less than $\delta$" (de Finetti, 1974a, p. 233). Symbolically we write that $G: \mathbb{R} \to \mathbb{R}$ is continuous means that

$$\forall x \forall \varepsilon > 0 \, \exists \delta > 0 \ni |x' - x| < \delta \Rightarrow |G(x') - G(x)| < \varepsilon,$$

whereas to write that $G: \mathbb{R} \to \mathbb{R}$ is absolutely continuous means that

$$\forall x \forall \varepsilon > 0 \exists \delta > 0 \ni \sum_{i=1}^{\infty} |x_i' - x| < \delta \Rightarrow \sum_{i=1}^{\infty} |G(x_i') - G(x)| < \varepsilon.$$

If you have not heard of this distinction, you may enjoy studying the standard introductory example of the "Cantor function," which is nondecreasing over the interval $[0,1]$ with $G(0)=0$ and $G(1)=1$, but it is also nonincreasing over a countable number of intervals in $[0,1]$ whose total length equals 1. Thus, the Cantor function is continuous everywhere and nonincreasing "almost everywhere," yet it increases from 0 to 1. Formal definitions and major results can be found in Kolmogorov and Fomin (1970, pp. 333–340) and many texts on real analysis.

By and large, distribution functions of the third type are irrelevant to the development of this text. One notational refinement that we do utilize from time to time (already in Sections 3.10, 3.12) derives mainly from the analytical awareness of distinctions between distribution functions of the second and third types. A *Stieltjes integral* of the form $\int g(x)\,dF(x)$ represents a general operation of integration that can represent in one symbolism an integral with respect to a distribution function $F(\cdot)$ of any of the three types mentioned.

In the case that $F(\cdot)$ is discrete, with its jump points in the countable set $\{x_1, x_2, \ldots, x_N, \ldots\}$, the integral $\int_{-\infty}^{\infty} g(x)\,dF(x)$ represents the series $\sum_{i=1}^{\infty} g(x)f(x)$, where the mass function value $f(x)$ represents the size of the jump at $x$, $f(x) \equiv F(x) - \lim_{a \to x(-)} F(a)$.

In the case that $F(\cdot)$ is absolutely continuous, the integral $\int_{-\infty}^{\infty} g(x)\,dF(x)$ represents the Riemann integral $\int_{-\infty}^{\infty} g(x)f(x)\,dx$ with respect to the derivative function $dF(x) = f(x)\,dx$.

Finally, the same Stieltjes integral symbol represents the result of a particular limiting summation operation even when the function $F(\cdot)$ is of the third type. Formalities can be better studied in an appropriate text on analysis.

We use the Stieltjes integration symbol only when it is convenient to represent mathematical results which pertain to any type of distribution function. You may interpret it as a sum, a series, or a Riemann integral, without conceptual confusion.

### 5.3.1 Applications of Distribution Functions in Statistical Inference

The operational subjective theory of statistical inference distinguishes and uses regularly four functions that can be recognized mathematically as distribution functions of the first or second type. Let us introduce them briefly.

A *probability distribution function* is, as we have noted, a discrete distribution function with jump points only at numbers in the realm of $X$. It is usually denoted by $F(\cdot)$, and its corresponding probability mass function (p.m.f.), by $f(\cdot)$. When several quantities are being discussed, say $X$, $Y$, and $Z$, their distribution functions are distinguished symbolically by subscripting the function names $F$ with the associated quantity symbol, $X$, $Y$, or $Z$. Thus, we may write $F_X(x)$, $F_Y(y)$, and $F_Z(z)$; similarly, their p.m.f.'s may be denoted by $f_X(x)$, $f_Y(y)$, and $f_Z(z)$.

A *statistical distribution function*, sometimes also called an *empirical distribution function*, is a function defined in terms of a specific vector of quantity observations. The statistical distribution function associated with the observed quantities $\mathbf{X}_N = (X_1, X_2, \ldots, X_N)$ is the function

$$F: \mathbb{R} \to [0,1] \quad \text{defined by} \quad F(x) = \sum_{i=1}^{N} (X_i \leqslant x)/N.$$

The numerical value of $F(x)$ equals the proportion of the observations that do not exceed $x$. A statistical distribution function is always a discrete distribution function. The mass function associated with a statistical distribution function, which identifies the proportion of the observations that equal $x$, is related to the histogram of the observations. The histogram identifies the number of observations that equal each specified value of $x$. We study applications of the histogram in Sections 5.5.1 and 5.5.2.

An *approximate probability distribution function* for $X$ *with approximation error less than* $\varepsilon$ is any distribution function $G(\cdot)$ with the property that $|G(x) - F(x)| < \varepsilon$ for all values of $x$ covering $\mathscr{R}(X)$ . The function $G(\cdot)$ may be discrete, in which case $g(x) \equiv G(x) - \lim_{a \to x(-)} G(x)$ is called your approximate probability mass function; or $G(\cdot)$ may be absolutely continuous, in which case $g(x) \equiv D_x G(x)$ is called your approximate probability density function; or $G$ may be a convex mixture of discrete and absolutely continuous distribution functions.

Suppose that a distribution function $F(\cdot)$ is representable in the form $F(x) = \int_\Theta F(x;\theta)\, dM(\theta)$, where $M(\theta)$ is a distribution function over a parameter space $\Theta$, and the functions denoted by $F(x;\theta)$ are members of a family of distributions that is parameterized by numerical values of the parameter $\theta$. At times we use the symbol $\mathscr{F}(\cdot;\Theta)$ to denote such a family of functions. In this context, the probability distribution function $F(x)$ is said to be a *mixture distribution* (sometimes called a "compound distribution") and the function $M(\theta)$ is said to be its *mixing distribution function.* The importance of mixture distributions in the operational subjective theory of statistical inference derives from their usefulness in representing judgments that regard quantities exchangeably or partially exchangeably, as we studied briefly in Chapter 3. We expand the details of those results in this chapter and in Chapter 7.

All four of these applications of mathematical distribution functions will be developed throughout this chapter and throughout this book. Let us now return to our study of distribution functions per se.

### 5.3.2 Moments of a Distribution Function

In Section 5.2 we studied how your probability distribution function for $X$ can be characterized by your prevision for powers of $X$. In the language of distribution functions, we established this characterization by using the fact that

$$P(X^\alpha) = \sum_{i=1}^{K} x_i^\alpha P(X = x_i) = \int_{-\infty}^{\infty} x^\alpha \, dF(x),$$

as required by coherency. It should not be surprising that this integral operation on a mathematical distribution function, as it appears on the right of the final equation, is generally meaningful and important in distribution theory. We commonly refer to the values of this integral operation on a distribution function as "the moments of the distribution." Let's write a formal definition.

***Definition 5.3.*** Let $F(\cdot)$ be a distribution function. Then for any real number $r$, the number $\mathbb{M}_r(F) \equiv \int_{-\infty}^{\infty} x^r \, dF(x)$, if it exists, is called *the rth moment of the distribution F*. If $\mathbb{M}_1(F)$ exists, then the number

$$\mathbb{M}_r^*(F) \equiv \int_{-\infty}^{\infty} [x - \mathbb{M}_1(F)]^r \, dF(x)$$

is called *the rth central moment* of the distribution $F$, whenever it exists. □

Notice that for any integer $r$, the $r$th central moment can be written as a function of the first $r$ moments of $F$, since

$$[x - \mathbb{M}_1(F)]^r = x^r - {}^rC_1 x^{(r-1)}\mathbb{M}_1(F) + {}^rC_2 x^{(r-2)}[\mathbb{M}_1(F)]^2 - \cdots + (-1)^i\, {}^rC_i x^{(r-i)}[\mathbb{M}_1(F)]^i + \cdots + (-1)^r[\mathbb{M}_1(F)]^r$$

according to the binomial expansion theorem. Thus, the linearity of the integral operator implies that

$$\mathbb{M}_r^*(F) = \mathbb{M}_r(F) - {}^rC_1 \mathbb{M}_{r-1}(F)\mathbb{M}_1(F) + {}^rC_2 \mathbb{M}_{r-2}(F)[\mathbb{M}_1(F)]^2 - \cdots + (-1)^i\, {}^rC_i \mathbb{M}_{r-i}(F)[\mathbb{M}_1(F)]^i + \cdots + (-1)^r[\mathbb{M}_1(F)]^r.$$

The qualifiers in Definition 5.3 concerning the existence of moments are required because of the fact that integrals of the form $\int_{-\infty}^{\infty} x^r \, dF(x)$ do not necessarily exist for every distribution function $F$. We shall come upon some examples in Chapter 7. The notation $\mathbb{M}_r(F_X)$ may be simplified to $\mathbb{M}_r(X)$.

Of course, if $X$ is a quantity and $F(\cdot)$ is your probability distribution for $X$, then coherency requires that, for each value of $r$, your prevision $P(X^r)$ must equal the $r$th moment of your probability distribution. Similarly, $\mathbb{M}_2^*(F)$ must equal your $V(X) \equiv P[(X - P(X))^2]$, as we defined your variance assertion in Chapter 4. Coherency requires that any asserted $P(X^r)$ must equal the corresponding $r$th moment of some discrete probability distribution. But not every sequence of moments, $\{\mathbb{M}_r(F)\}_{r=1}^{\infty}$, need represent previsions for the powers of some *quantity*. For there is no requirement that a generic distribution function specify probabilities for an observable quantity, for which objects alone the concept of prevision is relevant. The language of distribution functions and their moments is more general than the language of probability and prevision.

Within the general theory of distributions and their moments are two results that are worth stating for your general knowledge, but whose extensive discussion would distract us here. The first is an important bounding relation on the sizes of moments of different orders, called Lyapunov's inequality. The second is an interesting correspondence between all sequences of moments of distribution functions on $[0,1]$ and the family of completely monotone sequences, called Hausdorff's theorem.

Lyapunov's inequality on moments states that for any distribution function, $F(\cdot)$, for any real numbers $r$, $s$, and $t$ ordered by $r \leqslant s \leqslant t$, the powers of the moments $\mathbb{M}_r$, $\mathbb{M}_s$, and $\mathbb{M}_t$ must satisfy the inequality

$$\mathbb{M}_s^{(t-r)} \geqslant \mathbb{M}_r^{(s-r)} \mathbb{M}_t^{(t-s)}.$$

A derivation and discussion can be found in Marshall and Olkin (1979).

Hausdorff's theorem states that a nonincreasing positive sequence of real numbers, $\{1, m_1, m_2, \ldots, m_N, \ldots\}$, is a sequence of moments for a unique distribution $F(\cdot)$ on $[0,1]$ if and only if the sequence is a completely monotone sequence. Definitions, derivations, and discussion appear in the classic text of Hardy (1949) and the interesting article of Daboni (1982). Computable conditions for the extendability of a finite sequence to a completely monotone sequence, described in terms of Hankel determinants, appear in Karlin and Studden (1966).

Finally, a widely useful and conceptually important equality concerns the moments of distributions that are represented in the form of a mixture distribution, a feature relevant to the judgment of exchangeability. Suppose your distribution function $F(x)$ is the mixture distribution

$$F(x) = \int_{\Theta} F(x;\theta)\, dM(\theta).$$

Let $\mathbb{M}_1(X;\theta)$ denote the first moment of $X$ according to the distribution function $F(x;\theta)$. Thus, recognize $\mathbb{M}_1(X;\theta)$ as a function of $\theta$. Then coherency of your prevision requires that your

$$P(X) = \mathbb{M}_1(X) = \mathbb{M}_1[\mathbb{M}_1(X;\theta)]$$

and your

$$V(X) = \mathbb{M}_2^*(X) = \mathbb{M}_1[\mathbb{M}_2^*(X;\theta)] + \mathbb{M}_2^*[\mathbb{M}_1(X;\theta)].$$

A second use of this concept involving the structure of previsions, variances, conditional previsions, and conditional variances directly is the result we showed in section 4.2.1. Coherency requires that for any quantities $X$ and $Y$,

$$V(X) = P[V(X \mid \{\mathbf{Y} = \mathbf{y}\})] + V[P(X \mid \{\mathbf{Y} = \mathbf{y}\})],$$

where $\{\mathbf{Y} = \mathbf{y}\}$ denotes the partition generated by $Y$. Your variance for $X$ must equal your prevision for your conditional variance plus your variance for your partitioned conditional prevision.

### 5.3.3 Moment Generating Functions and Characteristic Functions

The relation established in Section 5.2 between a probability distribution function (always discrete) and its moments should motivate you to understand how a similar correspondence can be established between a general distribution function and its moments. The correspondence is generated in a rather different way, since the powers of a quantity $X$ were related to the constituents of a partition it generates by means of a finite linear combination which does not hold for countable or continuous distributions. Technically, the corresponding "moment generator" is constructed in these cases as follows.

Recall that for any real value of $t \neq 0$, the Maclaurin series expansion

$$\exp(tx) = 1 + tx + \frac{t^2x^2}{2!} + \frac{t^3x^3}{3!} + \frac{t^4x^4}{4!} + \cdots$$

The moment generating function, $M_F(t)$, for a distribution function $F(\cdot)$, when it exists, is defined by the corresponding series of integrals

$$M_F(t) \equiv \int_{-\infty}^{\infty} \exp(tx)\, dF(x)$$

$$= 1 + t\mathbb{M}_1(F) + \frac{t^2\mathbb{M}_2(F)}{2!} + \frac{t^3\mathbb{M}_3(F)}{3!} + \frac{t^4\mathbb{M}_4(F)}{4!} + \cdots$$

according to the linearity of the integral (summation) operator.

As a function of $t$, the moment generating function is continuous when it exists. Moreover, when it is differentiable at $t = 0$, its derivatives yield all the moments of $F(\cdot)$ through the following formula, which you can derive yourself by differentiating the representation in the previous paragraph:

$$D_t^{(r)} M_F(t)|_{t=0} = \mathbb{M}_r(F).$$

In words, this equation states that "the $r$th derivative of the moment generating function $M_F(t)$ with respect to $t$, evaluated at the value of $t = 0$, equals the $r$th moment of $F$. When we study specific distribution functions in the remainder of this text, we identify the associated moment generating function, allowing you to exemplify this identity.

A famous inversion formula derived in real analysis identifies a 1-1 correspondence between distribution functions and moment generating functions when the latter exist. More generally, even when the moments (and thus the moment generating function) of a distribution function do not exist, the complex function commonly denoted by $\chi_F(t) \equiv \int_{-\infty}^{\infty} \exp(itx)\, dF(x)$ always exists. The class of all such "characteristic functions" stands in 1-1 correspondence with the class of all distribution functions. The analytic study of characteristic functions is not a topic relevant to this text, but if you have tried grappling with them you may enjoy the beautiful geometrical expository article by Epps (1993).

Both moment generating and characteristic functions are related to the general theory of generating functions which are quite useful in determining the values of integrals and series. The amusing book of Wilff (1994), entitled *Generatingfunctionology*, develops the theory into broad numerical applications, and the text of Feller (1958), among others, discusses applications in probability.

### 5.3.4 The Distribution Function for a Transformation of Its Argument

Since your probability distribution for $X$ characterizes everything you know about $X$, it should also characterize everything you know about any quantity defined by a function of $X$, $Y \equiv G(X)$. In fact, we have derived a simple way to determine your probability distribution for any $G(X)$ from your probability distribution for $X$ when we specified Theorem 5.1. Coherency requires that

$$f_Y(y) \equiv P(Y=y) = \sum{}^* (G(x)=y)P(X=x) = \sum{}^* (G(x)=y)\, f_X(x),$$

where the summation $\sum^*$ runs over all $x \in \mathscr{R}(X)$. This result is corollary to the theorem, taking the function $G(\cdot)$ of the theorem in this case to be the event function $(G(X)=y)$.

When we consider an absolutely continuous distribution function, the relation between the density function $f_X(x)$ and the density $f_Y(y)$ can also be expressed easily, but under the restriction that the transformation $Y=G(X)$ be monotonic. The relation derives from an application of the chain rule of differential calculus. Let us state it with a brief proof, followed by two examples. The notation is based on the following result from calculus of inverse functions: if an invertible function $G(\cdot)$ is differentiable, with $dG(x) \equiv g(x)\,dx$, then its inverse function $G^{-1}(\cdot)$ is also differentiable, with $dG^{-1}(y) \equiv g^{-1}(y)\,dy = dy/g(G^{-1}(y))$.

**Theorem 5.3.** Suppose the function $G(\cdot)$ is strictly monotonic and differentiable, with $dG(x) \equiv g(x)\,dx$. If a distribution function $F_X(\cdot)$ is absolutely continuous, admitting the density $f_X(\cdot)$, then the corresponding distribution function for $Y \equiv G(X)$ is also absolutely continuous and admits the density function $f_Y(t) = f_X(G^{-1}(t))|g^{-1}(t)|$.

*Proof.* If $G(\cdot)$ is an increasing function, then $F_Y(t) \equiv F_X(G^{-1}(t))$ and $f_Y(t) = f_X(G^{-1}(t))g^{-1}(t)$ according to the chain rule for differentiation. In this case, $g^{-1}(t) > 0$, since the inverse function $G^{-1}(\cdot)$ is also increasing. Similarly, if $G(\cdot)$ is a decreasing function, then $F_Y(t) \equiv 1 - F_X(G^{-1}(t))$ and, thus, $f_Y(t) = -f_X(G^{-1}(t))g^{-1}(t)$, with $g^{-1}(t) < 0$. In either case the relation between the densities stated in the theorem using $|g^{-1}(t)|$ holds. □

The following two examples are misplaced in the sense that we have not yet formally defined either a Normal or a gamma density function. But you can understand at least the algebraical detail merely with a knowledge of calculus.

You may refer back to the distributional content of these examples once we specifically study these density functions.

***Example 1.*** A Normal(0, 1) density function for $X$ is defined by

$$f_X(x) = (2\pi)^{-1/2} \exp(-x^2/2) \qquad (x \in \mathbb{R}).$$

Defining $Y = G(X) \equiv X^2$, the symmetry of both $G(\cdot)$ and $f_X(\cdot)$ about 0 ensures that the density of $Y$ would be identical if the density for $X$ were doubled over only positive values of $x$:

$$f_X(x) = 2(2\pi)^{-1/2} \exp(-x^2/2) \qquad (x > 0).$$

Over this positive domain, the function $G(\cdot)$ is monotonic increasing. Its inverse function is specified by $G^{-1}(y) = y^{1/2}$, with $g^{-1}(y) = y^{-1/2}/2$. Thus, Theorem 5.3 ensures that

$$\begin{aligned} f_Y(y) &= f_X(G^{-1}(y))|g^{-1}(y)| = f_X(y^{1/2})y^{-1/2}/2 \\ &= (2\pi)^{-1/2} y^{-1/2} \exp(-y/2) \qquad (y > 0). \end{aligned}$$

This, as you will learn, is the formula for a $\chi^2(1)$ density. □

***Example 2.*** A gamma$(\alpha, \beta)$ density function for $X$ is defined for $\alpha > 0$ and $\beta > 0$ by

$$f_X(x) = [\beta^\alpha/\Gamma(\alpha)]x^{\alpha-1} \exp(-\beta x) \qquad (x > 0).$$

Defining $Y$ by the monotonic decreasing function $Y = G(X) \equiv 1/X$, the inverse function $G^{-1}(y) = 1/y$ has the derivative $g^{-1}(y) = -y^{-2}$. Thus, Theorem 5.3 ensures that

$$\begin{aligned} f_Y(y) &= f_X(G^{-1}(y))|g^{-1}(y)| = f_X(1/y)y^{-2} \\ &= [\beta^\alpha/\Gamma(\alpha)]y^{-\alpha-1} \exp(-\beta/y) \quad (y > 0). \end{aligned}$$

This, as you will learn, is the formula for an "inverted gamma" density.

□

***Example 3.*** This example has regular applications in the computer generation of sequences of numbers assessed independently with a particular distribution for use in Monte Carlo studies. Suppose $F_X(\cdot)$ is some generic distribution function, admitting the density function $f_X(\cdot)$. Consider the transformation $Y = F_X(X)$. Theorem 5.3 tells us that

$$f_Y(y) = f_X(F_X^{-1}(y))|f_X^{-1}(y)| = f_X(F_X^{-1}(y))/f_X(F_X^{-1}(y)) = 1 \quad (y \in [0, 1]).$$

That is, the distribution function value of a continuous distribution, itself has a uniform distribution over [0, 1], since its density is constant over this interval. If you would like to generate a sequence of quantities that are assessed independently with a gamma($\alpha, \beta$) distribution, for example, you can first generate a sequence of quantities that are each assessed with a uniform distribution on [0, 1], $\{Y_1 = y_1, Y_2 = y_2, \ldots\}$, and then transform each of them to the value of $x$ whose cumulative gamma distribution function value equals that value of $y$: $\{X_1 = x_1 = F_X^{-1}(y_1),\ X_2 = x_2 = F_X^{-1}(y_2), \ldots\}$. Software such as MATLAB, which can routinely generate a sequence of quantities assessed with a variety of distributions, embeds this very procedure into its programming code. (To learn the syntax, type "help binornd" or "help gamrnd".) □

Another useful example of Theorem 5.3 concerns the lognormal density function. If $X = \log(Y)$ has a normal density function, then $Y$ is said to have a lognormal density. You might wish to derive this density function for practice. Theorem 5.3 generalizes easily to a statement relating joint probability density functions for transformations of several variables. It is time that we formalize our terminology on this subject.

### 5.3.5 Joint, Marginal, and Conditional Probability Distributions for Several Quantities

We noted immediately after Definition 5.1 that the concept of a probability distribution is applicable to a vector of quantities as well as to a single quantity. Let us write the applicable definition in full detail and then explore the extension of concepts it provides relevant to statistical inference.

***Definition 5.4.*** Let $\mathbf{X}_N$ be a vector of quantities with realm $\mathcal{R}(\mathbf{X}_N) = \{\mathbf{x}_{N,1}, \mathbf{x}_{N,2}, \ldots, \mathbf{x}_{N,K}\}$. If you assert your probabilities for every constituent of the partition generated by $\mathbf{X}_N$ as $P(\mathbf{X}_N = \mathbf{x}_{N,1})$, $P(\mathbf{X}_N = \mathbf{x}_{N,2}), \ldots, P(\mathbf{X}_N = \mathbf{x}_{N,K})$, then the function $F(\mathbf{x}_N) \equiv \sum_{i=1}^{K} P(\mathbf{X}_N = \mathbf{x}_{N,i})(\mathbf{x}_{N,i} \leqslant \mathbf{x}_N)$, defined for any real vector $\mathbf{x}_N$, is called your *joint probability distribution function* for the components of $\mathbf{X}_N$. The function $f(\mathbf{x}_N) \equiv P(\mathbf{X}_N = \mathbf{x}_N)$ is called your *joint probability mass function* for the components of $\mathbf{X}_N$. □

The inequality $\mathbf{x}_{N,i} \leqslant \mathbf{x}_N$ is satisfied if and only if every component of the vector $\mathbf{x}_{N,i}$ is less than or equal to the corresponding component of $\mathbf{x}_N$.

The following simple example is rich enough to illustrate a joint probability distribution for two variables (for a vector in two dimensions) as well as their marginal and conditional distributions, which we develop shortly. This rather trite example is proposed merely to focus on our understanding the formal aspects of a joint distribution function and the multivariate notation.

***Example 1.*** An urn contains six distinguishable colored balls, which are identified numerically according to the scheme: 1 = red, 2 = orange, 3 = yellow,

4 = green, 5 = blue, 6 = violet. You are to scramble the balls in the urn, and to draw successively two balls, replacing the first and rescrambling the balls after the first draw. Let $X_1$ denote the number of the ball selected on the first draw, and $X_2$ denote the number of the ball selected on the second draw. Finally, let $D$ denote the absolute difference, $D \equiv |X_1 - X_2|$, and let $M$ denote the product of the results of the two draws, $M \equiv X_1 X_2$. In most of this example we concentrate on $D$ and $M$. The realm of the vector of logically dependent quantities, $(D, M)$, contains 21 elements:

$$\mathscr{R}(D, M) = \{(0,1),(0,4),(0,9),(0,16),(0,25),(0,36),(1,2),(1,6),(1,12),(1,20),(1,30),$$
$$(2,3),(2,8),(2,15),(2,24),(3,4),(3,10),(3,18),(4,5),(4,12),(5,6)\}.$$

Under the conditions of this experiment, I (and I presume you) would regard the quantities $X_1$ and $X_2$ independently, asserting $P(X_1 = i)(X_2 = j) = 1/36$ for each pair $(i, j) \in \{1,2,3,4,5,6\}^2$. Coherency then requires that my (and your?) joint probability mass function and your joint probability distribution function for $D$ and $M$ are those in Table 5.1.

Notice that each entry in the right half of Table 5.1, for example $F_{M,D}(13,4) = 21/36$, equals the sum of all the mass function entries $f_{M,D}(m, d)$ entered for values of $m \leqslant 13$ and $d \leqslant 4$. In general, we write that for any pair $(m^*, d^*)$, $F_{M,D}(m^*, d^*) = \sum_{m \leqslant m^*} \sum_{d \leqslant d^*} f_{M,D}(m, d)$. The joint distribution function for two quantities equals the cumulative sum, in both directions, of the associated probability masses.

Having considered the joint probability distribution function for $M$ and $D$, you might question "what has happened to the simple distribution function and probability mass function for $M$ or for $D$ that we might have considered in Section 5.1, before we began to study their joint distribution?" As it turns out, the answer is a simple one, since

$$f_D(d) = P(D = d) = \sum_{m \in \mathscr{R}(M)} f_{M,D}(m, d).$$

This means that the value of $f_D(d)$, for example, equals the sum of all the joint probability masses that appear in the column of the table headed by that value of $d$. The complete mass function appears in the bottom margin of the left side of Table 5.1. A similar equation defines the probability mass function for $M$,

$$f_M(m) = P(M = m) = \sum_{d \in \mathscr{R}(D)} f_{M,D}(m, d),$$

which also appears in the table as the right margin of the joint probability mass function matrix. The distribution functions for $M$ and for $D$ also appear in the table as the final (marginal) column and the final row of $F_{M,D}(m, d)$, although they are not specifically labeled. You can recognize them easily, however, on the basis

**Table 5.1. The Joint Probability Mass Function, $f_{M,D}(m, d)$, and the Joint Probability Distribution Function, $F_{M,D}(m, d)$, for $M$ and $D$[a]**

| | $36f_{M,D}(m,d)$ | | | | | | | $36F_{M,D}(m,d)$ | | | | | |
|---|---|---|---|---|---|---|---|---|---|---|---|---|---|
| $m\backslash d$ | 0 | 1 | 2 | 3 | 4 | 5 | $f_M(m)$ | 0 | 1 | 2 | 3 | 4 | 5 |
| 1 | 1 | | | | | | 1 | 1 | 1 | 1 | 1 | 1 | 1 |
| 2 | | 2 | | | | | 2 | 1 | 3 | 3 | 3 | 3 | 3 |
| 3 | | | 2 | | | | 2 | 1 | 3 | 5 | 5 | 5 | 5 |
| 4 | 1 | | | 2 | | | 3 | 2 | 4 | 6 | 8 | 8 | 8 |
| 5 | | | | | 2 | | 2 | 2 | 4 | 6 | 8 | 10 | 10 |
| 6 | | 2 | | | | 2 | 4 | 2 | 6 | 8 | 10 | 12 | 14 |
| 7 | | | | | | | 0 | 2 | 6 | 8 | 10 | 12 | 14 |
| 8 | | | 2 | | | | 2 | 2 | 6 | 10 | 12 | 14 | 16 |
| 9 | 1 | | | | | | 1 | 3 | 7 | 11 | 13 | 15 | 17 |
| 10 | | | | 2 | | | 2 | 3 | 7 | 11 | 15 | 17 | 17 |
| 11 | | | | | | | 0 | 3 | 7 | 11 | 15 | 17 | 19 |
| 12 | | 2 | | | 2 | | 4 | 3 | 9 | 13 | 17 | 21 | 23 |
| 13 | | | | | | | 0 | 3 | 9 | 13 | 17 | 21 | 23 |
| 14 | | | | | | | 0 | 3 | 9 | 13 | 17 | 21 | 23 |
| 15 | | | 2 | | | | 2 | 3 | 9 | 15 | 19 | 23 | 25 |
| 16 | 1 | | | | | | 1 | 4 | 10 | 16 | 20 | 24 | 26 |
| 17 | | | | | | | 0 | 4 | 10 | 16 | 20 | 24 | 26 |
| 18 | | | | 2 | | | 2 | 4 | 10 | 16 | 22 | 26 | 28 |
| 19 | | | | | | | 0 | 4 | 10 | 16 | 22 | 26 | 28 |
| 20 | | 2 | | | | | 2 | 4 | 12 | 18 | 24 | 28 | 30 |
| 21–23 | | | | | | | 0 | 4 | 12 | 18 | 24 | 28 | 30 |
| 24 | | | 2 | | | | 2 | 4 | 12 | 20 | 26 | 30 | 32 |
| 25 | 1 | | | | | | 1 | 5 | 13 | 21 | 27 | 31 | 33 |
| 26–29 | | | | | | | 0 | 5 | 13 | 21 | 27 | 31 | 33 |
| 30 | | 2 | | | | | 2 | 5 | 15 | 23 | 29 | 33 | 35 |
| 31–35 | | | | | | | 0 | 5 | 15 | 23 | 29 | 33 | 35 |
| 36 | 1 | | | | | | 1 | 6 | 16 | 24 | 30 | 34 | 36 |
| $f_D(d)$ | 6 | 10 | 8 | 6 | 4 | 2 | | | | | | | |

[a] Entries in each matrix are the appropriate value of $f_{M,D}(m, d)$ or $F_{M,D}(m, d)$ multiplied by 36. Missing entries for the joint mass function should be read as 0's. To save space, consecutive rows with identical entries are represented only once, as noted.

of the equations

$$F_M(m) = P(M \leqslant m) = F_{M,D}(m, \max d \in \mathscr{R}(D)),$$

$$F_D(d) = P(D \leqslant d) = F_{M,D}(\max m \in \mathscr{R}(M), d).$$

With a terminology based on this "marginal" relation between these several pairs of functions, the individual distributions and mass functions for $M$ and for $D$ are referred to as their "marginal" distributions with respect to one another. Let us write a formal definition with a little more generality, and then comment on some details. Of course, the following extensive notation can be reduced when the context of discussion is apparent, but it is worth writing once in full.

***Definition 5.5.*** Suppose the vector $\mathbf{X}_N$ is partitioned into $\mathbf{X}_N = (\mathbf{X}_{N_1}, \mathbf{X}_{N_2})$ and the elements of $\mathscr{R}(\mathbf{X}_N)$ are partitioned conformably. Then the function

$$f_{\mathbf{X}_{N_1}}(\mathbf{x}_{N_1}) = \sum f_{\mathbf{X}_N}(\mathbf{x}_{N_1}, \mathbf{x}_{N_2}) \qquad (\mathbf{x}_{N_2} \in \mathscr{R}(\mathbf{X}_{N_2}))$$

is called your *marginal probability mass function for* $\mathbf{X}_{N_1}$. Similarly, the function

$$f_{\mathbf{X}_{N_2}}(\mathbf{x}_{N_2}) = \sum f_{\mathbf{X}_N}(\mathbf{x}_{N_1}, \mathbf{x}_{N_2}) \qquad (\mathbf{x}_{N_1} \in \mathscr{R}(\mathbf{X}_{N_1}))$$

is called your *marginal probability mass function for* $\mathbf{X}_{N_2}$.

The distribution functions corresponding to these mass functions, $F_{\mathbf{X}_{N_1}}(\mathbf{x}_{N_1})$ and $F_{\mathbf{X}_{N_2}}(\mathbf{x}_{N_2})$, are called your *marginal probability distribution functions.*

□

Let us first note that the categories of joint probability mass function and marginal probability mass function are not exclusive. The single quantities $D$ and $M$ of Example 1 correspond to the vectors $\mathbf{X}_{N_1}$ and $\mathbf{X}_{N_2}$ of Definition 5.5, so their marginal distributions are not joint distributions. However, in larger problems, when $\mathbf{X}_{N_1}$ and $\mathbf{X}_{N_2}$ are actually vectors of several quantities, then the mass function $f_{\mathbf{X}_{N_1}}(\mathbf{x}_{N_1})$, which is marginal with respect to the components of the vector $\mathbf{X}_{N_2}$, is also a joint p.m.f.—joint with respect to its own several components.

***Example 1 continued.*** In this context, notice that the joint p.m.f. for $D$ and $M$ that we have developed is a *marginal* p.m.f with respect to the vector of quantities $(X_1, X_2)$. As an exercise, you should identify the realm of the quantity vector $(X_1, X_2, D, M)$ and consider the characteristics of the associated joint probability distribution for this vector. □

By now we are well familiar with the concept of conditional prevision and the requirements that coherency imposes upon its assertion. The formalism of a *conditional* distribution merely subsumes this notion into the convenient and useful language of probability distributions. Let us merely write the appropriate definition and then make relevant comments.

***Definition 5.6.*** Again suppose $\mathbf{X}_N = (\mathbf{X}_{N_1}, \mathbf{X}_{N_2})$ is assessed with the joint p.m.f. $f_{\mathbf{X}_N}(\mathbf{x}_{N_1}, \mathbf{x}_{N_2})$, having the associated marginal p.m.f.'s $f_{\mathbf{X}_{N_1}}(\mathbf{x}_{N_1})$ and $f_{\mathbf{X}_{N_2}}(\mathbf{x}_{N_2})$. By the *conditional probability mass function for* $\mathbf{X}_{N_2}$ *given* $(\mathbf{X}_{N_1} = \mathbf{x}_{N_1})$ is meant the function

$$f_{\mathbf{X}_{N_2}}(\mathbf{x}_{N_2} | (\mathbf{X}_{N_1} = \mathbf{x}_{N_1})) \equiv P((\mathbf{X}_{N_2} = \mathbf{x}_{N_2}) | (\mathbf{X}_{N_1} = \mathbf{x}_{N_1})) \qquad (\mathbf{x}_{N_2} \in \mathscr{R}(\mathbf{X}_{N_2})).$$

Similarly, the *conditional probability mass function for* $\mathbf{X}_{N_1}$ given $(\mathbf{X}_{N_2} = \mathbf{x}_{N_2})$ is the function

$$f_{\mathbf{X}_{N_1}}(\mathbf{x}_{N_1} | (\mathbf{X}_{N_2} = \mathbf{x}_{N_2})) \equiv P((\mathbf{X}_{N_1} = \mathbf{x}_{N_1}) | (\mathbf{X}_{N_2} = \mathbf{x}_{N_2})) \qquad (\mathbf{x}_{N_1} \in \mathscr{R}(\mathbf{X}_{N_1})). \qquad \square$$

First notice that the multiplication rule for coherent conditional probabilities requires the identities

$$\begin{aligned} f_{\mathbf{X}_N}(\mathbf{x}_{N1}, \mathbf{x}_{N2}) &= f_{X_{N_2}}(\mathbf{x}_{N2} | (\mathbf{X}_{N_1} = \mathbf{x}_{N1}))\, f_{\mathbf{X}_{N_1}}(\mathbf{x}_{N1}) \\ &= f_{\mathbf{X}_{N_1}}(\mathbf{x}_{N1} | (\mathbf{X}_{N2} = \mathbf{x}_{N2}))\, f_{\mathbf{X}_{N_2}}(\mathbf{x}_{N2}), \end{aligned}$$

for every $(\mathbf{x}_{N_1}, \mathbf{x}_{N_2}) \in \mathscr{R}(\mathbf{X}_N)$. (This equation must hold irrespective of whether the marginal probabilities $f_{\mathbf{X}_{N_1}}(\mathbf{x}_{N_2})$ and $f_{\mathbf{X}_{N_2}}(\mathbf{x}_{N_2})$ equal 0 or not. Of course, division by these marginal probabilities is allowable only if they do not equal 0.)

Next notice that we have defined many distinct conditional probability distribution functions by this definition. For each $\mathbf{x}_{N_1} \in \mathscr{R}(\mathbf{X}_{N_1})$ there may be asserted a distinct conditional p.m.f., $f_{\mathbf{X}_{N_2}}(\mathbf{x}_{N_2} | (\mathbf{X}_{N_1} = \mathbf{x}_{N_1}))$; and similarly, for each $\mathbf{x}_{N_2} \in \mathscr{R}(\mathbf{X}_{N_2})$ there may be asserted a distinct conditional p.m.f., $f_{\mathbf{X}_{N_1}}(\mathbf{x}_{N_1} | (\mathbf{X}_{N_2} = \mathbf{X}_{N_2}))$.

Before turning to some problems, a final definition is in order while we are on the topic of conditional and joint distributions. It identifies the judgment to regard $N$ quantities independently in a stochastic sense. As noted, the concept of stochastic independence is not often directly useful in statistical theory, but it does surface at times through considering transformations of quantities.

***Definition** 5.7.* You are said to *regard the vector of quantities* $\mathbf{X}_N$ *independently* if for every subvector of $K$ components your joint p.m.f. for the $K$-component vector equals the product of your marginal mass functions for the components:

$$f(\mathbf{x}_K) = \prod_{i=1}^{K} f_{X_i}(x_i) \qquad \text{for each } \mathbf{x}_K \in \mathscr{R}(\mathbf{X}_K). \qquad \square$$

This definition is the simplest way to express formally the motivating feature that your conditional distribution for any component of $\mathbf{X}_N$ conditional on any information specified by the values of the remaining components is the same as your marginal distribution for that component.

You can consolidate your understanding by considering the following problems. The first continues our example based on drawing two balls from an urn. The second states Bayes' theorem in the terminology of conditional and marginal mass functions. The third suggests how a single distribution function or p.m.f. can be simultaneously a marginal, joint, and conditional function. The final two concern features of independent judgments.

## PROBLEMS

1. Examining Table 5.1 depicting the joint distribution for $D$ and $M$, identify the six relevant conditional p.m.f.'s $f_M(m|D=d))$ and the 18 relevant conditional p.m.f.'s $f_D(d|(M=m))$.

2. Prove Bayes' theorem for p.m.f.'s, focusing on recognizing the role of the condition that $P(X_1 = x_1) > 0$ in the following statement:

$$f_{X_2}(x|(X_1 = x_1)) = f_{X_1}(x_1|(X_2 = x))\, f_{X_2}(x) \Big/ \sum_{x_2 \in \mathscr{R}(X_2)} f_{X_1}(x_1|(X_2 = x_2))\, f_{X_2}(x_2).$$

3. Returning to the context of Example 1 in which we draw twice from an urn of distinguishable colored balls, make a table that identifies the values of the function $f_{X_1,D}(x_1, d|(M = 12))$. Notice that this p.m.f. can be described simultaneously as joint with respect to $X_1$ and $D$, as conditional upon the event $(M = 12)$, and as marginal with respect to $X_2$.

4. Again in Example 1, show that you regard $X_1$ and $X_2$ independently in a stochastic sense, but of the four quantities $X_1, X_2, D$, and $M$, there is no other pair that you regard independently.

5. Review Definition 3.4 of the judgment to regard events independently and Example 2 of Section 3.7 to recall that the simple condition of $f(\mathbf{x}_N) = \prod_{i=1}^{N} f_{X_i}(x_i)$ for each $\mathbf{x}_N \in \mathscr{R}(\mathbf{X}_N)$ is not sufficient to characterize independent judgments.

### *5.3.5.1 Properties of Distribution Functions for Vectors*

This section is merely a notice that the concept of a distribution function, for which we studied many details in Sections 5.3.1 through 5.3.4, extends directly to a distribution function of many variables, called a *multivariate distribution function.* The technical details expand as follows.

For an absolutely continuous distribution function in $N$ dimensions, the *multivariate density function* is the cross partial derivative of $F(\mathbf{x}_N)$ with respect to each of its arguments:

$$f_{\mathbf{X}_N}(x_1, x_2, \ldots, x_N)\, dx_1\, dx_2 \cdots dx_N = dF(\mathbf{x}_N),$$

and

$$F_{\mathbf{X}_N}(t_1, t_2, \ldots, t_N) = \int_{-\infty}^{t_1} \int_{-\infty}^{t_2} \cdots \int_{-\infty}^{t_N} f_{\mathbf{X}_N}(x_1, x_2, \ldots, x_N)\, dx_1\, dx_2 \cdots dx_N.$$

The concept of the moments of a distribution applies to each of the types of distribution we have studied. First, there are marginal moments for each component of $\mathbf{X}_N$. These are precisely the moments which we studied for one variable, generated by each of the marginal distributions of components of $\mathbf{X}_N$. Then there are also *joint-product moments* for the distribution function $F_{\mathbf{X}_N}(\cdot)$, which are defined by

$$\mathbb{M}_{i_1,i_2,\ldots,i_N}(F_{\mathbf{X}_N}) \equiv \int_{-\infty}^{\infty}\int_{-\infty}^{\infty}\cdots\int_{-\infty}^{\infty} x_1^{i_1}x_2^{i_2}\cdots x_N^{i_N} f_{\mathbf{X}_N}(\mathbf{x}_N)\,dx_1\,dx_2\cdots dx_N.$$

Features such as the covariance between two arguments of a distribution, and the correlation between two arguments are defined in terms of these cross-product moments:

$$\sigma_{i,j} \equiv \mathbb{M}_{1,1}(F_{X_2}) - \mathbb{M}_1(F_{X_1})\mathbb{M}_1(F_{X_2}) \qquad \text{and} \qquad \rho_{i,j} \equiv \sigma_{i,j}/\sigma_i\sigma_j.$$

Similarly there are *conditional joint-product moments* associated with the several conditional distributions that can be specified for elements of $\mathbf{X}_N$. Moreover, there are joint moment generating functions, joint characteristic functions, and joint probability generating functions. In the absolutely continuous case, the joint moment generators, for example, are defined, using the column vectors $\mathbf{t}_N$ and $\mathbf{x}_N$, as

$$M_{\mathbf{X}_N}(\mathbf{t}_N) = \int_{-\infty}^{\infty}\int_{-\infty}^{\infty}\cdots\int_{-\infty}^{\infty} \exp(\mathbf{t}_N^{\mathrm{T}}\mathbf{x}_N)\, f_{\mathbf{X}_N}(x_1, x_2, \ldots, x_N)\,dx_1\,dx_2\cdots dx_N.$$

Finally, the formula we developed for the distribution of a function of a quantity translates to the multivariate context in a form that is well known in the calculus of many variables.

#### *5.3.5.2 Sequential Forecasts of Macroeconomic Statistics: An Example Involving Joint, Conditional, and Marginal Distributions*

Consider the problem of macroeconomic forecasting. Many financial, economic, and public policy decisions are made with an eye to what is happening in the economy of a nation. Measurements of what is happening include the national income accounts, recorded in the United States by the Department of Commerce, labor force statistics recorded by the Bureau of Labor Statistics, financial statistics recorded by the Federal Reserve Board, and national demographic statistics recorded by the Census Bureau, and myriad others. The frequency of measurement of these quantities ranges from daily to monthly to quarterly to annually to decennially. Accounts are kept in such extensive detail that any forecaster's attention necessarily focuses on a chosen subcollection of all the measurements. Nonetheless, a large number of measurements may be relevant concerns at any point in time. Typically we identify *vectors* of measurements that are of interest each time period by the denotation $\mathbf{X}_t, \mathbf{X}_{t+1}, \ldots$, where each vector $\mathbf{X}_t$ is composed of several (say, $K$) components.

For example,

$X_{1t}$ = nominal GNP in quarter $t$
$X_{2t}$ = real GNP in quarter $t$
$X_{3t}$ = civilian labor force in quarter $t$
$X_{4t}$ = civilian unemployment rate in quarter $t$
$X_{5t}$ = $M1$ component of money supply in quarter $t$
$\vdots$
$X_{Kt}$ = business fixed investment in quarter $t$

An economic forecaster is usually in the state of having available recorded observations $\mathbf{X}_{t-1}, \mathbf{X}_{t-2}, \ldots$ for several previous quarters, as well as other current information in various forms such as corporate news, political news, and geophysical news. Yet the forecaster, along with everyone else, is uncertain about the values of the vector quantities $\mathbf{X}_t$ for the current quarter and the values $\mathbf{X}_{t+1}$, $\mathbf{X}_{t+2}, \ldots$ for subsequent quarters.

The paradigmatic statistical questions are what conditional prevision assertions $P(\mathbf{X}_t, \mathbf{X}_{t+1}, \ldots | (\mathbf{X}_{t-1} = \mathbf{x}_{t-1})(\mathbf{X}_{t-2} = \mathbf{x}_{t-2}) \cdots)$ or what conditional distribution function $F(\mathbf{x}_t, \mathbf{x}_{t+1}, \ldots | (\mathbf{X}_{t-1} = \mathbf{x}_{t-1})(\mathbf{X}_{t-2} = \mathbf{x}_{t-2}) \cdots)$ represent best the forecaster's uncertainty about yet unobserved measurements given past measurements. Notation is simplified in the following discussion by writing $\mathbf{I}_{t-1} \equiv ((\mathbf{X}_{t-1} = \mathbf{x}_{t-1})(\mathbf{X}_{t-2} = \mathbf{x}_{t-2}) \cdots)$ to denote all the information available for the forecaster's (your) prevision about $\mathbf{X}_t, \mathbf{X}_{t+1}, \mathbf{X}_{t+2}, \ldots$. This may be available either in the form of recorded observations of previous quarters' $\mathbf{X}$ components or otherwise.

Suppose that the forecaster considers formulating an entire probability distribution for unknown values $\mathbf{X}_{t,T} \equiv (\mathbf{X}_t, \mathbf{X}_{t+1}, \ldots, \mathbf{X}_T)$ for the period of measurements covering the time interval $[t, T]$ via some conditional probability mass function (or approximating density function), $f: \mathscr{R}(\mathbf{X}_{t,T}) \rightarrow [0, 1]$, where

$$f(\mathbf{X}_{t,T} = \mathbf{x}_{t,T} | \mathbf{I}_{t-1}) \equiv P[(\mathbf{X}_t = \mathbf{x}_t)(\mathbf{X}_{t+1} = \mathbf{x}_{t+1}) \cdots (\mathbf{X}_T = \mathbf{x}_T) | \mathbf{I}_{t-1}],$$

specified for each $\mathbf{x}_{t,T} = (\mathbf{x}_t, \mathbf{x}_{t+1}, \ldots, \mathbf{x}_T) \in \mathscr{R}(\mathbf{X}_{t,T})$. The rules of coherent conditioning specify two useful ways in which this conditional probability mass function over so many quantities can be factored and analyzed.

First, the joint mass function for vectors over time can be factored into the product of conditional mass functions as

$$\begin{aligned} f(\mathbf{X}_{t,T} = \mathbf{x}_{t,T} | \mathbf{I}_{t-1}) = f(\mathbf{x}_t | \mathbf{I}_{t-1}) & \\ \cdot f(\mathbf{x}_{t+1} | (\mathbf{X}_t = \mathbf{x}_t) \mathbf{I}_{t-1}) & \\ \cdot f(\mathbf{x}_{t+2} | (\mathbf{X}_{t+1} = \mathbf{x}_{t+1})(\mathbf{X}_t = \mathbf{x}_t) \mathbf{I}_{t-1}) & \\ \cdots & \\ \cdot f(\mathbf{x}_T | (\mathbf{X}_{T-1} = \mathbf{x}_{T-1}) \cdots (\mathbf{X}_{t+1} = \mathbf{x}_{t+1})(\mathbf{X}_t = \mathbf{x}_t) \mathbf{I}_{t-1}). & \end{aligned}$$

Using this factorization, a forecast distribution for quantities $T$ periods ahead can be broken down into a product of $T$ distinct one-period-ahead conditional forecast assessments.

Second, the forecast mass function for any vector $\mathbf{X}_t$ given $\mathbf{I}_{t-1}$ can itself be factored into the products of conditional mass functions as

$$\begin{aligned} f(\mathbf{X}_t = \mathbf{x}_t | \mathbf{I}_{t-1}) = {} & f(x_{1t} | \mathbf{I}_{t-1}) \\ & \cdot f(x_{2t} | (X_{1t} = x_{1t}) \mathbf{I}_{t-1}) \\ & \cdot f(x_{3t} | (X_{2t} = x_{2t})(X_{1t} = x_{1t}) \mathbf{I}_{t-1}) \\ & \ldots \\ & \cdot f(x_{Kt} | (X_{(K-1)t} = x_{(K-1)t}) \cdots (X_{2t} = x_{2t})(X_{1t} = x_{1t}) \mathbf{I}_{t-1}). \end{aligned}$$

Forecasting $K$ distinct quantities one period ahead can be broken down into the product of conditional one-period-ahead forecasts, each conditioned on knowledge of one additional quantity among those to be forecast. Notice that both these factorizations mimic, structurally, the factorization rule for conditional probability that we noted as Theorem 3.3 in Section 3.4.

Knowledge of these ways that the joint uncertainty distribution can be coherently factored is very helpful in the specification and computation of economic forecasts. Yet this factorization structure, which honors the restrictions of coherency, is not sufficient to produce any particular forecast. Every coherent distribution function can be factored in these ways.

In identifying a specific forecast distribution to represent one's knowledge, there are two broad routes which can be followed. One is based on the representation of a discrete mass function over the large grid of points in the realm of $\mathbf{X}_t, \mathbf{X}_{t+1}, \ldots$. Numerical computation methods could be used to generate sequential forecast distributions as each successive observation $(\mathbf{X}_t = \mathbf{x}_t)$, $(\mathbf{X}_{t+1} = \mathbf{x}_{t+1})$, and so on, is made. Such methods demand a formidable amount of computing space to be operational for realistic problems. The development of supercomputers has only recently made these procedures even worthy to consider.

An alternative tack is to use convenient functional forms to approximate one's forecasting distribution. The specialized forms allow for an algebraic solution to many of the computational problems incurred in producing sequential forecasts. Of course, attention to the magnitude of approximation error is important in the interpretation and use of the computed results. The recent development of Monte Carlo and Gibbs sampling integration techniques have reduced greatly the crudeness of approximation to acceptable formulations. The papers of Casella and George (1992), Gelfand and Smith (1992), the text of Tanner (1993), and the compilation by Gilks et al. (1996) provide readable introductions to such technique.

Nonetheless, parametric families of distribution functions and their mixtures have widespread use in econometric forecasting as well as in every other scientific field. In the remainder of this chapter and in Chapter 7, we develop the mathematical structure of the operational subjective theory for using these forms. The topics in the present chapter use structures of joint and conditional distributions mainly in characterizing coherent inferences for vectors of quantities whose components are regarded exchangeably. Detailed characterizations involving vectors that are regarded exchangeably but whose components are not regarded exchangeably are studied in Chapter 7 in the context of the multivariate normal distribution and its mixtures.

## 5.4 THE PRACTICAL IMPORTANCE OF A PROBABILITY DISTRIBUTION: FAMILIES OF DISTRIBUTIONS FOR EVENTS REGARDED EXCHANGEABLY

The practical importance of distribution functions derives from the facts that they represent a complete specification of uncertain knowledge about quantities and that they are easy to conceptualize, if not to assess in every detail. Your assessing a limited number of values of the function and then evaluating the regularity of your opinions over possible values of $X$ in the domain of $F(\cdot)$ covering $\mathscr{R}(X)$ may be sufficient for you to determine that some specific function is a reasonable approximation to your probability distribution function for $X$. Thus, limited introspection about your knowledge of $X$, along with your general knowledge of the properties of some useful functional forms of distribution functions, may allow you to characterize approximately the full extent of your uncertain knowledge of $X$. No mean achievement! The purpose of the next few sections is to introduce some particular functional forms that have been useful in representing considered scientific opinion about quantities in many applied fields.

We have already studied some distribution functions rather extensively in Chapter 3, although our discussion there was not couched in this terminology. Let us begin by reviewing briefly what we know but use the language of distributions to express it.

### 5.4.1 The Hypergeometric Distribution and Its Mixtures

Suppose you regard the events $E_1, E_2, \ldots, E_N$ exchangeably. Throughout this section we use the symbol $S_N \equiv \sum_{i=1}^{N} E_i$ to denote their sum. Similarly, $S_n$ will denote the sum of the first $n$ of them, $S_n \equiv \sum_{i=1}^{n} E_i$. Now we know from Section 3.9, specifically from Corollary 3.11.1, that for any positive integer $n \leqslant N$, your conditional probability $P(S_n = x | S_N = A)$ must equal

$$P(S_n = x | S_N = A) = [{}^{A}C_x\, {}^{(N-A)}C_{(n-x)} / {}^{N}C_n] \qquad (x \in \mathscr{R}(S_n; A, N)), \qquad (*)$$

where

$$\mathscr{R}(S_n; A, N) \equiv \{\text{integers } x | 0 \leqslant x \leqslant n \leqslant N,\ x \leqslant A \leqslant N, \quad \text{and } n - x \leqslant N - A\}$$

denotes the realm of $S_n$ under this condition. We need to refer to this awkward set at several places in the following.

According to the general definition of a conditional probability mass function, the function $f_{S_n}(x|(S_N = A)) \equiv P(S_n = x|S_N = A)$ constitutes your conditional probability mass function for $S_n$ conditional upon the event $(S_N = A)$. To remind you of the motivation for equation (*), notice that the denominator ${}^N C_n$ is the number of ways in which $n$ events can be selected from $N$, and the numerator ${}^A C_x{}^{(N-A)}C_{(n-x)}$ is the number of those possible selections that would involve $x$ events that equal 1 and $n - x$ events that equal 0. The function $F_{S_n}(x|S_N = A) \equiv \sum_{a=0}^{[\![x]\!]} P(S_n = a|S_N = A)$ is the associated conditional probability distribution function for $S_n$ given $(S_N = A)$. (The symbol $[\![x]\!]$ denotes the greatest integer that does not exceed $x$.) In this circumstance we say that your asserted conditional probability distribution for $S_n$ given $(S_N = A)$ is hypergeometric, according to the next definition, and we write $S_n|(S_N = A) \sim H(n, A, N)$ to denote that you assess your knowledge in this form.

***Definition 5.8.*** To write $X \sim H(n, A, N)$ for nonnegative integers satisfying $n \leqslant N$ and $A \leqslant N$ denotes that someone asserts probabilities supporting the *hypergeometric probability mass function*, specified by

$$f(x) = [{}^A C_x{}^{(N-A)}C_{(n-x)}/{}^N C_n] \qquad \text{for } x \in \{0, 1, \ldots, n\} \text{ as long as}$$

$$x \leqslant A \quad \text{and } n - x \leqslant N - A,$$

$$\equiv 0 \qquad \text{for other } x.$$

The function $F(x) = \sum_{a=0}^{[\![x]\!]} f(a)$ is a *hypergeometric probability distribution function.* □

Now suppose you regard the events $E_1, E_2, \ldots, E_N$ exchangeably, but, as would be usual, you do not know precisely the value of their sum, $S_N$. Then for each positive integer $n \leqslant N$, the function

$$\begin{aligned} f_{S_n}(x) &\equiv P(S_n = x) \qquad (x \in \{0, 1, \ldots, n\}) \\ &= \sum_{A=0}^{N} P(S_n = x|S_N = A)P(S_N = A) \qquad (x \in \{0, 1, \ldots, n\}) \\ &= \sum_{A=0}^{N} ({}^A C_x{}^{(N-A)}C_{(n-x)}/{}^N C_n)P(S_N = A) \end{aligned}$$

constitutes your (unconditional) probability mass function for $S_n$. This is merely a restatement of Theorem 3.11.

Formally, this p.m.f. for $S_n$ can be written as a Stieltjes integral

$$f_{S_n}(x) = \int_{-\infty}^{\infty} ({}^A C_x{}^{(N-A)}C_{(n-x)}/{}^N C_n)\, dM_N(A),$$

where $M_N(A) \equiv P(S_N \leqslant A)$ denotes your probability distribution function for $S_N$. (When a mixing distribution function such as $M_N(\cdot)$ is discrete, the Stieltjes integral reduces to the sum that appears in the third equation for $f_{S_n}(x)$. Thus, the function $f_{S_n}(x)$ is itself a convex combination of hypergeometric mass functions. On this account we call the distribution for $S_n$ a *mixture hypergeometric distribution* function according to the following definition.

***Definition 5.9.*** Suppose that for each real number $\theta \in \Theta$, $F_X(x;\theta)$ is a distribution function for $X$ that is identified by a parameter $\theta$. The family of distributions $\mathscr{F}(X;\theta) = \{F_X(x;\theta) \mid \theta \in \Theta\}$ is called a *parametric family of distributions*. The set $\Theta$ of identifiers of the family members is called its *parameter space*. Furthermore, let $M(\theta)$ be any distribution function. Then a distribution function for $X$ that can be written as $F_X(x) = \int_\Theta F_X(x;\theta)\, dM(\theta)$ is said to be a *mixture distribution*, and the distribution function $M(\theta)$ is said to be its *mixing distribution function*. We write $X \sim M\text{-}F(\theta)$ to denote that your knowledge of $X$ is representable in this form. If the members of the parametric family of distributions are discrete (or absolutely continuous) distributions, then their associated mass (or density) functions would be denoted by $f_X(x;\theta)$. The *mixture mass* (*mixture density*) *function* associated with the mixture distribution $F_X(x)$ is denoted by $f_X(x) \equiv \int_\Theta f_X(x;\theta)\, dM(\theta)$.

□

This definition would remain formally the same if the variable $X$ were replaced by a vector of variables and/or if the parameter $\theta$ were replaced by a vector of parameters.

We study numerous examples of mixture distributions throughout this chapter and throughout the remainder of this book. To get your bearings on the terminology, consider the following review of the contents of this section, described as results concerning mixture hypergeometric distributions.

***Example 1.*** Any coherent probability distribution for the sum of $n$ events chosen from among $N$ events that are regarded exchangeably provides an example of a mixture distribution. Examine the p.m.f. for $S_n$:

$$f_{S_n}(x) = \sum_{A=0}^{N} ({}^{A}C_x\, {}^{(N-A)}C_{(n-x)} / {}^{N}C_n) P(S_N = A) \qquad (x \in \{0, 1, \ldots, n\}).$$

For any choice of $N$ and $A \in \{0, 1, \ldots, N\}$, the parenthetical expression ${}^{A}C_x\, {}^{(N-A)}C_{(n-x)} / {}^{N}C_n$ would itself constitute a hypergeometric probability mass function for $S_n$. In the notation of Definition 5.9, we denote it by $f_{S_n}(x; n, A, N)$. The integers $n$, $A$, and $N$ constitute a parameter vector specifying a family of parametric distributions, the hypergeometric family. The parameter space is the set of positive integer vectors $\{(n, A, N) \mid n \leqslant N, \text{ and } A \leqslant N\}$. The values of $n$ and $N$ are typically specified by an inference problem. Your probability distribution for $S_N$ constitutes the mixing distribution function over $A$:

$$M(A) = F_{S_N}(A) = \sum_{x=0}^{[\![A]\!]} P(S_N = x).$$

Making these substitutions, it is apparent that the p.m.f. $f_{S_n}(x)$ has the form of a mixture hypergeometric mass function, as specified in the definition. Symbolically we can write $S_n \sim M\text{-}H(n, A, N)$. □

## PROBLEMS

1. Show that asserting $X \sim H(n, A, N)$ implies via coherency that your $P(X) = n(A/N)$, that your $P(X^2) = n(A/N)(n-1)[(A-1)/(N+1]$, and that your variance for $X$ is $V(X) = n(A/N)[(N-A)/N][(N-n)/(N-1)]$. Thus, these three expressions constitute, respectively, the first moment, the second moment, and the second central moment of the hypergeometric distribution $H(n, A, N)$.

2. Show that the hypergeometric probability mass function sums to 1 over all values of $x$ from 0 to $n$. [*Hint*: Use the binomial expansion theorem to expand $(z+1)^A$, $(z+1)^B$, and $(z+1)^{A+B}$. Then compare the coefficient on $z^N$ in the final expansion with the sums of the coefficients on $z^N$ in the first two expansions.]

3. Show that if you regard the events $E_1, E_2, \ldots, E_N$ exchangeably and your probability distribution for $S_N$ is $S_N \sim H(N, \chi, \eta)$, then your probability distribution for $S_n$ is $S_n \sim H(n, \chi, \eta)$. In other words, *hypergeometric mixtures of hypergeometric distributions are hypergeometric.* (This problem and the next have already been discussed in detail in Example 3 of Section 3.11.2.1.)

4. Show further that under these same conditions your conditional probability distribution for $(S_N - S_n | S_n = x)$ is also hypergeometric; that is, $(S_N - S_n | S_n = x) \sim H(N-n, \chi - x, \eta - n)$. Because the conditional distribution for $(S_N - S_n | S_n)$ is in the same form as the marginal distribution of $S_N$ (both distribution forms are hypergeometric), we say that this form of mixing distribution is *natural conjugate* for a finite sequence of events that you regard exchangeably. We shall make a formal definition of a natural conjugate distribution after we see another example in the next section.

### 5.4.2 Infinite Exchangeable Extendability: Binomial, Beta, and Polya Distributions

We learned in Section 3.10 that if you regard the events $E_1, E_2, \ldots, E_N$ exchangeably, and your opinions are infinitely exchangeably extendable, then your probability mass function for the sum of any $n$ of them can be represented as

$$f_{S_n}(x) = {}^nC_x \int_0^1 \theta^x(1-\theta)^{n-x}\, dM(\theta).$$

where $M(\cdot)$ is some distribution function. It may either be discrete, absolutely continuous, or merely continuous. We say that your probability distribution for $S_n$ is mixture-binomial according to Definition 5.10, and we write $S_n \sim M\text{-}B(n, \theta)$. Notice that $M(\cdot)$ need not be a *probability* distribution function according to Definition 5.1, since there is no requirement that $M(\theta)$ exhibit a finite number of jump points.

***Definition 5.10.*** To write $X \sim B(n, \theta)$ for $n$ an integer and $\theta \in [0, 1]$ denotes that someone asserts probabilities supporting the *binomial probability mass function*, specified by

$$\begin{aligned} f(x) &= {}^nC_x\, \theta^x(1-\theta)^{n-x} \qquad && \text{for } x \in \{0, 1, \ldots, n\}, \\ &= 0 && \text{for other } x. \end{aligned}$$

The function $F(x) = \sum_{a=0}^{[\![x]\!]} f(a)$ is a *binomial probability distribution function.*

□

## PROBLEMS

1. Show that the binomial probability mass function of Definition 5.10 sums to 1 over integer values of $x$ from 0 to $n$. [*Hint*: Use the binomial expansion theorem that for any numbers $a$ and $b$ and for any integer $n$,

$$(A + B)^n = A^n + {}^nC_1 A^{n-1}B + {}^nC_2 A^{n-2}B^2 + \cdots + {}^nC_{n-1}AB^{n-1} + B^n.$$

Then replace $A$ by $\theta$ and $B$ by $1 - \theta$.]

2. Show that if you assert $X \sim B(n, \theta)$ then coherency requires your $P(X) = n\theta, V(X) = n\theta(1-\theta)$, and $P(e^{tX}) = [1 + \theta(e^t - 1)]^n$. Thus, these three expressions constitute the first moment, the second central moment, and the moment generating function for a binomial distribution, respectively. For practice with moment generating functions, evaluate $D_t(M_X(t))|_{t=0}$ and show that it equals $n\theta$. How about $D_t^2(M_x(t))|_{t=0}$?

3. Show the following relation between the limiting members of the hypergeometric family of distributions and the family of binomial distributions: If

$X \sim H(n, A, N)$, then

$$\lim_{\substack{N \to \infty \\ A \to \infty \\ \ni (A/N) = \theta}} f_X(x) = {}^nC_x \theta^x (1 - \theta)^{(n-x)}.$$

(*Hint*: Refer to Section 3.10.3.)

It is trite but true to say that once you assert your probability for any event as $P(E)$, then your probability distribution for that event is binomial, $B(n, \theta)$ with $n = 1$ and $\theta = P(E)$. With historical deference to James Bernoulli (*Ars Conjectandi*, 1713) the $B(1, \theta)$ distribution is commonly called a *Bernoulli distribution.*

Just as we found a "natural conjugate" form of mixing function for a mixture-hypergeometric distribution in problem 4 of Section 5.4.1, there is a simple form for the natural conjugate mixing function for a mixture-binomial distribution. This mixing distribution is called a beta distribution. Let us first study the mathematical properties of this family of distributions and its role in a mixture-binomial distribution. We formalize the concept of a natural conjugate distribution at the end of this section.

***Definition 5.11.*** The function $f: [0, 1] \to \mathbb{R}^+$ specified by

$$f(\theta) = [\Gamma(\alpha + \beta)/\Gamma(\alpha)\Gamma(\beta)]\theta^{\alpha - 1}(1 - \theta)^{\beta - 1} \qquad (\theta \in (0, 1))$$

for some specific values of $\alpha > 0$ and $\beta > 0$ is said to be the *beta density function* identified by the parameter values $\alpha$ and $\beta$. The associated cumulative function, $F(t) = \int_0^t f(\theta)\, d\theta$, is called a *beta distribution function.* We write $\theta \sim \text{beta}(\alpha, \beta)$ to denote that this distribution applies to $\theta$. □

The expressions $\Gamma(\alpha + \beta)$, $\Gamma(\alpha)$, and $\Gamma(\beta)$ appearing in the proportionality constant for the beta density function are evaluations of the gamma function, defined by $\Gamma(z) \equiv \int_0^\infty x^{z-1} \exp(-x)\, dx$. Notice that $\Gamma(1) = 1$. Integration by parts yields the characteristic recursive property of gamma functions, that $\Gamma(z) = (z - 1)\Gamma(z - 1)$. Thus, for integer values of $z$, $\Gamma(z) = (z - 1)!$. Approximate values of the gamma function over the interval $(0, 1]$ are computed by numerical methods. They appear in *Tables of Mathematical Functions* (1964), and they are now available in many computing packages. One notable value of the function is $\Gamma(1/2) = \sqrt{\pi}$, where $\pi$ is the irrational number pi. This is the reason that $\sqrt{\pi}$ appears so frequently in several related distribution functions we study. Graphically, the gamma function declines over $z \in (0, 1]$ from a vertical asymptote at 0 to the value of $\Gamma(1) = 1$. The function then continues to decline, achieving its minimum at $\Gamma(1.5)$, before increasing without bound thereafter.

Finally, the integral result that identifies the value of the proportionality constant of a beta density

$$\int_0^1 \theta^{\alpha-1}(1-\theta)^{\beta-1}\,d\theta = \Gamma(\alpha)\Gamma(\beta)/\Gamma(\alpha+\beta)$$

can be derived using the transformation $t = \theta/(1-\theta)$. This integral is studied in most texts on advanced calculus, for example Fulks (1979). We defer further study of the gamma function per se until we study the related *gamma distribution* in Chapter 7. All you need to realize for now is that the integral $\int_0^1 \theta^{\alpha-1}(1-\theta)^{\beta-1}\,d\theta = \Gamma(\alpha)\Gamma(\beta)/\Gamma(\alpha+\beta)$, which ensures that the beta density function integrates to 1 over its entire domain. You will use this fact in determining the moments of the beta distribution.

## PROBLEMS

**4.** Show that for $\theta \sim \text{beta}(\alpha, \beta)$, the first two moments of the distribution are $\mathbb{M}_1(F_\theta) = \alpha/(\alpha+\beta)$ and $\mathbb{M}_2(F_\theta) = \alpha(\alpha+1)/(\alpha+\beta)(\alpha+\beta+1)$. Thus, $\mathbb{M}_2^*(F_\theta) = \alpha\beta/(\alpha+\beta)^2(\alpha+\beta+1)$.

**5.** Differentiate the beta density function twice and study the shape of the function for various parameter configurations of $(\alpha, \beta)$. *Hint*: Notice that

$$D_\theta f(\theta) = f(\theta)\left[\frac{\alpha-1}{\theta} - \frac{\beta-1}{1-\theta}\right] = 0 \qquad \text{when } \theta = \frac{\alpha-1}{\alpha+\beta-2} \equiv \theta^*(\alpha,\beta)$$

if $(\alpha > 1)(\beta > 1)$ or $(\alpha < 1)(\beta < 1)$. Check the relevant second derivative evaluations for yourself, deriving the second derivative function as

$$D_\theta^2 f(\theta) = f(\theta)\left\{\left[\frac{\alpha-1}{\theta} - \frac{\beta-1}{1-\theta}\right]^2 - \frac{\alpha-1}{\theta^2} - \frac{\beta-1}{(1-\theta)^2}\right\}.$$

The graphical partition of $(\mathbb{R}^+, \mathbb{R}^+)$ in Figure 5.1 depicts four regions of the parameter space in which the density function has similar shapes. The shape of a typical function for parameter values in each of these regions is exemplified in Figure 5.2. You should also recognize algebraically, that for $\alpha = 1$ and $\beta = 1$ the beta density function is constant over the interval $[0, 1]$. This is the density defining a *uniform distribution on* $[0, 1]$.

**6.** Numerical values of beta distribution functions for various parameter specifications are available in many statistical software packages, including *Minitab* and *Matlab*. Using whatever computing package you have available, examine some beta density functions associated with $(\alpha, \beta)$ parameter configurations in

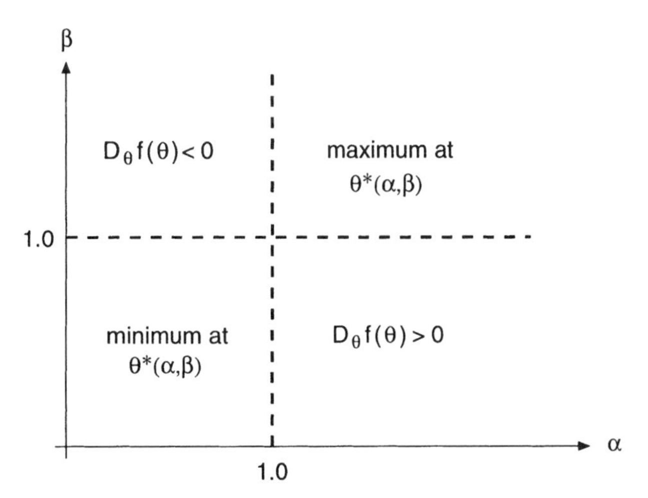


**Figure 5.1.** Four regions of $(\mathbb{R}^+, \mathbb{R}^+)$, the parameter space for $(\alpha, \beta)$, in which the beta $(\alpha, \beta)$ density exhibits similar shapes. For appropriate configurations, the maximum or minimum density value occurs at $\theta = \theta^*(\alpha, \beta) \equiv (\alpha - 1)/(\alpha + \beta - 2)$.

each of the four regions identified in problem 5. For some historical perspective, it is worth examining the classic *Tables of the Incomplete Beta Function* developed under the direction of Karl Pearson (1979).

#### *5.4.2.1 An Example Assessing Milk Yields from Commercial Dairy Cows*

We have developed enough mathematical apparatus to appreciate the details of an extended expository example. How much milk does a commercial dairy cow in the United States yield during a year? Before you begin reading this example, you might want to think a bit about the question yourself. To fix your ideas specifically, and to get your bearings on what you know about dairy cows, what is your prevision for the average annual milk yield of commercial dairy cows in the United States? The yield is measured by weight.

Now to stimulate your imagination for this example, let me tell you a few items of general information which you may or may not already know. As of the mid-1980s the commercial dairy herd in the United States of America numbered some 10 or 11 million cows. Of these, more than 90 percent were Holstein cows, either registered (pure bred) or grade. About 1 million cows lived in each of the states of California, Minnesota, and New York, and about 1.8 million lived in Wisconsin. On geophysical grounds alone, California is quite different from these other three leading dairy states, which are more homogeneous with respect to latitude, geography and weather patterns. Economically, the structure of dairy herds is quite different in California than the other three states, too. About 15 percent of California dairy cows live in herds of more than 950 cows, and about 80 percent of the cows live in herds exceeding 250 cows. Herd sizes in the other three states are typically much smaller.

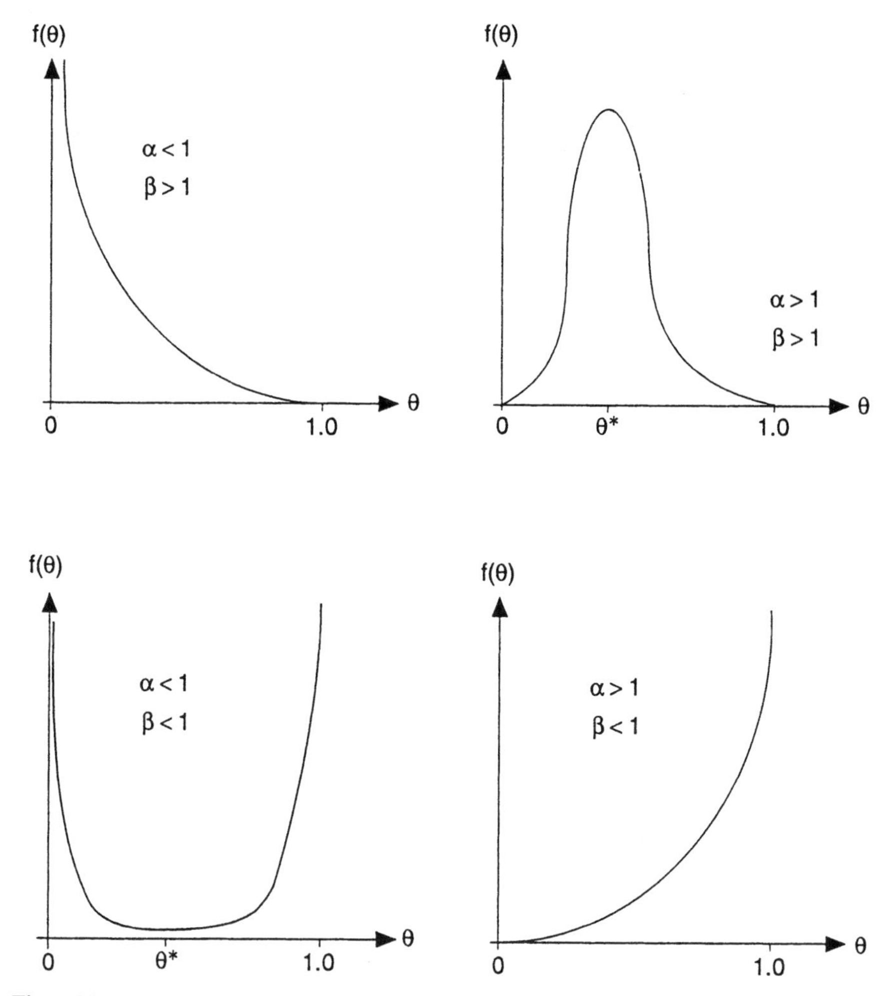


**Figure 5.2.** Four typical beta density functions for parameter configurations in each of the regions in Figure 5.1.

Such differences aside, the development of animal husbandry through artificial insemination has produced a national commercial dairy herd of surprising uniformity. The book of H. A. Herman (1981), *Improving Cattle by the Millions: NAAB and the Development and Worldwide Application of Artificial Insemination*, contains an informative historical account. The photograph on page 53 of a row of two-year-old Guernsey cows provides a good example of a group of cows whose annual milk yields almost anyone might well regard exchangeably. The gestation period of a dairy cow is roughly nine months. Commercial dairy cows are typically milked on 305 days per year, twice per day. This milking schedule allows each cow only 60 "dry days" during the final two months of pregnancy, which is induced about three months after calving. In fact, a 305-day milk yield by weight is the standard unit of milk yield measurement for a dairy cow. (Historically, measurement by weight provided a compensation for comparisons of yields across breeds. The smaller breeds of cow yield a higher percentage of butterfat and nonfat protein solids in their milk than do the larger breeds.) The timing of this example is the year 1984, before the commercial production of the controversial milk-stimulating hormone bovine somatatropine (BST), which has been approved by the USDA in 1994 for use in commercial dairying for the stimulation of so-called supercows.

Let us turn to a specific question. Consider a California dairy herd of some 1200 Holstein cows, which includes 76 registered heifers who first "freshened" (delivered their first calf and began milking) in February 1984. How many kilograms of milk will (did) these cows yield during their first 305-day milking period? I'm asking you. What do you know about such things? To pull a number out of the air, might any of them yield as much as 8000 kilos of milk during the year? none of them? all of them? How many? What is your *probability* that a specific one of these cows yields more than 8000 kg? What is your *probability distribution* for the vector of events, $\mathbf{E}_{76}$, which identifies the answer to this question for each of these 76 cows? Finally, to raise a further question which we answer only by the end of this section, what are your conditional probabilities for, say, the tenth event $E_{10}$ conditional upon the event that the sum of the first nine events, $S_9$, equals 0? or 1, or 2,..., or 9?

To provide notation for this conversation, suppose we denote the annual milk yields from these 76 cows (denominated in kilograms) by $X_1, X_2, \ldots, X_{76}$; and let's denote by $E_1, E_2, \ldots, E_{76}$ the corresponding events that their yields exceed 8000 kg. That is, $E_i \equiv (X_i > 8{,}000)$ for each $i = 1, 2, \ldots, 76$.

In this example, I develop an answer for "you," characterized in two different ways. I develop "your" answer, first supposing that you proclaim yourself as fairly ignorant about milk yields from cows. Second, I suppose that you are a fairly well informed agronomist from Minnesota whose primary area of expertise is not in dairying but in the area of grain production, hay, and animal feeds. One thing that I presume about each of "you" is that you regard the yields from these 76 cows exchangeably. [For any vector of positive numbers $\mathbf{x}_{76}$, your $P(\mathbf{X}_{76} = \mathbf{x}_{76})$ is identical over all the events identified by permutations of the components of $\mathbf{x}_{76}$. We dwell at greater length on the judgment to regard a sequence of general quantities (as opposed to mere events) exchangeably in Section 5.5.] Moreover, I presume you regard these events as exchangeably extendable to at least $N = 15{,}000$. This is approximately the number of commercial purebred Holstein heifers in California who first freshened in February 1984, all of whose milk yields you regard exchangeably. Thus, each of your probability mass functions for the sum of these 76 events, $S_{76} = \sum_{i=1}^{76} E_i$, can be expressed precisely as

$$f_{S_{76}}(x) = \sum_{A=0}^{15{,}000} [{}^{A}C_{x}\,{}^{(15{,}000-A)}C_{(76-x)}/{}^{15{,}000}C_{76}]\, f_{S_{15{,}000}}(A) \qquad (x \in \{0, 1, \ldots, 76\}),$$

where $f_{S_{15,000}}(A)$ is your probability mass function for the sum of the 15,000 events of the form $(X_i > 8000)$ that you regard exchangeably.

I presume further that you regard the 76 events as *infinitely* exchangeably extendable. That is, your p.m.f. for $S_{76}$ is representable as

$$f_{S_{76}}(x) = {}^{76}C_x \int_0^1 \theta^x(1-\theta)^{76-x}\, dM(\theta) \qquad (x \in \{0, 1, \ldots, 76\}),$$

where $M(\theta)$ represents your probability distribution function for the proportion of such events that would equal one out of a numberless collection of events that you regard exchangeably. It is this representation, with $M(\theta)$ selected from the family of beta distributions, that this example is designed to illustrate.

To begin, let us suppose that you aver that you know very little about cows and their milk yields. In fact, "Why are you asking me?" you challenge. "I have seen cows on television and perhaps occasionally from a passing car. But to be honest, I wouldn't know a Holstein from a Jersey from a Guernsey if I saw them. I haven't been really close to a cow since I was taken to a County Fair at the age of 8. I have no idea how much a dairy cow weighs, nor how large her udder might be!" We should answer this challenge before we begin to answer the questions about milk yields. On the one hand, your challenge is very appropriate and thoughtworthy. For there are many people who do know a lot about cows. If I really wanted to know the answers to these questions, surely it would be much better to ask one of these "experts" than to ask you. On the other hand, the operational subjective theory of probability purports to be able to express the logical requirements of anyone's uncertain knowledge about measurable quantities, even someone such as you, who knows "next to nothing."

As a matter of fact, however, in the course of the development of any scientific field, many very important and interesting new questions emerge each year, about which even "the experts" feel just as you do about milk yields from cows. Even the experts avow they know "next to nothing". It will be very worthwhile to study in detail the form of probability distribution asserted by someone who is in this particular state of knowledge. For it would be useful to identify intuitions regarding coherent patterns of learning (at least in the small) from experimentation (measurement) on questions about which we know very little to begin with. Finally, this example is meant to exhibit the positive role of the statistician in helping a client elicit his or her uncertain knowledge about unknown measurable quantities.

Let us progress now to answer the question, "Might any of these cows yield as much as 8000 kg of milk during the year?" Even knowing next to nothing about cows, it is easy to realize that the answer to this question hinges on how many liters of milk are contained in a cow's udder when it is full for milking. If you knew the "ballpark" weight of milk in one full udder, the milk yield per 305 days would be roughly 610 times that weight, since cows are milked twice per day. I presume you know that 1 liter of water weighs 1 kilogram. You might guess that the weight of milk is at least in the same ballpark as that of water, containing as it does components of water, butterfat, and nonfat solids. So how many liters of milk would a cow's full udder contain? "Well, surely it contains at least one liter," you begin. "Even 3 or 4 liters, as a low guess. That does sound like a lot of milk to produce and carry around with you twice per day. But an udder could contain as much as 10, 20, or even 50 liters for all I know. A cow is a pretty large animal after all; and realizing that they must be bred specifically for large udders, who knows? Come to think of it, a calf is fairly large too. The calf could probably drink a lot of milk if it were allowed to. How old are calves before they are weaned, by the way? As I've told you, there are many things to consider, and I don't know much about any of them."

"Fair enough! Relax!" I (the statistican) say. "No one is berating you for what you don't know. So your guess of the volume of an udder full of milk is somewhere between 4 and 50 liters? At roughly 1 kg per liter that would translate into roughly anywhere from 2440 kg to 30,500 kg milk yield per year as your ballpark guess for a single cow. Would you be willing then to assert the interval prevision $P(X_i) \in [2440, 30{,}500]$?"

"Yes! That's not saying too much," you allow. "Now that I'm thinking about it in fact, I'll even go further. For an average single udderful, I think I'll settle on 10 liters as my modal assertion, with my probability mass function relatively gentle between 5 and 20, and again between 25 and 50. Translated to 610 milkings per year, this means my p.m.f. for a single cow's milk yield would look something like Figure 5.3, at least to a crude degree of approximation. For having played around with similar sketches a bit, I think now that I'll say my $P(X)$ is somewhere around 12,000 and my median is 11,000. I'll also assert my $P(X > 15{,}000) = .3$, and my $P(X > 25{,}000) = .1$.

"As a result of all this, I now see that the number 8000 kg which was 'pulled out of the air' is a rather interesting one. Apparently my probability that the milk

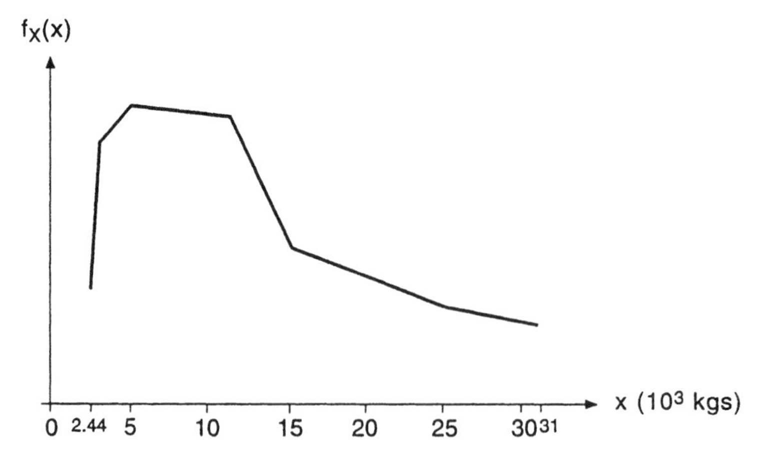


**Figure 5.3.** A crude sketch of "your" p.m.f. for the 305-day milk yield of a newly freshened California Holstein heifer in February 1984.

yield from any one of these cows under consideration exceeds 8000 kg is approximately .7."

We have discussed to this point your opinion about the milk yield from a single cow. It is evident that you are rather hazy in your knowledge about milk yields. Let us now expand our considerations to your knowledge about several individual cows, in particular the specified group of 76. Of course, since you regard these cows' milk yields exchangeably, coherency requires that your prevision for the average 305-day yield from the 76 cows equal your prevision for the yield from any one of them. $P(X_i) = P(\bar{X}_{76})$. Let us now pursue your opinions about differences among the milk yields of cows in this specific group. To focus on a quantity that is easy to think about, suppose we think about the average of the absolute differences of each of these cows' milk yields from their average yield, a quantity represented algebraically as $\sum_{i=1}^{76} |X_i - \bar{X}_{76}|/76$. How large do you think this number might be?

Even knowing rather little about milk yields, you do not expect the differences among individual yields to be too great. For these cows are all members of a large commercial herd, suggesting that they are all treated in a similar fashion. They are all approximately two-year-old Holsteins ("Whatever that means," you think to yourself) who have just given birth to a calf. By and large they are all healthy, since presumably any members of their own birth cohort who have shown problems of any sort have already been culled and sold for veal. Whatever the average yield these heifers produce, you would be quite surprised if the average absolute difference among their yields exceeds 10% say. Minimally, you might convey your attitude by the upper conditional prevision assertions

$$P\left( \sum_{i=1}^{76} |X_i - \bar{X}_{76}|/76 \,\middle|\, (\bar{X}_{76} = x) \right) \leqslant x/10,$$

whatever be the value of $x$ between 2.44 thousand and 31 thousand. You are expecting roughly that the 76 milk yields will be either bunched within 250 units around 2500; or within 3000 units around 30,000; or similarly bunched somewhere in between these locations. Of course, there may well be extreme observations. Any of these cows may die accidently early after freshening, yielding thus an observation $X_i$ near to zero. Some others, identified early as potential high-yielders, may be treated specially by their keepers, in the hopes of grooming "the county champion milker."

Enough! The amount of introspection you have aired to this point does support your disclaimer of being fairly ignorant about cows, though it does show you to be a fairly seasoned thinker. Congratulations! Let us return again now to the question, "Might any of these cows yield as much as 8000 kg of milk during the year?", a question you have recognized as an interesting one. A surprisingly strong implication of your admittedly limited knowledge of milk yields is that you are fairly certain that the events we've denoted by $E_1, E_2, \ldots, E_{76}$, where $E_i \equiv (X_i > 8{,}000)$, are either all 0 or all 1! Let's consider briefly the logic that yields this conclusion.

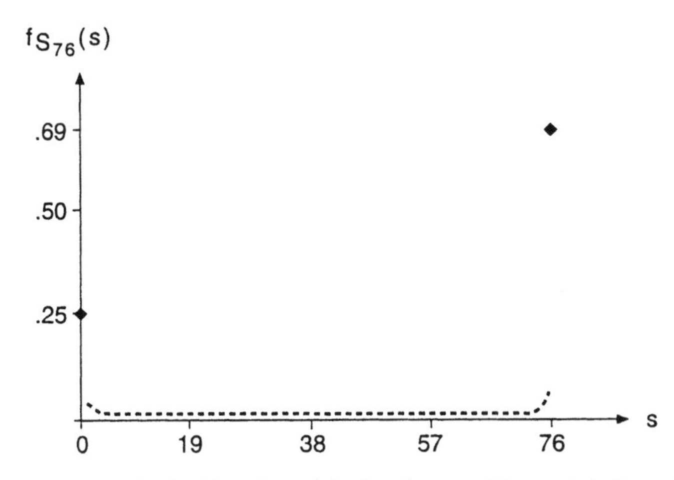


**Figure 5.4.** A crude sketch of "your" p.m.f. for $S_{76}$, the sum of the events in the vector $\mathbf{E}_{76}$.

At the lower extreme, you are expecting the milk yields to be bunched essentially within 250 units on either side of 2500 kg, while at the upper extreme you are expecting the yields to be bunched within 3000 units on either side of 30,000 kg. In these extreme cases you are expecting the specified events $\mathbf{E}_{76}$ to all equal 0 or all equal 1, respectively. Between these extreme imaginations, even the event that yields are bunched as low as within 700 units either side of 7000 kg or as high as within 900 units either side of 9000 kg would specify the general conclusion that the specified events $\mathbf{E}_{76}$ all equal 0 or all equal 1. It is only for possible average milk yields from the 76 cows ranging between 7500 and 8500 that you would expect to find a sizable proportion of these events equaling 0 and a sizable proportion equaling 1. Thinking a bit more about the crude p.m.f. you sketched in Figure 5.3, you recognize your $P(7500 < X < 8500)$ as being in the vicinity of .06; your $P(X < 7500) \approx .25$, and your $P(X > 8500) \approx .69$ . Roughly speaking, in the context of your expressed attitude regarding variation among the yields, this translates into assertions on the order of $P(S_{76} = 0) \approx .25$ and $P(S_{76} = 76) \approx .69$. Your probability mass function for $S_{76}$ would appear something like that displayed in Figure 5.4.

We need consider now only two more steps to identify a mixing function in the beta family of distributions that will adequately represent your uninformed opinions about the cows' milk yields, as they pertain to $\mathbf{E}_{76}$. First, realize that your judgment of exchangeability for components of $\mathbf{E}_{76}$ is surely extendable to $\mathbf{E}_{15,000}$, since there actually exist some 15,000 cows in California whose milk yields you regard exchangeably with the yields from these specific 76 cows. Thus, exchangeability would imply that your p.m.f. for $S_{15,000}$ is some appropriately refined extension of your p.m.f. for $S_{76}$ sketched in Figure 5.4. Now if this distribution were applied to the transformed quantity $S_{15,000}/15,000$, it would be supported only on the integer fractions 0, 1/15,000, 2/15,000, ..., 14,999/15,000, 1.

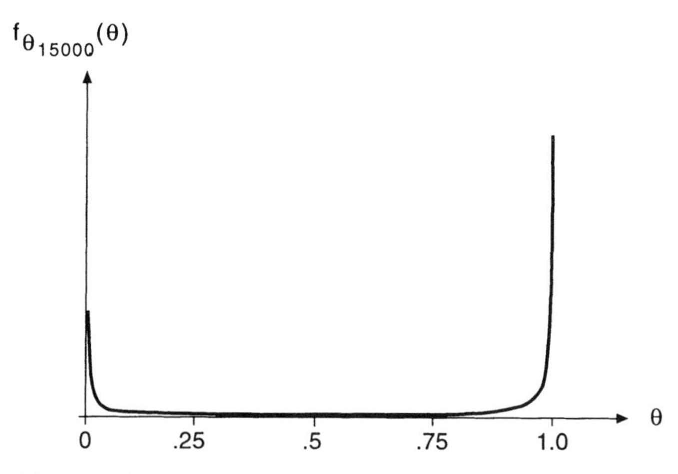


**Figure 5.5.** A crude sketch of "your" p.m.f. for $\theta_{15,000} \equiv S_{15,000}/15,000$.

The shape of such a p.m.f. is sketched in Figure 5.5, without reference to specific function values.

Finally, the infinitely extendable exchangeability that characterizes your judgment of $\mathbf{E}_{76}$ allows that your mixing distribution $M(\theta)$ may be specified by a continuous density function that merely "smooths" the p.m.f. we have sketched as $f_{\theta 15,000}(\cdot)$ in Figure 5.5. The conclusion you derived in problem 5 of Section 5.4.2 suggests that a beta($\alpha, \beta$) density function with parameters $\alpha$ and $\beta$ settled at values less than 1 would fit this bill. Studying functions of this form as you did in problem 6 of Section 5.4.2 yields a final conclusion that your opinions about $\mathbf{E}_{76}$ are well represented as

$$f_{S_{76}}(x) = {}^{76}C_x \int_0^1 \theta^x(1-\theta)^{76-x}\, dM(\theta) \qquad (x \in \{0, 1, \ldots, 76\})$$

where the distribution function $M(\theta)$ is the beta family member beta($\alpha = .0194$, $\beta = .0076$). Some values of this distribution function are

| $\theta$ | .01 | .1 | .2 | .3 | .4 | .6 | .7 | .8 | .9 | .99 | 1.0 |
|---|---|---|---|---|---|---|---|---|---|---|---|
| $F(\theta)$ | .2575 | .2698 | .2740 | .2769 | .2793 | .2836 | .2860 | .2889 | .2932 | .3060 | 1.0 |

Since the family of beta densities is characterized by only two parameters, an assertion of only two probabilities, say the lower and upper quartiles of your distribution for the proportion, would be sufficient to identify parameters $\alpha$ and $\beta$. In the give and take of assessing your approximate mixing distribution, you could check the implications of the resulting beta distribution against some of your other assertions.

Having laboriously worked through "your" introspection within your caricature as being relatively uninformed about dairying, we can be more brief in identifying "your" mixing function when "you" are portrayed as a fairly-well-informed Minnesota agronomist whose main field of expertise is in the area of grain production, hay, and animal feeds. The bottom line of the introspection needed to assess your mixing distribution $M(\theta)$ is to consider your opinion about the *proportion* of a large ("infinite") group of cows in the specified category whose milk yields would exceed 8000 kg for their 305-day milking period.

The Minnesotan muses: "I've seen a figure in the *USDA Bulletin* that says the national average milk yield for all Holstein cows is around 18,500 pounds. That would be roughly 8500 kg. Now I think that California dairies incur somewhat lower production costs than do dairies in north-central and northeastern states, since they do not experience such severe winters in which cows must be sheltered in controlled-temperature barns. Now I imagine that milk yields in California are somewhat lower to compensate for their lower production costs, though I'm not so sure about this. Come to think of it, 8000 kg is just about equal to my prevision for any of the milk yields $\mathbf{X}_{76}$ under consideration here. Moreover, my opinions are roughly symmetric about this number and heavily concentrated within 1000 kg on either side. As a first stab at my mixing distribution function for $\theta$, then, I'll settle on a beta density with parameters $\alpha = 10$ and $\beta = 10$. This density, symmetric about $\theta = .5$, allows a probability of .1861 that $\theta$ is less than .4, and the same probability that $\theta$ exceeds .6. □

Perhaps you, the enduring reader, can conclude this example for yourself by expanding the Minnesotan's considerations in the following problem.

## PROBLEM

1. Now suppose the Minnesotan ponders the first-year milk yield from 36 Holstein heifers in a herd that freshens in Minnesota in the spring of 1995. He is aware that BST has been approved for commercial use in 1994, but is also aware that in Minnesota several dairies are currently (July 1994) paying a premium for milk that is produced without the use of BST injections, and selling such milk in retail groceries in specially marked containers at a premium price. The research lore regarding milk yield gains from BST is that the daily increase in yield is between 3 and 7 kg, presuming the cow's food intake is increased appropriately (*Hoard's Dairyman* **139**(3), 1994, p. 89). If such were the case, the 305-day-yield from a cow would increase by some 1500 kg, easily pushing the yields above the 8000-kg mark. The Minnesotan's uncertainty regarding the events $\mathbf{E}_{36}$ is thus reduced to a quandary concerning whether the herd in question uses BST injections. In either case, the events $\mathbf{E}_{36}$ are regarded exchangeably. Assessing his probability of BST use by the herd as .25, the Minnesotan settles on a mixing distribution, $M(\theta)$, which is itself a mixture: $\theta \sim [.25 \times \text{beta}(\alpha = 99,\ \beta = 1) + .75 \times \text{beta}(10, 10)]$. Using com-

puter software that can access the family of beta distributions, determine the values of $M_\theta(t)$ at $t = .01, .1, .2, .3, \ldots, .9, .99$. You should see how the problem is becoming complicated. Observations of components of $\mathbf{E}_{36}$ should provide entangled information regarding this problem: regarding milk yields among BST using cows and among non-BST using cows, as well as whether this herd of 36 heifers uses BST injections. The theory of conditional probabilities can be used to sort out these information components.

In developing this extensive example, we have been sometimes casual in our sketching of probability mass functions and in our talk of probabilities being "in the vicinity of ... ." The degree of casualness allowable in such an elicitation depends of course on your purpose in specifying your opinions about any considered quantities. You may be concerned with a decision of great moment to yourself and your friends. Or you may be concerned with a decision of minor importance for which a "ballpark" answer would be sufficient.

#### 5.4.2.2 *The Mixture Distribution for $S_n$ Is Polya$(n, \alpha, \beta)$*

A technical development will complete this discussion. Supposing that you regard $N$ events exchangeably and your opinions are infinitely exchangeably extendable, your probability mass function for $S_n$ is representable as

$$f_{S_n}(x) = {}^nC_x \int_0^1 \theta^x(1-\theta)^{n-x}\, dM(\theta) \qquad (0 \leqslant x \leqslant n \leqslant N),$$

for some mixture distribution $M(\theta)$. Now if $M(\theta)$ is specified as some member of the family of beta distributions, beta$(\alpha, \beta)$, this representation specifies to

$$f_{S_n}(x) = {}^nC_x \int_0^1 \theta^x(1-\theta)^{n-x}[\Gamma(\alpha+\beta)/\Gamma(\alpha)\Gamma(\beta)]\theta^{\alpha-1}(1-\theta)^{\beta-1}\, d\theta,$$

which simplifies to

$$= {}^nC_x[\Gamma(\alpha+\beta)/\Gamma(\alpha)\Gamma(\beta)] \int_0^1 \theta^{x+\alpha-1}(1-\theta)^{n-x+\beta-1}\, d\theta,$$

and which integrates according to the beta integral as

$$= {}^nC_x[\Gamma(\alpha+\beta)/\Gamma(\alpha)\Gamma(\beta)]\,[\Gamma(\alpha+x)\Gamma(\beta+n-x)/\Gamma(\alpha+\beta+n)].$$

The probability distribution defined by this p.m.f. for $S_n$ has such widespread usefulness in problems of inference that we identify it by name as the Polya distribution according to the following definition.

***Definition 5.12.*** To write $X \sim \text{Polya}(n, \alpha, \beta)$ for $n$ an integer and $(\alpha, \beta) \in (\mathbb{R}^+, \mathbb{R}^+)$ denotes that someone asserts probabilities supporting the *Polya* *probability mass function for* $X$, specified by

$$f_X(x) = {}^nC_x \frac{\Gamma(\alpha+\beta)\Gamma(\alpha+x)\Gamma(\beta+n-x)}{\Gamma(\alpha+\beta+n)\Gamma(\alpha)\Gamma(\beta)} \qquad (x \in \{0, 1, \ldots, n\}),$$

$$= 0 \qquad \text{for other } x.$$

The corresponding function $F_X(x) \equiv \sum_{a=0}^{[x]} f(a)$ is a *Polya probability distribution function.* □

Using this terminology, we can summarize our technical development by the following theorem.

**Theorem 5.4.** A beta mixture of a binomial distribution is a Polya distribution: $X \sim$ beta-binomial$(n, \theta; \alpha, \beta)$ is equivalent to $X \sim$ Polya$(n, \alpha, \beta)$.

You should find the following problems helpful in consolidating your familiarity with the Polya distribution. In the first problem you will derive its first two moments. The second problem will remind you of your previous experience with the Polya distribution. In the third problem you will determine the Polya probabilities for the two "yous" of the dairy example in Section 5.4.2.1 and for the problem that concluded it.

## PROBLEMS

1. Show that if you regard $\mathbf{E}_N$ exchangeably, and your probability distribution function for $S_N$ is Polya$(N, \alpha, \beta)$, then your $P(S_N) = N\alpha/(\alpha+\beta)$ and your $V(S_N) = [N\alpha\beta/(\alpha+\beta)^2](\alpha+\beta+N)/(\alpha+\beta+1)$. You can do this problem in two ways: you can derive the first two moments of the distribution directly by appropriate summation, or you can use the known moments of the beta distribution and the binomial distribution, and the general result regarding moments of mixture distributions that if $X \sim M\text{-}F_x(\cdot;\theta)$ then $V(X) = \mathbb{M}_1[\mathbb{M}_2^*(X;\theta)] + \mathbb{M}_2^*[\mathbb{M}_1(X;\theta)]$. In either case once you have derived the solution, think about the size of the variance for large values of $N$ and fixed values of $(\alpha+\beta)$, and for large values of $(\alpha+\beta)$ for fixed $N$. What does this tell you?

2. Review problem 2 of Section 3.8 and show that the distribution for the total number of red balls drawn in $N$ draws from that urn process is Polya$(N, R, B)$.

3. Determine the Polya distributions that represent the assertions of the two "yous" in the dairy example of Section 5.4.2.1. Identify the distribu-

tions parametrically (just specify the values of $n$, $\alpha$, and $\beta$) and compute the associated mass functions. Compute the probability mass function for the mixture-Polya opinions described in the problem that concludes that section.

### 5.4.2.3 *The Conditional Distribution for $S_N|(S_n = x)$ Is Polya$(N-n, \alpha+x, \beta+n-x)$*

Suppose you are interested by $N$ events, whose values are unknown to you. You might well assess your previsions for these events, $P(\mathbf{E}_N)$, or even your joint probability distribution for them. Now presumably, if you observe the values of $n$ of the events, you would learn something about the value of the next one, and in fact about the values of all the rest of them. What would you want to learn? What information would the observation of $n$ events provide you concerning the values of the rest of them? What would you "infer" from the observation of $n$ events about the value of an $(n+1)$st? about the values of all $N-n$ subsequent events?

Your answer to the simplest of these questions is contained in the difference between your current prevision assertion $P(E_{n+1})$ and your current conditional prevision assertions $P(E_{n+1}|\mathbf{E}_n = \mathbf{e}_n)$ as specified for various values of the vector $\mathbf{e}_n \in \mathscr{R}(\mathbf{E}_n)$. For these conditional probabilities, remember, represent your assertion regarding your valuation of $E_{n+1}$ under the stipulation that your conditional prevision defining "transaction commitment" is called off unless the conditioning product event actually equals 1.

Similarly, your answer to the more complicated question (what would you learn about all the remaining events, $E_{n+1}, E_{n+2}, \ldots, E_N$?) is represented by the difference between your asserted marginal probability distribution for the events $E_{n+1}, E_{n+2}, \ldots, E_N$ and your asserted conditional distribution for these same $N-n$ events, conditioned again upon the various observation possibilities, $(\mathbf{E}_n = \mathbf{e}_n)$.

The following theorem states the coherent solution to both of these questions under the imagination that you regard the events $\mathbf{E}_N$ exchangeably, with your infinitely exchangeably extendable mixture distribution specifying opinions about the sum as Polya$(N, \theta; \alpha, \beta)$ or, equivalently, beta-binomial $(N, \theta; \alpha, \beta)$.

**Theorem 5.5.** Suppose you regard $N$ events exchangeably, and you assert a Polya$(N, \alpha, \beta)$ distribution for their sum, $S_N$. Denote by $S_n$ the sum of the first $n$ of these events, and by $S_{N-n}$ the sum of the remaining $N-n$ of them. Then

**(1)** your marginal distribution for $S_{N-n}$ is Polya$(N-n, \alpha, \beta)$, and

**(2)** your conditional distribution for $S_{N-n}|(S_n = x)$ is Polya$(N-n, \alpha+x, \beta+(n-x))$ for each $x = 0, 1, \ldots, n$.

Thus, the corresponding previsions and variances compare as

$$\begin{aligned}
P(S_{N-n}) &= (N-n)\alpha/(\alpha+\beta), \\
V(S_{N-n}) &= [(N-n)\alpha\beta/(\alpha+\beta)^2](\alpha+\beta+N-n)/(\alpha+\beta+1), \\
P(S_{N-n}|S_n = x) &= (N-n)(\alpha+x)/(\alpha+\beta+n), \\
V(S_{N-n}|S_n = x) &= [(N-n)(\alpha+x)(\beta+n-x)/(\alpha+\beta+n)^2] \\
&\quad \times (\alpha+\beta+N)/(\alpha+\beta+n+1), \\
&< V(S_{N-n}).
\end{aligned}$$

*Proof*. The easiest way to prove these results is through the Polya urn characterization of the distribution, which you reviewed in problem 2 of Section 5.4.2.2. In the first place, since the events are regarded exchangeably, any $N-n$ of them have the same joint distribution. So the marginal distribution for $S_{N-n}$ is generated by the Polya urn scheme in $N-n$ steps, beginning with $\alpha$ red and $\beta$ black balls in the urn. As to the conditional distribution of $S_{N-n}|(S_n = x)$, it is generated by a Polya urn scheme in $N-n$ steps, too, but beginning with $\alpha+x$ reds and $\beta+n-x$ blacks. The moment formulas are direct applications of the formulas derived in problem 1 of Section 5.4.2.2.

Alternatively, these results can be derived using standard margining and conditioning probability formulas applied to the Polya distribution, remembering that the sequence of basic events, $\mathbf{E}_N$, are regarded exchangeably. Thus, all event sequences yielding the same sum are assessed with the same probability. The results derive from applying standard counting procedures to this judgment. The developments of Sections 3.9 and 3.10 are directly applicable.

The following problem will confirm your understanding of the Polya distribution. We conclude this section with still another derivation of Theorem 5.5, using Theorem 5.4, which motivates a philosophical comment.

## PROBLEM

1. Show that a Polya mixture of a Polya distribution is a Polya distribution. To be precise, if conditionally $S_N|(\alpha, b)$ is Polya$(N, \alpha, \beta)$, and marginally $\alpha|(\alpha_0, \beta_0)$ is Polya$((\alpha_0+\beta_0), \alpha_0, \beta_0)$, where $\alpha_0$ and $\beta_0$ are integers, then $S_N|(\alpha_0, \beta_0)$ is Polya$(N, \alpha_0, \beta_0)$. (*Hint*: This result is immediately corollary to Theorem 5.5, that the marginals and conditionals for a Polya distribution are Polya. Notice the similarity of this result to the results of problems 3 and 4 of Section 5.4.1, stating that hypergeometric mixtures of hypergeometric distributions are hypergeometric distributions.)

***Philosophical Comment***

There is another way to derive the mathematical results regarding a Polya distribution that we have been studying. Since the alternative derivation allows for interesting philosophical comment, and it also has widespread algebraic application in other contexts, let us study it here.

First, examining the definition of a mixture distribution, we see that mathematically a mixture density (or mass) function $f_X(x) = \int_\Theta f_X(x;\theta)\,dM(\theta)$ can be recognized as representing a marginal distribution for $X$ with respect to a joint distribution for $X$ and $\theta$. The parametric mass function $f_X(x;\theta)$ surely sums to 1 over its domain of support (values of $x$) for each specification of the index parameter, $\theta$, and $M(\theta)$ is a distribution function. Thus, each mass function $f_X(x;\theta)$ can be considered as a conditional mass function for $X|(\theta=\theta)$, and, thus, we can denote it meaningfully by $f_{X|\theta}(x) \equiv f_X(x;\theta)$ or even by $f_X(x|\theta)$. On the basis of the standard relation between a joint distribution function and the associated marginal and conditional distributions, at least in the case of discrete and absolutely continuous distributions, we can write

$$f_{X,\theta}(x,\theta) = f_{X|\theta}(x)\, f_\theta(\theta) = f_{\theta|X=x}(\theta)\, f_X(x),$$

and thus

$$f_{\theta|X=x}(\theta) = f_{X|\theta}(x) f_\theta(\theta)/f_X(x) \qquad \text{as long as } f_X(x) > 0. \tag{*}$$

In the case of the beta-binomial mixture distribution for $S_N$, it is implied that $\theta \sim \text{beta}(\alpha,\beta)$ and $S_n|\theta \sim B(n,\theta)$. Thus, the conditional density for $\theta|(S_n=a)$ can be derived from product formula (*) as

$$f_{\theta|(S_n=a)}(\theta) = {}^nC_a\theta^a(1-\theta)^{n-a}[\Gamma(\alpha+\beta)/\Gamma(\alpha)\Gamma(\beta)]\theta^{\alpha-1}(1-\theta)^{\beta-1}/f_{S_n}(a),$$

where

$$f_{S_n}(a) = \int_0^1 {}^nC_a\theta^a(1-\theta)^{n-a}[\Gamma(\alpha+\beta)/\Gamma(\alpha)\Gamma(\beta)]\theta^{\alpha-1}(1-\theta)^{\beta-1}\,d\theta$$

in the denominator acts as a proportionality constant, ensuring that this posterior density function on $\theta$ integrates to 1 over its entire domain. Clearing the algebra should convince you of the result that the posterior mixing distribution on $\theta|(S_n=a)$ is $\text{beta}(\alpha+a, \beta+n-a)$.

But the beta-binomial mixture distribution for $S_N$ also implies that $S_{N-n}|(\theta, S_n=a) \sim B(N-n,\theta)$. Thus, since we have determined that the posterior mixing distribution on $\theta$ given $S_n=a$ is $\text{beta}(\alpha+a, \beta+n-a)$, we may conclude that the implied posterior distribution of $S_{N-n}|(S_n=a)$ is $\text{Polya}(N-n, \alpha+a, \beta+n-a)$, as stated in the Theorem 5.5.

We have recognized since our derivation of the mixture distribution formula for $S_N$ in Section 3.10.3, that the mixing distribution function on $\theta$ (either "prior" or "posterior" to $S_n = a$) which appears in these developments is virtually identical to one's probability distribution for the proportion of 1's occurring in a long sequence of $N$ events that you regard exchangeably, $S_N/N$. To be precise, the representation result is the limiting result as $N \to \infty$, but your assessing your mixing p.m.f. $f(\theta)$ can be achieved approximately by assessing your opinions regarding the proportion of 1's in a long string of such events.

Theorem 5.4, specifying that a beta mixture of a binomial distribution is algebraically equivalent to a Polya distribution, is worthy of philosophical and historical comment. The twentieth-century tradition of objectivist statistical practice has largely honored the characterization of events that we regard exchangeably, as events "generated independently by an unknown but constant probability, $\theta$, for success in each particular instance." Thus, the paradigmatic statistical program has been to "estimate the true probability of occurrence of an event" by means of observations from a sequence of independent events that occur with the same probability. Indeed, many Bayesian proponents suggest that "the posterior distribution for $\theta$" provides a subjectivist basis for statements about the "unknown probability," too. Such Bayesians worry that "the correct" prior distribution must be used to validate the procedure. We discuss these worries specifically in Chapter 6.

It is interesting in this context that the equivalence of the Polya distribution with the beta-binomial mixture identifies a representation of the problem that does not involve any parameter $\theta$ at all! Formulating the probability assessments for observable events in terms of the Polya distribution accentuates the recognition that the judgment of exchangeability regarding a sequence of events is a judgment that is in no way required for everyone. We can easily imagine vectors of events that someone regards exchangeably (who thus would find relevant a parameterized distribution in terms of a mixture over $\theta \in [0, 1]$), but that someone else does not regard exchangeably at all, knowing specific different things about different events in the sequence. For this latter, person, there is no concise representation of opinion in terms of a mixture over a parameter $\theta \in [0, 1]$.

The paradigmatic objectivist (and sometimes Bayesian) statistical program of estimating the true value of the probability that generates a sequence of observables is chimerical. The mixing parameter $\theta$ used in the representation of someone's infinitely extendable exchangeable opinions regarding a sequence $\mathbf{E}_N$ plays no role in representing the opinions of anyone who does not regard them exchangeably. It does not "exist" in itself, so it is not there to be estimated. The operationally defined measurement values identifying $\mathbf{E}_N$, while unknown, at least are observable numbers that are fit objects of our prevision assertions.

Let us conclude this section by formalizing some terminology. The fact that the marginal distribution for $S_{N-n}$ and the conditional distribution for $S_{N-n}|(S_n = x)$ are both Polya distributions, and that in the beta-binomial form of the corresponding mixing distributions for $\theta$ and for $\theta|(S_n = x)$ are both beta, allows for algebraic simplicity and for simple computational determination of their implications. This closure property of this family of distributions earns it the name of the "natural conjugate" family of distributions for quantities that are regarded in this way. Since this terminology has application in many distributional contexts, let us state here a formal definition.

***Definition 5.13.*** A family of distributions for a vector of quantities $\mathbf{X}_N$ is said to be a *natural conjugate family of distributions* if, for any choice of $n < N$, the joint distribution for the final $N - n$ components, $\mathbf{X}_{N-n}$, and the conditional joint distribution for these same components, conditioned on the values of the first $n$ components, $(\mathbf{X}_n = \mathbf{x}_n)$, are both members of this family. If the joint distribution for $\mathbf{X}_N$ is a mixture distribution $M\text{-}F(\cdot;\theta)$, and both of the mixing distributions $M(\theta)$ and $M(\theta|(\mathbf{X}_n = \mathbf{x}_n))$ are members of the same family, that family is called the natural conjugate family of mixing distributions. □

Using this terminology, we have learned thus far that the hypergeometric family of distributions is natural conjugate for the sum of events regarded exchangeably, and that the Polya family of distributions is natural conjugate for the sum of events regarded exchangeably and as infinitely exchangeably extendable. In the latter case, the beta family of distributions is the natural conjugate family of mixing distributions.

## 5.5 FAMILIES OF DISTRIBUTIONS FOR QUANTITIES REGARDED EXCHANGEABLY

To this point we have focused on the algebraic forms of probability distribution functions for events and for sums of events . In many applications, however, we are concerned with quantities whose realms are greater than merely the realm of an event $\{0, 1\}$. Let us now study some applicable forms of probability distribution functions for quantities whose realms we denote generically by $\mathscr{R}(X_i) = \{x_1, x_2, \ldots, x_R\}$.

The first part of the study involves a straightforward extension of the analysis we have just concluded to larger realms. First we identify the family of multiple-category hypergeometric distributions. Then we study a limiting subfamily, the family of multinomial distributions, along with its natural conjugate mixing distribution, the Dirichlet distribution. Finally, an examination of the abstract form of the multinomial distribution will lead us to a characterization of the exponential family of distributions, in Section 5.6. This large family includes many of the most commonly useful families of distributions in applied statistical inference. A specification of some of its subfamilies concludes this chapter.

Once again, the judgment to regard quantities exchangeably will be found central to the applied theory of statistical inference about quantities. Let us review the concept of exchangeability, here stating the definition and the major results as they pertain to general quantities rather than merely to events.

***Definition 5.14.*** Suppose that $X_1, X_2, \ldots, X_N$ are distinct quantities with the same realms, $\mathscr{R}(X_i) = \{x_1, x_2, \ldots, x_R\}$. You are said to *regard the quantities*

$X_1, X_2, \ldots, X_N$ *exchangeably* if, for any selection of $n \leqslant N$ of these quantities, your prevision for the product events of the form $(X_1 = y_1)(X_2 = y_2)\cdots(X_n = y_n)$ is constant for every permutation of the numbers $y_1, y_2, \ldots, y_n$. (Each of these numbers $y_i$ is presumably a member of the common realm of the quantities, $\mathscr{R}(X_i)$. Some of them may be identical, specifying repeated observation values.) Symbolically, we write that your $P(\mathbf{X}_N = \mathbf{x}_N) = P(\mathbf{X}_N = \mathbf{y}_N)$ whenever the vector $\mathbf{y}_N$ is a permutation of the vector $\mathbf{x}_N \in \mathscr{R}(\mathbf{X}_N)$. □

***Example 1.*** Suppose the quantities $X_1, \ldots, X_{1000}$ identify the measured yields of milk on a test day in April 1989 (in units of 10 kg) from 1000 Holstein cows who were born in the calendar year 1985, who were then living in a specified county of New York state, and whose milking records have been retrieved by scrambling the DHIA (Dairy Herd Improvement Association) list of records for participating dairies in that county. The measurement scale of the $X_i$ is based upon $\mathscr{R}(X_i) = \{0, 1, 2, \ldots, 9\}$, where $X_i = 0$ means the yield is less than 10 kilos, $X_i = 1$ means the yield is between 10 and 20 kilos, ..., and $X_i = 9$ means the yield exceeds 90 kilos. Knowing nothing more about these particular cows (such as when each had last given birth to a calf, which are purebred cows and which are grade, etc.), you would be said to regard the quantities exchangeably if you decide that your previsions are identical for product events of the form $(X_1 = y_1)(X_2 = y_2)\cdots(X_N = y_N)$ for every reordering of the numbers $y_1, y_2, \ldots, y_N$. For example, among many other possibilities, you would assert

$$P[(X_1 = 6)(X_2 = 3)\cdots(X_{1000} = 4)] = P[(X_1 = 3)(X_2 = 4)\cdots(X_{1000} = 6)]$$
$$= P[(X_1 = 6)(X_2 = 4)\cdots(X_{1000} = 3)].$$

For any prospective strings of observation values for $X_1, \ldots, X_{1000}$, you assert equal probabilities that the string of quantities may equal the observation string in each of its permuted orderings. We continue to expand on this example as we develop more extensive results about your probability distribution for such a sequence $X_1, X_2, \ldots, X_N$. □

Other examples of quantities that you might regard exchangeably are the yields (in bushels or in kilograms) of corn to be harvested from $N$ distinct acre (hectare) plots, or the prices (in eighths of US$) of the final trades of a named stock on the New York Stock Exchange on $N$ different days. Of course, there is no requirement that you regard any such quantities exchangeably. Alternatively, for example, you might be expecting the stock price sequence generally to rise during the coming five-day trading week before it falls a little the following week. In the context of such expectations you might well prefer to assert, among other previsions,

$$P[\mathbf{X}_{10} = (24, 24.5, 25.375, 25.25, 25.75, 25.25, 25.25, 23.75, 24.25, 23)]$$
$$> P[\mathbf{X}_{10} = (24.5, 24.25, 23.75, 24, 23, 25.375, 25.25, 25.75, 25.25, 25.25)].$$

Such an assertion would defy the definition of judging a sequence of quantities exchangeably, since the two enumerated sequences of $\mathbf{X}_{10}$ are permutations of one another. Harboring such opinions you might choose to engage speculative transactions on this stock! If you regard the prices exchangeably, you would not.

### 5.5.1 The Multiple-Category Hypergeometric Distribution and Its Mixtures

Suppose you regard the quantities composing the vector $\mathbf{X}_N = (X_1, \ldots, X_N)^{\mathrm{T}}$ exchangeably, each having the common realm $\mathscr{R}(X_i) = \{x_1, x_2, \ldots, x_R\}$. Supposing that $N > R$, it is evident that several of the quantities must be identical in value. For there are not enough possibilities in $\mathscr{R}(X_i)$ for each of them be distinct. One helpful summary of the quantities, which will be central to statistical inference regarding quantities that you judge exchangeably, is the *histogram*, or *category sum of the quantities.* This summary identifies of how many of the quantities $X_i$ equal each of the possible members of their common realm. Formally, let us define the notation

$$\begin{aligned}\mathbf{S}_R(\mathbf{X}_N) &\equiv \left[\sum_{i=1}^{N} (X_i = x_1), \sum_{i=1}^{N} (X_i = x_2), \ldots, \sum_{i=1}^{N} (X_i = x_R)\right] \\ &\equiv (S_1(N), \ldots, S_R(N))^{\mathrm{T}} \equiv \mathbf{S}_R(N).\end{aligned}$$

This vector $\mathbf{S}_R(N)$ is composed of $R$ nonnegative integers that sum to $N$. We call the vector $\mathbf{S}_R(\mathbf{X}_N)$ the *category sum* of the quantities $\mathbf{X}_N$, since it specifies how many of the quantities fall in each category of their realm.

The category sum is a helpful "summary" of the quantities because it identifies the values of $\mathbf{X}_N$ precisely, *except for their order.* If a vector of quantities $\mathbf{X}_N$ is a permutation of the vector $\mathbf{Y}_N$, then $\mathbf{S}_R(\mathbf{X}_N) = \mathbf{S}_R(\mathbf{Y}_N)$. However, if $\mathbf{S}_R(\mathbf{X}_N) \neq \mathbf{S}_R(\mathbf{Y}_N)$, then $\mathbf{X}_N$ and $\mathbf{Y}_N$ must differ by more than a mere permutation of their components.

Now for any selection of $n$ quantities from $N$, your joint conditional probability distribution for the $R$ components of their category sum, $\mathbf{S}_R(\mathbf{X}_n)$, conditional upon the event $(\mathbf{S}_R(\mathbf{X}_N) = \mathbf{s}_R(N))$ can be written in terms of the probability mass function

$$\begin{aligned}f_{\mathbf{S}_R(\mathbf{X}_n)}(\mathbf{s}_R(n)|\mathbf{S}_R(\mathbf{X}_N) = \mathbf{s}_R(N)) &\equiv P[\mathbf{S}_R(\mathbf{X}_n) = \mathbf{s}_R(n)|\mathbf{S}_R(\mathbf{X}_N) = \mathbf{s}_R(N)] \qquad (\mathbf{1}^{\mathrm{T}}\mathbf{s}_R(n) = n) \\ &= \left[\prod_{j=1}^{R} {}^{s_j(N)}C_{s_j(n)}\right] \Big/ {}^{N}C_n.\end{aligned}$$

For the denominator, ${}^{N}C_n$, is the number of distinct groups of $n$ quantities that can be selected from among the components of $\mathbf{X}_N$, while the numerator is the number of these possible selection groups with the category sum $\mathbf{S}_R(\mathbf{X}_n) = \mathbf{s}_R(n)$, under the supposition that the category sum of $\mathbf{X}_N$ is

$\mathbf{S}_R(\mathbf{X}_N) = \mathbf{s}_R(N)$. We say your joint conditional probability distribution for the $R$ components of $\mathbf{S}_R(\mathbf{X}_n)$ is "multiple-category hypergeometric" according to the following definition.

***Definition 5.15.*** To write $\mathbf{Y} \sim H(n, S_1, S_2, \ldots, S_{R-1}, N)$ for $n \leqslant N$ an integer and $S_1, S_2, \ldots, S_{R-1}$ all nonnegative integers whose sum does not exceed $N$ denotes that someone asserts probabilities supporting the *multiple-category hypergeometric probability mass function for* $\mathbf{Y}$, with parameters $n, S_1, S_2, \ldots, S_{R-1}$, and $S_R \equiv N - \sum_{i=1}^{R-1} S_i$, specified by

$$f(y_1, y_2, \ldots, y_R) = \left[\prod_{i=1}^{R} {}^{S_i}C_{y_i}\right] \Big/ {}^{N}C_n \qquad [(y_1, y_2, \ldots, y_R) \in \mathscr{R}(\mathbf{Y})]$$

$$= {}^{n}C_{(y_1, \ldots, y_R)} \, {}^{(N-n)}C_{(S_1 - y_1, \ldots, S_R - y_R)} / {}^{N}C_{(S_1, \ldots, S_R)},$$

where $\mathscr{R}(\mathbf{Y}) = \{\mathbf{y}_R \mid \text{each } y_i \text{ is a nonnegative integer, and } \mathbf{1}_R^{\mathrm{T}}\mathbf{y}_R = n\}$. □

You may prove the equivalence of these two algebraic expressions for the multiple-category p.m.f. using the multiple-combinatoric notation we reviewed in Section 3.8.1 note 6. Do notice that the multiple-category hypergeometric distribution is a joint distribution for the component quantities of the vector $\mathbf{Y}$. We shall specify some of its associated marginal and conditional distributions shortly.

Using this notation, our discussion has proved the following theorem.

**Theorem 5.6.** If you regard the quantities $\mathbf{X}_N$ exchangeably, where the components have the common realm $\mathscr{R}(X_i) = \{x_1, x_2, \ldots, x_R\}$, then coherency requires that your conditional distributions for the category sum of $n$ of them, $\mathbf{S}_R(n)$, are

$$\mathbf{S}_R(\mathbf{X}_n) \mid (\mathbf{S}_R(\mathbf{X}_N) = \mathbf{s}_R(N)) \sim H(n, \mathbf{s}_{R-1}(N), N),$$

for each $\mathbf{s}_R(N) \in \{\mathbf{s}_R \mid \text{ each } s_i \text{ is a nonnegative integer, and } \mathbf{1}_R^{\mathrm{T}}\mathbf{s}_R = N\}$.

***Example 1.*** Consider again the dairy cows whose milk yields you regarded exchangeably in Example 1 of Section 5.5. Now condition your considerations on the event that the category sum for the yields from the 1000 cows is $\mathbf{S}_{10}(\mathbf{X}_{1000}) = (93, 11, 20, 366, 290, 211, 9, 0, 0, 0)$. Your probability that of any three cows selected from the group there is one in each of the categories 2, 3, and 4 equals (remember the 10 category values are 0 through 9)

$$f_{\mathbf{S}_{10}(\mathbf{X}_3)}[(0,0,1,1,1,0,0,0,0,0) \mid (\mathbf{S}_{10}(\mathbf{X}_{1000}) = (93, 11, 20, 366, 290, 211, 9, 0, 0, 0))]$$

$$= \frac{{}^{20}C_1 \, {}^{366}C_1 \, {}^{290}C_1}{{}^{1000}C_3} = \frac{20 \cdot 366 \cdot 290}{1000 \cdot 999 \cdot 998/6}.$$

Your probability that all three are in category 3 equals

$$f_{\mathbf{S}_{10}(\mathbf{X}_3)}[(0,0,0,3,0,0,0,0,0,0)|(\mathbf{S}_{10}(\mathbf{X}_{1000})=(93,11,20,366,290,211,9,0,0,0))]$$
$$=\frac{{}^{366}C_3}{{}^{1000}C_3}=\frac{366\cdot 365\cdot 364}{1000\cdot 999\cdot 998}.$$

Symbolically, we summarize these probability assertions, along with many others, when writing

$$\mathbf{S}_{10}(\mathbf{X}_3)|(\mathbf{S}_{10}(\mathbf{X}_{1000})=(93,\ 11,\ 20,\ 366,\ 290,\ 211,\ 9,\ 0,\ 0,\ 0))$$
$$\sim H(3,93,11,20,366,290,211,9,0,0,1000).$$ □

This example of the multiple-category hypergeometric distribution has involved categories based upon numerical measurements of milk yields. But there are other ways that measurement categories could distinguish quantities in this example as well. The DHIA keeps records on many characteristics of each cow: its breed (1 = Ayrshire, 2 = Guernsey, 3 = Holstein, 4 = Jersey, 5 = Milking Shorthorn, 6 = Red Poll, 7 = other); whether the cow has been bred through artificial insemination (1 = yes, 0 = no); whether the cow is purebred or grade (1 = yes, 0 = no); if the cow is a grade, whether or not the dairy has a record of its mother (1 = yes, 0 = no); and so on. Each of these identifications provides the basis for defining a quantity, just as did the cows' measured milk yields on the specified day. You would, for example, regard exchangeably the "breed" quantity for each of the scrambled cow units of any 1000 cows under consideration. In this case, thus your conditional probability distribution for the breeds of any scrambled subsample of $n$ cows, conditional upon the breed makeup of the 1000 $(\mathbf{S}_7(\mathbf{X}_{1000})=(s_1,s_2,\ldots,s_7))$, would be $H(n,s_1,s_2,\ldots,s_6,1000)$.

The following two theorems on characteristics of a multiple-category hypergeometric distribution are worth noting for reference and discussion. You may prove them for yourself rather easily. The first identifies the mean vector and the variance-covariance matrix for this joint distribution; the second identifies several important associated marginal joint distributions.

**Theorem 5.7.** If you assert $\mathbf{Y}\sim H(n,S_1,S_2,\ldots,S_{R-1},N)$, then your prevision for individual components of $\mathbf{Y}$ is specified by the vector $P(\mathbf{Y})=(n/N)(S_1,S_2,\ldots,S_R)$, where $S_R\equiv N-\sum_{i=1}^{R-1}S_i$. Your variance-covariance matrix for $\mathbf{Y}$ shows diagonal elements of

$$\operatorname{Var}(Y_i)=n(S_i/N)[(N-S_i)/N](N-n)/(N-1)$$

and off-diagonal elements of

$$\operatorname{Cov}(Y_i,\ Y_j)=-n(S_i/N)(S_j/N)(N-n)/(N-1).$$

**Theorem 5.8.** If you assert $\mathbf{Y} \sim H(n, S_1, S_2, \ldots, S_{R-1}, N)$ then your marginal distribution for any component $Y_i$ is $Y_i \sim H(n, S_i, N)$. More generally, your marginal distribution for the sum of any $r$ components (call them the first $r$ components, without loss of generality) is $\sum_{i=1}^{r} Y_i \sim H(n, \sum_{i=1}^{r} S_i, N)$. As another generalization, your marginal joint distribution for any $r$ subcomponents, call them the first $r$ components $(Y_1, \ldots, Y_r)$, is a hypergeometric mixture of multiple-category hypergeometric distributions:

$$(Y_1, \ldots, Y_r) \left| \left( \sum_{i=r+1}^{N} Y_i = y \right) \sim H\left( n - y, S_1, \ldots, S_{r-1}, \sum_{i=1}^{r} S_i \right), \right.$$

$$\left( \sum_{i=r+1}^{N} Y_i \right) \sim H\left( n, \sum_{i=r+1}^{R} S_i, N \right).$$

Before considering an example and some simple computational problems, let us remind ourselves that your probability mass function for a particular *ordered sequence* of quantities that you regard exchangeably is proportional to your probability mass function for the category sum the sequence provokes. In particular, your

$$\begin{aligned} &f_{\mathbf{X}_n}(y_1, y_2, \ldots, y_n | [\mathbf{S}_R(\mathbf{X}_N) = \mathbf{s}_R(N)]) \\ &\quad \equiv P((X_1 = y_1)(X_2 = y_2) \cdots (X_n = y_n) | [\mathbf{S}_R(\mathbf{X}_N) = \mathbf{s}_R(N)]) \\ &\quad = [{}^nC_{(s_1(n), \ldots, s_R(n))}]^{-1} f_{\mathbf{S}_R(\mathbf{X}_n)}(s_1(n), \ldots, s_R(n) | \mathbf{S}_R(\mathbf{X}_N) = \mathbf{s}_R(N)), \end{aligned}$$

where $s_j(n) \equiv \sum_i^n (y_i = x_j)$ is the $j$th component of the category sum $\mathbf{S}_R(\mathbf{y}_n)$. For your judging $\mathbf{X}_N$ exchangeably means that you regard as equilikely all ordered sequences of components $X_1, \ldots, X_n$ whose category sums are identical. The multinomial coefficient ${}^nC_{(s_1(n), \ldots, s_R(n))}$ is the number of distinct sequences of $n$ quantities chosen from $\mathbf{X}_N$ that have the category sum $S_R(\mathbf{X}_n) = (s_1(n), \ldots, s_R(n))$. Thus, your

$$\begin{aligned} &f_{\mathbf{X}_n}(y_1, y_2, \ldots, y_n | [\mathbf{S}_R(\mathbf{X}_N) = \mathbf{s}_R(N)]) \\ &\quad = [{}^nC_{(s_1(n), \ldots, s_R(n))}]^{-1} \left[ \prod_{j=1}^{R} {}^{s_j(N)}C_{s_j(n)} \right] \Big/ {}^NC_n. \end{aligned}$$

***Example 1, continued.*** Conditioning again upon the category sum event, $\mathbf{S}_{10}(\mathbf{X}_{1000}) = (93, 11, 20, 366, 290, 211, 9, 0, 0, 0) = \mathbf{s}_{10}(1000)$, your probability that the first of three cows selected from the 1000 is in category 4, that the second is in category 2, and that the third is in category 3 equals

$$\begin{aligned} &f_{\mathbf{X}_3}(4, 2, 3 | \mathbf{S}_{10}(\mathbf{X}_{1000} = (93, 11, 20, 366, 290, 211, 9, 0, 0, 0))) \\ &\quad = (3!)^{-1} (290 \cdot 20 \cdot 366)/(1000 \cdot 999 \cdot 998) \end{aligned}$$

since there are $3! = 6$ distinct sequences of $X$'s pertaining to those three cows which have the category sum (0, 0, 1, 1, 1, 0, 0, 0, 0, 0). Your conditional p.m.f. evaluated at any permutation of the triple (4, 2, 3) is identical, for precisely the same p.m.f. formula applies.

Similarly, your conditional probability that the first cow selected is in category 3, the second is in category 3, and the third is in category 3 equals

$$f_{\mathbf{X}_3}(3, 3, 3 | \mathbf{S}_{10}(\mathbf{X}_{1000} = (93, 11, 20, 366, 290, 211, 9, 0, 0, 0)))$$
$$= (366 \cdot 365 \cdot 364)/(1000 \cdot 999 \cdot 998)$$

since there is only one distinct permutation of the triple (3, 3, 3). □

## PROBLEMS

**1.** Operational Setup. You are considering your opinions regarding 10000 possums who live in a certain 3500 hectares of forest. Each possum can be distinguished according to its sex: $S_i \equiv$ (possum $i$ is male); its pregnancy status: $PG_i \equiv$ (possum $i$ is pregnant); and its status as a carrier of the bovine tuberculosis virus: $V_i \equiv$ (possum $i$ carries the virus for bovine tuberculosis). A standard way to represent such multiple categorizations of an item is via the quantity that equals the base-10 numerical representation of the vector that identifies the value of component events. For example, in this case the vector $(S_i, PG_i, V_i)$ could be translated into a quantity $X_i$ defined by

$$X_i \equiv 4S_i + 2PG_i + V_i.$$

Recognizing the logical restriction that males are not pregnant, denoted by the restriction $(S_i PG_i = 0) = 1$, the realm for $X_i$ is seen as $\mathscr{R}(X_i) = \{0, 1, 2, 3, 4, 5\}$. Suppose that you regard the quantities $\mathbf{X}_{10,000}$ exchangeably.

**a.** Identify the form of your implied conditional probability distribution for the category sum of 25 captured possums, $\mathbf{S}_6(\mathbf{X}_{25})$, conditional upon $\mathbf{S}_6(\mathbf{X}_{1000}) = (9, 108, 493, 3874, 112, 5404)$.

**b.** Try to assess some broad features of your personal probability distribution for $\mathbf{S}_6(\mathbf{X}_{1000})$.

[*Hint*: Begin this problem by constructing the realm matrix for the vector $(S_i, PG_i, V_i, X_i)$.]

**2.** Computational Applications. Determine the numerical values of the following four probabilities in the context of the captured possums just described:

**a.** $P(\mathbf{S}_6(\mathbf{X}_{25}) = (1, 2, 5, 8, 1, 8) | \mathbf{S}_6(\mathbf{X}_{1000}) = (9, 108, 493, 3874, 112, 5404))$.

**b.** $P(\mathbf{S}_6(\mathbf{X}_{25}) = (1, 1, 4, 7, 1, 11) | \mathbf{S}_6(\mathbf{X}_{1000}) = (9, 108, 493, 3874, 112, 5404))$.

**c.** $P(\mathbf{S}_6(\mathbf{X}_{25}) = (1, 2, 5, 8, 1, 8) | \mathbf{S}_6(\mathbf{X}_{1000}) = (19, 98, 490, 3980, 108, 5305))$.
**d.** $P(\mathbf{S}_6(\mathbf{X}_{25}) = (1, 1, 4, 7, 1, 11) | \mathbf{S}_6(\mathbf{X}_{1000}) = (19, 98, 490, 3980, 108, 5305))$.

This analysis of a judgment to regard the quantities $X_1, X_2, \ldots, X_N$ exchangeably obviously parallels our analysis in Section 5.4.1 of a judgment of exchangeability regarding $N$ events. To this point we have studied only your *conditional* distribution for the category sum of $n$ quantities given the category sum of all $N$ quantities. The completion of the study demands the formulation of the joint, marginal, and conditional distributions for $n$ and $N-n$ quantities that are *not conditioned* on the category sum of all $N$ of them. Since the algebraic and conceptual parallels with the context of events continue, let me merely state the following results, which complete the analysis for quantities that you regard exchangeably. We continue to suppose that the realm of each of the quantities $X_1, \ldots, X_N$ is $\mathscr{R}(X_i) = \{x_1, \ldots, x_R\}$.

**Theorem 5.9.** If you regard the quantities $X_1, \ldots, X_N$ exchangeably, then your probability mass function for the category sum of any $n$ quantities that are selected from among them is a probability mixture of multiple-category hypergeometric distributions:

$$
\begin{aligned}
& f_{\mathbf{S}_R(\mathbf{X}_n)}(s_1(n), \ldots, s_R(n)) \\
&\quad = \textstyle\sum^* f_{\mathbf{S}_R(\mathbf{X}_n)}(s_1(n), \ldots, s_R(n) | [\mathbf{S}_R(\mathbf{X}_R) = \mathbf{s}_R(N)]) P[\mathbf{S}_R(\mathbf{X}_N) = \mathbf{s}_R(N)]) \\
&\quad = \textstyle\sum^* \left[ \prod_{j=1}^{R} {}^{s_j(N)}C_{s_j(n)} \right] f_{\mathbf{S}_R(\mathbf{X}_N)}(\mathbf{s}_R(N)) \Big/ {}^{N}C_n ,
\end{aligned}
$$

where the summations $\sum^*$ are over all $R$-dimensional vectors of nonnegative integers, $\mathbf{s}_R(N)$, whose components sum to $N$. Thus, your prevision for a sequence of quantities you regard exchangeably specifies a *mixture-multiple-category hypergeometric distribution*, characterized by your p.m.f. for their category sum, the mixing mass function in this expression. □

**Theorem 5.10.** Under the same conditions as Theorem 5.9, if your probability distribution for the category sum $\mathbf{S}_R(\mathbf{X}_N)$ is itself a multiple-category hypergeometric distribution,

$$\mathbf{S}_R(\mathbf{X}_N) \sim H(N, \eta_1, \eta_2, \ldots, \eta_{R-1}, \mathbb{N}),$$

then your marginal distribution for $\mathbf{S}_R(\mathbf{X}_n)$ is also hypergeometric:

$$\mathbf{S}_R(\mathbf{X}_N) \sim H(n, \eta_1, \eta_2, \ldots, \eta_{R-1}, \mathbb{N}).$$

Furthermore, your conditional distribution for the category sum of the remaining quantities, $X_{n+1}, \ldots, X_N$, given the event $\mathbf{S}_R(X_n) = \mathbf{s}_R(n)$, is multiple-category hypergeometric as well:

$$\mathbf{S}_R(X_{n+1},\ldots,X_N)|(\mathbf{S}_R(\mathbf{X}_n)=\mathbf{s}_R(n))$$
$$\sim H(N-n,\eta_1-s_1(n),\eta_2-s_2(n),\ldots,\eta_{R-1}-s_{R-1},\mathbb{N}-n).$$

Thus, the multiple-category hypergeometric distribution is the natural conjugate mixing function for mixture-multiple-category hypergeometric distributions.

***Definition 5.16.*** Your opinions regarding the quantities $X_1,\ldots,X_N$ are said to be *exchangeably extendable to degree* $K>N$ if there exists a probability distribution for the category sum of quantities $Y_1,\ldots,Y_K$ with the property that

$$P[\mathbf{S}_R(\mathbf{Y}_N)=\mathbf{s}_R(N)]=P[\mathbf{S}_R(\mathbf{X}_N)=\mathbf{s}_R(N)]$$

for any selection of $\mathbf{Y}_N$ from $\mathbf{Y}_K$, for every $\mathbf{s}_R(N)\in\mathscr{R}(\mathbf{S}_R(\mathbf{X}_N))$. □

Notice that if your opinions are exchangeably extendable on the basis of of the probability distribution specified by $P[\mathbf{S}_R(\mathbf{Y}_K)=\mathbf{s}_R(K)]$ for various $\mathbf{s}_R(K)$, then

$$f_{\mathbf{S}_R(\mathbf{X}_n)}(s_1(N),\ldots,s_R(N))=P[\mathbf{S}_R(\mathbf{Y}_N)=\mathbf{s}_R(N)]$$
$$=\sum{}^* f_{\mathbf{S}_R(\mathbf{X}_N)}[(\mathbf{s}_R(N))|\mathbf{S}_R(\mathbf{Y}_K)=\mathbf{s}_R(K)]P[\mathbf{S}_R(\mathbf{Y}_K)=\mathbf{s}_R(K)],$$

where the summation $\sum^*$ runs over all $R$-dimensional vectors of nonnegative integers, $\mathbf{s}_R(K)$, whose components sum to $K$. The definition of extendable exchangeability implies that the first function in the summation expression is a multiple-category hypergeometric p.m.f.,

$$f_{\mathbf{S}_R(\mathbf{X}_N)}[(s_R(N))|\mathbf{S}_R(\mathbf{Y}_K)=\mathbf{s}_R(K)]=\left[\prod_{j=1}^{R}{}^{s_j(K)}C_{s_j(N)}\right]\Big/{}^{K}C_N.$$

Thus, any probability distribution for a sequence of $N$ quantities that are regarded as exchangeably extendable to $K>N$ must be a mixture of multiple-category hypergeometric distributions, characterized by some mixing function in the $({}^{K+R-1}C_{R-1}-1)$-dimensional simplex. The components of the mixing function are the probabilities $P[\mathbf{S}_R(\mathbf{Y}_K)=\mathbf{s}_R(K)]$ for appropriate values of the vector $\mathbf{s}_R(K)$.

The characterization of probability distribution functions representing the assessment of $X_1,X_2,\ldots,X_N$ as exchangeably extendable to degree $K>N$ but no further (for each value of $K=N,N+1,\ldots$) requires an algebraically messy analogue to our discussion concerning events regarded exchangeably, which was presented in Section 3.10. It would lead to naturally to the concept of an assessment of the quantities $\mathbf{X}_N$ as exchangeably infinitely extendable, defined as follows.

***Definition 5.17.*** Your opinions regarding the quantities $X_1, \ldots, X_N$ are said to be *infinitely exchangeably extendable* if they are exchangeably extendable to degree $K$ for any positive integer $K$. □

An argument parallel to that of Section 3.10 uses the definitions and yields the conclusions presented in the next section.

### 5.5.2 The Multinomial Distribution and Its Dirichlet Mixture, the Multiple-Category Polya Distribution

We continue to analyze the algebraic form of your opinions about a sequence of quantities that you regard exchangeably. The formulation of probability distributions that are infinitely exchangeably extendable, presented in this section, continues as a straightforward generalization of the results of Section 5.4.2, which pertained to the special case of events. The presentation of mathematical results is direct, supposing that you can by now easily imagine several examples for yourself.

The context of the development is this: consider the quantity vector $\mathbf{X}_N = (X_1, \ldots, X_N)$, where each $X_i$ is a quantity with $\mathscr{R}(X_i) = \{x_i, \ldots, x_R\}$. For any selection of $n$ quantities from the components of $X_N$, we focus on the vector

$$\mathbf{S}_R(\mathbf{X}_n) = \left[ \sum_{i=1}^{n} (X_i = x_1), \sum_{i=1}^{n} (X_i = x_2), \ldots, \sum_{i=1}^{n} (X_i = x_R) \right],$$

which is the category sum of the $n$ selected quantities. Thus, it is a vector of quantities in $R$-dimensional space. Each of its components is a nonnegative integer, and the sum of its components necessarily equals $n$. If such vectors were proportioned by their sum, the elements of the realm $\mathscr{R}(n^{-1}\mathbf{S}_R(\mathbf{X}_n))$ would be representable geometrically as rational vector elements of the $(R-1)$-dimensional unit simplex, denoted simply by $\mathbf{S}^{R-1}$.

***Definition 5.18.*** Suppose $\mathbf{Y}_R$ is a vector of quantities $(Y_1, \ldots, Y_R)$, with realm $\mathscr{R}(\mathbf{Y}_R) = \{\mathbf{y}_R \equiv (y_1, y_2, \ldots, y_R) |$ each $y_i$ is a nonnegative integer, and $\sum_{i=1}^{R} y_i = N\}$. Then a probability mass function for $\mathbf{Y}_R$ of the form

$$f(\mathbf{y}_R) = {}^N C_{y_1, \ldots, y_R} \prod_{i=1}^{R} \theta_i^{y_i} \qquad (\mathbf{y}_R \in \mathscr{R}(\mathbf{Y}_R))$$

for some $\boldsymbol{\theta}_R = (\theta_1, \ldots, \theta_R)$ in the unit simplex $\mathbf{S}^{R-1}$ is said to be a *multinomial probability mass function* for $\mathbf{Y}$, identified by parameter values $\boldsymbol{\theta}_R \in \mathbf{S}^{R-1}$ and $N > 0$. We write $\mathbf{Y}_R \sim M(N, \theta_1, \theta_2, \ldots, \theta_{R-1})$ to denote that you assert probabilities regarding $\mathbf{Y}_R$ that are representable in this form. □

The proportionality constant ${}^N C_{y_1,\ldots,y_R}$ ensures that the mass function sums to 1 over all $\mathbf{y} \in \mathscr{R}(\mathbf{Y}_R)$. It arises from the generalization of the binomial expansion

$$(a+b)^N = \sum_{i=0}^{N} {}^N C_i \, a^i b^{n-i}$$

to the multinomial expansion

$$\left( \sum_{i=1}^{R} a_i \right)^N = \sum{}^* \, {}^N C_{n_1,\ldots,n_R} \prod_{i=1}^{R} a_i^{n_i},$$

where the summation $\sum^*$ runs over all vectors of nonnegative integers $(n_1, \ldots, n_R)$ that sum to $N$. If you replace the summands $a_1, \ldots, a_R$ in this expansion by nonnegative real numbers that sum to 1, you will see immediately that the multinomial p.m.f. sums to 1 over all its arguments.

We have developed enough terminology to state the following generalization of de Finetti's representation theorem as it applies to general quantities rather than events.

**Theorem 5.11.** Let $\mathbf{X}_N$ denote a vector of quantities, each with the same realm $\mathscr{R}(X_i) = \{x_1, \ldots, x_R\}$. If you regard the components of $\mathbf{X}_N$ exchangeably and your opinions are infinitely exchangeably extendable, then for any choice of $n$ quantities from these $N$, your p.m.f. can be written

$$f_{\mathbf{X}_n}(y_1, \ldots, y_n) = \int_0^1 \cdots \int_0^1 \prod_{i=1}^{R} \theta_i^{s_i(\mathbf{y}_n)} \, d_1 \cdots d_{R-1} M(\theta_1, \ldots, \theta_R),$$

where $\mathbf{s}(\mathbf{y}_n) = (s_1(\mathbf{y}_n), \ldots, s_R(\mathbf{y}_n))$ is the category sum of $\mathbf{y}_n$, and $M(\theta_1, \ldots, \theta_R)$ is a distribution function over the $(R-1)$-dimensional simplex, $\mathbf{S}^{R-1}$. Moreover, your associated probability mass function for the category sum of any subvector $\mathbf{X}_n$ is mixture multinomial, $M\text{-}M(n, \theta_1, \ldots, \theta_R)$, specifying your p.m.f. as

$$f_{\mathbf{S}_R(\mathbf{X}_n)}(s_R(\mathbf{y}_n)) = {}^n C_{s_1,\ldots,s_R} \int_0^1 \cdots \int_0^1 \prod_{i=1}^{R} \theta_i^{s_i(\mathbf{y}_n)} \, d_1 \cdots d_{R-1} M(\theta_1, \ldots, \theta_R).$$

This theorem can be proved easily using arguments parallel to those we developed in Chapter 3. You need merely analyze the limiting expression of the p.m.f. for any sequence of quantities that are regarded exchangeably and exchangeably extendable to degree $K$ (limit as $K \to \infty$). The theorem is discussed in detail, and with extensive historical references, by Good (1965, Chap. 4.)

As stated, the representation equations are exact. The representation for the p.m.f.'s $f_{\mathbf{X}_n}(y_1, \ldots, y_n)$ and $f_{S_R(\mathbf{X}_n)}(s_R(\mathbf{y}_n))$ are only approximate if you regard the components of $\mathbf{X}_N$ exchangeably but as exchangeably extendable only to degree $K$. The closeness of the approximation would increase, of course, as $K$ increases.

An important general conclusion of all our discussion to this point is that if you regard $X_1, \ldots, X_N$ exchangeably (no matter what the degree of extendability), then for any choice of $n$ quantities among them the category sum, or histogram, of $X_1, \ldots, X_n$ is sufficient for your inference about the remaining $N - n$ components of $\mathbf{X}_N$. Thus, no matter how many quantities are involved, if you regard them exchangeably there is a specified finite number of statistics that are sufficient to summarize their evidential content for one another. Using the language of Definition 3.10 in Section 3.9.1, we say that the category sum is a *sufficient statistic of fixed dimension* for quantities that are regarded exchangeably.

Let us now develop the natural conjugate form of mixing function for the mixture-multinomial distribution. This is a multivariate extension of the beta distribution, called a Dirichlet distribution.

***Definition 5.19.*** The function $f: \mathbf{S}^{R-1} \to \mathbb{R}^+$ defined by

$$f(\theta_1, \ldots, \theta_R) = \left\{ \Gamma\left( \sum_{i=1}^{R} \alpha_i \right) \Big/ \prod_{i=1}^{R} \Gamma(\alpha_i) \right\} \prod_{i=1}^{R} \theta_i^{\alpha_i - 1} \qquad (\boldsymbol{\theta}_R \in \mathbf{S}^{R-1})$$

for some vector of parameters $\boldsymbol{\alpha}_R = (\alpha_1, \ldots, \alpha_R) \in (\mathbb{R}^+)^R$ is said to be a *Dirichlet density function* with parameters $\boldsymbol{\alpha}_R$. By $\mathbf{S}^{R-1}$ is meant the $(R-1)$-dimensional unit simplex. The associated function

$$F(\theta_1, \ldots, \theta_R) = \int_0^{\theta_1} \cdots \int_0^{\theta_R} f(t_1, \ldots, t_R)\, dt_1 \cdots dt_R$$

is called the *Dirichlet distribution function.* We denote this by $\boldsymbol{\theta}_R = (\theta_1, \ldots, \theta_R) \sim D(\alpha_1, \ldots, \alpha_R)$. □

**Theorem 5.12.** If $\boldsymbol{\theta}_R \sim D(\alpha_1, \ldots, \alpha_R)$, then the marginal distributions of the vector components are each $\theta_j \sim \text{beta}[\alpha_j, (\sum_{i=1}^{R} \alpha_i) - \alpha_j]$, for $j = 1, \ldots, R$. Thus, the first two moments of each $\theta_j$ are known from the moment formulas of the beta distribution:

$$\mathbb{M}_1(\theta_i) = \alpha_i \Big/ \sum_{i=1}^{R} \alpha_i,$$

$$\mathbb{M}_2(\theta_i) = \alpha_i(\alpha_i + 1) \Big/ \left( \sum_{i=1}^{R} \alpha_i \right) \left[ \left( \sum_{i=1}^{R} \alpha_i \right) + 1 \right],$$

$$\mathbb{M}_2^*(\theta_i) = \alpha_i \left[ \left( \sum_{i=1}^{R} \alpha_i \right) - \alpha_i \right] \Big/ \left( \sum_{i=1}^{R} \alpha_i \right)^2 \left[ \left( \sum_{i=1}^{R} \alpha_i \right) + 1 \right].$$

The direct cross-product moments are

$$\mathbb{M}_{1,1}(\theta_i, \theta_j) = \alpha_i \alpha_j \Big/ \left( \sum_{i=1}^{R} \alpha_i \right) \left[ \sum_{i=1}^{R} \alpha_i + 1 \right];$$

the central cross-product moments are

$$\mathbb{M}^*_{1,1}(\theta_i, \theta_j) = -\alpha_i \alpha_j \bigg/ \left( \sum_{i=1}^{R} \alpha_i \right)^2 \left[ \sum_{i=1}^{R} \alpha_i + 1 \right];$$

and thus the correlations are

$$\mathrm{Cor}(\theta_i, \theta_j) = -(\alpha_i \alpha_j)^{1/2} \bigg/ \left[ \left( \sum_{i=1}^{R} \alpha_i \right) - \alpha_i \right]^{1/2} \left[ \left( \sum_{i=1}^{R} \alpha_i \right) - \alpha_j \right]^{1/2}.$$

For any value of $K < R - 1$, the conditional distribution of $\theta_{K+1} | \boldsymbol{\theta}_K$ is beta$(\alpha_{K+1}, \sum_{i=K+2}^{R} \alpha_i)$. The conditional distribution of $\theta_R | \boldsymbol{\theta}_{R-1}$ is degenerate on $1 - \sum_{i=1}^{R-1} \theta_i$. □

Analogous to your assertion of opinions about events via mixture-binomial distributions, you can assess your mixing distribution for the vector $\boldsymbol{\theta}_R$ by considering your opinion regarding the proportions of a large number of observations that fall in the various categories. Thus, your willingness to use a Dirichlet-multinomial mixture distribution to represent your opinions should be based upon the similarity of some Dirichlet family member to your opinions about the proportions of many observations that fit into each of the categories. In these terms, the negativity of covariance between any two components of the $\boldsymbol{\theta}_R$ vector,

$$\mathbb{M}^*_{1,1}(\theta_i, \theta_j) = -\alpha_i \alpha_j \bigg/ \left( \sum_{i=1}^{R} \alpha_i \right)^2 \left[ \left( \sum_{i=1}^{R} \alpha_i \right) + 1 \right],$$

is sometimes annoying. The Dirichlet form embeds the conditional reaction that if some component of $\boldsymbol{\theta}_R$ is unduly large, all others are prevised to be proportionally smaller. Such a reaction may well represent someone's knowledge about proportions in some instances, but this should not be presumed routinely without thought.

In many applications, positive covariance between $(\theta_i, \theta_j)$ for values of $i$ near $j$ would seem to be a desirable characteristic of distributions that are to represent mixing functions, allowing negative covariance between pairs with subscripts farther apart. Dickey (1983) has developed an interpretable family of special functions that can exhibit this character, showing how to construct a generalized Dirichlet family which allows positive covariance among nearby components. Aitchison (1986) presents another family of logistic Normal distributions on the unit simplex allowing this characteristic. Lad and Brabyn (1993) analyze an application that induces positive covariance in another way, by imposing mixed-cyclic relations among components of $\boldsymbol{\theta}_R$ as motivated by the application. Today, an array of Monte Carlo integration procedures allow for the numerical computations relevant to most any appropriate form of mixing distribution that can be designed on the unit simplex. Readable sources on these computational matters are available in Gelfand and Smith (1992), Casella and George (1992), Tanner (1993), and Gilks et al. (1996). Their application in the mixture-binomial context requires the specifications of the conditional distributions of $\theta_{K+1}|\boldsymbol{\theta}_K$ for $K = 1, \ldots, R-1$.

The following definition and theorems state the multivariate analogue of the beta-binomial mixture relation to the Polya distribution, for general reference. The parameters $\boldsymbol{\theta}_R$ disappear from the Dirichlet mixture multinomial distribution when it is recognized as equivalent to a multiple-category Polya distribution, defined as follows.

***Definition 5.20.*** Suppose $\mathbf{X}_R$ is a vector of quantities with realm $\mathscr{R}(\mathbf{X}_R) = \{\mathbf{x}_R | \mathbf{1}^T\mathbf{x}_R = N$, and each component $\mathbf{x}_i$ is a nonnegative integer$\}$. Then the function $f\colon \mathscr{R}(\mathbf{X}_R) \to \mathbb{R}^+$ defined by

$$f(\mathbf{x}_R) = {}^N C_{\mathbf{x}_R} \frac{\Gamma\left(\sum_{i=1}^{R} \alpha_i\right) \prod_{i=1}^{R} \Gamma(\alpha_i + x_i)}{\Gamma\left(N + \sum_{i=1}^{R} \alpha_i\right) \prod_{i=1}^{R} \Gamma(\alpha_i)} \qquad (\mathbf{x}_R \in \mathscr{R}(\mathbf{X}_R))$$

for some $\boldsymbol{\alpha}_R = (\alpha_1, \ldots, \alpha_R) \in \mathbb{R}^{+R}$ and $N > 0$ is said to be the p.m.f. for a *multiple-category Polya distribution*, identified by parameter values $N, \alpha_1, \ldots, \alpha_R$. We write $\mathbf{X}_R \sim \text{Polya}(N, \alpha_1, \ldots, \alpha_R)$ to denote that you assert probabilities regarding $\mathbf{X}_R$ that are representable in this form. □

The following statement of the equivalence of the Dirichlet mixture- multinomial distribution and the Polya distribution can be derived algebraically by integrating the Dirichlet-multinomial probability mass function over the parameter space of $\boldsymbol{\theta}_R \in \mathbf{S}^{R-1}$.

**Theorem 5.13.** The mixture distribution $\mathbf{S}_R(\mathbf{X}_N) \sim D\text{-}M(N, \theta_1, \ldots, \theta_{R-1}; \alpha_1, \ldots, \alpha_R)$ is identical to the distribution $\mathbf{S}_R(\mathbf{X}_N) \sim \text{Polya}(N, \alpha_1, \ldots, \alpha_R)$.

Implied prevision and conditional prevision assertions can be derived easily from the moments of the multinomial and the Dirichlet distributions, using again the familiar moment representations, now in multivariate form: If the joint distribution of the vector $\mathbf{Y}$ is the mixture $M\text{-}F(\cdot; \boldsymbol{\theta}_R)$ then the prevision for $\mathbf{Y}$ is $P(\mathbf{Y}) = \mathbb{M}_1(\mathbb{M}_1(\mathbf{Y}; \boldsymbol{\theta}_R))$, and the variance-covariance matrix for $\mathbf{Y}$ is Var-Cov$(\mathbf{Y}) = \mathbb{M}_1(\mathbb{M}_2^*(\mathbf{Y}; \boldsymbol{\theta}_R)) + \mathbb{M}_2^*(\mathbb{M}_1(\mathbf{Y}; \boldsymbol{\theta}_R))$. We give them here for reference:

$$P[\mathbf{S}_R(\mathbf{X}_{N-n})] = (N-n)(\alpha_1, \ldots, \alpha_R) \Big/ \sum_{i=1}^{R} \alpha_i,$$

$$P[\mathbf{S}_R(\mathbf{X}_{N-n}) | (\mathbf{S}_R(\mathbf{X}_n) = \mathbf{s}_R(n))] = (N-n)(\alpha_1 + s_1(n), \ldots, \alpha_R + s_R(n)) \Big/ \sum_{i=1}^{R} (\alpha_i + s_i(n)).$$

As to the variance-covariance matrix for the components of the category sum vector, Var-Cov$[\mathbf{S}_R(\mathbf{X}_{N-n})]$, it is composed of diagonal components $(i = 1, 2, \ldots, R)$

$$V[S_i(\mathbf{X}_{N-n})] = \frac{(N-n)\alpha_i\left[\sum_{i=1}^{R}\alpha_i - \alpha_i\right]\left[\sum_{i=1}^{R}\alpha_i + N - n + 1\right]}{\left(\sum_{i=1}^{R}\alpha_i\right)^2\left[\sum_{i=1}^{R}\alpha_i + 1\right]},$$

and off-diagonal components $(1 \leqslant i < j \leqslant R)$

$$\text{Cov}[S_i(\mathbf{X}_{N-n}), S_j(\mathbf{X}_{N-n})] = \frac{-(N-n)\alpha_i\alpha_j\left[\sum_{i=1}^{R}\alpha_i + N - n\right]}{\left(\sum_{i=1}^{R}\alpha_i\right)^2\left[\sum_{i=1}^{R}\alpha_i + 1\right]}.$$

The formulas for the conditional variance-covariance matrix, Var-Cov$[\mathbf{S}_R(\mathbf{X}_{N-n}) | (\mathbf{S}_R(\mathbf{X}_n) = \mathbf{s}_R(n))]$ are similar, requiring only the replacement of each $\alpha_i$ by $\alpha_i + s_i(\mathbf{x}_n)$.

The simplicity of these sequential conditioning formulas derives from the fact that the Dirichlet-multinomial, or Polya, distribution is natural conjugate for quantities regarded as infinitely exchangeably extendable. We conclude this section with a formal statement.

**Theorem 5.14.** If you regard the quantities $X_1, \ldots, X_N$ exchangeably and your probability distribution for their category sum is the infinitely exchangeably extendable Dirichlet-multinomial distribution,

$$\mathbf{S}_R(\mathbf{X}_N) \sim D\text{-}M(N, \theta_1, \ldots, \theta_{R-1}; \alpha_1, \ldots, \alpha_R),$$

then your conditional distribution for the category sum of the remaining quantities, $\mathbf{S}_R(\mathbf{X}_{N-n})$, given the observation of any $n \leqslant N$ quantities, $(\mathbf{X}_n = \mathbf{x}_n)$, is also Dirichlet-multinomial:

$$(X_{n+1}, \ldots, X_N | (\mathbf{X}_n = \mathbf{x}_n)) \sim D\text{-}M(N - n, \theta_1, \ldots, \theta_{R-1}; \alpha_1 + s_1(n), \ldots, \alpha_R + s_R(n)),$$

where $\mathbf{s}_R(n)$ is the category sum $\mathbf{S}_R(\mathbf{x}_n)$.

## 5.6 DISTRIBUTIONS ADMITTING SUFFICIENT STATISTICS OF FIXED DIMENSION: THE MIXTURE EXPONENTIAL FAMILY AND ITS USES

This concluding section motivates the study of a large family of parametric mixture distributions, called the exponential family, through its connections to the theory of sufficient statistics of fixed dimension. Although proofs of details are beyond the scope of this text, you may like to be familiar with the content of this theory as you study the many useful distributions that are subfamilies of this family. We study some of them in great detail in Chapter 7 on the Normal distribution function and its gamma mixtures.

First we should notice that any p.m.f. for a quantity $X$ with the realm $\mathscr{R}(X) = \{x_1, x_2, \ldots, x_R\}$ defined by

$$f(x) = P(X = x) = \prod_{i=1}^{R} \theta_i^{(x = x_i)} \qquad [x \in \mathscr{R}(X)],$$

can be rewritten in a form based on $\exp[\ln(f(x))]$ as

$$f(x) = \exp\left[\sum_{i=1}^{R} (x = x_i) \log(\theta_i)\right] \qquad [x \in \mathscr{R}(X)],$$

a functional form which has widespread application.

For example, we know that any distribution function for a sequence of quantities regarded as infinitely exchangeably extendable can be written as

$$dF(\mathbf{x}_N) = \int_0^1 \cdots \int_0^1 \exp\left[\sum_{i=1}^{R} S_i(\mathbf{x}_N) \log(\theta_i)\right] \; d_1 \cdots d_{R-1} F(\theta_1, \ldots, \theta_{R-1}).$$

This representation is achieved by rewriting the mixture-multinomial mass function $f(\mathbf{x}_N)$ as $\exp[\ln(f(\mathbf{x}_N))]$. In this form, the mixture-multinomial distribution can be recognized as a member of the mixture-exponential family, defined as follows.

***Definition 5.21.*** Any distribution function of $N$ variables, representable by

$$dF(\mathbf{x}_N) = \int_{\Theta_1} \cdots \int_{\Theta_R} g(\theta_1, \ldots, \theta_R) h(\mathbf{x}_N) \exp\left[\sum_{i=1}^{R} \theta_i T_i(\mathbf{x}_N)\right] \; d_1 \cdots d_R M(\theta_1, \ldots, \theta_R)$$

is called a member of the *mixture-exponential family of distributions.* We write $\mathbf{X}_N \sim M\text{-}\exp[T_1, \ldots, T_R; \theta_1, \ldots, \theta_R]$ to denote this. The vector of statistics $\mathbf{T}_R(\mathbf{X}_N) = [T_1(\mathbf{X}_N), \ldots, T_R(\mathbf{X}_N)]$ is called a *sufficient statistic of fixed dimension.*

□

The importance of the exponential family of distributions derives from its role in characterizing distributions that specify sufficient statistics of fixed dimension. Let us state this formally as an unproved theorem in two parts, and then comment on its relevance for the use of mixture-exponential forms in specifying approximately your distribution for a vector of quantities $\mathbf{X}_N$.

**Theorem 5.15.** If $\mathbf{X}_N \sim M\text{-exp}[\theta_1, \ldots, \theta_R, T_1(\cdot), \ldots, T_R(\cdot)]$, then for any $n \leqslant N$, the $R$-dimensional vector of statistics $\mathbf{T}_R(\mathbf{X}_n)$ is sufficient for inference about the remaining $N - n$ quantities $X_{n+1}, \ldots, X_N$.

The proof of this part is easy. The precise statement, as well as the proof, of the following converse is more difficult.

**Theorem 5.16.** If a joint probability mass function or approximating density function can be written in terms of some parametric mixture

$$f(\mathbf{x}_N) = \int_{\boldsymbol{\Theta}} f(\mathbf{x}_N; \boldsymbol{\theta})\, \mathbf{d}M(\boldsymbol{\theta}),$$

where the domain of support for $f(\cdot)$ does not vary with $\boldsymbol{\theta} \in \boldsymbol{\Theta}$, and certain other regularity conditions hold, then sufficient statistics of fixed dimension, $\mathbf{T}_R(\cdot)$, can be determined only if $f(\mathbf{x}_N; \boldsymbol{\theta})$ is a member of the exponential family.

Details of the regularity conditions mentioned for the discrete mass functions can be found in Andersen (1970). Continuous results of this type were developed almost simultaneously (early 1930s) in France, Australia, and the United States by Darmois, Pitman, and Koopman. Recent important related works are those of Diaconis and Freedman (1984) and Lauritzen (1988).

In some applications, even where the assessment of exchangeability is quite reasonable, the data reduction afforded by sufficiency of the category sum, described in Section 5.5 is not too helpful. For example, if you regard the annual milk production of California's grade Holstein cows exchangeably, you can reduce the information contained in observations of any subgroup of them into no more than 30,000 statistics via the histogram of their yields. For in the United States today, milk output is still measured in units of pounds. A Holstein who gives less than 10,000 pounds milk in a year would be extremely lucky to survive the butcher's block, while yields of more than 40,000 pounds are rare indeed. [Though supercow cometh. The record 365-day milk yield of 60,380 pounds (27,445 kg) was produced during 1994–1995 by Bell-Jr Rosabell-ET, a Holstein cow in the Miramont Dairy herd of Wayne and Cathy Gillespie of Calhan, Colorado. *Hoard's Dairyman*, **140**(16).] Thus, a range of 30,000 categories would exhaust the relevant realm of a cow's milk yield, if the highest category is understood to represent any observation of 40,000 pounds or more. In this way, any number of observations can be reduced to 30,000 statistics without any loss of information to those who regard the yields exchangeably. But this is little comfort. For assessing one's probability distribution for the classification of 300,000 milk yields (the approximate number of grade Holstein dairy cows in California) into even 30,000 categories would be a formidable project. Can't we reduce still further the number of statistics that could sufficiently summarize the data?

One tack to follow would be to merge irrelevant category distinctions, keeping track only of the number of cows in yield categories to the nearest 100 pounds. This would reduce reduce the number of statistics to a mere 300. (Though mentioned merely as an aside here, this is an important feature of any applied statistical analysis that is worthy of another sentence. Statistical analysis of substantive problems can often be simplified greatly by ignoring digits of measurement that are insignificant relative to the issues at hand. Think, for example of the U.S. gross national product, which runs in the trillions ($10^{12}$ of dollars and which is recorded to the nearest 100 millions of dollars.) But can we reduce the dimension of the sufficient statistic still further?

A second tack in achieving greater reduction would be by admitting further characteristics of your probability distribution. One characteristic, for example, which implies exchangeability and more, is the property of spherical symmetry.

***Definition 5.22.*** A distribution function of $N$ variables is said to be *spherically symmetric* about the center, $C$, if it specifies a mass or density function that is constant over all vectors in its domain lying on the same sphere centered at $C$: $f(\mathbf{x}_N) = g[\sum_{i=1}^{N} (x_i - C)^2]$. □

Spherical symmetry would feature in your opinions regarding a vector of quantities $\mathbf{X}_N$ in any case where you expect them to be "somewhere around $C$," and your probability that the quantity vector equals any of its possibilities lying on the same sphere centered at $C$ is the same. Such an attitude may well apply to your opinions about grade Holstein milk yields in California.

If you recognize your opinions about $\mathbf{X}_N$ as spherically symmetric, Theorem 5.14 implies that under the further conditions alluded to, your p.m.f. or approximating p.d.f. must have the form

$$f(\mathbf{x}_N) = \int_{\Theta_1, \Theta_2} g(\theta_1, \theta_2) h(\mathbf{x}_N) \exp\left[ \theta_1 \left( \sum_{i=1}^{N} x_i \right) + \theta_2 \left( \sum_{i=1}^{N} x_i^2 \right) \right] d_1 d_2 M(\theta_1, \theta_2).$$

The exponential component of this function can be rewritten in a way that identifies the distribution as the digital analogue of a mixture-Normal distribution, a member of the exponential family that we study in detail in Chapter 7. Further remarks are deferred until then. The sum and the sum of squares of the observations are the sufficient statistics of fixed dimension, 2.

We study the Normal and gamma subfamilies of the exponential family in Chapter 7. We conclude this chapter with a brief presentation of the mixture-Poisson family as a subfamily, and the mixture-Weibull family of distributions which is not a subfamily of the exponential family. The presentations shall be mainly technical, accentuating the mixing feature of the analysis which identifies the relevant probability distribution for observable quantities. The inclusion of this material is largely to provide a technical completion of mathematical details of distributions that are commonly studied in a second-year university course in mathematical probability. A compilation of similar results in various distributional contexts appears in the work of Bernardo and Smith (1994), which is worth examining on may other counts as well. The text of Lee (1989) presents the distributions at a useful, more elementary level.

### 5.6.1 Gamma Mixtures of Poisson Distributions Are Negative Binomial

How many customers will enter your retail jewelry shop between 4 P.M. and 5 P.M. on a Friday afternoon before a nonholiday weekend? How many errors of information transmission will occur in a megabyte of boosted signals at a relay station? How many trout will you catch by fly-fishing between 7 A.M. and 7:10 A.M. in Lake Seneca of the Wind River Range on a summer morning? How many times will a recognizable activity occur during a specified duration of time under conditions that do not seem to change very much? Considered opinions about such questions are often approximated as mixtures of Poisson distributions, whose complete analysis involves extensive study. You can gain an idea of the extent of it by perusing the text of Johnson, Kotz, and Kemp (1992). This volume along with two others by Johnson and Kotz (1970, 1972) constitute a classic reference on distribution functions, discrete and continuous. We shall be brief. The honest undergraduate may well need a teacher as a guide, both to help you through some details and to embellish the presentation by drawing some comparisons between various functions.

***Definition 5.23.*** $X \sim \text{Poisson}(\lambda > 0)$ denotes that a probability mass function for $X$ is asserted as

$$f_X(x) = \exp(-\lambda)\lambda^x/x! \qquad x \in \{\text{nonnegative integers}\}. \qquad \square$$

The Poisson distribution is perhaps the simplest member of the exponential family of distributions. The Poisson probability mass function is generated by the terms of the Maclaurin series expansion of the function $\exp(\lambda) = \sum_{x=0}^{\infty} \lambda^x/x!$. You can determine for yourself that it has the moments $\mathbb{M}_1(F) = \lambda$, $\mathbb{M}_2(F) = \lambda^2 + \lambda$, and $\mathbb{M}_2^*(F) = \lambda$ and the moment generating function $M_F(t) = \exp[\lambda(e^t - 1)]$.

The Poisson probability mass function can be generated as the solution to a system of differential equations for the function

$$f(x; t) \equiv P(x \text{ instances of the activity occur in a time interval of length } t).$$

The conditions on the function are that for any $t > 0$ and for any nonnegative integer $x$, for small $\Delta t$,

1. $f(x + 1; t + \Delta t) - f(x; t) = \alpha \Delta t$ for some specified value of $\alpha$.
2. $P(\text{more than one occurrence between } t \text{ and } t + \Delta t) \approx 0$.

3. Occurrences of the activity during nonoverlapping time intervals are regarded independently (having specified the value of $\alpha$).

The Poisson($\lambda = \alpha t$) mass function satisfies these conditions. See Freund (1971).

Let us next introduce the generalized negative binomial distribution, whose relevance to the Poisson distribution will be made apparent shortly. There are several forms in which the negative binomial distribution appears in statistical literature. We define it here in the functional form that is most appropriate to our application.

***Definition 5.24.*** $N \sim$ negative binomial($r > 0$, $\theta \in [0, 1]$) denotes that a probability mass function for $X$ is asserted as

$$f_N(n) = [\Gamma(n)/\Gamma(r)\Gamma(n - r + 1)]\theta^r(1 - \theta)^{n-r} \qquad \text{for } n \in \{\text{nonnegative integers} \geqslant r\}$$

$$= {}^{(n-1)}C_{(r-1)}\theta^r(1 - \theta)^{n-r} \qquad \text{in the special case that } r \text{ is an integer.}$$

□

The negative binomial probability mass function is usually developed (in the special case of its second form) as the mass function associated with the number of events regarded independently (with probability $\theta$) that will be observed before the sum of the said number of events equals $r$. (Thus, $N$ may not be less than $r$.) But, as we shall soon see, this mass function (in its first form) can represent more interesting opinions as well. You can easily determine its moments, $\mathbb{M}_1(F) = r/\theta$, $\mathbb{M}_2(F) = r(r + 1 - \theta)/\theta^2$, and $\mathbb{M}_2^*(F) = r(1 - \theta)/\theta^2$. The moment generating function is more difficult to derive, but the $k$th moment can be found directly: $\mathbb{M}_K(F) = (r/\theta)^{K-1}[(r + 1 - \theta)/\theta]$.

The gamma distribution will be presented in brief detail in Section 7.4.5.1. You should study that section now to prepare for the following theorem.

**Theorem 5.17.** A gamma-Poisson mixture distribution, $\Gamma$-Poisson($\lambda; \alpha, \beta$) is equivalent to a negative binomial[$r = \alpha, \theta = \beta/(\beta + 1)$] distribution.

*Proof*. Using the result the $f(x) = \int_0^\infty f(x|\lambda)f(\lambda)\,d\lambda$, it is apparent that for nonnegative integers $x$,

$$\begin{aligned}
f(x) &= [\beta^\alpha/\Gamma(\alpha)] \int_0^\infty [\exp(-\lambda)\lambda^x/x!]\lambda^{\alpha-1}\exp(-\beta\lambda)\,d\lambda \\
&= [\beta^\alpha/\Gamma(\alpha)x!] \int_0^\infty \lambda^{x+\alpha-1}\exp(-(\beta+1)\lambda)\,d\lambda \\
&= [\beta^\alpha/\Gamma(\alpha)x!]\,\Gamma(x+\alpha)/(\beta+1)^{(x+\alpha)} \\
&= [\Gamma(x+\alpha)/\Gamma(\alpha)x!]\,[\beta/(\beta+1)]^\alpha[1/(\beta+1)^x].
\end{aligned}$$

This can be recognized as the negative binomial mass function

$$f_Y(y) = [\Gamma(y)/\Gamma(\alpha)\Gamma(y-\alpha+1)]\theta^\alpha(1-\theta)^{y-\alpha}, \qquad y \in \{\text{nonnegative integers} \geqslant \alpha\},$$

using the transformation $y = x + \alpha$ and $\theta = (\beta/(\beta+1))$. □

In using a Poisson distribution to approximate your opinions in applied problems, one is typically in a quandary as to an appropriate specification of $\lambda$. In such a case, it may well be that a mixture-Poisson distribution is more appropriate. It turns out that if a quantity $X$ is regarded exchangeably with a large number of other quantities, and assessed with a gamma-Poisson mixture distribution (or if your opinions about $X$ are merely exchangeably extendable to a large extent) then your probability distribution for the average of the quantities (or your imagination regarding the extended average) is indistinguishable from your mixing function on the parameter structure $\lambda$. (An argument to this effect will be made in the context of mixture-Normal distributions in Chapter 7. It mimics the argument that your mixing distribution on $\theta$ for events you regard exchangeably is indistinguishable from your probability distribution for the proportion of the events that equal 1.) If a member of the gamma family of densities approximates well enough your opinion regarding the average of these quantities, you would find Theorem 5.17 appropriate to your opinion about $X$. A complete specification of such opinions regarding the entire sequence $\mathbf{X}_N$ is contained in the following definition and discussion.

***Definition 5.25.*** Suppose $\mathbf{X}_N$ is a vector of quantites whose realms are the nonnegative integers. To write $\mathbf{X}_N \sim$ gamma-Poisson$(N, \lambda; \alpha, \beta)$ denotes that you regard them exchangeably, asserting your joint probability mass function as

$$f(\mathbf{x}_N) = [\beta^\alpha/\Gamma(\alpha)] \int_0^\infty \lambda^{\alpha-1} \exp(-\beta\lambda) \prod_{i=1}^N [\exp(-\lambda)\lambda^{x_i}/x_i!]\, d\lambda$$

for any vector $\mathbf{x}_N$ whose components are nonnegative integers. □

In this form, the gamma-Poisson mixture is recognizable as a member of the exponential family of distributions. With some algebraic simplification and the recognition of the gamma integral, the p.m.f. can be simplified to

$$f(\mathbf{x}_N) = \frac{\Gamma(\alpha + \sum \mathbf{x}_N)\beta^\alpha}{\Gamma(\alpha) \prod_{i=1}^N x_i!(\beta+N)^{(\alpha+\sum \mathbf{x}_N)}},$$

where $\sum \mathbf{x}_N$ denotes the sum of the vector components. Evidently, this function is constant over permutations of its arguments, since the sum of the $\mathbf{x}_N$ vector components and the product of their factorials are both constant over permutations.

Because of the exchangeability feature of this distribution, any subvector of the $N$ quantities (the final $N-n$ of them, for example) are also assessed with a gamma-Poisson($N-n, \lambda; \alpha, \beta$) distribution. [Merely replace the $N$'s in the mass function $f(\mathbf{x}_N)$ with $(N-n)$'s.] Furthermore, the conditional mass function for the final $N-n$ components given the values of the first $n$ can be derived as

$$f(\mathbf{x}_{N-n}|\mathbf{X}_n=\mathbf{x}_n)=\frac{\Gamma(\alpha+\sum\mathbf{x}_n+\sum\mathbf{x}_{N-n})(\beta+n)^{(\alpha+\Sigma\mathbf{x}_n)}}{\Gamma(\alpha+\sum\mathbf{x}_n)\prod_{i=n+1}^{N}x_i!(\beta+N)^{(\alpha+\Sigma\mathbf{x}_n+\Sigma\mathbf{x}_{N-n})}},$$

which is the form of a gamma-Poisson($N-n, \lambda$; $\alpha+\sum\mathbf{x}_n, \beta+n$) mass function. Thus, the gamma-Poisson distribution is a natural conjugate form for opinions regarding quantities of this sort. In the case that $N=n+1$, both the prior and posterior (conditional on $\mathbf{X}_n=\mathbf{x}_n$) p.m.f. for $X_{n+1}$ can be recognized as negative binomial. Notice that in any case, the sum of the conditioning quantities is sufficient for inference regarding the remaining quantities.

## PROBLEM

**1.** Consider the lake-trout-fishing question that introduced this section. Suppose you regard exchangeably the number of fish caught between 7 A.M. and 7:10 A.M. on the mornings after full moons in July during the decade of the 1990s. Suppose further that you assess your joint p.m.f. for the numbers of fish caught in the form gamma-Poisson($N, \lambda; \alpha, \beta$). Considering your opinions regarding the average number of fish caught on many days of the sort you are imagining, you assess your mixing function as gamma($\alpha=2.5, \beta=1.087$).

- **a.** Sketch the density function for your mixing distribution over $\lambda$.
- **b.** Determine the values of your p.m.f. for the number of fish you will catch on the first attempt.
- **c.** Determine the values of your p.m.f. for your (i) fourth attempt conditional on the first three attempts yielding 1, 3, and 0, and then conditional on their yielding 4, 2, and 7; (ii) seventh attempt conditional on the first six yielding 1, 3, 0, 0, 1, 3, and then conditional on their yielding 4, 2, 7, 7, 4, 2, and finally conditional on their yielding 4, 0, 2, 1, 7, 3. If each person in your class computes two of these proposals, you can merge your computational results for interesting comparisons.
- **d.** Sketch the density function for your mixing distribution over $\lambda$ conditional on any two of the catch conditions mentioned in part c.
- **e.** Determine the values of your joint p.m.f. for the fish caught on your first two attempts, over some small array of possibilities. Recognize that the function values will be symmetric over this array about its main diagonal.

### 5.6.2 Mixture-Weibull Distributions Are *Not* in the Exponential Family

Of course, not every useful probability distribution is a member of the exponential family. One of these which merits your familiarity on account of its frequent application in assessing the reliability of machinery is the family of Weibull distributions.

***Definition 5.26.*** The function $f: \mathbb{R}^+ \to \mathbb{R}^+$ defined by

$$f(t) = (\beta/\alpha)(t/\alpha)^{\beta-1} \exp[-(t/\alpha)^\beta] \qquad (t \geqslant 0),$$

is said to be a *Weibull density function* identified by numerical values of the parameters $\alpha > 0$ and $\beta > 0$. The associated function

$$F(t) = \int_0^t f(x)\,dx = 1 - \exp[-(t/\alpha)^\beta] \qquad (t \geqslant 0)$$

is called the *Weibull distribution function.* We denote this by $T \sim \text{Weibull}(\alpha, \beta)$. □

The Weibull family members, one for each pair $(\alpha, \beta) \in (\mathbb{R}^+, \mathbb{R}^+)$, are continuous distribution functions whose mixtures are commonly used to represent opinions regarding the duration of the reliable lifetime of a manufactured item. How many complete washing cycles will this dishwasher provide before it fails (for, say, leaking gaskets, broken water pump, burned-out motor, short circuit in the electrical system, damaged rotating water sprinkling blade, etc.)? The text of Martz and Waller (1982) provides a useful introduction to statistical problems of reliability, with specific attention to the mixture-Weibull family in these applications.

Technically, one feature of the Weibull family to recognize is that it shares some family members with the family of gamma distributions. The density functions shared by both families are the simple exponential densities, parameterized in the Weibull case by specifying by $\beta = 1$. But whereas the entire gamma family of distributions is a subfamily of the mixture-exponential distributions of Definition 5.15, the Weibull family members with $\beta \neq 1$ are not in this class. The consequence of this is that mixtures of Weibull distributions therefore do not afford sufficient statistics of fixed dimension. The following definition will make this clear.

***Definition 5.27.*** Suppose $\mathbf{T}_N$ is a vector of nonnegative quantities. To write $\mathbf{T}_N \sim \text{mixture-Weibull}(N, \alpha, \beta; M(\cdot))$ denotes that you regard these quantities exchangeably, asserting your approximate joint density function as

$$f(\mathbf{t}_N) = (\beta/\alpha)^N \int_0^\infty \int_0^\infty \prod_{i=1}^N (t_i/\alpha)^{\beta-1} \exp[-(t_i/\alpha)^\beta]\,dM(\alpha, \beta)$$

for any vector $\mathbf{t}_N$ whose components are nonnegative integers. □

Evidently, this density function is constant over permutations of its arguments, since the product is constant over permutations of components of $\mathbf{t}_N$. Algebraic simplification yields the alternative form

$$f(\mathbf{t}_N) = (\beta/\alpha)^N \alpha^{-N(\beta-1)} \int_0^\infty \int_0^\infty [\pi(\mathbf{t}_N)]^{\beta-1} \exp\left[ -\alpha^{-\beta} \sum_{i=1}^{N} t_i^\beta \right] dM(\alpha, \beta)$$

where $\pi(\mathbf{t}_N)$ denotes the product of the components of $\mathbf{t}_N$. In this form it is evident that the density is not in the mixture-exponential form: not only is the functional form not an exponential function of a linear function of the parameters and their sufficient statistics, but the the density does not even admit sufficient statistics. While the product function $\pi(\mathbf{t}_N)$ surely defines a statistic, the remaining factor in the density, $\exp[-\alpha^{-\beta}\sum_{i=1}^{N} t_i^\beta]$, cannot be decomposed into separable functions of statistics and parameters. The value of the statistic $\sum_{i=1}^{N} t_i^\beta$ depends algebraically on the parameter value, $\beta$. Thus, the mixture-Weibull family of distributions provides a counterexample of how the exponential family of distributions and the theory of sufficient statistics go hand in hand.

Computationally, the absence of sufficient statistics for the Weibull family requires some sophistication in computational strategies to provide sequential forecasting distributions based on data accumulation, in an efficient way. Again, the text of Martz and Waller (1982) provides a readable introduction.

Our study of probability distributions and their approximations will conclude in Chapter 7 with an extensive examination of the multivariate Normal distribution and its mixtures, focusing on exchangeable judgments as a unifying theme. Theoretical, philosophical, and computational issues that have been presented tersely here will receive more explicit attention in that chapter. By presenting distribution theory in the way I have I hope you will be able to make use of the extensive technical analyses of probability distributions that are typically described from another viewpoint, that of observing independent outcomes of random variables generated according to some distributional process. Our focusing on exchangeable judgments leads us to dwell more extensively on the theory of mixture distributions than is customary.

CHAPTER 6

# Proper Scoring Rules

*Some see the flesh before the bones, and some see the bones before the flesh, and some never see the bones at all, and some never see the flesh at all, never never see the flesh at all.* *Watt*, SAMUEL BECKETT

## 6.1 INTRODUCTION

We refer to prevision and probability in this text exclusively as a representation of your personal uncertain knowledge about the numerical values of measured quantities. To a scientist who is interested in using the statistical methods of inference supported by this operational subjective theory, several questions may well arise: "Probability only represents my state of knowledge about the world? You mean I may think anything I want to think—and that is fine as long as I am coherent? What about public scientific knowledge? Is this merely a matter of opinion? Aren't there scientific facts that we know to be the case from objective scientific procedures of experimentation and conclusion via "scientific method" —for example, Boyle's law of gases that temperature equals pressure times volume, or Ohm's law that voltage equals current times resistance? What about 'causal laws' which allow us to design and build machinery to accomplish the tasks we set out for it?"

Operational subjective statistical theory indeed answers these questions. Although the answers defy the common contemporary understanding of "scientific method" alluded to, supporters of the operational subjective theory conclude that the resolutions it provides to these issues actually synthesize the understanding gained during this century from our major scientific achievements.

The viewpoint of the operational subjective theory is that scientific method is a procedure in which scientists take some care in gauging their activities in the world, in recording measured features of them, and in reflecting on their observations. The use of operationally defined measurements of gauged activities and observations allows scientists to achieve a mode of objectivity in reporting their results by means of statistics. Once the measurements are recorded, anyone who believes that they were not created as a hoax will at least conclude that whatever was happening, its measurement according to the specified operationally defined procedures is represented by the recorded numbers. But what to infer from known recorded measurements about the values of yet unrecorded measurements is the responsibility of each "knowing" person.

In fact, people often agree to a great extent on what is to be inferred. Moreover, for particular quantities we are sometimes willing to turn over our responsibility for knowing to particular people, to "scientists," whose experience in the relevant domain far exceeds our own, and whose judgment we trust. This happy state of affairs is called a scientific consensus. There is general agreement in evaluating the evidence that the measured observations provide for yet unmeasured experiences. Even then there may be disagreement on minor details among those who largely agree.

Nonetheless, in the history of science, as well as on notable contemporary matters, there have been numerous well-known controversies among groups of very respected scientists regarding what conclusions are to be drawn from our observational experience. Thus, it is a crucial question whether any distinction can be made between the *validity* of the uncertain knowledge proclaimed by different people who assert different previsions for particular quantities?

In addition to providing an understanding of scientific consensus, the operational subjective statistical theory provides a method for evaluating competing theories empirically according to objective procedures that can be regarded as "fair" by anyone who subscribes to the possibility of resolving theoretical disputes by empirical observation. This chapter describes the logic of this method, called scored sequential forecasting. Readers who are familiar with statistical theory as currently practiced should be aware from the start that the developments in this chapter motivate computational evaluative procedures that are meant to replace the activities of "hypothesis testing" and "model fitting" or "parameter estimation" in which statisticians have engaged for the past half-century or so. The seriousness and care with which you study this chapter ought to be weighted by the recognized importance these activities are still accorded in most practical scientific work.

For despite the appeal of theoretical work on proper scoring rules and their properties, specific applications in empirical work have not been extensive to date. In addition to the experimental applications reviewed by Cooke (1991) and Lichtenstein, Fischhoff, and Phillips (1982), the applied papers by Blattenberger (1996), Blattenberger and Lad (1986, 1988), Hill (1994b), use them explicitly in the empirical assessment of scientific controversies. There is still much to learn about the procedures that can only derive from practice. I hope that the presentation here will motivate their broader application. We study an example in Chapter 8.

## 6.2 SCORING RULES

Operational subjective statistical theory insists that there are no "objectively correct" previsions for unknown quantities. There are only *your* and *my* and *our* previsions, representing each of our uncertainties in our personal states of knowledge. To the extent it could be said that that there exists a "correct" prevision for a quantity, this correct prevision must surely identically equal the quantity value itself. How could such a prevision assertion be considered to be incorrect? If $X = 6$, for example, but its value is unknown to you for certain, is there any sense in which the assertion $P(X) = 6$ could be considered incorrect? Thus, your prevision for any supposed "correct prevision" would have to be the same as your prevision for the quantity itself, rendering the concept of "correct" prevision as operationally useless. Rather than dwell on it, let us address an operationally meaningful question: How can we fairly distinguish better-informed states of knowledge from more poorly informed states?

The answer to this question lies in using proper scoring rules for evaluating states of uncertain knowledge. We are familiar by now with several forms of specified knowledge about $X$, which we might denote generically by $K(X)$. So $K(X)$ is your numerical assertion of knowledge about $X$ in some well defined form. For example, $K(X)$ may be your prevision $P(X)$, your variance $V(X)$, some higher moment assertion $\mathbb{M}_r(X) = P(X^r)$, your Mode($X$), your $p$th quantile $Q_p(X)$, or a vector of several of these. In fact, you may have asserted numerically even your complete distribution function, $F_X(\cdot)$. This amounts to your specifying your prevision for a vector of quantities, the events of the form $(X = x_i)$ for each $x_i \in \mathscr{R}(X)$. Let us begin by describing scoring rules generally as they pertain to any form of specified knowledge about a quantity, $K(X)$. Specific application to distributions and to vectors of quantities shall be made apparent when it is needed.

A scoring rule is a function of two variables that accords a real number (a score) to each pair of numbers that could represent the value of $X$ and the assertion value, $K(X)$. Thus, aside from approximate computations that we describe, the function is considered to be discrete relative to its first argument, the possible values of $X$, and continuous as a function of the second, the possible coherent assertion values, $K(X)$. Requirements on a scoring rule are that it achieves its maximum for each $x \in \mathscr{R}(X)$ when $K(X) = x$, and it is nonincreasing as $x$ departs from $K(X)$ for each $K(X)$, and as $K(X)$ departs from $x$ for each $x$. Formally, we make the following definition, which allows a wide range to the form of qualifying functions.

***Definition 6.1.*** Let $X$ be a quantity whose value is unknown to you. Let $K(X)$ be a number that you assert as your numerical assessment of your knowledge about $X$ in some generic form. Any function $S: \mathscr{R}(X) \otimes [\min \mathscr{R}(X), \max \mathscr{R}(X)] \to \mathbb{R}$ is called a *scoring rule* for $K(X)$ as long as for each $x \in \mathscr{R}(X)$, $S(x, K(X))$ achieves its maximum at $K(X) = x$, and for each $K(X) \in [\min \mathscr{R}(X), \max \mathscr{R}(X)]$, $S(x, K(X))$ does not increase as $x$ departs from $K(X)$ in either direction. □

Graphically, a scoring function of two variables exhibits a ridge along the line $x = K(X)$.

Scoring functions are defined here so that a large score for an assertion is preferred to a small score. According to this convention, scores are to be interpreted as "gains" rather than as "losses." The best score is achieved when knowledge of $X$ is asserted as the value of $X$ itself, whatever that might be. Most of the scoring rules we study also exhibit an arbitrary normalizing feature, that $S[x, K(X) = x] = 0$ for each $x \in \mathscr{R}(X)$. These simplifications will ensure that we do not get tripped up in theoretical statements by confusing whether someone should assert knowledge about $X$ in a way that would maximize or minimize their expected (prevised) score. In practice, reported scores are sometimes published in a form where the scoring rule represents not a positive gain, but a punitive loss. "Squared error loss" is a common example. At any rate, results are interpretable as gains or as losses simply by algebraically negating losses to treat them as gains, and vice versa.

Mathematically, many functions satisfy the criteria for a scoring rule specified by Definition 6.1, but not every scoring rule would be regarded as providing a fair assessment of the assertion $K(X)$. If you were asked to assert your prevision, for example, knowing that your assertion were to be scored according to the rule $S(X, P(X))$, is it possible that you might privately assess your prevision value numerically as $P(X) = p$ but desire to announce publicly instead some other number as your "$P(X) = q$," in the expectation that you would achieve a better score by announcing this number rather than your actual assessment? In order to preclude such misleading intrigue in the meaningful use of prevision assertions in communicating our knowledge in science and commerce, the operational subjective theory provides that knowledge assertions should be scored according to "proper" or "fair" scoring rules, defined as follows.

***Definition 6.2.*** A scoring rule for the assertion $K(X)$ is said to be a *proper scoring rule*, or fair scoring rule, if your prevision for your numerical score $S[X, K(X)]$ is the greatest when you assert as $K(X)$ your honest assessment of that value; that is, if your $P(S[X, K(X)]) \geqslant P[S(X, Q)]$ for every $Q \neq K(X)$. It is presumed that your prevision, upon which you base your determination that the scoring rule is proper, is coherent. □

The use of a proper scoring rule for evaluating the relative quality of different states of uncertain knowledge is said to encourage scientists' honesty and accuracy in assessing numerical values for their $K(X)$'s. When scored according to a proper rule, it is to everyone's advantage to assert your $K(X)$ accurately as the numerical value of your own assessed knowledge about $X$. Every person announcing a $K(X)$ would regard the scoring procedure as fair. Proper scoring rules are devised so that no one previses (expects) that intrigue in falsely announcing some other number as "$K(X)$" would be advantageous for achieving a better score.

In the context of general decision problems, the use of proper scoring functions is equivalent to the use of utility functions in evaluating your experience of $X$ while in the state of uncertain knowledge represented by $K(X)$. If you were a farmer, for example, your planting preparations for the current growing season would be different if your prevision for the 90-day rainfall measured in millimeters is $P(X) = 150$, than they would be if you were prevising $P(X) = 100$ or $P(X) = 200$. Furthermore, in each of these three proclaimed states of information regarding rainfall, your experience would be quite different if you would subsequently live through a history summarized, say, by the measurement $X = 57$. The particular scoring functions we study in the generic form of $S(X, K(X))$ can also be considered as utility functions, $U(X, K(X))$. The principle of asserting numerical values for $K(X)$ in such a way as to maximize your prevision for your score is equivalent to acting in a decision problem so as to maximize your expected utility.

To the extent that you are familiar with the mathematics of decision theory, you will recognize some technical relations between scoring rules and utilities in the specific algebraic forms of scoring rules we shall study. Connections between the concepts of utility, scoring rules, and information have been discussed extensively throughout the writings of Good (1950, 1957, 1971, 1983) and Buehler (1971). The articles by Bernardo (1979) and DeGroot (1984) add technical discussion. Before addressing details of specific functions, let's think a bit.

## 6.3 A BRIEF DETOUR: ON DEFERRING TO AN EXPERT

Let us not be misunderstood at the start. Consider the following objection: "Myself, I know very little of value about the measurable conductive capacity of a cylinder of single silicon crystal that has been constructed using a 'Bridgeman technique.' Thus, I would certainly prefer to specify not my own assessment $K(X)$, but rather I would announce the evaluation of my brother as my $K(X)$, since he is an expert on these matters. You could never devise a scoring rule that would make me expect to achieve a better score when announcing my prevision, say, for this quantity than in announcing my brother's prevision as my own."

The subjectivist response to such a challenge should clarify the purpose of proper scoring procedures for the empirical evaluation of uncertain knowledge. If your brother would tell you his prevision, for example, then you are perfectly welcome to specify his prevision as your prevision too. For the numerical value of his asserted prevision would then be part of your state of knowledge. This is what it means to rely on "experts" for all sorts of details when you are making decisions.

But suppose you call your brother on the phone, and he is not available to tell you his $P(X)$. Again I ask you, "What is your prevision for the conductive capacity of this semiconductor?"

Using the notation $P_Y$ to represent your previson assertions and $P_B$ to represent your brother's, you have allowed as how your $P_Y[X \mid P_B(X) = b] = b$, no matter what the value of $b$. Thus, your $P_Y(X)$ must equal your $P_Y[P_B(X)]$, as required by the theorem on total prevision. In brief, this coherency condition requires that your

$$P_Y(X) = \sum_{b \in \mathscr{R}[P_B(X)]} P_Y[X|(P_B(X) = b)]P_Y(P_B(X) = b)$$
$$= \sum_{b \in \mathscr{R}[P_B(X)]} bP_Y(P_B(X) = b) = P_Y[P_B(X)].$$

Your prevision for what your brother would tell you as his $P_B(X)$ must be the same as your own previson for $X$!

The point of this detour is that your merely realizing that there is an expert whom you would trust with specifying a prevision gives you no help in identifying your own prevision unless you know the expert's asserted value.

This resolution of the issue of deferring to an expert can be applied further to the following conundrum. Suppose you had said, "my prevision is the same as the prevision of either my brother or my father, who is also an 'expert' on synthetic crystal products and their properties". Coherency would then require your previsions for their two previsions to be equal, even though you may be fairly sure that their two previsions are not actually equal. Now what if you are unable to contact either of them, but when they return your phone calls, each gives you a different number as his prevision for $X$? Then what would you say is your $P(X)$ in this state of knowledge?

A complete discussion of this interesting question would be distracting just now, so I merely offer you a provocative technical problem below. But it has been discussed extensively in statistical literature for practical reasons. What should you do when several "experts" on whom you rely disagree in their assessments of a problem that concerns you? Happily, the resolution to this question would eventually turn you back to the topic of this chapter, which is what we do now. You would want to study a record of your experts' assessments of related quantities as scored by proper scoring rules.

Rather than continue this discussion here, let us conclude this detour with a list of interesting, though nonexhaustive, references on group decision making: de Finetti (1951, 1952, 1954), Stone (1961), Raiffa (1968), DeGroot (1974), McConway (1978, 1981), French (1980, 1985), R. L. Berger (1981), Lindley (1982) Cooke (1991), and Genest, McConway, and Schervish (1986). This final reference is somewhat synthetic, and contains important technical results.

## PROBLEMS

**1.** Suppose the components of $\mathbf{X}_3$ are logically independent, $\mathscr{R}(X_i) = \{1, 2, 3\}$, and that you regard them exchangeably. If you assert further that $P(X_1|(X_2 = r)) = P(X_1|(X_3 = r)) = r$ for each $r \in \mathscr{R}(X_i)$, determine the bounds coherency would require of your assertions $P(X_1|(X_2 = r_2)(X_3 = r_3))$ for each pair $(r_2, r_3) \in \{1, 2, 3\}^2$. Use the fundamental theorem of prevision to solve the problem as a linear programming problem. You can think of this problem as a small caricature of the problem of disagreeing experts. Think of $X_1$ as a quantity of interest and of $X_2$ and $X_3$ as the previsions $P_2(X_1)$ and $P_2(X_1)$ to be asserted by two experts, where the units of prevision assertion are distinguished to the same scale as the units of $X_1$. The suppositions $P(X_1|(X_2=r))=r=P(X_1|(X_3=r))$ may be taken to represent the fact that you would agree with either of the experts singly. The previsions $P(X_1|(X_2=r_2)(X_3=r_3))$ under investigation would represent your conditional previsions supposing their disagreement in the form of $(X_2,X_2)=(r_2,r_3)$. Why would the presumption that you regard the three components of $X_3$ exchangeably be reasonable for this problem?

**2.** Expand problem 1 to address the situation of three experts and a larger realm for $X_i$, $R(X_i)=\{1,2,\ldots,r\}$.

## 6.4 RETURN TO THE TRAIL: A MULTIPLICITY OF PROPER SCORING RULES

As it turns out, there are many functions that qualify as proper scoring rules for any assertion about $X$, and each has its own distinguishing features. We study them in two groups: proper scores for previsions and their variants, and proper scores for distributions. In this section we study the mathematical forms of several scoring rules. In Section 6.5 we identify their distinguishing properties.

### 6.4.1 The Quadratic Score of Previsions

***Definition 6.3.*** The function $S(X,K)=-(X-K)^2$ is called the *quadratic score of* $K$, or the *Brier score of* $K$. □

The quadratic scoring function is clearly concave in $K$ and achieves is maximum at $S(x,K=x)$ for any $x\in\mathcal{R}(X)$. The name "Brier score" is in deference to the work of G. W. Brier (1950), who applied this score to probabilistic forecasts in meteorology.

**Theorem 6.1.** The quadratic scoring rule is a proper scoring rule for $P(X)$.

*Proof.* The quadratic function $S(X,K)$ is continuous and differentiable in $K$. Your prevision for the score achieved by the assertion of $K$ is

$$P[S(X,K)]=-[P(X^2)-2KP(X)+K^2].$$

Thus, its derivative is $D_KP[S(X,K)]=2P(X)-2K=0$ if and only if $K=P(X)$. Moreover, the concavity of $S$ over $K$ ensures the second-order conditions that $P[S(X,K)]$ achieves a maximum at $P(S[X,P(X)])$. □

**Corollary 6.1.1.** Your coherent prevision for the quadratic score accorded to your asserted $P(X)$ must equal $-[P(X^2) - P^2(X)] = -V(X)$.

If your variance for $X$ is large, your prevision for the quadratic score achieved by your $P(X)$ assertion is small (that is, negative and large in absolute value). Conversely, if your variance for $X$ is small, then your prevision for your quadratics core would still be negative, but near 0, the maximum possible quadratic score for any prevision.

This raises an interesting prospect. Your variance for $X$ is itself a prevision for the quantity $[X - P(X)]^2$, where $P(X)$ is your prevision for $X$. If you specify your prevision and your variance for $X$, then your assertions about $X$ define a vector $\mathbf{K}_2(X) = [P(X), V(X)]$. A vector of quadratic scores could be used to score both of these previsions:

$$\mathbf{S}(X, \mathbf{K}_2(X) = [P(X), V(X)]) = (-[X - P(X)]^2, -[(X - P(X))^2 - V(X)]^2).$$

The first component of the vector score is the quadratic score of your prevision for $X$. The second component of the vector score is the quadratic score of your prevision for the first component of the vector score. This expansion of your assessed knowledge can, in principle, be continued indefinitely, at least until your complete distribution has been assessed. For the second component of this vector score is also a well-defined quantity. Thus, your prevision for it, $P(-[(X - P(X))^2 - V(X)]^2)$, is your prevision for the quadratic score of your variance. This prevision, if specified, could itself be scored by a quadratic score, and so on.

### 6.4.2 The Linear-Logarithmic Score of Previsions for Positive Quantities

We mentioned that proper scoring rules are not unique. In this section we study a second proper score of previsions for positive quantities.

***Definition 6.4.*** Suppose $X$ is a quantity whose realm contains only positive numbers. The function $S(X, K) = 1 - [X/K - \log(X/K)]$ is called the *linear-logarithmic ratio discrepancy score of* $K$.

***Theorem 6.2.*** The linear-logarithmic ratio discrepancy scoring rule is a proper scoring rule for $P(X)$.

*Proof.* The linear-logarithmic scoring function is continuous and differentiable in $K$. For each possible value of $X$, its derivative is $D_K S(X, K) = XK^{-2}(1 - K/X)$, which equals 0 only when $K = X$. Moreover, the second derivative of $S(X, K)$ is negative at $K = X$, so the function achieves a maximum at $K = X$. Similarly, the discrete functions $S(X, K)$ achieve their maximum at $X = K$, and decline as $X$ departs from $K$ in either direction. Thus, this function qualifies as a scoring rule. Furthermore, your prevision for your score as a function of your announcement of $K$ is $P[S(X,K)] = 1 - P(X)/K + P[\log(X)] - \log(K)$. Thus, its derivative function is $D_K P[S(X,K)] = P(X)/K^2 - 1/K = 0$ if and only if $K = P(X)$. Moreover, the second derivative value there is also negative, ensuring that $P[S(X,K)]$ achieves a maximum at $P(S[X,P(X)])$. □

Whereas the quadratic score of $P(X)$ is symmetric in $X$ about $P(X)$, the linear-logarithmic ratio discrepancy score is evidently nonsymmetric. As an exercise, you should construct graphs of this scoring rule as a function of each of its two components.

**Corollary 6.2.1.** Your coherent prevision for the linear-logarithmic ratio discrepancy score achieved by your own prevision would be

$$P(S[X,P(X)]) = P[\log(X)] - \log[P(X)].$$

The maximum prevision that might be asserted for this score would be 0, occurring when you feel certain of the value of $X$. For in this case you must also assert $P(\log(X)) = \log(P(X))$. (Jensen's inequality of Section 4.3 is relevant, since the logarithm function is convex.)

## PROBLEMS

1. Let $Q$ be any number that does not equal your prevision for $X$. Show that for both the quadratic scoring function and for the linear-logarithmic ratio scoring function, $P[S(X,Q)] = P[S(X,P(X)) + S[P(X),Q]$. Of course the domain of $S$ would need to be expanded to $[\min \mathscr{R}(X), \max \mathscr{R}(X)]^2$ for the expression $S[P(X),Q]$ to always make sense. If you are familiar with the results of classical statistics, you should recognize that this result parallels the result that "mean squared error equals variance plus the square of the bias." Coupled with our knowledge that scoring rules satisfy the requirement that for any value of $A$ the function $S(A,B)$ achieves a maximum at $B = A$, this result will prove helpful if you extend your understanding of scoring issues by studying the seminal article on scoring rules by Savage (1971). He characterizes a scoring rule as "proper," using a formulation of the problem based on these two results.

2. Even though proper scoring rules are defined as proper in terms of your prevision for the score you would achieve, proper scoring rules can also be used to evaluate features of your opinions about a quantity other than your prevision for the quantity. Show that the scoring rule

$$S(X,K) \equiv -b[e^{a(K-X)} - a(K-X) - 1], \qquad \text{defined for } b > 0 \text{ and } a \neq 0,$$

is a proper score for $K^* = -(1/a)\log[P(e^{-aX})]$. Graph this scoring function for several values of $a$, choosing $b = 1$. Notice that this function is not symmetric about $K^*$. This so-called LINEX scoring function was developed for an applied decision problem by Varian (1975). Properties of the rule have been studied by Zellner (1986a).

**3.** Show that $S(E, K) \equiv \exp[(1/2 - E)K]$ is a proper score for your log odds in favor of $E$; that is, for $K^* = \log[P(E)/(1 - P(E))]$.

**4.** Show that $S(X, K) \equiv -|X - K|$ is a proper scoring rule for your median of $X$.

**5.** Show that $S(X, K) \equiv (X = K)$ is a proper scoring rule for your mode of $X$.

### 6.4.3 Proper Scores of a Probability Distribution for $X$

Suppose you assert your entire probability mass function for a quantity whose realm is $\mathscr{R}(X) = \{x_1, x_2, \ldots, x_N\}$. That is, you assert your probabilities for the constituents of the partition generated by $X$:

$$P(\mathbf{E}_N(X)) \equiv P[(X = x_1), (X = x_2), \ldots, (X = x_N)] = \mathbf{p}_N.$$

Of course coherency requires this vector to reside in the $(N - 1)$-dimensional unit simplex. In the generic notation for asserted knowledge about $X$ that we have been using for general discussion, your specification of knowledge about $X$ would be denoted by $K(\mathbf{E}_N(X)) = \mathbf{p}_N$. Theorem 6.3 will identify four distinct proper scoring rules for your probability mass function thus specified, on the basis of observing $X$, and thus of observing the vector $\mathbf{E}_N(X)$.

The motivation for the names of these scoring rules should be fairly obvious. The quadratic score of your distribution is the sum of the quadratic scores of the probabilities you assert for the constituents of the partition generated by $X$. Remember that one of those constituents will equal 1, and the rest will equal 0. The log score associates with your distribution only the logarithm of your probability for the constituent event that happens to equal 1. The denominator of the spherical score is identical for all vectors $\mathbf{Q}_N$ in $\mathbf{S}^{N-1}$ that lie on the surface of the same $N$-dimensional Euclidean sphere. The spherical score is a special case of the $\beta$-pseudospherical score for the specification $\beta = 2$.

Each function is represented in two algebraic forms. The first form, $S(X, \mathbf{Q}_N)$, represents the scoring function as computed from the value of X, whatever that might be. The second form, $S(X = x_i, \mathbf{Q}_N)$, represents the same function in terms of the specification that that $X$ equals this particular element of its realm, $x_i$.

**Theorem 6.3.** Let $X$ be a quantity with realm $\mathscr{R}(X) = \{x_1, x_2, \ldots, x_N\}$. Each of the following named scoring rules for any vector $\mathbf{Q}_N = (Q_1, Q_2, \ldots, Q_N)$ in the unit simplex $\mathbf{S}^{N-1}$ is a proper scoring rule for your probability mass function, $f_X(\cdot)$.

**Quadratic Score (Brier Score) of a Distribution**

$$S(X, \mathbf{Q}_N) \equiv -\left[\sum_{i=1}^{N} (X = x_i)(1 - Q_i)^2 + \sum_{i=1}^{N} (X \neq x_i) Q_i^2\right]$$

$$= -\left[\sum_{i=1}^{N} Q_i^2 - 2\sum_{i=1}^{N} (X = x_i) Q_i + 1\right];$$

equivalently,

$$S(X = x_i, \mathbf{Q}_N) \equiv -\left[\sum_{i=1}^{N} Q_i^2 - 2Q_i + 1\right].$$

**Logarithmic Score of a Distribution**

$$S(X, \mathbf{Q}_N) \equiv \sum_{i=1}^{N} (X = x_i) \log(Q_i);$$

equivalently,

$$S(X = x_i, \mathbf{Q}_N) \equiv \log(Q_i).$$

**Spherical Score of a Distribution**

$$S(X, \mathbf{Q}_N) \equiv \sum_{i=1}^{N} (X = x_i) Q_i \Bigg/ \left(\sum_{i=1}^{N} Q_i^2\right)^{1/2};$$

equivalently,

$$S(X = x_i, \mathbf{Q}_N) \equiv Q_i \Bigg/ \left(\sum_{i=1}^{N} Q_i^2\right)^{1/2}.$$

**$\beta$-Pseudospherical Score of a Distribution.** For each $\beta > 1$,

$$S_\beta(X, \mathbf{Q}_N) \equiv \left\{\sum_{i=1}^{N} (X = x_i) Q_i \Bigg/ \left[\sum_{i=1}^{N} Q_i^\beta\right]^{1/\beta}\right\}^{\beta - 1};$$

equivalently,

$$S_\beta(X = x_i, \mathbf{Q}_N) \equiv \left\{Q_i \Bigg/ \left[\sum_{i=1}^{N} Q_i^\beta\right]^{1/\beta}\right\}^{\beta - 1}.$$

*Proof*. The structure of the proof that each of these scoring rules is proper for your mass function is the same. It relies on Lagrangian methods to determine the vector $\mathbf{Q}_N$ that maximizes your prevised score, $P[S(X, \mathbf{Q}_N)]$, subject to the constraint that $\mathbf{Q}_N$ lies in the unit simplex, $\mathbf{S}^{N-1}$. In the following details of the proof for the quadratic score, we denote by $\mathbf{p}_N$ your presumed probability masses for the constituents of the partition generated by $X$: $p_i \equiv P(X = x_i)$ for $i = 1, \ldots, N$.

To begin, your prevision for the quadratic score achieved by the specification $\mathbf{Q}_N$ is

$$P[S(X, \mathbf{Q}_N)] = -\left[\sum_{i=1}^{N} Q_i^2 - 2\sum_{i=1}^{N} Q_i p_i + 1\right].$$

Thus, the Lagrange function, including the constraint component, is

$$L(P[S(X, \mathbf{Q}_N)], \lambda) = P[S(X, \mathbf{Q})] + \lambda\left(1 - \sum_{i=1}^{N} Q_i\right).$$

The $N$ partial derivative functions of $L$ with respect to $Q_i$ are

$$D_{Q_i} L(P[S(X, \mathbf{Q}_N)], \lambda) = -(2Q_i - 2p_i) - \lambda, \qquad \text{for } i = 1, \ldots, N,$$

and the partial derivative with respect to $\lambda$ is $D_\lambda L(P[S(X, \mathbf{Q})], \lambda) = 1 - \sum_{i=1}^{N} Q_i$. *Setting each of the* $N$ partial derivatives $D_{Q_i} = 0$ requires each $p_i - Q_i = \lambda/2$. Summing these $N$ equations yields $\sum_{i=1}^{N} p_i - \sum_{i=1}^{N} Q_i = N\lambda/2$, which implies that $\lambda = 0$, since both the sums of the $p_i$ and of the $Q_i$ must equal 1. Thus, for each $i$, $Q_i = p_i$, which was to be proved.

As to second-order conditions, the $(N+1) \times (N+1)$ matrix of cross partial derivatives is entirely composed of 0's, except that the first $N$ diagonal elements each equal $-2$, the last diagonal element equals 0, and the final bordering row and column elements are otherwise equal to $-1$. This matrix is negative-definite, since the determinants of its principle minors alternate in sign, beginning with a negative. Thus, the solution $Q_i = p_i$ for each $i = 1, \ldots, N$ yields a maximum value for your prevised score, $P(S[X, \mathbf{Q}_N])$. □

Details of the proofs that the remaining scoring rules are also proper scores of your probability mass function are uneventful. You may derive the results yourself as an exercise in applied Lagrangian methods, to assure your understanding. We should notice one peculiarity of the specific result just proved for the quadratic score, that the value of the Lagrange multiplier is $\lambda = 0$. Generally, the value of a Lagrange multiplier can be interpreted as the cost incurred by subscribing to the constraint with which it is associated. The fact that $\lambda = 0$ for the quadratic scoring rule means that the use of the quadratic score itself encourages you to specify as $\mathbf{Q}_N$ a *coherent* vector of probability masses. For there is no cost involved in meeting the constraining condition that $\sum_{i=1}^{N} Q_i = 1$. Although this is true using the quadratic and the spherical scores of distributions, it is not true for the other scores mentioned in the theorem.

#### 6.4.3.1 *Coherent Previsions for Proper Scores of Your Distribution*

For each of the scoring rules for a distribution we have defined, it is a simple matter to determine your prevision for the score to be achieved by your asserted probability mass function. Rather than present them formally en masse as a corollary to Theorem 6.3, we discuss them informally in turn.

*Coherent Prevision for the Quadratic Distribution Score: Repeat Rate*
Since the quadratic score of your distribution is

$$S[X, \mathbf{p}_N] = -\left[\sum_{i=1}^{N} p_i^2 - 2\sum_{i=1}^{N} (X = x_i)p_i + 1\right],$$

your coherent prevision for this score must be

$$P(S[X, \mathbf{p}_N]) = -\left[\sum_{i=1}^{N} p_i^2 - 2\sum_{i=1}^{N} p_i^2 + 1\right] = \sum_{i=1}^{N} p_i^2 - 1.$$

This prevised score can be interpreted in terms of your prevision for the probability you assess for the constituent $(X = x_i)$ that happens to equal 1, because $\sum_{i=1}^{N} p_i^2$ also equals $P[\sum_{i=1}^{N}(X = x_i)P(X = x_i)]$. Of course, when asserting your probability distribution for $X$ you have specified your probability for every constituent of the partition it generates: $P(X = x_1), \ldots, P(X = x_N)$. But by asserting these probabilities, you avow that you are not certain which of these constituents equals 1 (except in the extreme case of a degenerate distribution). The value of $\sum_{i=1}^{N} p_i^2$ is a measure of the amount of information about $X$ proclaimed in the assertions of $\mathbf{p}_N$. It achieves its maximum value of 1 for any extreme point of the simplex, $\mathbf{p}_N = \mathbf{e}_{i,N}$ for some value of $i$; this is a mass function proclaiming *certain* knowledge that $X = x_i$. It achieves its minimum value for the uniform distribution, $\mathbf{p}_N = (1/N)\mathbf{1}_N$. (Using Lagrangian methods, choose $\mathbf{p}_N$ from the unit simplex $\mathbf{S}^{N-1}$ to minimize $\sum_{i=1}^{N} p_i^2$.)

This measure of a distribution has many applications in statistical theory and in statistical physics. In a completely different context from scoring rules, Good (1965) refers to the measure $\sum_{i=1}^{N} p_i^2$ as the *repeat rate* of a distribution. It would equal the probability of a repeated value of $X_1 = x_i$ and $X_2 = x_i$ if $X_1$ and $X_2$ were regarded independently and with the same distribution, $\mathbf{p}_N$.

In the first place, the possible values of a quadratic score for distributions range from $-2$ to 0. The value of $-2$ is scored by any distribution $\mathbf{p}_N \in \mathbf{S}^{N-1}$ that associates probability 1 with an event $(X = x_i)$ when in fact $X$ does not equal $x_i$. The largest possible quadratic score, 0, is accorded to the distribution that associates probability 1 with the event $(X = x_i)$ for the realm element $x_i$ that actually obtains, and 0 with all other possibilities. Any other distribution achieves a score between $-2$ and 0.

In the second place your coherent prevision for the quadratic score achieved by your own distribution, $\sum_{i=1}^{N} p_i^2 - 1$, can range only from $1/N - 1$ to 0. The maximum prevised score of 0 is expected by everyone who asserts any of the $N$ degenerate distributions associating probability 1 with one of the $N$ possibilities, and associating probability 0 with each of the other $N-1$ possibilities. But only one of these degenerate distributions actually achieves this maximum possible score! The remainder of these maximum score prevising distributions all in fact achieve the lowest possible score of $-2$. The lowest possible prevised score, $1/N - 1$, is expected by assertors of a uniform distribution, $\mathbf{p}_N = (1/N)\mathbf{1}_N$. Moreover, the score actually achieved by this minimum prevised scoring distribution is certain to equal $1/N - 1$, no matter what value of $X$ is found to obtain. So when compared to other asserted distributions in the simplex $\mathbf{S}^{N-1}$, the minimum score prevising distribution can actually achieve a better score than those with which it is compared. In fact, if you do assert a uniform distribution for $X$, you must expect (previse) that this will be the case. The use of proper scoring rules to evaluate distribution assertions ensures this. Otherwise you would assert the distribution whose score you expect to be the largest.

*Coherent Prevision for the Logarithmic Distribution Score: Entropy*
Since the logarithmic score of your distribution is

$$S[X, \mathbf{p}_N] = \sum_{i=1}^{N} (X = x_i) \log(p_i),$$

your coherent prevision for this score must be

$$P(S[X, \mathbf{p}_N]) = \sum_{i=1}^{N} p_i \log(p_i).$$

Your prevision for the logarithmic score of your distribution is your prevision for the log of your prevision for whichever constituent it is that equals 1. This function of a distribution is another important measure of the *amount* of information about $X$ that is proclaimed by anyone who asserts it. You may recognize this measure as the (negative) *entropy* of your distribution for $X$. We discuss it in more detail in Section 6.5.2. For now, notice only that similar to the situation of the quadratic score, the distribution $\mathbf{p}_N \in \mathbf{S}^{N-1}$ that entails the lowest possible prevised score is the uniform distribution. (Again, you can prove this easily using Lagrangian methods.)

Moreover, the uniform distribution surely achieves its prevised score of $\log(1/N)$, no matter what be the value of $X$. Computed logarithmic scores of distributions can range from $-\infty$ to 0, depending both on the distribution and on the value of $X$. The extreme possible values are achieved by the $N$ degenerate distributions, represented by the unit vectors $\mathbf{e}_{1,N}, \ldots, \mathbf{e}_{N,N}$: one of these achieves the score of 0, while the others achieve the score of $-\infty$. Of course, you expect to achieve the maximum score if you assert any one of these degenerate distributions. In fact, the proponents of all of these degenerate distributions feel sure of achieving it. Nonetheless, only one of these distributions can achieve it. All the other $N-1$ of them achieve the worst possible score of $-\infty$.

*Coherent Prevision for the Spherical and $\beta$-Pseudospherical Scores*
Since the spherical score of your distribution is

$$S[X, \mathbf{p}_N] = \sum_{i=1}^{N} (X = x_i)p_i \Bigg/ \left[ \sum_{i=1}^{N} p_i^2 \right]^{1/2},$$

your coherent prevision for this score must be

$$P(S[X, \mathbf{p}_N]) = \sum_{i=1}^{N} p_i^2 \Bigg/ \left[ \sum_{i=1}^{N} p_i^2 \right]^{1/2} = \left[ \sum_{i=1}^{N} p_i^2 \right]^{1/2}.$$

This number is the Euclidean length of your distribution defining vector, $\mathbf{p}_N$. It equals the square root of the so-called repeat rate of your distribution, which has already appeared in your prevision for the quadratic score of your prevision.

Finally, your prevision for the $\beta$-pseudospherical score of your distribution is $P(S_\beta[X, \mathbf{p}_N]) = \{\sum_{i=1}^{N} p_i^2 / [\sum_{i=1}^{N} p_i^\beta]^{1/\beta}\}^{\beta-1}$, involving an obvious generalization of the repeat rate.

### 6.4.3.2 *A Graphical Display of Constant Score Contours*

The implications of using a particular one of these scoring rules can be understood by graphically displaying the contours of constant scores and contours of constant prevised scores over distributions in the two-dimensional unit simplex. Suppose $\mathscr{R}(X) = \{1, 2, 3\}$, so the event vector $\mathbf{E}_3 = [(X=1), (X=2), (X=3)]$ constitutes the partition generated by $X$. As we learned in the prelude to Section 3.10, the unit simplex representing coherent probability distributions over this partition is an equilateral triangle. Figures 6.1–6.3 exhibit on their left side the contours of vectors in the simplex that would be accorded the same quadratic, logarithmic, and spherical scores, respectively, on the basis of the presumed observation that $X$ is observed to equal 2. On the right side of each figure appear the contours of distributions whose proponents would all expect to receive the same score based on their uncertainty before observing $X$.

Algebraically the contours of constant quadratic scores, $S(X=2, \mathbf{p}_3) = \sum_{i=1}^{3} p_i^2 - 2p_2 + 1$, can be recognized as arcs of circles centered at the observation point $\mathbf{E}_3 = (0, 1, 0)$. Comparatively, the contours of $\mathbf{p}_3$ vectors with constant prevised quadratic scores, $\sum_{i=1}^{3} p_i^2 - 1$, displayed on the right side of Figure 6.1 are circles centered at the center of the simplex (1/3, 1/3, 1/3). The contours displayed are continued outside the boundary of the simplex only as a visual aid, for all coherent distributions are represented by vectors within the simplex.

In the left side of Figure 6.2, the contours of $\mathbf{p}_3$ vectors that achieve an identical logarithmic score are displayed as parallel lines through the simplex, depending only on the value of $P(X=2)$. This represents the fact that the logarithmic score accords $S(X=2, \mathbf{p}_3) = \log(p_2)$. The contours of constant prevised logarithmic

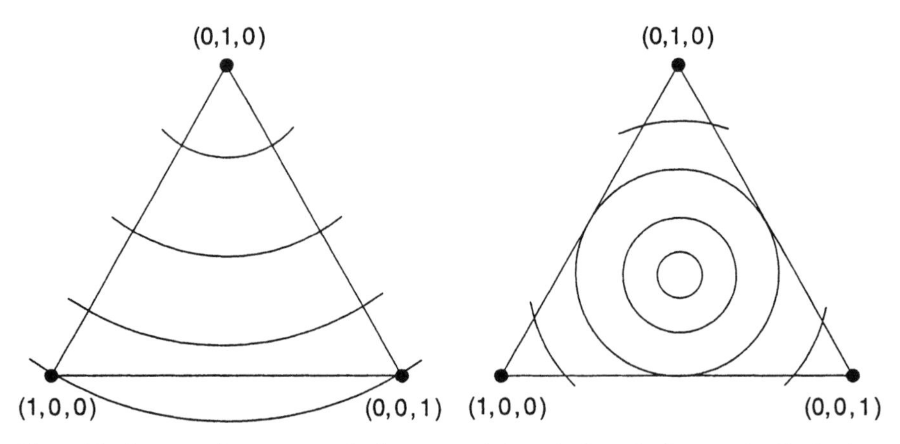


**Figure 6.1.** Contours of constant quadratic scores and of prevised quadratic scores for distributions in a two-dimensional simplex, supposing that the observed partition vector is $\mathbf{E}_2 = (0, 1, 0)$, deriving from the observation $X = 2$.

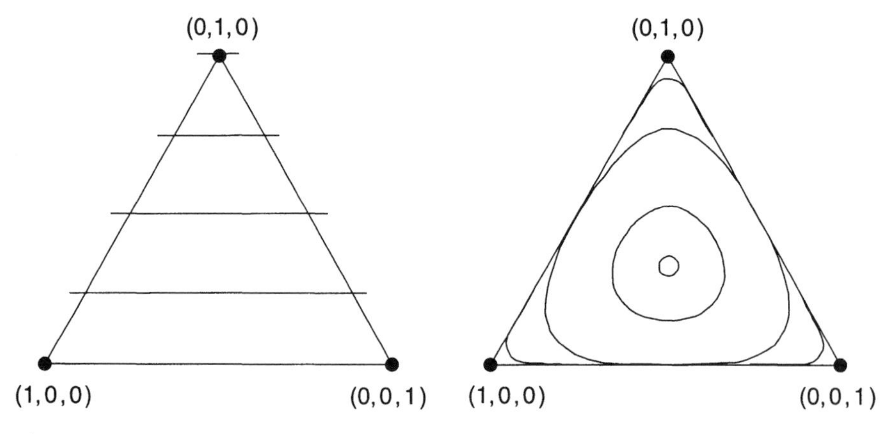


**Figure 6.2.** Contours of constant logarithmic scores and of prevised logarithmic scores for distributions in a two-dimensional simplex, supposing that the observed partition value is $\mathbf{E}_3 = (0, 1, 0)$.

scores, shown on the right side of the figure, are somewhat more difficult to describe and to compute. The components of any contour can be described only implicitly by the equation

$$p_1 \log(p_1) + p_2 \log(p_2) + (1 - p_1 - p_2)\log(1 - p_1 - p_2) = C \in [\log(1/3), 0].$$

One feature is that if a contour includes an edge point of the simplex, then the edge of the simplex is tangent to the contour at that point. A second is that

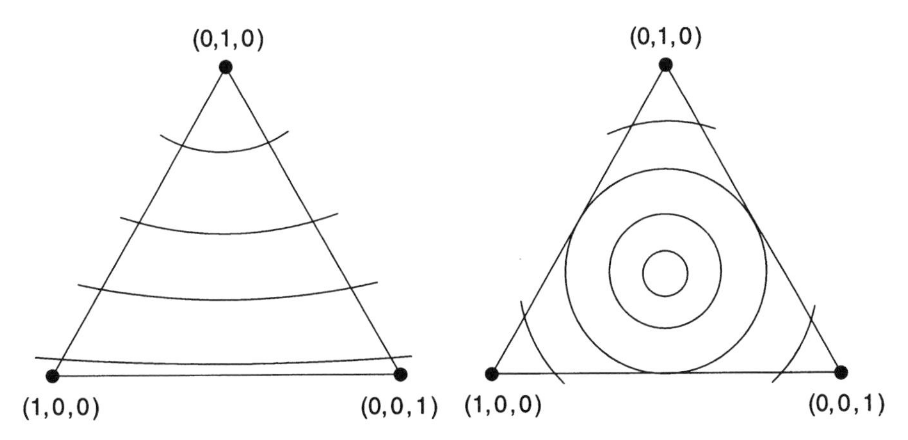


**Figure 6.3.** Contours of constant spherical scores and of prevised spherical scores for distributions in a two-dimensional simplex, supposing that the observed partition value is $\mathbf{E}_3 = (0, 1, 0)$.

wherever a contour is intersected by the bisector of any vertex angle, the tangent to the contour is perpendicular to that bisecting line. Finally, the shapes of the contours approach that of a triangle as the maximum component of all points on the contour approaches 1 (the maximum prevision contour), whereas they approach the shape of a circle as the maximum component approaches 1/3 (the minimum prevision contour).

Finally, in Figure 6.3 the contours of vectors that are accorded the same spherical score appear similar to the constant quadratic score contours for $\mathbf{p}_3$ providing $p_2 \equiv P(X = 2)$ is near to 1; but they appear similar to the constant logarithmic score contours when $P(X = 2)$ is near to 0. The contours of $\mathbf{p}_3$ vectors are defined by the equation $p_2 = K(p_1^2 + p_2^2 + p_3^2)^{1/2}$, where the constant $K \in [0, 1]$ is the value of the score achieved by vectors on the contour. Also rather complicated, this equation says that the value of $p_2$ for each point on the contour must be proportional to the radius of the circle centered at (1/3, 1/3, 1/3) on which the point lies. The function identifying any particular contour is an implicit function that is awkward to present in detail. It is easy to show that the minimum value of $p_2$ is achieved on each contour when $p_1 = p_3$. Moreover, it is evident that the contour for the minimum score is the line at the base of the simplex, $p_2 = 0$, just as is the minimum logarithmic score contour. The associated contours of constant prevised scoring vectors are arcs of circles, just as for the quadratic score.

As a general conclusion, this geometrical exposition suggests that the choice between scoring rules should be based upon your relative interest in punishing distributions that spread the total probability for the unobserved but "possible" observation values uniformly and those that assess some of those "possibilities" with probabilities with very small values compared to the others.

Contemplating a specific application to educational testing will reinforce this idea, accentuating how the choice of a scoring rule really revolves on your utility valuations for the consequences of experiencing the observed value of $X$ in the state of uncertainty represented by $\mathbf{p}_3$. Suppose you construct a three-possibility "multiple-choice question," so that the first possibility is incorrect and would be favored only by a respondent who harbors a gross misconception about the question's content; the second possibility is correct and is favored by a respondent who can make some very fine desirable distinctions in thinking about the content; and the third possibility is not precisely correct, but only on account of some minor feature involving a very fine distinction. In such a case, you would presumably want to score a respondent's elicited $\mathbf{p}_3$ response vector, for a given value of $p_2$, much more severely if the value of $1 - p_2$ were more heavily placed on $E_1$ relative to $E_3$. Moreover, for a given sum of $p_2 + p_3$, the penalty for the size of $p_3$ would not increase too drastically as $p_3$ increases relative to $p_2$. The article of de Finetti (1965) on assessing partial knowledge of a test item is stimulating.

Mathematically enviable properties of proper scores to be studied shortly, such as the approximation property of the quadratic score, seem more persuasive for motivating their use, the greater are the differences in utility valuations among those whose probabilities are being scored.

#### *6.4.3.3 Proper Scores for Continuous Approximating Distributions*

Suppose $f_X(\cdot)$ is the density for a continuous and differentiable function $F_X(\cdot)$ which you judge to approximate your own personal probability distribution function to an adequate degree of accuracy. The following functions are proper scoring functions for $f_X(\cdot)$, where propriety is determined in terms of the approximating distribution function $F_X(\cdot)$.

Quadratic score: $$S[X, f_X(\cdot)] \equiv 2f_X(X) - \int_D f_X^2(x)\,dx - 1,$$

Logarithmic score: $$S[X, f_X(\cdot)] \equiv \log[f_X(X)],$$

Spherical score: $$S[X, f_X(\cdot)] \equiv f_X(X) \bigg/ \left[\int_D f_X^2(x)\,dx\right]^{1/2}.$$

These functions are the continuous analogues of the three scores of distributions that we have already studied. The notation $\int_D$ denotes integration over the appropriate domain of $f_X(\cdot)$. Previsions for the scores of distributions based on approximating density functions are defined by $\int_D S[x, f_X(x)] f_X(x)\,dx$. Performing this integral on these three density scores yields approximate previsions for the scores of a distribution that are the continuous analogues of the previsions for scores of discrete mass functions. For continuous approximating densities, the repeat rate is defined as $\int_D f_X^2(x)\,dx$, and the entropy is defined as $\int_D \log[f_X(x)] f_X(x)\,dx$. A general analysis of the forms of proper scoring rules for continuous approximating distributions is presented in a readable and detailed article by Matheson and Winkler (1976).

## PROBLEM

1. Applying scoring rules to many of the parametric distributions we have studied involves straightforward numerical computation of the repeat rate or the entropy, based on details of the probability mass functions available in most software. You can easily derive closed-form expressions for the repeat rates of the beta and gamma distributions, and both the repeat rate and the entropy for a Normal distribution (presented here in Chapter 7). The entropy for the beta and gamma distributions can be represented in closed form, but these are more difficult to derive.

## 6.5 DISTINGUISHING PROPERTIES OF VARIOUS PROPER SCORING RULES

All of the scoring rules for previsions and for distributions that we have studied are proper rules. What properties distinguish one proper scoring rule from another? On what grounds should we decide which scoring rule to use when comparing assertions of different people, and how are they to be used? In this section we address these questions theoretically. Proofs for some of the theorems discussed are beyond the scope of this book, but references are made to literature where they can be found.

### 6.5.1 Properties of Proper Scoring Rules for Prevision

We begin with the quadratic scoring function, which has been used in applied scientific statistical computations for centuries, albeit from a different perspective on probability and for different proclaimed statistical purposes than we are imagining. In the first place, the quadratic score is widely useful because it provides a good approximation to many proper scoring rules for $P(X)$. Perhaps the most important feature of the quadratic score, this precludes undue concern over the choice of a scoring rule when there are no obvious motivating factors for another specific choice, and it provides a possible compromise when the choice of the appropriate rule is controversial.

Suppose $S(X, Q)$ is a differentiable function of $Q$ for which $P[S(X, Q)]$ achieves its maximum at $Q = P(X)$. The function $S(X, Q)$ can be expressed by its Taylor series expansion in $Q$ about $X$ (whatever that value may be) as

$$S(X, Q) = S(X, X) + a_1(Q - X)D_Q S(X, X) + a_2(Q - X)^2 D_Q^2 S(X, X) \\ + a_3(C - X)D_Q^3 S(X, X)$$

for some number $C$ between $Q$ and $X$. Thus, your $P[S(X, Q)]$ is your prevision for this polynomial in $X$. Considering the summands of this expansion shows why the quadratic component can be expected (by anyone who proclaims substantive knowledge of $X$) to dominate the others. First, scoring functions can typically be scaled so the maximum score, $S(X, X)$, equals 0. Furthermore, because of the concavity of the function $S$ at its maximum, the derivative $D_Q S(X, X) = 0$ as well. Thus for any differentiable proper scoring rule, if your $P(C - X)^3$ is sufficiently small, the quadratic score, $K(X - P(X))^2$, provides an adequate approximation to the scoring function $S[X, P(X)]$. But there are two other appealing properties specific to the quadratic rule. Let us state them in a theorem.

**Theorem 6.4.** The quadratic scoring rule, $S(X, Q) = -(X - Q)^2$, is the unique (up to positive linear transformations) proper scoring rule for $P(X)$ that has either of the following two properties:

**(a)** the score is a function only of the difference between $P(X)$ and $X$; that is,

$$S[X, P(X)] = g(X - P(X));$$

**(b)** the scoring function is a symmetric function of its arguments; that is,

$$S[X, P(X)] = S[P(X), X].$$

*Proof.* The proof can be found in Sections 5.1 and 5.2 of Savage (1971).

A comment is needed to ensure your understanding of this theorem. In problems 4 and 5 of Section 6.4.2 you showed that $S(X, Q) = -|X - Q|$ is a proper scoring rule for your median of $X$, and that $S(X, Q) = (X = Q)$ is a proper scoring rule for your mode of $X$. Thus, you may be puzzled by this theorem, since both of these scoring rules are symmetric functions only of the difference between $X$ and $Q$. To clarify the situation, notice that these scoring functions provide proper scores not of your $P(X)$, but of your median for $X$ and your mode for $X$, respectively. The quadratic score is the unique proper score of your prevision, $P(X)$, that has these properties. This is the burden of Theorem 6.4.

The uniqueness of the quadratic score specified in the theorem is only up to positive linearity. This qualification is of some consequence, for it allows that only the relative sizes of quadratics scores are worthy of comparison. Recalibrating the function from $S(X, P) = -(X - P)^2$ to $S_{a,b}(X, P) = a - b(X - P)^2$ with $b > 0$ would preserve the ordering of the scores for any two previsions. That is, $-(X - P)^2 > -(X - Q)^2$ if and only if $a - b(X - P)^2 > a - b(X - Q)^2$. Moreover, as long as the coefficient $a = 0$, the relative scale of these pairs of scores is unchanged as well. The standard practice of choosing the coefficients as $a = 0$ and $b = 1$ scales the maximum achievable score at 0 and interpretably scales your prevised score as the value of your variance.

In some situations, the property of the quadratic score as a symmetric function only of the difference between $X$ and $P(X)$ is desirable only as an approximation when $X$ is expected to be near to $P(X)$. Alternatively, we might prefer the score accorded to a prevision that exceeds $X$ by a specific amount to be greater than the score accorded a prevision that is smaller than $X$ by the same amount. The linear-logarithmic ratio discrepancy scoring function is nonsymmetric about $X$ in this way, being a function of the ratio of $X$ to $P(X)$. In fact it exhibits its own uniqueness property in this regard, again proved in Savage (1971, Section 5.3).

**Theorem 6.5.** If $X$ is a quantity whose realm contains only positive numbers, the linear-logarithmic ratio discrepancy rule, $S[X, P(X)] = 1 - X/P(X) + \log(X/P(X))$, is the unique (up to positive linearity) proper scoring rule for $P(X)$ that is a function only of the ratio of $X$ to $P(X)$: $S[X, P(X)] = g(X/P(X))$.

The linear-logarithmic scoring function also preserves the ordering of scores accorded to differing previsions when it is translated by a positive linear function: $S_{a,b}(X, P(X)) \equiv a + bS(X, P(X))$. But as opposed to the quadratic score, the relative scales of the linear-logarithmic scores depend on the magnitude of $X$ for any choice of coefficients $a$ and $b$. The standard determination of coefficients as $a = 0$ and $b = 1$ locates the maximum achievable score at 0, and scales your prevision for your score as $P[\log(X)] - \log(P(X))$.

The uniqueness properties of the quadratic and the linear-logarithmic scoring functions within the class of proper scoring rules are even more extensive than specified in these theorems. Again, details can be found in Savage (1971).

### 6.5.2 Properties of Proper Scoring Rules for Distributions

Two distinguishing properties of the logarithmic scoring function for probability distributions suggest that it ought to be used more commonly in statistical practice than is the case today. The first provides that the logarithmic score of a distribution is the unique proper score of a distribution that is a function only of $P(X = x^0)$, where $x^0$ denotes the observed value of $X$ that actually obtains. This unique locality property of the logarithmic score was identified by Shuford, Albert, and Massengill (1966) and Bernardo (1979). It specifies that if two asserted distributions agree in their specified value of $P(X = x^0)$, they should achieve the same score, regardless of how they differ in their assertions concerning the possibility that $X$ might be different from $x^0$ by any specified amount. (Recall the contours of constant logarithmic scores displayed in Figure 6.2.) The motivation for this property of a scoring rule is that the empirical observation of $X = x^0$, taken at face value, tells us nothing about the possibilities that $X$ "could have been" values other than $x^0$, whether near or far from $x^0$. Thus, the score for a distribution should not depend on what it asserts about those possibilities. (You may consider a challenge to this motivation in a problem at the end of this section.) Bernardo's proof uses techniques of functional equations in the style of Savage's (1971) work. The simpler proof of the theorem presented here is based on the properties of Lagrangian methods.

**Theorem 6.6.** Suppose $\mathscr{R}(X) = \{x_1, \ldots, x_N\}$, and a probability distribution for $X$ has been asserted via the previsions denoted by $\mathbf{p}_N \equiv P[(X = x_1), \ldots,$

$(X = x_N)]$. The logarithmic scoring rule, $S(X, \mathbf{p}_N) \equiv \sum_{i=1}^{N}(X = x_i)\log(p_i)$, is the unique (up to positive linear transformations) proper scoring rule that can be expressed in the form $S(X, \mathbf{p}_N) = \sum_{i=1}^{N}(X = x_i)g(p_i)$ for some differentiable function $g(\cdot)$.

*Proof.* When the scoring function has this form, your prevision for your score must equal $\sum_{i=1}^{N} p_i g(p_i)$. The presumed propriety of the scoring rule $S(X, \mathbf{q}_N)$ requires that the Lagrange function, $L(\mathbf{q}_N, \lambda) \equiv \sum_{i=1}^{N} p_i g(q_i) + \lambda(\sum_{i=1}^{N} q_i - 1)$, achieve a maximum when $\mathbf{q}_N = \mathbf{p}_N$. But since the partial derivatives

$$D_{q_i} L(\mathbf{q}_N, \lambda) = p_i D_{q_i} g(q_i) + \lambda = 0 \qquad \text{at } q_i = p_i \text{ for every } i$$

if and only if $D_{p_i} g(p_i) = \lambda / p_i$, the function $g(\cdot)$ is identified uniquely as the logarithm function, qualified by the linear transformations allowed by the value of the Lagrange multiplier, $\lambda$, and the constant of integration. □

Bernardo noted that this property of the logarithmic rule parallels the property of the likelihood principle, which we mentioned briefly in Section 3.11.1. The logarithmic scoring function is a function of the data only through the probability asserted for the event $(X = x^0)$ that actually occurs, not on other values of $x$ that might have equaled $X$ but didn't. Suppose that the 10 possible values of $X$ specify $\mathscr{R}(X)$ as $\{1, 2, \ldots, 10\}$. Would you like to ascribe the same score to the distribution $\mathbf{p}_{10} = (0, 0, .9, .1, 0, 0, 0, 0, 0, 0)$ and the distribution $\mathbf{q}_{10} = (.1, .1, .1, .1, .1, .1, .1, .1, .1, .1\}$ based on the observation $X = 4$? The locality property of the logarithmic score would be relevant as long as you would be happy to give the same score to any two distributions that do not differ in their probabilities ascribed to the event $(X = x^0)$.

A second property of the logarithmic scoring rule for distributions may be even more generally appealing. Suppose we want to score a joint distribution for several quantities, represented by the joint p.m.f. $f_{\mathbf{X}}(\mathbf{x})$, and compare it with the cumulative score accorded to its associated sequence of conditional p.m.f.'s,

$$f_{X_1}(x_1), f_{X_2}(x_2 | (X_1 = x_1)), f_{X_3}(x_3 | \mathbf{X}_2 = \mathbf{x}_2), \ldots, f_{X_N}(x_N | \mathbf{X}_{N-1} = \mathbf{x}_{N-1}).$$

This would be to compare the numerical values of the score

$$S(\mathbf{X}, f_{\mathbf{X}}(\mathbf{x})) \qquad \text{with } \sum_{i=1}^{N} S(X_i, f_{X_i}(x_i | (\mathbf{X}_{i-1} = \mathbf{x}_{i-1}))).$$

It is easy to see that these two scores would be the same using the logarithmic scoring function. For any coherent joint p.m.f. must equal the product of the conditional p.m.f.'s in this sequence. Thus, the logarithmic score of the joint p.m.f. (evaluated at whatever the observed values of $\mathbf{X}$ may be) must equal the sum of the logarithmic scores for the sequence of conditional p.m.f.'s at these same values.

My conjecture is that the logarithmic scoring rule is the unique scoring rule for joint distributions that has this property.

## PROBLEM

1. One could argue that the observation of $X = x_0$, amounts to the observation of $N$ events that occur: $(X = x^0)$ along with $N - 1$ events of the form $(X \neq x_i)$ for the elements in $\mathscr{R}(X)$ that are different from $x^0$. A logarithmic score for the probabilities assessed for all of these events that obtain would be

$$S(X,\mathbf{q}_N) \equiv \sum_{i=1}^{N} (X = x_i)\log(q_i) + \sum_{i=1}^{N} (X \neq x_i)\log(1 - q_i).$$

Show that this is a proper scoring rule for your distribution of $X$, identified by the partition probabilities $\mathbf{p}_N$. Study the constant score contours for $\mathbf{p}_3$, presuming $X = 2$, as in our graphical exposition, satisfying yourself that the pattern of the contours is similar to that for the spherical scoring rule.

### 6.5.3 Previsions for Scores of Distributions: Measuring Information

Perhaps the most widely known properties of scoring functions are properties not of the scores themselves, but of coherent previsions for the scores to be accorded to one's asserted distribution by various scoring functions. This topic, which we are now well prepared to discuss, establishes a connection between scores of distributions and measures of the amount of information about $X$ that is proclaimed by someone who asserts a probability distribution.

When you assert your probability distribution, you are displaying the information that you hold regarding the quantity in question. In asserting distinct probability distributions for a quantity, two people proclaim different information about it. To provide some bearings for our discussion, consider the case of claims to certain (sure) knowledge. Suppose, for example, I tell you that I have four living brothers. In fact, they are sitting with me just now in this room. I'm telling you some information. Suppose again I tell you that the official world land speed record as of 1 April 1978 was 1312 kilometers per hour. I'm telling you some information. Now suppose my brother tells you that the world land speed record as of 1 April 1978 was 1164 kilometers per hour. He is telling you some information. But surely, if one of us in fact is telling you something that is correct, then the other is not. You do not know which, if either, of us is correct, although you may well assert your previsions that each of us is correct or that neither of us is correct. In considering a numerical measure for the amount (not the different content) of information that each of us has proclaimed to you, does it not seem reasonable to say that the amount of information asserted by each is numerically equal?

The *amount* of information contained in an assertion about a quantity does not depend on the unknown value of the quantity, but only on the structure of the assertion. Now you, in your own state of uncertain knowledge about the world land speed record as of 1 April 1978 may well regard one of these two assertions as more surprising and less believable than the other. If you assert positive probabilities for the truth of each of the claimants' assertions, the amount of information in your assertion would be less than the amount in either of theirs, even though only one of their assertions could possibly be true.

Our use of the words "information," "amount of information," and "surprise," in the context of proclaimed certain knowledge transfers easily to the context of uncertain knowledge expressed via previsions and probability distributions. Two distinct distributions may express different information content, even while proclaiming the same *amount* of information.

The scoring rules for distributions we have been studying amount to an assessment of the *value* of the information about $X \in \{x_1, \ldots, x_N\}$ contained in the probability vector $\mathbf{p}_N$, measured by a function $S(X, \mathbf{p}_N)$ that is continuous in all components of $\mathbf{p}_N$, achieves a maximum value at $S(X = x_i, \mathbf{e}_{i,N})$ and achieves a minimum value at $S(X = x_i, \mathbf{e}_{j,N})$. Here $\mathbf{e}_{i,N}$ denotes the vector $\mathbf{p}_N$ for which $p_i = 1$ with the remaining components equal to 0.

In contrast, the *amount* of information asserted about $X \in \{x_1, \ldots, x_N\}$ via the probability vector $\mathbf{p}_N$ is measured by a function $H(\mathbf{p}_N)$ that does not depend on the value of $X$ at all. The following central result on this topic has provided the basis for the subsequent development of mathematical information theory.

**Theorem 6.7.** (Shannon, 1948). The function defined by $H(\mathbf{p}_N) = \sum_{i=1}^{N} p_i \log(p_i)$, called the *negentropy* of $\mathbf{p}_N$, is the unique function $H(\cdot)$ with the properties that

**(a)** $H(\cdot)$ is continuous in all its arguments,
**(b)** $H(N^{-1}\mathbf{1}_N)$ is a monotonic decreasing function of $N$, and
**(c)** $H(tp_1, (1-t)p_1, p_2, \ldots, p_N) = H(\mathbf{p}_N) + p_1 H(t, 1-t)$.

A beautiful exposition of this result appears in Khinchin (1953). Deference is made in this definition to Maxwell's and Boltzmann's earlier definition of the entropy in a distribution $\mathbf{p}_N$ as $-\sum_{i=1}^{N} p_i \log(p_i)$. Thus, their *entropy* is a positive measure of the disorder specified by a distribution, maximized over all distributions by the uniform distribution. As our negentropy is a measure of the amount of information asserted about $X$ by someone who asserts a probability distribution $\mathbf{p}_N$, it is defined as the negative of their entropy, and is minimized over all distribution assertions by someone who asserts a uniform distribution.

Property (c) in Shannon's theorem identifies the negentropy function of a probability distribution as appropriately additive over an asserted joint distribution for several quantities. Suppose $X$ and $Y$ are quantities with $\mathscr{R}(X, Y) \subseteq \{x_1, \ldots, x_N\} \otimes \{y_1, \ldots, y_M\}$. An asserted joint distribution is representable by a matrix $\mathbf{p}_{N,M} \equiv [p_{i,j}]$, where $p_{i,j} \equiv P[(X = x_i)(Y = y_j)]$. The marginal mass functions are specified by the row and column sums of $\mathbf{p}_{N,M}$, denoted by $\mathbf{p}_{N\cdot} = (p_{1\cdot}, p_{2\cdot}, \ldots, p_{N\cdot})$ and $\mathbf{p}_{\cdot M} = (p_{\cdot 1}, p_{\cdot 2}, \ldots, p_{\cdot N})$; and the conditional mass functions are defined according to $p_{i|j} \equiv p_{i,j}/p_{\cdot j}$ and $p_{j|i} \equiv p_{i,j}/p_{i\cdot}$. So $\mathbf{p}_{N|j} \equiv [p_{1|j}, p_{2|j}, \ldots, p_{N|j}]$. Then $H(\mathbf{p}_{N,M})$ can be seen to equal

$$\begin{aligned} H(\mathbf{p}_{N,M}) &= \sum_{i=1}^{N} \sum_{j=1}^{M} p_{i,j} \log(p_{i,j}) = \sum_{i=1}^{N} \sum_{j=1}^{M} p_{i\cdot} p_{j|i} \log(p_{i\cdot} p_{j|i}) \\ &= \sum_{i=1}^{N} \sum_{j=1}^{M} p_{i\cdot} p_{j|i} [\log(p_{i\cdot}) + \log(p_{j|i})] \\ &= \sum_{i=1}^{N} p_{i\cdot} \log(p_{i\cdot}) + \sum_{i=1}^{N} p_{i\cdot} \sum_{j=1}^{M} p_{j|i} \log(p_{j|i}) \\ &= H(\mathbf{p}_{N\cdot}) + \sum_{i=1}^{N} p_{i\cdot} H(\mathbf{p}_{N|i}). \end{aligned}$$

The symmetrical treatment of $X$ and $Y$ in this development also provides that

$$H(\mathbf{p}_{N,M}) = H(\mathbf{p}_{\cdot M}) + \sum_{j=1}^{M} p_{\cdot j} H(\mathbf{p}_{M|j}).$$

In either case, the amount of information asserted about two quantities via a joint distribution equals the amount of information asserted marginally about one of them plus the assertor's prevision for the amount of information asserted about the other conditional upon the observed value of the one.

By now, many characterizations of (negative) entropy as a measure of the information in a distribution have been developed. For example, the properties proposed by Fadeev, as reported by Renyi (1961), replace Shannon's property (b) with

**(b′)** $H(\cdot)$ is symmetric with respect to permutations of the components of its argument vector, $\mathbf{p}_N$; and

**(d)** $H(.5, .5) = 1$.

This final property merely scales the logarithm to be represented in base 2.

More detailed study of information measures can best be continued elsewhere, as a vast literature has developed. Renyi's (1984) expository essay provides many ideas as well as details at an introductory level. Kullback (1959) is a now classic exposition. His work is noted especially for identifying the amount of information about a quantity $X_2$ that is considered to be contained in the observation of $X_1 = x$ as the increase in information in the conditional distribution for $X_2$ given $X_1 = x$ relative to the information measure of the marginal distribution for $X_2$. Rather than discuss such ideas here, we conclude our theoretical consideration of scoring rules with a presentation of "the calibration question," which continues to arouse critics of subjectivism in statistics to the present day.

## 6.6 THE CALIBRATION QUESTION

Throughout the twentieth century, developments of the operational subjective theory of probability have regularly been challenged by the appealing notion that good assertions of probabilities must somehow be "borne out" by the frequency with which events occur. A spirited statement of the challenge can be found in several influential writings of Maurice Fréchet. Let's listen to his argument against the subjective foundation for probability assessments in some detail, relishing its emotional appeal:

> ... Suppose that a degree of belief is small for events that are not rare. If someone acts upon that real degree of belief as the base of his conduct, we could observe the considered individual out walking in his shirt sleeves in the middle of winter and all bundled up in summer, or promenading tranquilly upon the guardrail of a bridge, for example, because he finds himself to have attributed a small degree of belief to the possibility that it is cold in winter, hot in summer, or that one falls when climbing the guardrail. If people would act this way they would disappear rapidly from the face of the earth, and with them this unfortunate* interpretation of probability. ... either the degree of belief is estimated badly enough to lead to their death those who made it the foundation of their conduct, or else the degree of belief is not at all subjective and is estimated approximately by a frequency. (Fréchet, 1946)

Fréchet concluded that only subjective probabilities that are equivalent to frequencies can be regarded as valid, and thus the subjective character of probability assertions is more apparent than real. He was explicit on this criterion for the "correctness" of a probability assessment at the International Congress on Philosophy of Science in Paris:

> ... But how would one judge if his estimations of probability are sufficiently correct? Would it be when an event of which he judges the probability greater than 1/2 does not occur? No, certainly not if this circumstance occurs only one time. It only incites him to revise his estimation if the frequency of this event remains small in a great enough number of trials.
>
> Thus this frequency which our subjectivist would like to ignore returns to him like a boomerang to astonish him and lead him to repentance. (Fréchet, 1949)

An earlier variation on this argument by Bonferroni (1925) had apparently pleased Fréchet, as an annotated copy of that paper can be found among his professional papers which are deposited at the University of Paris VI Department of Statistics. Bonferroni wrote "that the a-priori determination of probability is based upon equal frequencies which can be verified by experience—and thus the notion of a-priori probability is based either consciously or unconsciously on this experience. Thus, the determination of probability is only apparently a matter of a-priori judgment."

*The French word translated here as "unfortunate" is "facheuse," which connotes "odious," "vexatious," or "troublesome."

It was perhaps in response to Bonferroni's argument that de Finetti (1937) had outlined the relationship between frequencies and probabilities in his famous Paris lectures. More extensive discussion of Fréchet's arguments by de Finetti can be found in Section 5.9 of his treatise (de Finetti, 1974a) entitled "Frequency and Wisdom after the Event." Since probability assertions regarding a sequence of quantities represent the avowed state of uncertainty of their assessor, there is no point in comparing the probabilities with the outcomes as if it made sense to ask whether they had been "correct" or not. Posterior probabilities of subsequent events are not corrections of previous probability assertions. They are simply additional assertions, cohering with the previous ones, corresponding to uncertainties conditioned upon a new state of information.

Because he outlived Bonferroni, de Finetti (1961) had the last word on this topic between them, explicitly chiding the idea that subjective probabilities are really frequencies, in an otherwise laudatory obituary essay. But de Finetti (1974a) mentioned on the first page of his treatise that his own consternation with the uncomprehending attitudes displayed at the 1949 Congress motivated him to attempt a complete exposition of the subjectivist viewpoint, a project which engaged him for some 20 years, along with his other work.

Nonetheless, de Finetti himself was still struggling with the issue of evaluating probability assessments in his contribution to Good's (1962) *Anthology of Half Baked Ideas*, entitled with the question "Does It Make Sense to Speak of Good Probability Appraisers?" About the same time, an important contribution to a mathematical formulation of the calibration question appeared in a seminar paper by Pratt (1962), one of the most-cited unpublished papers in modern statistical literature. Also entitled with a question, "Must Subjective Probabilities be Realized as Relative Frequencies?", Pratt's discussion set the tone for most of the analysis of the calibration issue conducted during the subsequent 30 years. Virtually all the analysis has considered good calibration as a property of probabilities relative to their limiting frequency of instantiation among events for which they are asserted. Dawid (1982, 1985) brought up the issue again, claiming to show theoretically that good calibration is a property of only some subjective probability distributions, and that two sequences of forecasts which both meet this criterion must be in "asymptotic agreement." This result was proposed to substantiate the notion of a "correct" probability assessment and to raise a frightful specter for any truly subjective theory of probability. Oakes (1985) proposed a contrary result. Schervish (1985) also rejected Dawid's claims, providing an array of examples in which the limiting frequency definition of calibration appears silly.

Yet the concept of probability calibration continues to be taken seriously by many, even in the critical commentary of Cooke (1991) and in applied work such as Clemens and Murphy (1986, 1987). Empiricists have also been involved in the act. Lichtenstein, Fischhoff, and Phillips (1982) surveyed the large body of empirical psychological experiments and applied statistical forecasting results through 1980 that were oriented toward studying whether the subjective probabilities asserted by forecasters exhibit the property of good calibration over finite sequences of events.

My own attitude is to dismiss the usual concept of probability calibration as ill-posed and misleading, as I argued in Lad (1984) and detailed further in Blattenberger and Lad (1985). In the remainder of this section I present my objections and offer a resolution that recognizes good calibration as equivalent to the coherence of one's assertions.

### 6.6.1 Defining Probability Calibration

As defined by most critics, a person's probability assertions are said to be well calibrated at the value $p$ if the limiting frequency of occurrence of events that are assessed with probability $p$, equals $p$. Their recourse to a *limiting* frequency definition has been necessitated by the fact that to assess each of $N$ events with the same probability $p$ does not at all amount to a claim that the proportion of these events that occur, $S_N/N$, equals $p$. In the simple case to which a binomial distribution applies, for example, to regard $N$ events independently with the same probability $p$ is to assert positive probabilities for each of the $N + 1$ frequencies 0, $1/N, 2/N, \ldots, (N-1)/N$, and 1. It is only when $N$ is infinite that these assertions would equate with the unequivocal claim that the proportion of the occurrences equals $p$, as would be shown by the strong law of large numbers.

But a limiting frequency criterion for the calibration quality of probability assertions is objectionable. First of all, the fact that an infinity of events cannot be observed in a finite time renders the limiting frequency concept of good calibration operationally useless for determining whether any real person's probabilities satisfy the definition. Furthermore, even if we could identify two people whose probability assertions were somehow known to be well calibrated according to this definition (at every value of $p \in [0, 1]$), there is nothing that would prevent the two of them from asserting different probability values for any particular event. No matter whether the event in fact occurs or not, and no matter what are the distinct numerical values of their probabilities, both of these persons would remain well calibrated according to the limiting frequency definition. For appending a single 0 or 1 to an infinite string of 0's and 1's for which the proportion of 1's equals $p$ would not change the proportion of 1's in the enlarged string. The supposed calibration quality of both of these forecasters, based on a limiting frequency, would be operationally useless for aiding you to determine what you should assert as your probability for the event in question.

Even in the context of presumably infinite lists of probabilities, a final objection to calibration procedures can lead us to an appropriate definition of calibration that will settle many arguments. Suppose you would challenge that I am not well calibrated at probability 3/5 according to the limiting frequency definition by exhibiting an infinite sequence of events that I have avowedly assessed with probability 3/5 and showing that in fact only 1/10 of them have occurred. If I could produce another list (an infinitely long one) of events that I also assess with probability 3/5 but which you have not considered in your analysis of my "bad calibration," would I not have a right to be incensed? Why should you call my probability assessments "uncalibrated" when you have checked out only a portion of the events that I have assessed at this probability level?

If there is anything substantive to the concept of calibration of one's probabilities, or, generally, previsions, it surely must be a global property of all one's assertions, not merely some selected assertions. For purposes of further discussion, let us propose a definition of the calibration of probability assessments that makes this feature explicit. Moreover, this definition refrains from metaphysical allusion to limiting frequencies, confining its domain to the finite world of experience in which we find ourselves.

***Definition 6.5.*** Probability assertions are said to be *well calibrated at the level of probability p* if the observed proportion of *all* propositions that are assessed with probability $p$ equals $p$. □

Similarly, remember, coherency is a property that is relevant to the collection of all the assertions that you are willing to make about quantities. To the extent that you feel comfortable in your understanding that coherence is the driving property of logical thinking in the context of uncertain knowledge, the conclusion of our next argument should not be surprising.

### 6.6.2 Coherent Previsions Are Well Calibrated at Every Probability

Could anyone possibly be uncalibrated in assertions of probability at the level of $P(\cdot) = 1/2$? Suppose I assert my probabilities as 1/2 that the land area of New Zealand exceeds the land area of Australia, that by 1998 the men's world record in the 100-meter dash will be less than 7.00 seconds, and that the maximum recorded temperature in Spain during 1990 exceeded 30°C. Would you worry that my probabilities are uncalibrated? The question is not whether you would be shocked by my assertions or would assert different probabilities for these events, but whether my probability assessments at the level of 1/2 are uncalibrated.

Remember that when anyone asserts $P(A) = 1/2$, coherency also requires the concomitant assertion of $P(\tilde{A}) = 1/2$. Thus, since events assessed with probability 1/2 necessarily accumulate in exclusive pairs, the proportion of all events coherently assessed with probability 1/2 necessarily equals 1/2. Essentially the same argument extends to probabilities different than 1/2.

Consider once again the first-year's milk yield from a registered California Holstein heifer. In particular, we consider two different people's assessment of their 20, 40, 60, and 80th percentiles for her milk yield, $M$, denominated in kilograms. Suppose the president of the California Holstein Breeders' Association asserts these quantiles as 8500, 8700, 9000, and 9800; and suppose a frazzled student in Christchurch, under my grilling, asserts her quintiles as 150, 300, 500, and 1000. Would you worry that her probability assertions are uncalibrated?

Notice that the student has identified, via coherency, five exclusive and exhaustive events that she assesses with probabilities 1/5:

$$E_1 = (M \leqslant 150); \quad E_2 = (150 < M \leqslant 300); \quad E_3 = (300 < M \leqslant 500);$$
$$E_4 = (500 < M \leqslant 1000); \text{ and } E_5 = (1000 < M).$$

Again, the question is not whether you think she is less informed about cows or a poorer judge of this cow than the president. The question is whether her probability assertions are uncalibrated. In the words of Fréchet, might she be giving high probabilities (1/5) to events that are quite "rare," such as $E_1$, $E_2$, $E_3$, and $E_4$, and low probabilities (1/5) to events that are quite common, such as $E_5$?

To be sure, the student has assessed five exclusive and exhaustive events each with the probability 1/5, and no matter what happens to the milk yield from this heifer, exactly one of these five events will equal 1. Furthermore, by adding any two of these five events, we can define ${}^5C_2 = 10$ events that coherency requires her to assess with probability 2/5; and exactly 4 of these 10 events will equal 1, while the other six will equal 0, no matter what happens to the milk yield from this cow. The four events that equal 1 will be those defined by the sum of the one event of $\mathbf{E}_5$ that equals 1 and each of the other four events. The six events that equal 0 are the ${}^4C_2$ events defined by summing any two of the four that equal 0. Do you get it? How many events does coherency require the student to assess with probability 3/5, and with probability 4/5? How many of these collections will equal 1? What difference does it make if we analyze the assertions of the Breeders' Association president instead of the student?

You can see what is happening. As far as the calibration properties of the probabilities asserted by these two very differently informed assessors goes, precisely the same thing can be said. Let us state the result as a theorem.

**Theorem 6.8.** Suppose $N$ coherent $N$-tile assertions regarding a quantity, $X$, are denoted by $X_{(1,N)} < X_{(2,N)} < \cdots < X_{(N,N)} = \max \mathscr{R}(X)$, meaning $P(X \leqslant X_{(i,N)}) = i/N$ is asserted for each $i = 1, 2, \ldots, N$. Then no matter what the measured value of $X$, the assessed probabilities are well calibrated at each $P(\cdot) = p = i/N$ for $i = 1, 2, \ldots, N$.

*Proof.* For each value of $i$, the elicited $N$-tiles specify by coherency ${}^NC_i$ distinct event sums that are assessed with probability $i/N$. Of these summed events, those that equal 1 are all those that include among their summands the single constituent of the partition generated by the $N$-tile assertions that equals 1, and any $i-1$ other constituents as summands. Thus, there are ${}^{(N-1)}C_{(i-1)}$ of these events that equal 1. Thus, the proportion of events assessed with probability $i/N$ that in fact equal 1 is ${}^{(N-1)}C_{(i-1)}/{}^NC_i = i/N$, no matter what is the value of $X$. □

Coherency of the asserted $N$-tiles implies the good calibration of all the asserted probabilities they imply. In a certain sense this theorem is more symbolic than practical, since we are not required to and often cannot assert $N$-tiles for a quantity on account of the discreteness of its realm. But even in these cases, something similar can be said. Suppose for example that I assert only $P(E) = .8$, and thus $P(\tilde{E}) = .2$. What can be said about the calibration of these coherent assertions? Practically speaking, I would aver that whatever I assert about any event such as this $E$, I can always identify two events that I regard independently of $E$ and of each other, such as $H_1$ and $H_2$ that I flip a head with each of the two coins in my pocket. Thus, I can always embed an assertion such as $P(E) = .8$ into an array of five assertions:

$$P(EH_1H_2) = P(EH_1\tilde{H}_2) = P(E\tilde{H}_1H_2) = P(E\tilde{H}_1\tilde{H}_2) = P(\tilde{E}) = 1/5.$$

The same argument we applied to the quantiles for the milk yield can be applied to these five coherent and equal probabilities over a partition. They imply via coherency the good calibration of my probability assessments at the values of $P(\cdot)$ equal to .2, .4, .6, and .8.

Companion to this image that coherent assertions imply the good calibration of all probabilities in the global sense of Definition 6.5 is the awareness that incoherent assertions entail bad calibration in some sense. As there are so many ways to be incoherent, it is difficult to propose a precise statement of this result, since incoherent assertions need obey no rules. One example would be that if the $N$ constituents of a partition were incoherently assessed each with the probability $1/(N-1)$, say, then the frequency with which these specific $N$ events would equal 1 would be $1/N \neq 1/(N-1)$.

However, another understanding can be made quite clear. If only a subgroup of all the events assessed with a probability $P(\cdot) = p$ is examined, it is almost certain that the proportion of them that are found to equal 1 will be different than $p$, perhaps even quite different. As far as the coherent assessor is concerned, however, such an occurrence is just a matter of luck and constitutes no criticism of the assessed probabilities. We shall return to this matter.

### 6.6.3 Separating Proper Scores into Calibration and Refinement Components

A calibration assessment of a limited number of prevision assertions does play a specific role in the general theory of proper scoring rules for probabilities. This is easiest to identify by analyzing the quadratic scoring rule, but the role is the same in any proper score of probabilities whatsoever. For simplicity of exposition here, we consider a vector $\mathbf{E}_N$ of logically independent events that you assess with the probabilities $P(\mathbf{E}_N) = \mathbf{p}_N$. There are no coherency restrictions on the components of $\mathbf{p}_N$ other than they each lie within [0, 1]. In this context, a simple transformation of our quadratic scoring rule is commonly called the "Brier score" of your probabilities:

$$\mathrm{BS}(P(\mathbf{E}_N) = \mathbf{p}_N, \mathbf{E}_N) \equiv N^{-1} \sum_{i=1}^{N} (E_i - p_i)^2.$$

The Brier score of a vector of probability assertions is simply the average squared difference between the probabilities and the event values. Notice that a low Brier score achievement would be preferable to a high Brier score. It was in this form that our quadratic score was studied by Brier (1950) in the context of weather forecasts.

Algebraically, the Brier score for an arbitrary collection of probability assertions can be partitioned into two components, as shown by Sanders (1963). One component represents a calibration score of the various asserted probability levels, and the other represents the refinement of the subgroups of events that happen to be assessed with the same probabilities, via their homogeneity. Suppose that among the $N$ asserted probabilities, there are $K < N$ distinct assertion values: $N_1$ events assessed with probability $p_1^*$, $N_2$ events assessed with probability $p_2^*, \ldots,$ and $N_K$ assessed with probability $p_K^*$, where $\sum_{i=1}^{N} N_i = N$. Defining $r_j$ to equal the observed proportion of the events that equal 1 among those assessed with probability $p_j^*$, it is a simple matter to show that the Brier score can be expressed as

$$\begin{aligned} \mathrm{BS}(P(\mathbf{E}_N) = \mathbf{p}_N, \mathbf{E}_N) &= N^{-1} \sum_{j=1}^{K} N_j (p_j^* - r_j)^2 + N^{-1} \sum_{j=1}^{K} N_j (r_j)(1 - r_j) \\ &\equiv C(P(\mathbf{E}_N) = \mathbf{p}_N, \mathbf{E}_N) + R(P(\mathbf{E}_N) = \mathbf{p}_N, \mathbf{E}_N). \end{aligned}$$

These two components of the score are called the calibration component and the refinement component, respectively. The calibration component is a weighted average of the squared differences between each distinct probability value, $p_j^*$, and the observed proportion of occurrences among events assessed with that probability. The refinement component is a measure of the extent to which the group of events that are assessed with the same probability is homogeneous in their occurring. At one extreme, if the observed values of all events assessed with the same probability turn out to be identical, either all 1's or all 0's, then the refinement component of the Brier score equals 0. At the opposite extreme, the greatest dissimilarity among them is exhibited if half of them equal 1 and the other half 0.

Notice that while the Brier score is a proper scoring rule for the assessed probabilities of a sequence of events, the calibration score component alone is not. An easy counterexample to the propriety of the calibration score arises when the components of $\mathbf{E}_N$ constitute a partition. You can be sure of achieving a calibration score of 0 if you announce your probabilities for the events as the uniform distribution, each $P(E_i) = 1/N$. For in this case, of course, exactly $1/N$ of the $N$ events would necessarily equal 1. Even if your honest probability assessments are not all equal, you can be sure of achieving a better calibration score by pretending that your probabilities are uniform. Thus, the calibration score itself is not a proper score of asserted probabilities.

A numerical example will highlight just why a "good" calibration score component is not even particularly desirable for forecasters who want to minimize their achieved Brier scores. The exhibition here comes from the work of Blattenberger and Lad (1985), which contains graphical depictions and more extensive details. To begin, suppose that the $N$ events are all assessed with the identical probability, $p$, though not necessarily independently. In this case, in the partition formula for the Brier score, the value of $K = 1$, so the components of the score reduce to

$$C(p,r) = (r-p)^2 \qquad \text{and} \qquad R(p,r) = r(1-r);$$

and thus,

$$\text{BS}(p,r) = (1-2p)r + p^2.$$

Together, these three equations identify a curve of possible ($C$, $R$, BS) scores that can be achieved when *any N events* are assessed with the same probability $p$. Remember that every score triple must satisfy the restriction $\text{BS} = C + R$, so these points do trace out a curve in three dimensions.

Projected into the spaces of (BS, $C$) pairs and (BS, $R$) pairs, the curve traces out parabolas, specified by the equations

$$C(\text{BS}) = (1-2p)^{-2}[\text{BS} - p + p^2]^2,$$

$$R(\text{BS}) = (1-2p)^{-2}[-\text{BS}^2 + \text{BS}(1-2p+2p^2) + 2p^3 - p^4].$$

The projection into ($C$, $R$)-space is defined by the pair of functions

$$R^+(C) = p - p^2 - C + C^{1/2}(1-2p) \quad \text{and} \quad R^-(C) = p - p^2 - C - C^{1/2}(1-2p),$$

which together define a parabola symmetric about the line $R = .25 - C$.

Table 6.1 displays all the possible score configurations ($C$, $R$, BS) that could possibly result from observing 10 events that are all assessed with the probability .2 and with the probability .4, when the proportion $r$ of them equal 1. Each triple represented lies on a ($C$, $R$, BS) curve for the appropriate value

**Table 6.1. Eleven Possible ($C$, $R$, BS) Score Configurations Resulting from 10 Events Assessed with the Same Probabilities, .2 and .4, when the Proportion $r$ of them Equal 1**

| $r$ | 0 | .1 | .2 | .3 | .4 | .5 | .6 | .7 | .8 | .9 | 1.0 |
|---|---|---|---|---|---|---|---|---|---|---|---|
| $C(.2, r)$ | .04 | .01 | .00 | .01 | .04 | .09 | .16 | .25 | .36 | .49 | .64 |
| $R(.2, r)$ | .00 | .09 | .16 | .21 | .24 | .25 | .24 | .21 | .16 | .09 | .00 |
| BS(.2, $r$) | .04 | .10 | .16 | .22 | .28 | .34 | .40 | .46 | .52 | .58 | .64 |
| $C(.4, r)$ | .16 | .09 | .04 | .01 | .00 | .01 | .04 | .09 | .16 | .25 | .36 |
| $R(.4, r)$ | .00 | .09 | .16 | .21 | .24 | .25 | .24 | .21 | .16 | .09 | .00 |
| BS(.4, $r$) | .16 | .18 | .20 | .22 | .24 | .26 | .28 | .30 | .32 | .34 | .36 |

of $p$, either .2 or .4. If 20 events were assessed, 10 with probability .2 and 10 with .4, the resulting score triple would be one of the 121 possibilities derived from averaging any column from the top bank of ($C$, $R$, BS) scores possible for $p = .2$ and any column from the bottom bank of scores possible for $p = .4$.

The first important point to notice is that the best (lowest) Brier score is achieved not when the calibration score equals 0 but when the refinement score equals 0. (The same is true when the worst Brier score is achieved.) When the calibration score equals 0, an intermediate value of the Brier score is achieved, both when the 10 events are assessed with the probability .2 and with .4.

Second to consider is how the achieved score decomposition might relate to the assertor's prevision for the score configuration. This depends on the interconnections among one's attitudes toward the several considered events. I merely state and discuss the following results, which you may prove as an exercise if you wish:

**a.** In any case, as long as $N$ events are assessed with the same probability, $p$, the assertor's prevision for the Brier score is $P(\mathrm{BS}(p, r)) = p(1 - p)$, which is the implied variance for any one of the events. (When the events are assessed with different probabilities, the prevision for the Brier score is the average of their implied variances.)

**b.** In the extreme case that the $N$ considered events are also regarded independently, the assertor's prevision for the calibration score is $P(C(p, r)) = p(1 - p)/N$, which is the implied variance for the proportion of 1's among the assessed events.

Thus, in this case, one's prevision for the calibration score converges to 0 as the number $N$ of events assessed increases. Moreover, this assessor becomes more and more certain that the calibration score will equal 0, even though there is no guarantee that it will equal 0. Do notice that in the example of finite $N = 10$ displayed in the table, it is only when the calibration score equals 0 that the expected value of the Brier score is achieved. But in these cases, the expected value of the calibration score, $p(1 - p)/N$, is not achieved.

**c.** In one extreme case of dependence, suppose the assertor is certain that the sum of the 10 events equals 2. (It may even be the case that this is a logical restriction on the events.) In such a case, this person must be certain of achieving a calibration score of 0, the entire Brier score then equaling the refinement score.

In the case of the logical restriction on the sum, this result is guaranteed since there is only one possibility for the value of $r = 2/10$. But in the case that the assessor is merely certain about something that is not logically required, a calibration score of 0 is not guaranteed.

**d.** In another extreme case of dependence, suppose logical restrictions require that the sum of the $N$ events equal 0 or $N$. To cohere with the assertion of each $P(E_i) = p$, it must be asserted concomitantly that $P(\sum_{i=1}^{N} E_i = N) = p$ and $P(\sum_{i=1}^{N} E_i = 0) = 1 - p$. In such a case, this person must be certain of achieving nonzero calibration score and a refinement score of 0.

In this case, the achieved Brier score is composed exclusively of its calibration component, and it might equal either its lowest possibility, $(0 - p)^2$, or its highest possibility, $(1 - p)^2$. (These would be the lowest and highest, respectively, presuming $p < .5$, as in our example.) The only possible score columns in Table 6.1 would be those for which $r = 0$ and $r = 1$.

We have examined an array of situations in which several events are assessed by a proper score, the Brier Score, for the purpose of displaying how differently the calibration component might figure into the total score. Because the calibration score alone is not a proper scoring rule, it surely does not rank as an important evaluation device in assessing the validity of probability assessments, as has sometimes been proposed. However, in view of what we have seen, when the Brier score is near to the prevised score of an assessor, the calibration component of the score will be minimal. It plays a larger component in achieved scores that turn out to be much worse or much better than expected by the person whose previsions are being evaluated.

We have examined the separation of the Brier score into a calibration component and a refinement component, but the principle behind such a separation is much more general. DeGroot and Fienberg (1982, 1983) have proved that every proper scoring rule for a sequence of probability distributions can be decomposed into a calibration component and a refinement component. Concern with calibration on its own is known to motivate the distortion of prevision pronouncements, for it does not lead to a proper scoring procedure. The only question requiring further comment is how ought one respond to comparative empirical evaluations of competing scientific theories represented by different prevision assertions.

### 6.6.4 Coherent Responses to a Proper Score Comparison between Different Opinion Structures over Selected Quantities

Consider the extreme possibility that you assert your joint probability distribution regarding a sequence of quantities. Embedded within your asserted distribution is the sequence of conditional distributions that are scored according to a specified scoring rule, say the logarithmic score. How should you react to the observed values of your score as it accumulates? When your opinions are being compared with someone else's who thinks differently from you, how should you react if your accumulating score appears to be better or worse than this person?

A surprising but defensible answer is "not at all." The sequence of conditional distributions that are being scored purport to express your conditional probability assessments for each subsequent observation conditioned upon the specific data that has determined your accumulated score. The score you achieve on the basis of this data is merely a summary of the data. Thus, it cannot be any more informative regarding subsequent quantities than is the data itself which you are already conditioning on when moving through the sequence of conditional distributions being scored. In assessing your joint distribution which implies these conditional distributions, you have already assessed your conditional attitude toward the next observation based upon the observations which determine your score. Presuming that you are learning "in the small," your reaction to the score is already encompassed in the computation of your current conditional distribution via Bayes' theorem.

Having said that, it may be an interesting project to derive a form of a distribution as a mixture of two different distributions where the mixing weights over the two distributions are specified functions of the scores achieved by the two distributions separately. When margined over your opinions about these two separate scores, this construction strategy would yield a joint distribution of its own for the data sequence, which itself could be scored against an appropriate, even the same, scoring rule.

But there is more to say about scientific reaction to a sequence of comparative accumulating scores. It is quite evident in the array of scores displayed in Table 6.1, and it is true very generally, that the score one achieves for any specific collection of assertions may be smaller, about the same, or larger than one had expected. Suppose, for example, that all 10 of the events $\mathbf{E}_{10}$ that are assessed identically with $P(E_i) = .2$ are observed to equal 1. In this case, you the proponent would receive the score $(C, R, \text{BS}) = (.64, 0, .64)$, the worst of the scores you could possibly achieve. Two reactions are possible, or some mixture of the two.

You might feel rather unlucky, assured that your assessment of your probabilities was quite appropriate. In the case of considered independence, for example, you would have assessed the event that all 10 events equal 1 with a probability of only $(.8)^{10} = .1024(10^{-6})$. You could be left wondering "Why did I have such bad luck?" Content with such a reaction, you would be insulated from learning from this experience and could happily continue to learn in the small via Bayes' theorem, computed with respect to your current conditional distribution function for whatever you are going to observe next.

On the other hand, you might feel rather stupid, wondering why you had assessed this occurrence with such small probability. Hoping to learn in the large, you might seek out someone who had assessed them with high probabilities, to find out what you might learn about life as it pertains to still other quantities that concern you. (You may even seek out a proponent of an alternative scientific viewpoint, whose assertion scores are being compared with yours, trying to understand better what you may be missing in that viewpoint.) Such a venture would lead you into the possibility we have termed "learning in the large," in which you completely reassess your opinions about many quantities on the basis of some new understanding. There can be no numerical rule identifying your new probability distribution from your old one. What is needed in such real learning is not some numerical procedure for changing your old conditional distribution into a new one, but the clarifying process of understanding nature in a new, albeit still uncertain, way that motivates you to assert different previsions than you did before.

Learning in this sense is a matter not of a mechanical transformation of distribution functions but the creation of a new understanding. The easiest case to describe occurs when you find you decide to share an understanding and agree with the prevision assertions of someone else, perhaps again your scientific challenger, whom you could not understand before. More complex is the learning that occurs when you begin to understand things in a way that no one else has already described.

Having recognized that anyone may always regard a bad score as due to luck, we should also notice that a good score may well be attributed to luck, too. One thing that comparative sequential scoring procedures allows is the comparison of your achieved score with your prevised score as well. This can aid in your assessment of the appropriate reaction to your achieved score, balancing your sense of luck with your desire to learn something that you don't already know. Whereas there can be no dismissal of the "just luck" defense of a scientific point of view out of hand, at some point the proponent of a poorly scoring forecast will be laughed out of court.

Data analytic studies that are meant to distinguish the evidence regarding two (several) different scientific viewpoints, whose proponents assert different distributions or previsions, should have the goal of displaying every feature of the scoring experience that would be relevant to both proponents' determining how to balance the two dimensions of their personal reactions to their scores. Thus, the relevant comparisons would include not merely the accumulating scores of proponents A's and B's assertions but also each of their accumulating previsions for their own scores and for the other's scores. The relevant sequences of accumulating computations should include the score of proponent A's assertions; A's prevision for the score of A; A's prevision for the score of B; B's score; B's prevision for the score of B; and B's prevision for the score of A:

$$S(X, P_{\mathrm{A}}(X)), \quad P_{\mathrm{A}}[S(X, P_{\mathrm{A}}(X))], \quad P_{\mathrm{A}}[S(X, P_{\mathrm{B}}(X))],$$
$$S(X, P_{\mathrm{B}}(X)), \quad P_{\mathrm{B}}[S(X, P_{\mathrm{B}}(X))], \quad P_{\mathrm{B}}[S(X, P_{\mathrm{A}}(X))].$$

The result you proved in problem 1 of Section 6.4.2 implies, interestingly, that

$$P_{\mathrm{A}}[S(X, P_{\mathrm{B}}(X))] = P_{\mathrm{A}}[S(X, P_{\mathrm{A}}(X))] + S(P_{\mathrm{A}}(X), P_{\mathrm{B}}(X)),$$
$$P_{\mathrm{B}}[S(X, P_{\mathrm{A}}(X))] = P_{\mathrm{B}}[S(X, P_{\mathrm{B}}(X))] + S(P_{\mathrm{B}}(X), P_{\mathrm{A}}(X)).$$

Attention should be given not merely to the cumulative totals of these six series but to the pattern of their accumulation. In the cumulative total a very good score for a single observation can mask an otherwise poor record of fairly regular bad scores.

Despite the theoretical understandings of scoring procedures displayed in this chapter, our experience with evaluating close scores of competing distribution assertions in the ways suggested is rather limited. The article of Blattenberger and Lad (1988) is the most easily available example of an interesting comparison. One distribution (of six compared) which one might be tempted to rule out of worthy consideration on the basis of very poor scores early in a sequence of forecasts subsequently achieves a fine record of scored forecasts, rating it as a serious contender for the leading score later in the sequence. Hill (1994b) provides another example. In other research, Fountain, Macfarland, and McCosker (1995) and Fountain and McCosker (1995) use comparisons of probabilities assessed that scores would be better or worse than actually achieved, in their considerations of economic experiments on utility assessments.

In Chapter 8 we compute a simple example of the sequential logarithmic scores of three distributions and the quadratic scores for their associated previsions in a problem of forecasting New Zealand's dollar exchange rate against the U.S. dollar during the first five years of New Zealand's floating exchange rate regime, beginning in March 1985.

CHAPTER 7

# The Multivariate Normal Distribution and Its Mixtures

*This shoe and this boot were so close in colour, the one to the other, and so veiled, as to their uppers, in the first place by the trousers, and in the second by the greatcoat, that they might almost have been taken not for a shoe on the one hand, and on the other for a boot, but for a true pair, of boots, or of shoes, had not the boot been blunt, and the shoe sharp, at the toe.*
*In this boot, a twelve, and in this shoe, a ten, Watt, whose size was eleven, suffered, if not agony, at least pain, with his feet, of which each would willingly have changed places with the other, if only for a moment.*

*Watt*, SAMUEL BECKETT

## 7.1 INTRODUCTION

In this chapter, we methodically develop the structure of multivariate Normal distributions and their extension to Normal-Wishart-Normal mixtures, producing the multivariate $t$-distribution and the matrix-$t$-distribution. The theory of forecasting using the mixture normal family of distributions provides insight into the general structure of coherent statistical inference, as well as practical procedures for application. Through its inherent linear regression structure, the distribution provides an insightful introduction to more general forms of linear regression analysis, which we discuss in Chapter 8.

By the conclusion of the present chapter, we shall have developed a convenient functional representation of probability distributions for large arrays of quantities, such as the example of macroeconomic forecasting we described in Section 5.3.5.2. There we described a temporal sequence of vectors $\mathbf{X}_1, \ldots, \mathbf{X}_\tau$, each of which represents $K$ distinct quantities measured during the subscripted time period. Algebraically, such a sequence can be represented either as a single "stacked" vector $\mathbf{X}_{K\tau} = (\mathbf{X}_1^T, \ldots, \mathbf{X}_\tau^T)^T$ or as a matrix $\mathbf{M}_{K,\tau}$ whose $\tau$ columns are the $K$-dimensional vectors $\mathbf{X}_1, \ldots, \mathbf{X}_\tau$. As we shall see, the most convenient way to think about such a sequence of vectors depends on what you are willing to assert about them in terms of your own uncertain knowledge. Since it will take us this entire chapter and even part of the next to formulate representative probability distributions for such a sequence of vectors, we shall be patient in the development, addressing some practical inference problems about less complicated groups of observations along the way. Our early applications will focus on output yields in grain and dairy production, which provide natural examples of quantities assessed exchangeably with mixture-Normal distributions.

We begin with a specification of the multivariate Normal distribution and its properties, proceeding subsequently to an analysis of the family of mixture distributions.

## 7.2 THE MULTIVARIATE NORMAL DISTRIBUTION AND ITS PROPERTIES

Consider a $K$-dimensional vector of quantities, $\mathbf{X}$, for which you have assessed your prevision vector $P(\mathbf{X}) \equiv \boldsymbol{\mu}$, and your variance-covariance matrix $P[(\mathbf{X}-\boldsymbol{\mu})(\mathbf{X}-\boldsymbol{\mu})^{\mathrm{T}}] \equiv \boldsymbol{\Sigma}$. We presume throughout that the rank of the covariance matrix $\boldsymbol{\Sigma}$ equals $K$, and often we use the inverse of your variance-covariance matrix, which is called your *precision* matrix, denoted by $\boldsymbol{\Pi} \equiv \boldsymbol{\Sigma}^{-1}$. We don't subscript $\mathbf{X}$ by $K$. When we do use subscripts on $\mathbf{X}$, it will be to denote interesting partitions of the vector into subvectors. Unless stated otherwise, you should presume that a boldface vector without subscripts has dimension $K$.

Denoting asserted prevision values by $\boldsymbol{\mu}$, $\boldsymbol{\Sigma}$, and $\boldsymbol{\Pi}$ derives from historical precedents which have a motivation of their own. For our purposes, it serves as a reminder of the distinction that these are not measurements of historic activity of an economy or of any other natural process. Rather, these are numerical assessments made by you, a forecaster, in specifying your uncertain knowledge about measured quantities, which are denoted generically by bold Roman letters, $\mathbf{X}$.

The representation of your knowledge about $\mathbf{X}$ by a multivariate Normal distribution requires a much more extensive specification than the mere assertion of your previsions for the quantities and your assertion of covariance among them. For it would identify your probability for every event that is representable as a function of the measurements $\mathbf{X}$.

***Definition 7.1.*** Your uncertain knowledge about a quantity vector $\mathbf{X} = (X_1, \ldots, X_K)^{\mathrm{T}}$ is said to specify a *$K$-dimensional multivariate Normal distribution* with location vector $\boldsymbol{\mu} = P(\mathbf{X})$, variance-covariance matrix $\boldsymbol{\Sigma} \equiv P[(\mathbf{X}-\boldsymbol{\mu})(\mathbf{X}-\boldsymbol{\mu})^{\mathrm{T}}]$, and precision matrix $\boldsymbol{\Pi} \equiv \boldsymbol{\Sigma}^{-1}$ if your joint distribution function for $\mathbf{X}$ can be represented approximately by the integral of the density function

$$f(\mathbf{x}) = (2\pi)^{-K/2} |\boldsymbol{\Pi}|^{1/2} \exp\{-\tfrac{1}{2}(\mathbf{x}-\boldsymbol{\mu})^{\mathrm{T}} \boldsymbol{\Pi} (\mathbf{x}-\boldsymbol{\mu})\},$$

which is called the *K-dimensional multivariate Normal density function* for $\mathbf{X}$. (The symbol $\pi$ denotes the real number pi.) The distribution function is thus defined by

$$F(\mathbf{t}) = \int_{-\infty}^{t_1} \int_{-\infty}^{t_2} \cdots \int_{-\infty}^{t_K} f(\mathbf{x})\, dx_K \cdots dx_1$$

for any $\mathbf{t} = (t_1, \ldots, t_K)^{\mathrm{T}}$ in the realm of $\mathbf{X}$. For approximation purposes this realm $\mathscr{R}(\mathbf{X})$ is presumed to be real, $K$-dimensional space, denoted by $\mathbb{R}^K$. We write $\mathbf{X} \sim N_K(\boldsymbol{\mu}, \boldsymbol{\Sigma} = \boldsymbol{\Pi}^{-1})$ to denote that you assert such a distribution for $\mathbf{X}$.

□

Your willingness to assert your opinions about $\mathbf{X}$ approximately in the form of a multivariate Normal distribution can be motivated only by a full understanding of the structure of inferences that this form of distribution implies. Our purpose in this section is to develop this understanding.

At some point in our study (Section 7.4) it will simplify matters to parameterize the multivariate normal family members by the precision matrix; we then write $\mathbf{X} \sim N_K(\boldsymbol{\mu}, \boldsymbol{\Pi})$. But for now we identify the members via the usual parameterization of $N_K(\boldsymbol{\mu}, \boldsymbol{\Sigma})$.

## PROBLEMS

**1.** Invert $\boldsymbol{\Sigma}$ in a $2 \times 2$ case and show that $\pi_{12} = -\sigma_{12}/(\sigma_{11}\sigma_{22} - \sigma_{12}^2)$. Thus, the "coprecision" for two quantities considered together has the reverse sign of the covariance between them.

**2.** Invert $\boldsymbol{\Sigma}$ in a $3 \times 3$ case and note that $\pi_{12} = (\sigma_{13}\sigma_{23} - \sigma_{12}\sigma_{33})/|\boldsymbol{\Sigma}|$. (Remember that $\boldsymbol{\Sigma}$ is symmetric.) Thus, the value of the coprecision between two quantities depends on your assessment of variance and covariance among all the quantities included as components of the vector $\mathbf{X}$. In eliciting one's uncertain knowledge, it is much easier to *assess* $\boldsymbol{\Sigma}$ and *compute* $\boldsymbol{\Pi}$ as its inverse than vice versa. For the assessment of $\boldsymbol{\Sigma}$ demands that you think about quantities only two at a time, no matter what is the dimension of $\mathbf{X}$, whereas the direct assessment of any component of $\boldsymbol{\Pi}$ demands that you think jointly about complicated functions of all the quantities under consideration at once. However, in the computation of sequential forecasts, it is often more convenient to work algebraically in terms of the $\boldsymbol{\Pi}$ matrix, as we shall see.

**3.** For the case of $K = 1$, show that the normal density function achieves its peak at $x = \mu$ and that it inflects at $x = \mu \pm \sigma$. For $K = 2$ write the bivariate Normal density in the form of a function of two variables.

**4.** What is your prevision for $W$, the yield of wheat in bushels from a selected acre of Kansas farmland planted with wheat in 1982? What is your variance for $W$?

Assess your median, quartiles, and octiles for $W$. [N.B. A bushel is a unit of dry measurement equal to 4 pecks, or 2150.42 cubic inches. This constitutes a cube about 13 inches, or 33 centimeters on a side. One acre is defined as 1/640 of a square mile. It equals about two fifths of a hectare, about the size of a football field. How many bushels (cubes) of wheat grain would be harvested from one acre (one football field)?]

**5.** Using a well-known result, which we discuss in due time, that $X \sim N_1(\mu, \sigma^2)$ implies $Z = (X - \mu)/\sigma \sim N_1(0, 1)$, use a Normal distribution table to determine the octiles of a Normal distribution with your prevision and variance. How similar are the octiles of this Normal distribution to those which you assessed for the wheat yield from this acre?

**6.** Now let me tell you that in 1982 there were 77.9 million acres of wheat planted in the United States, and the total yield was 2.77 billion (that is, 1000 million) bushels. Thus, the U.S. national average yield of wheat per acre that year was 36 bushels. Repeat the assessment and approximation sequence described in problems 4 and 5. I comment on problems 4–6 after stating a final technical problem.

**7.** If you are in practice with multidimensional integrals, it is not hard to show that the moment generating function associated with the distribution of $\mathbf{X} \sim N_K(\boldsymbol{\mu}, \boldsymbol{\Sigma})$ is $M_{\mathbf{X}}(\mathbf{t}) \equiv P[\exp(\mathbf{t}^T\mathbf{X})] = \exp[\mathbf{t}^T\boldsymbol{\mu} + (1/2)\mathbf{t}^T\boldsymbol{\Sigma}\mathbf{t})$. For scoring rule purposes, the negative entropy is $P[\log f(\mathbf{X})] = -(K/2)[1 + \log(2\pi)] + (1/2)\log|\boldsymbol{\Pi}|$ and $\int_{\mathbb{R}^K} f^2(\mathbf{x})\,d\mathbf{x} = 2^{-1/2}(2\pi)^{-K/2}|\boldsymbol{\Pi}|^{1/2}$.

In objective statistical theory it is proposed that certain quantities, such as the wheat yield from an acre are *generated* by a Normal distribution. In applications, it is taken by an objectivist to be a very serious question whether such and such data are generated by a Normal distribution or by some other distribution. Although various curve-fitting procedures have been developed in an attempt to formulate an objective "probabilistic" answer, such a question is unanswerable in principle, because the "generating distribution" is not observable itself. Only the "randomly generated outcomes" are observable.

On the contrary, in the operational subjective theory there is no such unanswerable question as to whether a distribution is Normal or not. Either your opinions are expressible approximately in this form, or they aren't. If they are, you may use the properties of this distribution to simplify your analysis. If they aren't, don't. You decide. Your willingness to use approximating distributions of any parametric form should be based on your detailed understanding of the mathematical properties of the distribution. After considering your prevision for a quantity and some of your quantiles, you might decide that your probability distribution for $X$ is not very different from a Normal distribution. On the other hand, if you recognize that your distribution is not Normal, you might also identify what type of algebraic adjustments would be needed to represent your own distribution more accurately. In this spirit, let us begin to identify the salient features of the multivariate Normal distribution.

### 7.2.1 Marginal and Conditional Distributions for the Components of X

The following theorem and discussion involve the matrix partitioning of $\mathbf{X}$, $\boldsymbol{\mu}$, and $\boldsymbol{\Sigma}$. Suppose the $K$-dimensional vector $\mathbf{X}$ is partitioned into two subvectors as $(\mathbf{X}_1, \mathbf{X}_2)^{\mathrm{T}}$, where the dimension of each $\mathbf{X}_i$ is $K_i \times 1$, with $K_1 + K_2 = K$. The components of $\boldsymbol{\mu}$ are partitioned conformably as $(\boldsymbol{\mu}_1, \boldsymbol{\mu}_2)^{\mathrm{T}}$, and the matrix $\boldsymbol{\Sigma}$ is partitioned into component matrices $\boldsymbol{\Sigma}_{11}, \boldsymbol{\Sigma}_{12}, \boldsymbol{\Sigma}_{21}$, and $\boldsymbol{\Sigma}_{22}$, where the dimension of $\boldsymbol{\Sigma}_{ij}$ is $K_i \times K_j$.

**Theorem 7.1.** If you assert $\mathbf{X} \sim N_K(\boldsymbol{\mu}, \boldsymbol{\Sigma})$, then for any partitioning of $\mathbf{X}$ into $(\mathbf{X}_1, \mathbf{X}_2)^{\mathrm{T}}$, coherency requires that your marginal distributions for either subvector $\mathbf{X}_i$ be specified as $\mathbf{X}_i \sim N_{K_i}[\boldsymbol{\mu}_i, \boldsymbol{\Sigma}_{ii}]$ and that your conditional distribution for either $\mathbf{X}_j|(\mathbf{X}_i = \mathbf{x}_i)$ be specified as

$$\mathbf{X}_j|(\mathbf{X}_i = \mathbf{x}_i) \sim N_{K_j}[\boldsymbol{\mu}_j + \boldsymbol{\Sigma}_{ji}\boldsymbol{\Sigma}_{ii}^{-1}(\mathbf{x}_i - \boldsymbol{\mu}_i), \quad \boldsymbol{\Sigma}_{jj} - \boldsymbol{\Sigma}_{ji}\boldsymbol{\Sigma}_{ii}^{-1}\boldsymbol{\Sigma}_{ij}].$$

*Proof*. The proof involves showing that the joint multivariate Normal density for $\mathbf{X}$ can be factored into the product of the specified multivariate densities for $\mathbf{X}_i$ and for $\mathbf{X}_j|(\mathbf{X}_i = \mathbf{x}_i)$. You can provide it yourself. Merely write the product of the marginal density function for $\mathbf{X}_1$ at $\mathbf{x}_1$ with the conditional density function for $\mathbf{X}_2$ at $\mathbf{x}_2$ given $(\mathbf{X}_1 = \mathbf{x}_1)$. Perform the necessary algebraic simplification that equates this product function with the joint density function for $(\mathbf{X}_1, \mathbf{X}_2)$ at $(\mathbf{x}_1, \mathbf{x}_2)$. A complete algebraic derivation appears in Rao (1973, pp. 522–523).

□

Aside from its recognition that both the conditional and marginal distributions of subvectors are also multivariate Normal, the most important aspect of Theorem 7.1 is its specification of your prevision for the conditional quantities $\mathbf{X}_i|(\mathbf{X}_j = \mathbf{x}_j)$ as

$$P(\mathbf{X}_1|\mathbf{X}_2 = \mathbf{x}_2) = \boldsymbol{\mu}_1 + \boldsymbol{\Sigma}_{12}\boldsymbol{\Sigma}_{22}^{-1}(\mathbf{x}_2 - \boldsymbol{\mu}_2)$$

and

$$P(\mathbf{X}_2|\mathbf{X}_1 = \mathbf{x}_1) = \boldsymbol{\mu}_2 + \boldsymbol{\Sigma}_{21}\boldsymbol{\Sigma}_{11}^{-1}(\mathbf{x}_1 - \boldsymbol{\mu}_1).$$

Thus, the conditional prevision vectors are specified by linear functions of the conditioning quantities. These linear equations are commonly called the *linear regression equations* of $\mathbf{X}_i$ on $(\mathbf{X}_j = \mathbf{x}_j)$. Geometrically they specify hyperplanes in the $\mathbf{x}_j$ variables that identify each component of the vector $P(\mathbf{X}_i|\mathbf{X}_j = \mathbf{x}_j)$ as a function of the conditioning observation values. Denoting $\boldsymbol{\beta}_{i \cdot j} \equiv \boldsymbol{\Sigma}_{ij} \boldsymbol{\Sigma}_{jj}^{-1}$, the equations can be written $P(\mathbf{X}_i | \mathbf{X}_j = \mathbf{x}_j) = \boldsymbol{\mu}_i + \boldsymbol{\beta}_{i \cdot j}(\mathbf{x}_j - \boldsymbol{\mu}_j)$.

We study more general applications of a linear regression structure in the next chapter. It is worth noting here in the context of the multivariate Normal distribution that *both* of these regression equations, $P(\mathbf{X}_1 | \mathbf{X}_2 = \mathbf{x}_2)$ and $P(\mathbf{X}_2 | \mathbf{X}_1 = \mathbf{x}_1)$, are well defined. There is nothing in the defining opinion structure to denote either one as any more "valid" or "correct" than the other. Both $P(\mathbf{X}_1 | \mathbf{X}_2 = \mathbf{x}_2)$ and $P(\mathbf{X}_2 | \mathbf{X}_1 = \mathbf{x}_1)$ are meaningful assertions. The regression equations will be represented geometrically when we study the "elliptical symmetry" of the multivariate Normal density function shortly. For the moment, let us conclude our attention to the conditional distributions with a comment on the algebraic implications of assessing two quantities independently with a multivariate Normal distribution.

We have realized from the outset of our study of statistical inference that the direct assessment of quantities independently plays little role in operational subjective analysis. For the assessment of independence amounts to the assertion that nothing can be learned about one quantity from the observation of the other. However, as we expand our understanding, we shall find that some specific linear transformations of quantities assessed with a multivariate Normal distribution are *necessarily regarded independently*. For this reason, it is worthy of notice that quantities assessed with a multivariate Normal distribution are regarded independently if and only if their specified covariance matrix is diagonal. This is evident from Theorem 7.1, since every conditional distribution of $\mathbf{X}_i | (\mathbf{X}_j = \mathbf{x}_j)$ equals the marginal distribution of $\mathbf{X}_i$ if and only if all nondiagonal components of $\boldsymbol{\Sigma}$ equal 0.

## PROBLEMS

**1.** An example in two dimensions: Supposing the dimension of $\mathbf{X}$ is $2 \times 1$, reduce the notation for the conditional distribution of $X_2 | (X_1 = x_1)$ to see

$$X_2 | (X_1 = x_1) \sim N_1[\mu_2 + (\sigma_{12}/\sigma_{11})(X_1 - \mu_1), \sigma_{22}(1 - \rho^2)],$$

where $\rho \equiv \sigma_{12}/(\sigma_{11}\sigma_{22})^{1/2}$ is the correlation of $X_1$ with $X_2$ specified by the covariance matrix. The distribution of $X_i$ is $N_1(\mu_i, \sigma_{ii})$ for $i = 1$ or 2. Thus, the factor $1 - \rho^2$ in the conditional variance expression represents the proportional decline in your variance for $X_2$ that accrues from the conditioning of your assessment of $X_2$ upon $(X_1 = x_1)$, and vice versa.

**2.** The correlation between any two quantities is defined as the quotient of your assertions of their covariance and the product of their standard deviations: $\rho_{ij} \equiv \sigma_{ij}/(\sigma_{ii}\sigma_{jj})^{1/2}$. Show that the matrix of all correlations, $[\rho_{ij}]$, can be computed from $\boldsymbol{\Sigma}$ via the equation $[\rho_{ij}] = \boldsymbol{\Delta\Sigma\Delta}$, where $\boldsymbol{\Delta}$ is a diagonal matrix whose diagonal entries are $\delta_{ii} = 1/\sqrt{\sigma_{ii}}$.

### 7.2.2 Geometrical Features of a Multivariate Normal Density

The striking feature of the multivariate Normal density function,

$$f(\mathbf{x}) = (2\pi)^{-K/2}|\mathbf{\Pi}|^{1/2}\exp\{-\tfrac{1}{2}(\mathbf{x}-\boldsymbol{\mu})^{\mathrm{T}}\mathbf{\Pi}(\mathbf{x}-\boldsymbol{\mu})\},$$

is that it is constant for all vectors $\mathbf{x}$ that lie on an ellipse of the form

$$(\mathbf{x}-\boldsymbol{\mu})^{\mathrm{T}}\mathbf{\Pi}(\mathbf{x}-\boldsymbol{\mu}) = \sum_{i=1}^{K}\sum_{j=1}^{K}\pi_{ij}(x_i-\mu_i)(x_j-\mu_j) = c \geqslant 0.$$

Algebraic expressions of this form are called "quadratic forms in $\mathbf{x}-\boldsymbol{\mu}$ about $\mathbf{\Pi}$." Geometrically, all vectors $\mathbf{x}$ that satisfy such an equation lie on a multi-dimensional ellipse.

For various values of $c > 0$, the quadratic form equations determine a family of concentric ellipses, as shown in Figure 7.1. It displays three members of one such family of ellipses in two dimensions. This family is identified by the specification of $\mu_1 = 4$, $\mu_2 = 3$, $\pi_{11} = 1/5$, $\pi_{22} = 9/20$, and $\pi_{12} = -1/5$. (Equivalently, $\sigma_{11} = 9$, $\sigma_{22} = 4$, and $\sigma_{12} = 4$. Thus, $\rho_{12} = 2/3$.) Our first project, begun in the next subsection, is to study how the directions of the major and minor axes of a family of ellipses are determined by the components of its defining precision matrix, $\mathbf{\Pi}$ or, equivalently, by the covariance matrix, $\mathbf{\Sigma}$.

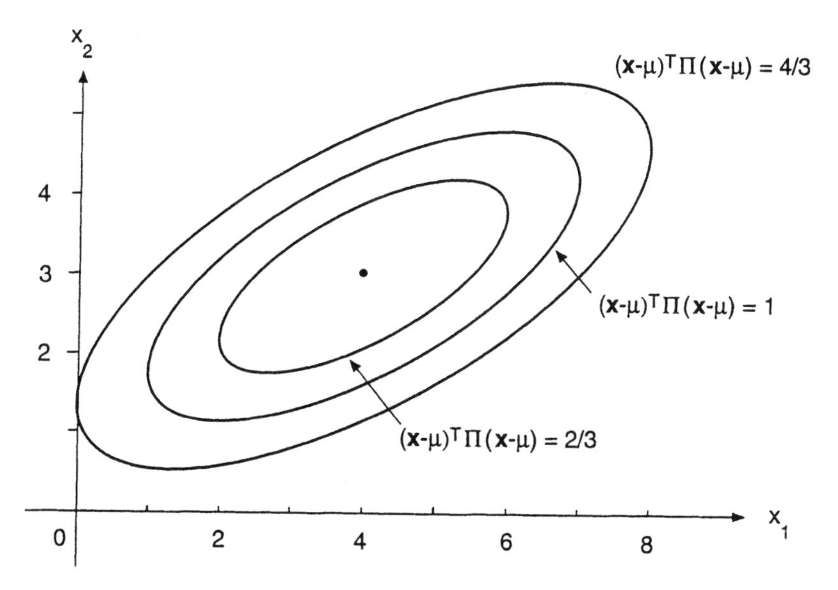


**Figure 7.1.** Equidensity contours at $c = 2/3$, $c = 1$, and $c = 4/3$ for the bivariate Normal density defined by $\mu_1 = 4$, $\mu_2 = 3$, $\sigma_{11} = 9$, $\sigma_{22} = 4$, and $\sigma_{12} = 4$. Thus, $\rho_{12} = 2/3$. Equivalently, $\pi_{11} = 1/5$, $\pi_{22} = 9/20$, and $\pi_{12} = -1/5$.

## PROBLEMS

1. Show the equivalence of the precisions and the variances stipulated beneath Figure 7.1.

2. Find the precision matrix that corresponds to the covariance matrix in which $\sigma_{11} = 9, \sigma_{22} = 4$, and $\sigma_{12} = 2$. Thus, $\rho_{12} = 1/3$. Notice that every component of this precision matrix differs from that in Figure 7.1, whereas the covariance matrices differ only in their $\sigma_{12}$ component.

3. Write a program in MATLAB that will generate a graphical display of the bivariate Normal density for various values of the variances $\sigma_{11}$ and $\sigma_{22}$ and the correlation $\rho_{12}$. Let $\sigma_{11} = 9$, $\sigma_{22} = 4$, and let $\sigma_{12}$ vary from $-5.7$ to 5.7 in steps of 1.9 to produce a variety of pictures. Notice that MATLAB also will display the elliptical contours of constant density. Use the commands "contour" and "subplot."

### *7.2.2.1 Major and Minor Axes for Ellipses of Specified Density*

The major and minor axes of the ellipses represented in Figure 7.1 are rotated away from the directions of the standard $(x_1, x_2)$ axes. We now examine how the angle of rotation for the ellipse is computable from the components of the covariance matrix $\mathbf{\Sigma}$. As it turns out, this algebraic development is neater when derived in terms of $\mathbf{\Sigma}$ rather than in terms of $\mathbf{\Pi}$. Figure 7.2 will provide a reference for our discussion. The single ellipse it displays is identical to the middle ellipse of Figure 7.1, except that it is translated to be centered at the origin (0, 0) instead of at (4, 3). Since the prevision vector $\boldsymbol{\mu}$ serves only to locate the center of the ellipse, the following algebraic development is simplified by presuming $\boldsymbol{\mu}$ to be the $\mathbf{0}$ vector.

Any ellipse specifies a unique circumscribed circle and a unique inscribed circle with their centers at the center of the ellipse. When the ellipse is centered at the origin, the equations describing these circles have the form $\mathbf{x}^T\mathbf{x} = c_1$ and $\mathbf{x}^T\mathbf{x} = c_2 < c_1$. The constant $c_i$ is the squared radius of the corresponding circle. The major and minor axes of the ellipse intersect these two circles at specific points. Any other vector on the ellipse lies on a circle with the same center, but with a squared radius value between $c_1$ and $c_2$.

Thus, the axis vectors of the ellipse can be determined algebraically as the solution to the following two extreme-value problems: Find the vectors $\mathbf{M}$ and $\mathbf{m}$ for which $\mathbf{M}^T\mathbf{M} = \max \mathbf{x}^T\mathbf{x}$ and $\mathbf{m}^T\mathbf{m} = \min \mathbf{x}^T\mathbf{x}$, both subject to the restriction that $\mathbf{x}^T\mathbf{\Sigma}^{-1}\mathbf{x} = c$.

The Lagrangian expressions for the two extreme-value problems have the same form: $L(\mathbf{x}, \lambda) = \mathbf{x}^T\mathbf{x} + \lambda(c - \mathbf{x}^T\mathbf{\Sigma}^{-1}\mathbf{x})$. The partial derivatives are

$$\partial_{\mathbf{x}} L(\mathbf{x}, \lambda) = 2\mathbf{x} - 2\lambda\mathbf{\Sigma}^{-1}\mathbf{x} \qquad \text{and} \qquad \partial_{\lambda} L(\mathbf{x}, \lambda) = c - \mathbf{x}^T\mathbf{\Sigma}^{-1}\mathbf{x},$$

where $\partial_{\mathbf{x}}(L(\mathbf{x}, \lambda)$ denotes the vector of partial derivatives of $L$ with respect to the

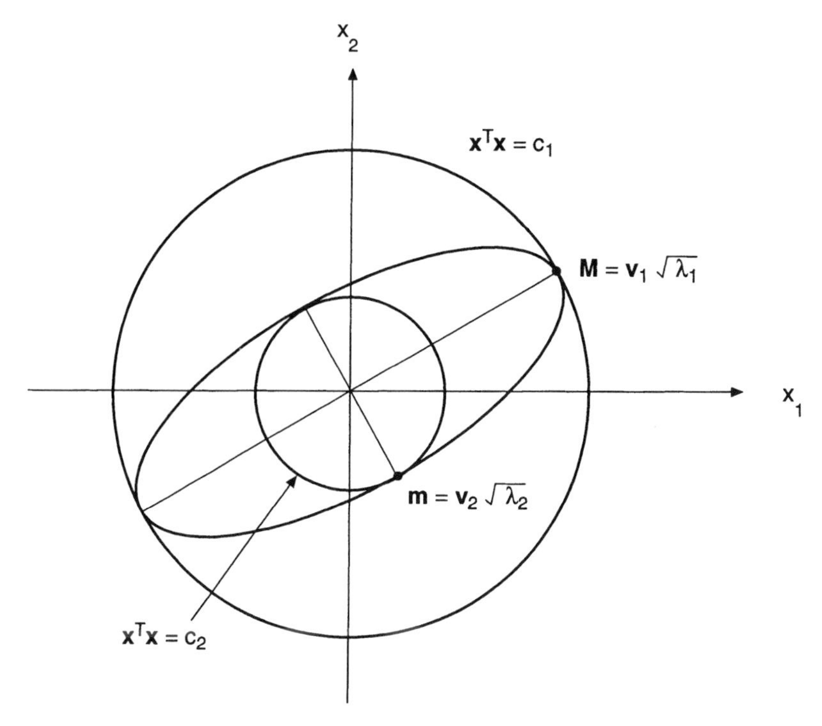


**Figure 7.2.** Circumscribed and inscribed circles for the ellipse $\mathbf{x}^{\mathrm{T}}\boldsymbol{\Sigma}^{-1}\mathbf{x} = 1$, which identify its major and minor axes.

components of $\mathbf{x}$. Setting $\partial_{\mathrm{X}} L(\mathbf{x}, \lambda) = \mathbf{0}$ requires that $\boldsymbol{\Sigma}^{-1}\mathbf{x} = \lambda^{-1}\mathbf{x}$ or, equivalently, $\boldsymbol{\Sigma}\mathbf{x} = \lambda\mathbf{x}$. Setting $\lambda_{\lambda} L(\mathbf{x}, \lambda) = 0$ requires that $\mathbf{x}^{\mathrm{T}}\boldsymbol{\Sigma}^{-1}\mathbf{x} = c$. Equivalently, then, $\mathbf{x}^{\mathrm{T}}\mathbf{x} = \lambda c$. The equation $\boldsymbol{\Sigma}\mathbf{x} = \lambda\mathbf{x}$ identifies the solution vectors $\mathbf{x}$ as the eigenvectors of $\boldsymbol{\Sigma}$, and the values of $\lambda$ as their corresponding eigenvalues. The constraining condition, $\mathbf{x}^{\mathrm{T}}\mathbf{x} = \lambda c$ specifies the normalization of the eigenvectors to have this squared length. We denote these normalized eigenvectors by $\mathbf{M} = \sqrt{\lambda_1 c}\,\mathbf{v}_1$ and $\mathbf{m} = \sqrt{\lambda_2 c}\,\mathbf{v}_2$, where $\lambda_1 > \lambda_2$ and $\mathbf{v}_1$ and $\mathbf{v}_2$ are the unit-length eigenvectors.

Thus, the *eigenvector* constituting the *major axis* of the ellipse for the bivariate Normal density is the one corresponding to the *larger eigenvalue* of $\boldsymbol{\Sigma}$. The minor axis of the ellipse is the eigenvector corresponding to the smaller eigenvalue of $\boldsymbol{\Sigma}$. Finally, the ratio of the lengths of the major and minor axes equals the square root of the ratio of the larger to the smaller eigenvalue. We review a numerical example shortly.

Although our geometrical example has been presented in only two dimensions, our algebraic analysis has identified characteristics of the quadratic form generally for $K$ dimensions. For a $K$-dimensional ellipsoid there are $K$ eigensolutions to the first-order conditions. In $K$ dimensions, each axis of the hyperellipse intersects a unique hypersphere at exactly two points of the ellipsoid, presuming that there are no repeated eigenvalues. The ordered sizes of the eigenvalues determine the ordered squared radii of these hyperspheres.

The following well-known, but seldom printed, formulae identify the eigenvalues and the unit-length eigenvectors of a $2 \times 2$ matrix $\mathbf{\Sigma}$:

$$\lambda_1 = (1/2)\{\sigma_{11} + \sigma_{22} + [(\sigma_{11} - \sigma_{22})^2 + 4\sigma_{12}^2]^{1/2}\},$$
$$\lambda_2 = (1/2)\{\sigma_{11} + \sigma_{22} - [(\sigma_{11} - \sigma_{22})^2 + 4\sigma_{12}^2]^{1/2}\},$$
$$v_{11} = |\sqrt{2}\sigma_{12}\{(\sigma_{22} - \sigma_{11})^2 + 4\sigma_{12}^2 + (\sigma_{22} - \sigma_{11})[(\sigma_{22} - \sigma_{11})^2 + 4\sigma_{12}^2]^{1/2}\}^{-1/2}|,$$
$$v_{12} = (2\sigma_{12})^{-1}\{(\sigma_{22} - \sigma_{11}) + [(\sigma_{22} - \sigma_{11})^2 + 4\sigma_{12}^2]^{1/2}\}v_{11},$$
$$v_{21} = |\sqrt{2}\sigma_{12}\{(\sigma_{22} - \sigma_{11})^2 + 4\sigma_{12}^2 - (\sigma_{22} - \sigma_{11})[(\sigma_{22} - \sigma_{11})^2 + 4\sigma_{12}^2]^{1/2}\}^{-1/2}|,$$
$$v_{22} = (2\sigma_{12})^{-1}\{(\sigma_{22} - \sigma_{11}) - [(\sigma_{22} - \sigma_{11})^2 + 4\sigma_{12}^2]^{1/2}\}v_{21}.$$

In this formulation, $\lambda_1$ is presumed to exceed $\lambda_2$. Thus, $\mathbf{v}_1 = (v_{11}, v_{12})^{\mathrm{T}}$ corresponds to the major axis of the ellipse, and $\mathbf{v}_2$ to the minor axis. Both $v_{11}$ and $v_{21}$ are positive in these formulas, arbitrarily. Thus, the quadrant of the major-axis eigenvector is determined by the sign of $\sigma_{12}$, as can be seen from the formula for $v_{12}$. Notice that $v_{22}$ has the opposite sign of $\sigma_{12}$. For the ellipse displayed in Figure 7.2, the major axis runs through in the first quadrant since $\sigma_{12}$ is positive.

Of course, all computations to generate eigenvalues and eigenvectors can be achieved quite easily with MATLAB. If Sigma is specified as a square matrix, then eig(Sigma) will produce the unit-length eigenvectors of Sigma along with their corresponding eigenvalues. The algebraic detail here and the problems suggested are produced merely to refresh your understanding of the concepts involved.

***Example 1.*** The ellipse displayed in Figure 7.2 reflects the following computations:

**a.** Remember that $\mathbf{\Sigma}$ is defined by $\sigma_{11} = 9$, $\sigma_{22} = 4$, and $\sigma_{12} = 4$. Thus, $\rho_{12} = 2/3$.

**b.** The eigenvalues of $\mathbf{\Sigma}$ are $\lambda_1 = [13 + \sqrt{89}]/2 \approx 11.217$ and $\lambda_2 = [13 - \sqrt{89}]/2 \approx 1.783$.

**c.** The corresponding unit-length eigenvectors of $\mathbf{\Sigma}$ are $\mathbf{v}_1(\lambda_1) = (v_{11}, v_{12})^{\mathrm{T}} \approx (.87464, .48477)$ and $\mathbf{v}_2(\lambda_2) = (v_{21}, v_{22})^{\mathrm{T}} \approx (.48477, -.87464)$. Multiplying these vectors by the square roots of their eigenvalues yields the major and minor axes of the ellipse displayed in the figure, for which $c = 1$.

**d.** Thus, the ratio of the lengths of the major and minor axes of any ellipse defined by $\mathbf{x}^{\mathrm{T}}\mathbf{\Sigma}^{-1}\mathbf{x} = c$ equals $(\lambda_1/\lambda_2)^{1/2} \approx (11.217/1.783)^{1/2} \approx 2.51$. □

Finally, we can specify precisely the rotation angle of the major axis of the ellipse away from the $x_1$-axis. Using the convention that $\mathbf{v}_1$ is the eigenvector defining the major axis of the ellipse (with $v_{11} > 0$), then $\theta \equiv \tan^{-1}(v_{12}/v_{11})$ is the angle that determines the rotation. To identify this angle, we can note from the formula for $v_{12}$ that the slope of the major axis of the ellipse is

$$v_{12}/v_{11} = (2\sigma_{12})^{-1}\{(\sigma_{22} - \sigma_{11}) + [(\sigma_{22} - \sigma_{11})^2 + 4\sigma_{12}^2]^{1/2}\}.$$

In the example displayed in Figure 7.2, $v_{12}/v_{11} \approx .55425$, and $\theta \approx 29°$.

## PROBLEMS

1. For comparative purposes, compute the major and minor axes of the ellipses specified by the covariance matrix $\mathbf{\Sigma}$ for which $\sigma_{11} = 9$, $\sigma_{22} = 4$, and $\sigma_{12} = 2$. Draw the resulting ellipses.

2. Using the given formulae, notice that the product $\lambda^1\lambda^2 = |\mathbf{\Sigma}|$ and that the sum $\lambda_1 + \lambda_2 = \text{trace}(\mathbf{\Sigma})$. These relations hold quite generally: the determinant of a matrix equals the product of its eigenvalues; the trace of a matrix equals the sum of its eigenvalues.

3. Notice that if $\sigma_{12} = 0$, then $\lambda_1 = \sigma_{11}$ and $\lambda_2 = \sigma_{22}$. (This presumes that $\mathbf{\Sigma}$ is constructed so that $\sigma_{11} > \sigma_{22}$, as in our example.)

4. Notice that if $\sigma_{11} = \sigma_{22}$, then $v_{21} = 1/\sqrt{2}$, and $v_{22}$ equals either $1/\sqrt{2}$ or $-1/\sqrt{2}$, depending on the sign of $\sigma_{12}$. Thus, the major axis of the ellipse is the 45° line either in the first quadrant (if $\sigma_{12}$ is positive) or in the fourth quadrant (if $\sigma_{12}$ is negative).

5. Referring to problem 1, determine the slope of the major axis of an isodensity ellipse, and determine the angle of its rotation relative to the $x$-axis. Notice that the correlation is lower and the degree of twist of the ellipse is smaller than for our displayed example. But this is not a general result, as the next problem shows.

6. Use your MATLAB program to generate the ellipses associated with the covariance matrix with $\sigma_{11} = \sigma_{22} = 1$ and $\sigma_{12} = 0, .3, .6$, and $.9$. The size of the correlation plays a role in rotating the ellipse, but it is modified by the relative size of the variances, as can be seen in the formula for the rotation angle.

#### *7.2.2.2 Situating the Regression Equations of $\mathbf{X}_i$ on $\mathbf{X}_j$*

We have identified the central axes of the ellipses of constant Normal density via the eigenvectors and eigenvalues of the covariance matrix. Our next project is to identify the location of the regression lines within the family of concentric ellipses.

The linear regression equations for $\mathbf{X}_i$ on $\mathbf{X}_j$ and for $\mathbf{X}_j$ on $\mathbf{X}_i$ intersect specific points on the ellipsoids of constant Normal density. We study the situation both

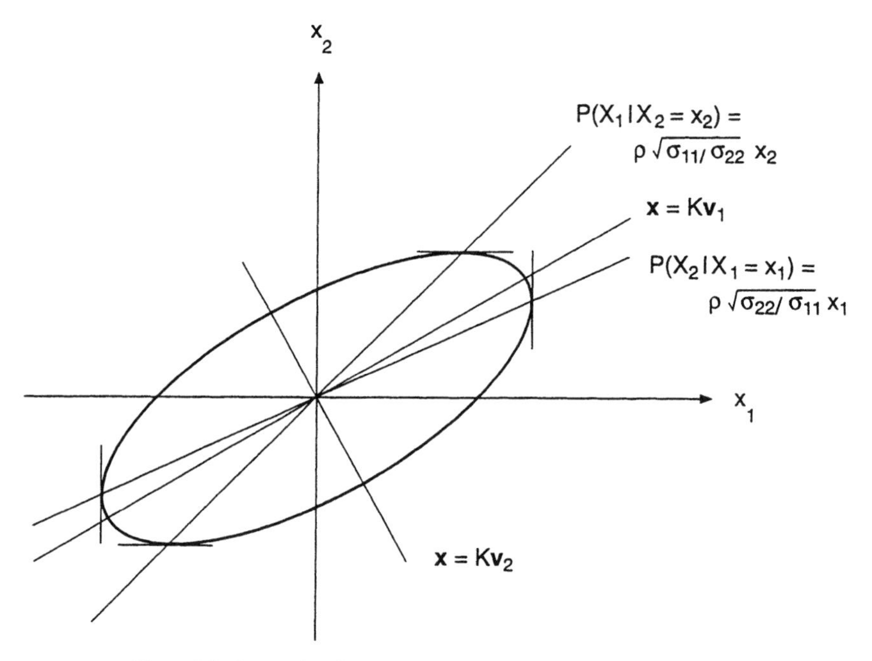


**Figure 7.3.** Regression lines situated in an ellipse of constant density.

algebraically and geometrically in only two dimensions, since the concepts and detail extend easily to many dimensions. Figure 7.3 provides a reference for the discussion. Again, the location of the bivariate Normal distribution is presumed to be the origin, $\boldsymbol{\mu} = \mathbf{0}$.

The quadratic form $\mathbf{x}^{\mathrm{T}}\boldsymbol{\Pi}\mathbf{x} = c$, which identifies the vectors on an ellipse, specifies implicitly the components of these vectors as functions of one another. Using the implicit function form

$$G(x_1, x_2) = \mathbf{x}^{\mathrm{T}}\boldsymbol{\Pi}\mathbf{x} - c = \pi_{11}x_1^2 + 2\pi_{12}x_1x_2 + \pi_{22}x_2^2 - c = 0,$$

we know that the derivative $dx_1/dx_2$, for example, is

$$dx_1/dx_2 = -\partial_{x2}G(x_1, x_2)/\partial_{x1}G(x_1, x_2),$$

where $\partial_{x2}G(x_1, x_2)$ denotes the partial derivative of $G$ with respect to $x_2$. Thus, $dx_1/dx_2 = 0$ when $\partial_{x2}G(x_1, x_2) = 0$. The points on the ellipse that satisfy this result are the points at which *vertical* lines are tangent to the ellipse. Assessing this differential equation for our implicit function $G(x_1, x_2)$ yields the equation

$$2\pi_{12}x_1 + 2\pi_{22}x_2 = 0, \quad \text{or equivalently,} \quad x_2 = -(\pi_{12}/\pi_{22})x_1 = (\sigma_{12}/\sigma_{11})x_1.$$

But this equation coincides with the regression equation of $X_2$ on $X_1$, which we have identified in Section 7.2.1 from the conditional distribution for $X_2$ given $(X_1 = x_1)$ as

$$P(X_2|X_1 = x_1) = (\sigma_{12}/\sigma_{11})x_1.$$

Thus, the regression line of $X_2$ on $X_1$ intersects any ellipse of constant density at the points where its tangent lines are vertical.

A similar analysis of the condition $dx_2/dx_1 = 0$ yields the regression line

$$P(X_1|X_2 = x_2) = (\sigma_{12}/\sigma_{22})x_2,$$

which intersects the isodensity ellipses at the points where *horizontal* lines are tangent to them. In order to recognize the regression line $P(X_1|X_2 = x_2)$ as a function of $x_2$ in Figure 7.3, you need to "look" at the graph from the perspective of standing behind it with your head tilted to the left at 90°, so the positive $x_1$-axis rises "upward" for you.

We can summarize the geometrical developments of this entire section by listing a general procedure for graphing an isodensity ellipse of a bivariate Normal distribution such as the one in Figure 7.3. It corresponds to our example specification of $\mathbf{\Sigma}$ by $\sigma_{11} = 9$, $\sigma_{22} = 4$, and $\sigma_{12} = 4$. Thus, the equation for the isodensity ellipse is

$$\mathbf{x}^{\mathrm{T}}\mathbf{\Sigma}^{-1}\mathbf{x} = \mathbf{x}^{\mathrm{T}}\mathbf{\Pi}\mathbf{x} = (1/5)x_{11}^2 - (2/5)x_{11}x_{12} + (9/20)x_{12}^2 = c.$$

**a.** Locate the center of the ellipse $(\mu_1, \mu_2)$: $(\mu_1, \mu_2) = (0, 0)$.

**b.** Determine the major and minor axes from the eigenvectors of the covariance matrix: $\mathbf{v}_1 \propto (.87464,\ .48477) \propto (1,\ .55)$, and $\mathbf{v}_2 \propto (.48477,\ -.87464) \propto (1, -2.2)$.

**c.** Determine the relative lengths of these axes from the square root of the ratio of their corresponding eigenvalues, $\sqrt{\lambda_1/\lambda_2} \approx (11.217/1.783)^{1/2} \approx 2.5$.

**d.** Draw the two regression lines through the center of the ellipse, with the appropriate slopes:

$$P(X_2|X_1 = x_1) = (\sigma_{12}/\sigma_{11})(x_1 - \mu_1) = (4/9)x_1,$$
$$P(X_1|X_2 = x_2) = (\sigma_{12}/\sigma_{22})(x_2 - \mu_2) = (4/4)x_2 = x_2.$$

**e.** Draw the ellipse so the lengths of its major and minor axes are in proper proportions, its tangent lines are orthogonal to these axes at the endpoints, and so its tangent lines have slope zero where they intersect the two regression lines.

Because the multivariate Normal distribution has constant density over any ellipsoid of the form $(\mathbf{x} - \boldsymbol{\mu})^{\mathrm{T}}\mathbf{\Pi}(\mathbf{x} - \boldsymbol{\mu}) = c$, it is said to display the feature of *elliptical symmetry*. This is an important general concept in our uncertain assessments of quantities that can have even broader application than multivariate Normal assessments. But its features are subsumed in the simpler concept of *spherical symmetry*, because a locus of points that form an ellipse in the space of pairs, $(x_1, x_2)$ identifies a sphere in a certain linear transformation of the space to $T(x_1, x_2)$. We study some details of this transformation in the following sections.

### 7.2.3 Implied Distributions for Linear Transformations of Quantities Assessed with a Multivariate Normal Distribution

Coherency requires another important property to your uncertain knowledge when you assert a multivariate Normal distribution for a vector $\mathbf{X}$—that you must assess any quantities defined by linear combinations of this vector with a multivariate normal distribution as well. Let us state the result succinctly as a theorem.

**Theorem 7.2.** If you assess your distribution for $\mathbf{X}$ as $N_K(\boldsymbol{\mu}, \boldsymbol{\Sigma})$, then for any matrix $\mathbf{A}$ with dimension $L \times K$, coherency of prevision demands that you assess your distribution for $\mathbf{Y} \equiv \mathbf{AX}$ as $N_L(\mathbf{A}\boldsymbol{\mu}, \mathbf{A}\boldsymbol{\Sigma}\mathbf{A}^T)$.

*Proof*. The proof is immediate, using the moment generating function for $\mathbf{AX}$, since $P[\exp(\mathbf{t}^T\mathbf{AX})] = \exp[\mathbf{t}^T\mathbf{A}\boldsymbol{\mu} + (1/2)\mathbf{t}^T\mathbf{A}\boldsymbol{\Sigma}\mathbf{A}^T\mathbf{t}]$, which defines the moment generating function for the specified multivariate normal distribution, $M_Y(\mathbf{t}) = M_{\mathbf{AX}}(\mathbf{t})$. (Refer to its display in problem 7 of Section 7.2.) □

In defining the moment generating function for a vector of quantities, the argument vector of, say, $M_X(\mathbf{t}) \equiv P[\exp(\mathbf{t}^T\mathbf{X})]$, is a vector $\mathbf{t}$ that is conformable for multiplication with $\mathbf{X}$. Thus, when considering the moment generating function for $\mathbf{AX}$, defined by $P[\exp(\mathbf{t}^T\mathbf{AX})]$, the coefficients of $\mathbf{X}$ specified by $\mathbf{t}^T\mathbf{A}$ can also be understood as an arbitrary vector that is conformable for multiplication with $\mathbf{X}$. That is why the prevision for $[\exp(\mathbf{t}^T\mathbf{AX})]$ can be written directly from our knowledge of the moment generating function form for the multivariate Normal distribution $N_K(\boldsymbol{\mu}, \boldsymbol{\Sigma})$. In most applications of Theorem 7.2, the rank of $\boldsymbol{\Sigma}$ equals $K$, and the rank of $\mathbf{A}$ equals $L \leqslant K$.

Theorem 7.2 has extensive applications, since in many situations we are interested in linear combinations of quantities: their sums, averages, or weighted sums such as the total price of a bundle of several commodities with distinct prices. At this point in our study, the theorem also serves to complete our understanding of the *spherical* symmetry of the multivariate Normal distribution as follows.

Since the covariance matrix $\boldsymbol{\Sigma}$ is symmetric and positive-definite, we know there is a unique matrix $\mathbf{T}$ having the property that $\mathbf{T}\boldsymbol{\Sigma}\mathbf{T}^T = \mathbf{I}_K$, the $K$-dimensional identity matrix. (Again, we are assuming that $\boldsymbol{\Sigma}$ has full rank.) This matrix $\mathbf{T}$ is composed by rows of the eigenvectors of $\boldsymbol{\Sigma}$, but with their lengths normed to equal the inverse square root of their corresponding eigenvalues. Theorem 7.2, then, implies that the vector quantity $\mathbf{Z} = (\mathbf{TX} - \mathbf{T}\boldsymbol{\mu})$ be assessed as $N_K(\mathbf{0}, \mathbf{I}_K)$. This result motivates the statement that the multivariate Normal distribution is *spherically symmetric*, since the density for $\mathbf{Z}$ is obviously constant on spheres. The vector quantity $\mathbf{Z}$ is said to be the *standard form* of the vector $\mathbf{X}$. Any vector $\mathbf{X}$ assessed with prevision $P(\mathbf{X}) = \boldsymbol{\mu}$ and covariance matrix $\boldsymbol{\Sigma}$ can be transformed into standard form via such a matrix $\mathbf{T}$. However, only if $\mathbf{X}$ is assessed with a multivariate Normal distribution can $\mathbf{Z}$ be assessed with the standard spherical *Normal* distribution.

A second aspect of this transformation is that in the context of multivariate Normal distributions, the components of the vector $\mathbf{TX}$ are necessarily regarded *independently*, even if the components of $\mathbf{X}$ are not regarded independently. (For the covariance matrix of $\mathbf{TX}$ is diagonal, even if the covariance matrix for $\mathbf{X}$ is not.)

### 7.2.4 Implied Distributions for Quadratic Transformations of Quantities Assessed with a Multivariate Normal Distribution

The algebraic simplicity of multivariate Normal distributions extends to simple forms of implied distributions for *quadratic* functions of a vector. In this section, we study the distribution for the inner product (squared length) of a vector. (We study the matrix defined by the outer product in due time in Section 7.5, wherein the Wishart distribution is defined.) The main result of this section can be stated directly as a theorem.

**Theorem 7.3.** If you assess your distribution for the $\mathbf{X}$ as $N_K(\boldsymbol{\mu}, \boldsymbol{\Sigma})$, then coherency requires that you assess your distribution for the value of the quadratic form $(\mathbf{x} - \boldsymbol{\mu})^{\mathrm{T}}\boldsymbol{\Pi}(\mathbf{x} - \boldsymbol{\mu})$ as $\chi^2_K$.

*Proof*. Using the matrix $\mathbf{T}$ of eigenvectors of $\boldsymbol{\Sigma}$, we know that $\mathbf{Z} = \mathbf{T}(\mathbf{X} - \boldsymbol{\mu})$ must be assessed as $N_K(\mathbf{0}, \mathbf{I}_K)$. Thus, the quantity $\mathbf{Z}^{\mathrm{T}}\mathbf{Z} = (\mathbf{x} - \boldsymbol{\mu})^{\mathrm{T}}\mathbf{T}^{\mathrm{T}}\mathbf{T}(\mathbf{x} - \boldsymbol{\mu})$ is a is a sum of squared quantities assessed independently, each with a $N_1(0, 1)$ distribution. Thus, each square is assessed independently with a $\chi^2_1$ distribution, according to Section 5.3.4. (We shall learn shortly why the sum of these squares must be asessed with a $\chi^2_K$ distribution, when we study the family of gamma distributions in detail.) Finally, the equivalence of $\mathbf{T}^{\mathrm{T}}\mathbf{T}$ with $\boldsymbol{\Pi}$ derives from standard properties of the eigenvector matrix $\mathbf{T}$. Since $\mathbf{T}^{\mathrm{T}}\boldsymbol{\Sigma}\mathbf{T} = \mathbf{I}$, $\mathbf{T}\mathbf{T}^{\mathrm{T}}\boldsymbol{\Sigma}\mathbf{T}\mathbf{T}^{\mathrm{T}} = \mathbf{T}\mathbf{T}^{\mathrm{T}}$, and thus $\mathbf{T}\mathbf{T}^{\mathrm{T}}\boldsymbol{\Sigma} = \mathbf{I}$. Thus, $\mathbf{T}\mathbf{T}^{\mathrm{T}} = \boldsymbol{\Sigma}^{-1} = \boldsymbol{\Pi}$. □

Our study of isodensity ellipsoids provides us with a geometrical understanding of Theorem 7.3. The family members of concentric ellipsoids defined by $(\mathbf{x} - \boldsymbol{\mu})^{\mathrm{T}}\boldsymbol{\Pi}(\mathbf{x} - \boldsymbol{\mu}) = c$ are indexed by $c$, which equals the squared radius of their corresponding hyperspheres in the space of the transformed quantities $\mathbf{TX}$. If you assess the vector $\mathbf{X}$ as multivariate Normal, you must assess the unknown value of the squared radius of its corresponding hypersphere as $\chi^2_K$. Given the hypersphere, of course, your opinion is uniform over all the vectors that compose it.

We have identified enough structure of the multivariate Normal distribution to begin a systematic study of its use in statistical inference about quantities that you regard exchangeably.

## 7.3 ASSESSING QUANTITIES EXCHANGEABLY WITH A MULTIVARIATE NORMAL DISTRIBUTION

We introduced our discussion of the multivariate Normal distribution by pondering your prevision for the yield of a nondescript acre of Kansas planted with wheat during the year 1982. The state of Kansas is part of the immense glacial Great Plains of North America. It is quite regularly the largest wheat-producing state in the United States, both in terms of acreage planted in wheat and in terms of total harvest. Its output is surpassed in some years by the state of North Dakota. Looking at a topographical map, you may be surprised by the extensive uniformity of Kansas topography if you have never traversed the state by car, by rail, or by air. You probably have not much reason to tender distinguishable opinions about the wheat yield from any two distinct acres that are planted with wheat. Armed with this much (little) knowledge, you may well regard exchangeably the wheat yields $X_1, X_2, \ldots, X_N$ from the $N$ acres of Kansas planted in wheat.

### PROBLEM

1. Of course there are differences in soil quality and natural irrigation among the acres of Kansas, but the differences are minimal. You would need to be an agricultural expert within the state to know their detail. Such an expert would not regard the yields from distinct acres exchangeably (unless their location identifying subscripts had been scrambled). Why would you, even in your situation of meager agricultural knowledge, not regard wheat yields from acres throughout the entire United States exchangeably? Examine the *Statistical Abstract of the United States* for any year, turning to the table entitled "Wheat–Acreage, Production, and Value, by State." Think about the statistics you find there and discuss the reasonability of the exchangeability assessment within various regions. You may enjoy the exercise more if you try to assess your own prevision and variance for wheat yields in various states before examining the statistics. Recall the definitions of measurement units discussed in problem 4 of Section 7.2.

Enough reflection on exchangeability per se! Suppose that you do regard wheat yields from the distinct acres of Kansas farmland exchangeably. Is your distribution for the yields from the several acres of Kansas wheat approximately multivariate Normal? Aware that the yield from a particular acre is the sum of the yields from a large number of plants, you may convince yourself that your distribution is approximately normal on account of the central limit theorem, which we shall discuss in Section 7.4.3. In this section we study the algebraic implications of such a prevision assessment. Once we are aware of the consequences, we may know better how to formulate "non-normal" opinions in adjusted forms.

### 7.3.1 Inferences Concerning a Vector of Quantities Based on the Observation of Some of Its Components

What is the algebraic form of the multivariate Normal distribution for $\mathbf{X}$ if you regard its components exchangeably? And what is the form of your conditional distribution for $\mathbf{X}_2|(\mathbf{X}_1 = \mathbf{x}_1)$ if your knowledge about $\mathbf{X}$ can be specified in this way?

The answer to the first question is easy. For the symmetric feature of previsions when you regard quantities exchangeably specifies not only that your prevision and variance for every quantity be the same as your prevision and variance for every other quantity but also that your covariance between any two of the quantities be equal to your covariance for any other pair. For the judgment of exchangeability implies that your joint distribution for any two quantities chosen from the $K$ be identical. Thus, we have the following result.

**Theorem 7.4.** To regard the components of a $K$-dimensional vector $\mathbf{X}$ exchangeably with a multivariate Normal distribution is to specify your distribution in the form $\mathbf{X} \sim N_K(\mu\mathbf{1}_K, \alpha\mathbf{I}_K + \beta\mathbf{1}_{K,K})$, where $\mathbf{1}_K$ is a $K \times 1$ vector with each component equal to 1, $\mathbf{I}_K$ is the identity matrix in $K$ dimensions, and $\mathbf{1}_{K,K}$ is a $K \times K$ matrix with each component equal to 1. The scalar $\mu$ may be any real number, but $\alpha$ is necessarily nonnegative, and $\beta \geqslant -\alpha/K$.

*Proof*. These forms for the prevision vector and variance-covariance matrix are the only forms that identify equal previsions and variances for every component of $\mathbf{X}$ and equal covariances between any two components. As to the restrictions on $\alpha$ and $\beta$, since the correlation $\rho_{i,j} = \beta/(\alpha + \beta) \leqslant 1$ and the variance $V(X_i) = \alpha + \beta \geqslant 0$, it is necessary that $\alpha \geqslant 0$. Furthermore, the requirement that the variance-covariance matrix, $\alpha\mathbf{I}_K + \beta\mathbf{1}_{K,K}$, be positive semidefinite requires that the determinant of each of its principle minors be nonnegative. You can show by induction (problem 1 at the end of this section) that $|\alpha\mathbf{I}_K + \beta\mathbf{1}_{K,K}| = \alpha^K + K\alpha^{K-1}\beta$ for any positive integer $K$. Thus, $|\alpha\mathbf{I}_K + \beta\mathbf{1}_{K,K}| \geqslant 0$ implies that $\beta \geqslant -\alpha/K$. □

The following corollary to Theorem 7.4 specifies further restrictions on the values of $\alpha$ and $\beta$ in the covariance matrix for $\mathbf{X}$ that are required if the distribution for $\mathbf{X}$ is exchangeably extendable as multivariate Normal.

**Corollary 7.4.1.** If you regard the components of the $K$-dimensional vector $\mathbf{X}$ exchangeably as multivariate Normal, and your distribution is exchangeably extendable to dimension $N$, then $\beta \geqslant -\alpha/N$. The distribution is infinitely exchangeably extendable if and only if $\beta \geqslant 0$.

*Proof*. The extendability of the exchangeable distribution on $\mathbf{X}$ to dimension $N$ requires the existence of a positive semidefinite matrix of the form $\alpha\mathbf{I}_N + \beta\mathbf{1}_{N,N}$, where $\beta \geqslant -\alpha/N$. The corollary follows as a consequence. □

With opinions about $\mathbf{X}$ asserted as exchangeable multivariate Normal, what should be learned about the component quantities of a subvector $\mathbf{X}_2 \equiv X_{K_1+1}, X_{K_1+2}, \ldots, X_{K_2}$ from the conditioning observation of the remaining quantities $\mathbf{X}_1 \equiv (X_1, X_2, \ldots, X_{K_1})^{\mathrm{T}}$? This question can be answered directly, since the conditional distribution of $\mathbf{X}_2 | (\mathbf{X}_1 = \mathbf{x}_1)$ in the multivariate Normal context is known to have the form $N_{K_2}(\boldsymbol{\mu}_2 + \boldsymbol{\Sigma}_{21}\boldsymbol{\Sigma}_{11}^{-1}(\mathbf{X}_1 - \boldsymbol{\mu}_1), \boldsymbol{\Sigma}_{22} - \boldsymbol{\Sigma}_{21}\boldsymbol{\Sigma}_{11}^{-1}\boldsymbol{\Sigma}_{22})$. Thus, the details of the proof of the following theorem are merely an exercise in linear algebra.

**Theorem 7.5.** Suppose you regard the components of the $K$-dimensional vector $\mathbf{X}$ exchangeably with a multivariate Normal distribution, $N_K(\mu\mathbf{1}_K, \alpha\mathbf{I}_K + \beta\mathbf{1}_K)$, and suppose $(\mathbf{X}_1, \mathbf{X}_2)^{\mathrm{T}}$ is any partition of $\mathbf{X}$ into its first $K_1$ and its remaining $K_2$ components. Then coherency requires that your conditional distribution for $\mathbf{X}_2 | (\mathbf{X}_1 = \mathbf{x}_2)$ be assessed as

$$N_{K_2}\left[(\alpha + K_1\beta)^{-1}\left(\alpha\mu + \beta\sum_{i=1}^{K_1} x_{1_i}\right)\mathbf{1}_{K_2}, \alpha\mathbf{I}_{K_2} + \alpha\beta(\alpha + K_1\beta)^{-1}\mathbf{1}_{K_2,K_2}\right].$$

*Proof*. Details are left as an exercise, described in problem 2. □

Before considering the proof exercises, let's discuss the content and meaning of the theorem. First, examine the conditional previsions, $P[\mathbf{X}_2 | (\mathbf{X}_1 = \mathbf{x}_1)]$, as they differ from the marginal previsions, $P[\mathbf{X}_2] = \mu\mathbf{1}_{K_2}$:

$$\begin{aligned} P[\mathbf{X}_2 | (\mathbf{X}_1 = \mathbf{x}_1)] &= (\alpha + K_1\beta)^{-1}\left(\alpha\mu + \beta\sum_{i=1}^{K_1} x_{1_i}\right)\mathbf{1}_{K_2} \\ &= (\alpha + K_1\beta)^{-1}(\alpha\mu + K_1\beta\bar{x}_{K_1})\mathbf{1}_{K_2}, \end{aligned}$$

where $\bar{x}_{K_1}$ is the average of the conditioning observations. The posterior (conditional) prevision for $\mathbf{X}_2$ given the observation $(\mathbf{X}_1 = \mathbf{x}_1)$ specifies a linear (and even convex when $\beta > 0$) regression equation for the components of $\mathbf{X}_2$ on all the components of $\mathbf{X}_1$ and on the constant $\mu$. The coefficients on each of the conditioning observations are identical, equal to $\beta/(\alpha + K_1\beta)$; and the coefficient on $\mu$, the prior (marginal) prevision for each component of $\mathbf{X}_2$, equals $\alpha/(\alpha + K_1\beta)$.

It is evident from examining the regression coefficient weights on each value of $x_{1_i}$ (or alternatively on $\bar{x}_{K_1}$) and on $\mu$ that in determining the posterior prevision for any component of $\mathbf{X}_2$,

1. the relative weight on each observation value, $x_{1_i}$, increases as $\beta$ increases, and diminishes as the observation size, $K_1$, increases;
2. the relative weight on the observation mean, $\bar{x}_{K_1}$, increases as $\beta$ and as $K_1$ increase;
3. the relative weight on the prior prevision, $\mu$, increases as $\alpha$ increases and diminishes as $\beta$ and $K_1$ increase.

In considering these relations, remember that the variance specified for any $X_i$ equals $\alpha + \beta$, while its covariance with any other $X_j$ equals $\beta$. The greater is the absolute value of the covariance relative to the variance, the greater is the weight on the content of any particular observation in informing the posterior prevision for any other quantity. This is quite appropriate, for the covariance assertion amounts to an assessment of how the value of the conditioning quantity codiffers with the prevised quantity from their common prevision. The relative information content of any *specific* observation value for the conditioned prevision value diminishes as the number of observations increase, but the relative information content of the *average* of all the observations increases as the number of observations increase.

Finally, since the posterior distribution for $\mathbf{X}_2$ given $(\mathbf{X}_1 = \mathbf{x}_1)$ is a function of $\mathbf{x}_1$ only through its dimension and through the sum of its components, we can also state the following corollary.

**Corollary 7.5.1.** If you regard the components of $\mathbf{X}$ exchangeably with a multivariate Normal distribution, then $\mathbf{S}(\mathbf{X}_1) = (K_1, \sum_{i=1}^{K_1} X_{1_i})$ are sufficient statistics for $\mathbf{X}_1$ relative to the components of $\mathbf{X}_2$, no matter which components of $\mathbf{X}$ are selected to specify $\mathbf{X}_1 = (X_{1_1}, \ldots, X_{1_{K_1}})$.

This result is pleasing, since it tells us, for example, that the information regarding the components of $\mathbf{X}_2$ contained in the observation of wheat yields from $K_1$ selected acres of Kansas is summarized by the number of observations and the average of their yields. The sufficiency characterization of the exchangeable multivariate Normal distribution also allows an understanding of the power the normality assertion entails. We learned in Chapter 5 that to regard $K$ quantities exchangeably, when each of them has the same realm $\mathscr{R}(X_i) = \{0, 1, \ldots, M\}$, say, the $M + 1$ histogram statistics are sufficient for representing the information contained in any number $K_1$ of them regarding any of the others. We noticed in Chapter 3 how the requirement of a larger number of sufficient statistics to characterize the information in the observations amounts to a weakening of the exchangeability assertion to one of partial exchangeability. In asserting a multivariate Normal exchangeable distribution for $\mathbf{X}$, the opposite relation holds. Fewer than $\mathbf{M} + 1$ statistics, only 2 in fact, are considered sufficient to summarize the information in the observation $(\mathbf{X}_1 = \mathbf{x}_1)$ that is relevant to inference about $\mathbf{X}_2$.

Finally, Theorem 7.5 identifies that the conditional covariance matrix reduces to $V(\mathbf{X}_2 | (\mathbf{X}_1 = \mathbf{x}_1)) = \alpha I_{K_2} + \alpha\beta(\alpha + K_1\beta)^{-1}\mathbf{1}_{K_2,K_2}$ from the marginal covariance $V(\mathbf{X}_2) = \alpha\mathbf{I}_{K_2} + \beta\mathbf{1}_{K_2,K_2}$. Thus, the conditioning observations reduce the variance for any further quantity from $(\alpha + \beta)$ to $[\alpha + \beta\alpha/(\alpha + K_1\beta)]$, which reduces toward $\alpha$ as the conditioning observation size, $K_1$, increases. Thus, there is a limit to the extent to which uncertainty regarding subsequent unknown quantities can be reduced by increasing the number of conditioning observations.

We extend our understanding of these relations in the next section, and find reason to have second thoughts about your exchangeable assessment of quantities as multivariate Normal. For now, the following directed algebra problems should convince you of the details of Theorems 7.4 and 7.5.

## PROBLEMS

**1.** Since we use matrices of this form extensively, it is a useful exercise to convince yourself of the following results:

**a.** $(a\mathbf{I}_K + b\mathbf{1}_{K,K})(c\mathbf{I}_K + d\mathbf{1}_{K,K}) = ac\mathbf{I}_K + (ad + bc + bdK)\mathbf{1}_{K,K}$.

**b.** $D_K \equiv |a\mathbf{I}_K + b\mathbf{1}_{K,K}| = a^K + Ka^{K-1}b$. [*Hint*: First notice that $D_{K+1} = (a + b)D_K - Kb(D_K - aD_{K-1})$. Then use induction.]

**c.** $(a\mathbf{I}_K + b\mathbf{1}_{K,K})^{-1} = a^{-1}\mathbf{I}_K - [b/(a^2 + Kab)]\mathbf{1}_{K,K}$. [*Hint*: Multiply this matrix by $(a\mathbf{I}_K + b\mathbf{1}_{K,K})$ and use result a to yield $\mathbf{I}_K$.]

**d.** A matrix of this form has one distinct eigenvalue, $\lambda_1 = a + Kb$, and $K - 1$ repeated eigenvalues, $\lambda_i = a$, for $i = 2, \ldots, K$. (Apply result b to the determinant equation that specifies the eigenvalues.) Since this root is repeated, there is some degree of arbitrariness in the determining the corresponding eigenvectors. One possible set of corresponding orthonormal eigenvectors consists of the first vector as $\mathbf{v}_1 = K^{-1/2}\mathbf{1}_K$, and the remaining vectors $\mathbf{v}_2$ through $\mathbf{v}_K$ specified generically for $i = 2, \ldots, K$ by

$$v_{i,j} = 0 \qquad \text{for } j = 1, \ldots, i - 2,$$

$$v_{i,i-1} = -[(K - i + 1)/(K - i + 2)]^{1/2},$$

$$v_{i,j} = [(K - i + 2)(K - i + 1)]^{-1/2} \qquad \text{for } j = i, \ldots, K.$$

**2.** Show that if $\boldsymbol{\Sigma} = \alpha\mathbf{I}_K + \beta\mathbf{1}_{K,K}$ is partitioned into $\boldsymbol{\Sigma}_{11}$ of dimension $K_1 \times K_1$, $\boldsymbol{\Sigma}_{12}$ of dimension $K_1 \times K_2$, $\boldsymbol{\Sigma}_{21} = \boldsymbol{\Sigma}_{21}^{\mathrm{T}}$, and $\boldsymbol{\Sigma}_{22}$ of dimension $K_2 \times K_2$, then

**a.** $\boldsymbol{\Sigma}_{21} = \beta\mathbf{1}_{K_2,K_1}$.

**b.** $\boldsymbol{\Sigma}_{11}^{-1} = (\alpha^2 + K_1\alpha\beta)^{-1}[(\alpha + K_1\beta)\mathbf{I}_{K_1} - \beta\mathbf{1}_{K_1,K_1}]$.

**c.** $\boldsymbol{\Sigma}_{21}\boldsymbol{\Sigma}_{11}^{-1} = (\alpha + K_1\beta)^{-1}\beta\mathbf{1}_{K_2,K_1}$.

**d.**

$$P(\mathbf{X}_2|\mathbf{X}_1 = \mathbf{x}_1) = \mu \mathbf{1}_{K_2} + \left( \alpha + K_1\beta \right)^{-1} \beta \sum_{i=1}^{K_1} (x_i - \mu)\mathbf{1}_{K_2}$$

$$= (\alpha + K_1\beta)^{-1}\left( \alpha\mu + \beta \sum_{i=1}^{K_1} x_{1_i} \right)\mathbf{1}_{K_2}$$

$$= (\alpha + K_1\beta)^{-1}(\alpha\mu + K_1\beta\bar{x}_{K_1})\mathbf{1}_{K_2},$$

where

$$\bar{x}_{K_1} \equiv K_1^{-1} \sum_{i=1}^{K_1} x_i.$$

**e.** $\mathbf{\Sigma}_{21}\mathbf{\Sigma}_{11}^{-1}\mathbf{\Sigma}_{12} = (\alpha + K_1\beta)^{-1}K_1\beta^2\mathbf{1}_{K_2,K_2}$.

**f.** $V(\mathbf{X}_2|\mathbf{X}_1 = \mathbf{x}_1) = \alpha\mathbf{I}_{K_2} + (\alpha + K_1\beta)^{-1}\alpha\beta\mathbf{1}_{K_2,K_2}$.

Show that any diagonal component of this conditional covariance matrix is smaller than the corresponding component of $V(\mathbf{X}_2)$, which equals $\alpha + \beta$. How does the posterior correlation between any two components of $\mathbf{X}_2$ compare to the prior correlation between the same two components?

3. Analyze the partial derivatives of the weights on $\mu$, on $x_{1_i}$, and on $\bar{x}_{K_1}$ in the posterior prevision $P(\mathbf{X}_2|\mathbf{X}_1 = \mathbf{x}_1)$, derived in problem 2d, with respect to the variables $\alpha$, $\beta$, and $K_1$, which motivate the discussion following Theorem 7.5.

### 7.3.2 The Implied Joint Distribution for $\bar{X}_K$, for the Differences $(X_i - \bar{X}_K)$, and for the Average of These Squared Differences

Our analysis in this section will motivate an investigation of more elaborate distributions to represent exchangeable assessments of $\mathbf{X}$.

Consider the transformation of the vector of quantities $\mathbf{X}$ to the vector of quantities $\mathbf{Y} = [(X_1 - \bar{X}_K), (X_2 - \bar{X}_K), \ldots, (X_{K-1} - \bar{X}_K), \bar{X}_K]$. Notice that final component of this vector $\mathbf{Y}$ is $\bar{X}_K \equiv K^{-1}\sum_{i=1}^{K} X_i$. This transformation is achieved via the linear transformation $\mathbf{Y} = \mathbf{T}_K\mathbf{X}$, where the matrix $\mathbf{T}_K$ has the form

$$\mathbf{T}_K = \begin{pmatrix} 1-K^{-1} & -K^{-1} & -K^{-1} & & \cdots & -K^{-1} \\ -K^{-1} & 1-K^{-1} & -K^{-1} & & & -K^{-1} \\ -K^{-1} & -K^{-1} & 1-K^{-1} & & & -K^{-1} \\ \vdots & & & \ddots & & \vdots \\ -K^{-1} & -K^{-1} & -K^{-1} & & 1-K^{-1} & -K^{-1} \\ K^{-1} & K^{-1} & K^{-1} & \cdots & K^{-1} & K^{-1}. \end{pmatrix}$$

Your assessment of $\mathbf{X} \sim N_K[\mu\mathbf{1}_K, \alpha\mathbf{I}_K + \beta\mathbf{1}_{K,K}]$ implies via coherency that you assess $\mathbf{Y} = \mathbf{T}_K\mathbf{X} \sim N_K[\mu\mathbf{T}_K\mathbf{1}_K, \mathbf{T}_K(\alpha\mathbf{I}_K + \beta\mathbf{1}_{K,K})\mathbf{T}_K^{\mathrm{T}}]$. It is merely a matter of grinding out the algebra (in problem 1, shortly) to yield the following theorem.

**Theorem 7.6.** If you regard the components of the $K$-dimensional vector $\mathbf{X}$ exchangeably with a multivariate Normal distribution, $N_K(\mu\mathbf{1}_K, \alpha\mathbf{I}_K + \beta\mathbf{1}_K)$, then coherency requires that you assess the average of the components of $\mathbf{X}$, $\bar{X}_K$, with distribution $N_1(\mu, \alpha K^{-1} + \beta)$. Moreover, you must regard $\bar{X}_K$ independently of the difference of any component of $\mathbf{X}$ from $\bar{X}_K$. Your joint distribution for the differences of any $K-1$ components of $\mathbf{X}$ from their average, $\bar{X}_K$, is $N_{K-1}[\mathbf{0}_{K-1}, \alpha(\mathbf{I}_{K-1} + K^{-1}\mathbf{1}_{K-1,K-1})]$.

Again, some discussion can precede your derivation of the details of this result. Its first implication is that your exchangeable assessment of the components of $\mathbf{X}$ as multivariate Normal means you must regard the average of all the quantities independently of the deviation of any quantity from that average. For the final row and column of the covariance matrix for $\mathbf{T}_K\mathbf{X}$, to be derived, designates 0 for the covariance between the average and any of these differences. This result is not so startling, and indeed could well be a characteristic of your approximate opinion about wheat yields from acres in Kansas. (It would surely not be precisely true, since, for example, being told that $\bar{X}$ were 0, you could be very sure that each deviation $(X_i - \bar{X})$ were also 0. This cautionary remark serves only to underline the extent to which you merely regard the multivariate Normal distribution as a practical approximation to your distribution for $\mathbf{X}$ during "usual" harvests.)

Second, it is apparent from the theorem that your variance for $\bar{X}$ is approximately equal to $\beta$ if $K$ is a large number, while your variance for each $(X_i - \bar{X})$ is approximately equal to $\alpha$. These two results are pleasing since they expose meaningful *single* quantities in terms of which you can assess your *joint* distribution for the entire vector $\mathbf{X}$. In the context of exchangeable multivariate normal opinions about the components of $\mathbf{X}$, you need specify your $P(X_i) \equiv \mu$, your $P[(\bar{X}_K - \mu)^2] = V(\bar{X}_K) \approx \beta$ when $K$ is large, and your $P[(X_i - \bar{X}_K)^2] = V(X_i - \bar{X}) \approx \alpha$, again for large $K$, as in the Kansas wheat example. (Some 13 million acres of Kansas were planted with wheat during 1982.)

Before we discuss an example and pursue some related matters, you may want to clinch the details of the derivation of Theorem 7.6.

## PROBLEM

1. Based on the matrix $\mathbf{T}_K$ specified in the beginning of this section, assure yourself of the following results:

   **a.** $\mathbf{T}_K\mu\mathbf{1}_K = (0, 0, \ldots, 0, \mu)^{\mathrm{T}}$.

   **b.** $\mathbf{T}_K(\alpha\mathbf{I}_K + \beta\mathbf{1}_{K,K})\mathbf{T}_K^{\mathrm{T}}$ is a matrix whose first $K-1$ rows by $K-1$ columns determine a matrix of the form $\alpha\mathbf{1}_{K-1} - \alpha K^{-1}\mathbf{1}_{K-1,K-1}$. Every component of its final row and column is 0 except for the last element (row $K$, column $K$) ,which equals $\alpha K^{-1} + \beta$.

***Example 1.*** An imaginative Use of Theorem 7.6. Suppose that you have made three very high fidelity radio receivers according to the same design, and you are interested in some measurement they each provide for a specific signal. Now suppose that your distribution for the measurement vector $\mathbf{X}_3$ is exchangeable multivariate Normal with positive covariance, $\text{Cov}(X_i, X_j) = \beta > 0$. Thus, your opinions are infinitely exchangeably extendable as a multivariate Normal distribution. You may, of course, assess and $\alpha$ and $\beta$ by thinking directly about the product quantity $X_i X_j$. But you could also use Theorem 7.6 by imagining that if you had constructed many such receivers, call the number of them $K$, what would be your prevision assessment of $P[(\bar{X}_K - \mu)^2]$ and of any $P[(X_i - \bar{X}_K)^2]$? Your answers to these questions would constitute your assessments of $\alpha$ and $\beta$. □

Returning to a discussion of our theorem, notice that if $\beta = 0$ (so you regard the components of $\mathbf{X}$ independently) and $K$ is large, you must be practically certain that $\bar{X}_K$ equals your asserted value of $\mu = P(\bar{X}_K)$, since your variance for $\bar{X}_K$ would then equal $\alpha K^{-1} \approx 0$. But if you merely regard the components of $\mathbf{X}$ exchangeably ($\beta \neq 0$), you are still avowedly uncertain about the value of $\bar{X}_K$, even if $K$ is large, since $V(\bar{X}_K) = \beta + \alpha K^{-1} \approx \beta$. We can pursue this idea to generate a surprising result.

**Corollary 7.6.1.** If you regard the components of the $K$-dimensional vector $\mathbf{X}$ exchangeably with a multivariate Normal distribution, $N_K(\mu \mathbf{1}_K, \alpha \mathbf{I}_K + \beta \mathbf{1}_K)$, then coherency requires that you assess the quantity $Q_{K-1} = \alpha^{-1} \sum_{i=1}^{K} (X_i - \bar{X}_K)^2$ with the distribution $\chi^2_{K-1}$. Moreover, if $K$ is large, you must be practically certain that $(K-1)^{-1} \sum_{i=1}^{K} (X_i - \bar{X}_K)^2$ approximately equals $\alpha$.

*Proof.* We have shown that the first $K-1$ components of $\mathbf{Y} = \mathbf{T}\mathbf{X}$, denoted by $\mathbf{Y}_{K-1}$, are assessed with distribution $N_{K-1}[\mathbf{0}_{K-1}, (\alpha \mathbf{I}_{K-1} - \alpha K^{-1} \mathbf{1}_{K-1,K-1})]$. Consider, then, the quadratic transformation $Q_{K-1} \equiv \mathbf{Y}^{\text{T}}_{K-1} (\alpha \mathbf{I}_{K-1} - \alpha K^{-1} \mathbf{1}_{K-1,K-1})^{-1} \mathbf{Y}_{K-1}$, which we know (from Section 7.2.4) must be assessed with the distribution $\chi^2_{K-1}$. The remainder of this proof uses the following algebraic results which you will derive as an exercise:

**(a)**

$$(\alpha \mathbf{I}_{K-1} - \alpha K^{-1} \mathbf{1}_{K-1,K-1})^{-1} = \alpha^{-1} [\mathbf{I}_{K-1} + \mathbf{1}_{K-1,K-1}];$$

**(b)**

$$Q_{K-1} \equiv \mathbf{Y}^{\text{T}}_{K-1} (\alpha \mathbf{I}_{K-1} - \alpha K^{-1} \mathbf{1}_{K-1,K-1})^{-1} \mathbf{Y}_{K-1} = \alpha^{-1} \sum_{i=1}^{K} (X_i - \bar{X}_K)^2.$$

Now recall the first two moments of a quantity assessed with a $\chi^2$-distribution: $X \sim \chi^2_N$ implies $P(X) = N$, and $V(X) = 2N$. Thus, in the application we are now considering, $P(Q_{K-1}) = K - 1$ and $V(Q_{K-1}) = 2(K - 1)$. Thus, multiplying $Q_{K-1}$ by $\alpha(K - 1)^{-1}$ and assessing prevision and variance yield the results

$$P[\alpha Q_{K-1}/(K-1)] = P[(K-1)^{-1} \sum_{i=1}^{K} (X_i - \bar{X}_K)^2] = \alpha,$$

$$V[\alpha Q_{K-1}/(K-1)] = V[(K-1)^{-1} \sum_{i=1}^{K} (X_i - \bar{X}_K)^2] = 2\alpha^2/(K-1).$$

Thus, this variance approaches 0 as the value of $K$ increases. Supposing that $K$ is a large number, Chebyshev's inequality (Section 4.2) implies you are fairly sure, then, that $(K-1)^{-1}\sum_{i=1}^{K}(X_i - \bar{X}_K)^2 \approx \alpha$.

Relaxing from the detail of this proof, the result is rather startling. We have already allowed as how your exchangeable Normal assessment of $\mathbf{X}$ implies that you are uncertain about the average of its many components, $\bar{\mathrm{X}}$, with your distribution assessed as $N(\mu, \beta + \alpha/K \approx \beta)$. Now we find you to be virtually certain that the average squared deviation of the many components of $\mathbf{X}$ about $\bar{X}$ is $(K-1)\sum_{i=1}^{K}(X_i - \bar{X}_K)^2 \approx \alpha$.!! To echo an old question, "where did you get such knowledge?"

Of course, if you do have some reason to be certain of the value of this quantity, it cannot be disputed that you feel that way. You may be happy to announce your assessed distribution for the components of $\mathbf{X}$ as exchangeable multivariate Normal, and I would refrain from mocking, even in jest. But do you have such knowledge—about the average variation in wheat yield per acre in Kansas, for example?

Our first pass at representing approximately some reasoned opinion about the vector $\mathbf{X}$ in multivariate Normal form has resulted, after long study, in a problem. In the next section we pursue a related tack to find a solution to the dilemma.

## 7.4 ASSESSING QUANTITIES EXCHANGEABLY VIA MIXTURES OF INDEPENDENT NORMAL DISTRIBUTIONS

De Finetti's theorem for distributions of general quantities that are regarded exchangeably provides the clue for how we can expand our practical ability to represent such uncertain knowledge. We ignored this theorem in the analysis of the previous section, and you may wonder how it is even applicable. To review, the theorem states that any infinitely extendable judgment to regard a sequence of quantities exchangeably can be represented by a *mixture* of distributions that assess the quantities *independently*. When the realm of the quantities has $R$ components, the exact form of the joint distribution is a mixture of the product of multinomial distributions. When $R$ is large, a mixture of independent Normal distributions, which we study in this section, may provide an adequate approximation to your uncertain assertions. By analyzing the algebraic details of such a mixture, we can come to understand when you would agree to such an approximate specification of your knowledge.

Consider a vector of quantities $\mathbf{X} = (X_1, \ldots, X_K)^T$, each component of which has the identical realm $\mathcal{R}(X_i) = \{y_1, y_2, \ldots, y_R\}$. When $R$ is large, it is common to consider continuous distribution functions over $K$-dimensional real space, $\mathbb{R}^K$, as approximating forms of one's assertions. This is the context in which we study the mixture of independent Normal distributions.

Since a multivariate Normal distribution that treats the components of $\mathbf{X}$ independently is identified by only two parameters, $\mu$ and $\sigma^2 \equiv \pi^{-1}$, it is easy to formulate the expression for a mixture of independent Normal distributions. We study the more general mixture-multivariate Normal distribution later in this chapter.

***Definition 7.2.*** The distribution function for a vector of quantities $\mathbf{X} \in \mathbb{R}^K$,

$$F(\mathbf{x}) \equiv \int_{-\infty}^{\mathbf{x}} \int_{0}^{\infty} \int_{-\infty}^{\infty} \prod_{i=1}^{K} (2\Pi)^{-1/2} \pi^{1/2} \exp\{-\pi(x_i - \mu)^2/2\}\, dM(\mu, \pi)\, d\mathbf{x}$$

$$= \int_{-\infty}^{\mathbf{x}} \int_{0}^{\infty} \int_{-\infty}^{\infty} (2\Pi)^{-K/2} \pi^{K/2} \exp\left\{-\pi \sum_{i=1}^{K} (x_i - \mu)^2/2\right\} dM(\mu, \pi)\, d\mathbf{x},$$

where $\Pi$ denotes the real number pi, is said to be a *mixture of conditionally independent Normal distributions* with mixing parameters $\mu$ and $\pi$ and *mixing distribution function* $M$. The associated density function is

$$f(\mathbf{x}) = \int_{-\infty}^{\infty} \int_{-\infty}^{\infty} (2\Pi)^{-K/2} \pi^{K/2} \exp\left\{-\pi \sum_{i=1}^{K} (x_i - \mu)^2/2\right\} dM(\mu, \pi).$$

If you would regard such a function as an approximation to your probability distribution for $\mathbf{X}$, we would denote your assertion by $\mathbf{X} \sim M - N_K(\mu, \pi)$. □

**Important notice:** Do notice that from this point on we are parameterizing the $N(\mu, \pi)$ family members in terms of their precisions, $\pi$, rather than their variances, $\sigma^2 = \pi^{-1}$.

It is evident that a distribution for $\mathbf{X}$ in this form entails a judgment to regard the components of $\mathbf{X}$ exchangeably. For the density function is obviously constant for any reordering of the arguments $x_i$. What are the further implications of specifying that your uncertain knowledge of $\mathbf{X}$ is representable approximately as a mixture of conditionally independent Normal distributions? Our analysis of this question will be aided by noting the following algebraic results, which you should prove as an exercise.

## PROBLEMS

1. Show that $\sum_{i=1}^{K} (X_i - \mu)^2$, which appears in Definition 7.2, can be written as $\sum_{i=1}^{K} (X_i - \mu)^2 = \sum_{i=1}^{K} (X_i - \bar{X}_K)^2 + K(\bar{X}_K - \mu)^2$. [*Hint*: Just express

$\sum_{i=1}^{K}(X_i-\mu)^2$ as $\sum_{i=1}^{K}[(X_i-\bar{X}_K)+(\bar{X}_K-\mu)]^2$, expand the square, and simplify the algebra, noting that $\sum_{i=1}^{K}(X_i-\bar{X}_K)(\bar{X}_K-\mu)=0$.]

**2.** Notice that if we denote the average squared deviation of the components of $\mathbf{X}$ from their own average, $\bar{X}_K$, by $S_K^2 \equiv K^{-1}\sum_{i=1}^{K}(X_i-\bar{X}_K)^2$, then $\sum_{i=1}^{K}(X_i-\mu)^2$ can be written conveniently as $K[S_K^2+(\bar{X}_K-\mu)^2]$. Thus, the sum of squares of components of $\mathbf{X}$ about $\mu$ is a function of these components only through the summary statistics $\bar{X}_K$ and $S_K^2$.

### 7.4.1 Inference Based on Mixture-Normal Distributions

Now let us partition $\mathbf{X}=(\mathbf{X}_1,\mathbf{X}_2)$ into two component vectors, where $\mathbf{X}_1$ has dimension $K_1\times 1$ and $\mathbf{X}_2$ has dimension $K_2\times 1$, with $K_1+K_2=K$. Remember that your regarding any vector of quantities exchangeably via a mixture of conditionally independent distributions with some specified mixing function $M$ implies that your marginal distribution for any subvector is representable in the same way with the identical mixing function $M$. Thus, your assessment of $\mathbf{X}$ as $M\text{-}N_K(\mu,\pi)$ implies your assessment of its component vectors $\mathbf{X}_1$ and $\mathbf{X}_2$ with a distribution in the same mixture-Normal form, and with the same mixing function $M$. That is, you must assess $\mathbf{X}_1$ as $M\text{-}N_{K_1}(\mu,\pi)$ and $\mathbf{X}_2$ as $M\text{-}N_{K_2}(\mu,\pi)$.

We are now prepared to address the general inferential question in the present context, that your uncertain knowledge about $\mathbf{X}$ is approximately representable as $\mathbf{X}\sim M\text{-}N_K(\mu,\pi)$: What is your avowed conditional distribution for the subvector $\mathbf{X}_2$ given the observation that $\mathbf{X}_1=\mathbf{x}_1$? Using Bayes' theorem to answer, we need only consult the conditional density function

$$f(\mathbf{x}_2|\mathbf{X}_1=\mathbf{x}_1)=f(\mathbf{x}_1,\mathbf{x}_2)/f(\mathbf{x}_1).$$

Notice that in studying the quotient of these two densities, we may ignore any proportionality constants that do not involve components of $\mathbf{x}_2$. For coherency requires that the function $f(\mathbf{x}_2|\mathbf{X}_1=\mathbf{x}_1)$ be a density function too. Thus, whatever its form as a function of $\mathbf{x}_2$, it must be scaled by whatever proportionality constant is needed to ensure that its integral over $\mathbf{x}_2\in\mathbb{R}^{K_2}$ cumulates to 1. Ignoring these constants will simplify our algebra in what follows.

Based on the algebraic results of problems 1 and 2 of Section 7.4, the joint density $f(\mathbf{x}_1,\mathbf{x}_2)$ can be seen to be proportional to

$$\int_0^\infty\int_{-\infty}^\infty \pi^{K/2}\exp\left\{-\frac{\pi}{2}\left[\sum_{i=K_1+1}^{K}(x_i-\mu)^2+K_1s_{K_1}^2+K_1(\bar{x}_{K_1}-\mu)^2\right]\right\}dM(\mu,\pi).$$

The denominator then, $f(\mathbf{x}_1)$, is merely proportional to the integral of this function over all values of $\mathbf{x}_2\in\mathbb{R}^{K_2}$. This integral yields

$$f(\mathbf{x}_1)\propto\int_0^\infty\int_{-\infty}^\infty \pi^{K_1/2}\exp\{-(K_1\pi/2)[s_{K_1}^2+(\bar{x}_{K_1}-\mu)^2]\}\,dM(\mu,\pi),$$

which we recognize as specifying the marginal mixture-Normal density for $\mathbf{X}_1 \sim M\text{-}N_{K_1}(\mu, \pi)$.

Examining the ratio $f(\mathbf{x}_1, \mathbf{x}_2)/f(\mathbf{x}_1)$, we reach the pleasing conclusion that your conditional distribution for $\mathbf{X}_2|(\mathbf{X}_1 = \mathbf{x}_1)$ is also mixture Normal and with a computable mixing function. Let us first state this as a theorem with a corollary. In Section 7.4.2 we discuss it along with some terminology it motivates.

**Theorem 7.7.** If you regard the components of $\mathbf{X}$ exchangeably, asserting the infinitely extendable distribution $\mathbf{X} \sim M\text{-}N_K(\mu, \pi)$, then coherency requires that your conditional distribution for $\mathbf{X}_2|(\mathbf{X}_1 = \mathbf{x}_1)$ also be a mixture of independent Normal distributions, $M^*\text{-}N_{K_2}(\mu, \pi)$, with a mixing distribution function $M^*(\cdot)$ denoted by $M(\mu, \pi|\mathbf{X}_1 = \mathbf{x}_1)$ and calculated as

$$dM(\mu, \pi|\mathbf{X}_1 = \mathbf{x}_1) \propto \pi^{K_1/2} \exp\{-(K_1\pi/2)[s^2_{K_1} + (\bar{x}_{K_1} - \mu)^2]\}\, dM(\mu, \pi).$$

**Corollary 7.7.1.** Under the conditions of Theoreom 7.7, the statistics defined by $\mathbf{S}(\mathbf{X}_1 = \mathbf{x}_1) = [K_1, \bar{x}_{K_1}, s^2_{K_1}]$ are sufficient summaries of the observation $\mathbf{X}_1 = \mathbf{x}_1$ for any inference about the components of $\mathbf{X}_2$. The information transfer function (ITF, or likelihood function for $\mu$ and $\pi$), both defined subsequently, has the form

$$\text{ITF}[\mu, \pi; (\mathbf{X}_1 = \mathbf{x}_1), \mathbf{X}_2] = \pi^{K_1/2} \exp\{-(K_1\pi/2)[s^2_{K_1} + (\bar{x}_{K_1} - \mu)^2]\}.$$

### 7.4.2 Information Transfer, or Likelihood, Functions

***Definition 7.3.*** If your probability distribution for a vector $\mathbf{X}_2$ is a mixture distribution $M\text{-}F(\theta)$ with mixing distribution $M(\theta)$, and your conditional probability distribution for $\mathbf{X}_2$ given the observation of another vector $\mathbf{X}_1 = \mathbf{x}_1$ is a distribution in the same mixture form but with a different mixing distribution, $M^*(\theta) \neq M(\theta)$, then we denote this conditional mixing distribution $M^*(\theta)$ by $M(\theta|\mathbf{X}_1 = \mathbf{x}_1)$, and we call it your *conditional mixing function over* $\theta$ given $\mathbf{X}_1 = \mathbf{x}_1$. The ratio of the derivatives of these mixing distributions defines your *information transfer function through* $\theta$ *from* $\mathbf{X}_1 = \mathbf{x}_1$ *to* $\mathbf{X}_2$:

$$\text{ITF}[\theta; (\mathbf{X}_1 = \mathbf{x}_1), \mathbf{X}_2] \equiv dM(\theta|\mathbf{X}_1 = \mathbf{x}_1)/dM(\theta).$$

Alternatively, it is called your *likelihood function for* $\theta$ *on the basis of* $\mathbf{X}_1 = \mathbf{x}_1$, denoted simply by $L(\theta; \mathbf{x}_1)$. □

The algebraic form of the information transfer function through $\theta$ is $\text{ITF}[\theta; (\mathbf{X}_1 = \mathbf{x}_1), \mathbf{X}_2] = f(\mathbf{x}_1; \theta)$, the density or probability mass value at the observation vector $\mathbf{x}_1$ specified by each of the $\theta$-parametric distributions whose mixture represents your opinions. Algebraically, when $M(\theta)$ admits a density $f(\theta)$, the same relations govern the conditional mixing function that Bayes' theorem specifies for probabilities: $f(\theta|\mathbf{X}_1 = \mathbf{x}_1) = f(\mathbf{x}_1; \theta) f(\theta)/\int f(\mathbf{x}_1; \theta)\, f(\theta)\, d\theta$.

The integral denominator merely defines a proportionality constant that ensures $\int f(\theta|\mathbf{X}_1 = \mathbf{x}_1)\,d\theta = 1$. Thus, the information transfer ratio $f(\theta|\mathbf{X}_1 = \mathbf{x}_1)/f(\theta)$ is proportional to $f(\mathbf{x}_1; \theta)$.

In understanding Theorem 7.7, and in applying it to statistical inference, it is important to be clear that the object of this analysis is your uncertain knowledge about the components of $\mathbf{X}$. Various aspects of your knowledge are represented in terms of your approximate probability distribution functions $F(\mathbf{x})$, $F(\mathbf{x}_1)$, $F(\mathbf{x}_2)$, $F(\mathbf{x}_2|\mathbf{X}_1 = \mathbf{x}_1)$, and $F(\mathbf{x}_1|\mathbf{X}_2 = \mathbf{x}_2)$ or their associated densities. These functions are appropriately termed "probability" distribution functions on account of their use in representing your approximate probabilities for various events involving components of $\mathbf{X}$. Such functions are distinct both in their purpose and in their name from the "mixing distribution functions" $M(\mu, \pi)$ and $M(\mu, \pi|\mathbf{X}_1 = \mathbf{x}_1)$.

The arguments of the mixing functions, $\mu$ and $\pi$, are not observable quantities defined by some operationally defined measurement. Rather, these are parameters of a particular family of functions whose *mixture* represents your knowledge about the specific measurable quantities, $\mathbf{X}$. These parameters are evidently useful in representing your state of knowledge about $X_1, \ldots, X_K$; moreover, other people may find them useful too, even when they do not agree with you in every detail of their mixing functions. But someone else in still another state of knowledge may have uncertain opinions about $\mathbf{X}$ that are not representable in this way. For example, someone else may not even regard the components of $\mathbf{X}$ exchangeably, not to speak of the detail that you assess $\mathbf{X}$ with the mixture-Normal distribution $\mathbf{X} \sim M\text{-}N_K(\mu, \pi)$.

It is true that the mixing functions $M(\mu, \pi)$ and $M(\mu, \pi|\mathbf{X}_1 = \mathbf{x}_1)$ do have the mathematical properties of a distribution function. That is, they are nondecreasing functions of each argument, bounded between 0 and 1. Moreover, the moments of such a function, defined by

$$\mathbb{M}_{i,j}(\mu, \pi) \equiv \int_0^\infty \int_{-\infty}^\infty \mu^i \pi^j \, dM(\mu, \pi)$$

and

$$\mathbb{M}_{i,j}(\mu, \pi|\mathbf{X}_1 = \mathbf{x}_1) \equiv \int_0^\infty \int_{-\infty}^\infty \mu^i \pi^j \, dM(\mu, \pi|\mathbf{X}_1 = \mathbf{x}_1),$$

are meaningful numbers as moment characteristics of the distribution functions $M(\mu, \pi)$ and $M(\mu, \pi|\mathbf{X}_1 = \mathbf{x}_1)$. However, the function $M$ is not appropriately called a *probability* distribution function. For coherent probabilities (previsions) are defined operationally only in terms of your assessed valuations of transactions denominated in units of *observable* quantities. The arguments $\mu$ and $\pi$ do not represent observable quantities. Rather they constitute the mixing structure framework of your opinions about observable quantities. Thus, the function $M$ is appropriately called a *mixing distribution function.*

Studying the transformation of $F(\mathbf{x}_2)$ to $F(\mathbf{x}_2|\mathbf{X}_1 = \mathbf{x}_1)$, we can also distinguish the role played by the observation values $\mathbf{X}_1 = \mathbf{x}_1$ in the inference. The conditional density $f(\mathbf{x}_2|\mathbf{X}_1 = \mathbf{x}_1)$ is a function of $\mathbf{x}_1$ only through the expression $\pi^{K_1/2} \exp\{-(K_1\pi/2)[\sum_{i=1}^{K_1}(x_i - \mu)^2]\}$ in the posterior mixing function. Thus, we designate this expression according to Definition 7.3 as the *information transfer function* through $(\mu, \pi)$ from $\mathbf{X}_1 = \mathbf{x}_1$ to $\mathbf{X}_2$. We also can identify the statistics $K_1$, $\bar{X}_{K_1} = \bar{x}_{K_1}$, and $S_{K_1}^2 = s_{K_1}^2$ as sufficient statistics of $\mathbf{X}_1$ for inference about $\mathbf{X}_2$.

An older terminology refers to this very same expression as specifying a *likelihood function*. Originating in the work of Ronald Fisher (1925) it is so widespread in statistical usage that it would be difficult to ignore it completely in deference to "information transfer function" which is more appropriate to our understanding of it. Fisher's terminology for the identical function expression makes sense within the objectivist supposition that an unobservable parametric distribution, specified by a particular family parameter value, $\theta$, generates the data. As a function of the "possible" parameter values, the likelihood of any possible $\theta$ on the basis of observing $\mathbf{X}_1 = \mathbf{x}_1$, denoted by $L(\theta; \mathbf{x}_1) \equiv f(\mathbf{x}_1; \theta)$, is a measurement of how likely (probable) this observation would be if the numerical value of the generating parameter were that value of $\theta$. The book of Edwards (1972) is devoted to foundational discussion of the "likelihood approach" to statistical inference, and the monograph by Berger and Wolpert (1984) develops a vast array of intriguing examples of "the likelihood principle."

Our alternative terminology of the information transfer function allows that there is no unobservable parameter value of a generating distribution for the data. Probability distributions representing the uncertainties of people who regard the data sequence, $\mathbf{X}$, in a particular way may well be representable algebraically as mixture distributions with respect to a family of distributions that are indexed by the parameterization $\theta$. The role this parameterization plays in the representation of $f(\mathbf{x}) = \int_\Theta f(\mathbf{x}; \theta)\, dM(\theta)$ is to weight the family member densities $f(\mathbf{x}; \theta)$ for various $\theta \in \Theta$ in determining the density $f(\mathbf{x})$ as a weighted average of the family members. Thus, the information conveyed regarding the value of $\mathbf{X}_2$ by the observation of $\mathbf{X}_1 = \mathbf{x}_1$ is transferred algebraically by the quotient of the two mixing functions, $dM(\theta|\mathbf{X}_1 = \mathbf{x}_1)/dM(\theta)$, which motivates the terminology.

We realize explicitly that this parameterization of the information transfer function is appropriate only for those people who agree generally with the "particular way" of thinking that allows this specific mixture representation. Other "ways of thinking" can be parameterized quite differently, not involving $\theta$ at all. Thus, this parameterization of the information transfer function would be irrelevant to them. They could not even dream of getting involved in a project of estimating the parameter $\theta$ within the objectivist imagination: their opinions about the observable quantities are not even representable in these terms.

We have lingered on a philosophical analysis of the meaning of our mathematical constructions for conceptual purposes that should make the practical developments which follow more easily understood. Before studying some algebraic details of mixture-Normal distribution theory, we first discuss briefly some theorems that may motivate your choice of a mixture-multivariate Normal distribution to approximate your opinions in specific instances. Then we outline some general procedures of assessment that can be followed in identifying an appropriate mixing function.

### 7.4.3 Motivations for the Mixture-Normal Distribution to Represent Uncertain Knowledge

Here we merely state and discuss three important theorems that serve to motivate the use of mixture-Normal distributions in a variety of applied contexts. References shall be made to supporting research literature for analysis and generalizations. We do not focus extensively on details, but we try to interpret the theorems as providing an abstraction of situations in which their conclusions are relevant.

**Theorem 7.8.** If your opinions regarding a sequence of quantities $X_1, X_2, \ldots, X_K$ are infinitely extendable to a distribution function that is symmetric on spheres centered at $M$, $(\mathbf{x}_N - M\mathbf{1}_N)^{\mathrm{T}}(\mathbf{x}_N - M\mathbf{1}_N)$, for any $N > K$, then your distribution for $\mathbf{X}_K$ must be representable as a mixture-Normal distribution, denoted $M\text{-}N_K(\mu, \pi)$.

A simple proof using characteristic functions can be found in Smith (1981). Results of this form have an interesting history of analysis dating back at least to Borel, as noted in Diaconis and Freedman (1987), Hill (1967), and Dawid (1981), and in the summary monograph by Fang, Kotz, and Ng (1990).

The full practical import of this theorem derives from its extension to symmetry on ellipsoids and to symmetric distributions over vectors of quantities. For interpretation now, let us merely notice a certain scenario of opinion that applies directly to the simple theorem as stated. In three dimensions the spherical symmetry of your opinions regarding $\mathbf{X}_3$ amounts to your specifying that your approximating probability density is constant on any argument vector $\mathbf{x}_3$ lying on the same sphere centered at $(M, M, M)$. For example, you might muse: "I think that each of the quantities will be somewhere around $M$, and my probabilities for vectors of the quantities are constant on spheres surrounding $M\mathbf{1}_3^{\mathrm{T}}$. Moreover, my probabilities are low on spheres tightly bound around this center, they increase as the sphere radius increases to some size, and then they decline over spheres with increasingly larger radii."

First, in such a case, your prevision for each component would need to be $P(X_i) = M$, on account of the spherical symmetry of your opinion around $M\mathbf{1}_3^{\mathrm{T}}$. Thus, your mixing function over $(\mu, \pi)$ must have the first moment $\mathbb{M}_1(\mu) = M$. Second, nonetheless, you are allowed any array of probability orderings over the spheres with varying radii centered at $(M, M, M)$. Uniformity is required only over vectors on each given sphere. Thus, your mixing function may allow any ordering of the sizes of mixture weight on various values of $\pi$. So your uncertainty about the $X_i$ measurements is reduced to your mixing function over the center of the spheres, parameterized by $\mu$, and over the radii of the spheres, parameterized by $\pi$.

Our next theorem is actually a corollary to a theorem in Section 5.6, relating the exponential family of distributions to the theory of sufficient statistics.

**Theorem 7.9.** If your joint distribution for the vector of quantities $\mathbf{X}_K$ admits the statistics $K_1$, $\bar{X}_{K_1}$ and $S^2_{K_1}$ as sufficient for inference about $\mathbf{X}_{K_2}$ no matter what the size of $K$, $K_1$, and $K_2$, then your probability distribution for $\mathbf{X}_K$ must be approximately mixture-Normal, $M\text{-}N_K(\mu, \pi)$.

Proof of the general theorem from which this result is derived appear in the original papers by Darmois, Koopman, and Pitman, which are referenced in the discussion by Lehman (1959). The paper of Anderson (1970) presents details of the associated theorem for discrete-value quantities, which implies the exact distribution for $\mathbf{X}_K$ would be a mixture of "digitized" normal distributions.

Technicalities and extensions aside, Theorem 7.9 is a powerful one. If you do regard merely the average and the average squared deviation of a vector of quantities as sufficient for your inference about some indeterminately large number of remaining quantities, then you must assess your distribution for the entire vector as essentially mixture-Normal. In deciding whether or not this willingness is applicable to a specific situation, you need only consider the extent to which you would really desire some further summary statistics to assess your conditional distribution for the components of $\mathbf{X}_{K_2}$.

Our final "theorem" is a variation of the famous central limit theorem, as it is relevant in the context of quantities judged exchangeably. I state it rather informally, as our considerations here are focused on general ideas of content rather than precise detail.

**Theorem 7.10.** Suppose you regard the quantities $\mathbf{X}_K$ exchangeably, and you realize that each of these components is itself a sum of many other quantities that you regard exchangeably among themselves. Then your joint distribution for the vector $\mathbf{X}_K$ must be approximately mixture-Normal, $M\text{-}N_K(\mu, \pi)$.

Several results of this type appears in the survey by Taylor, Daffer, and Patterson (1985), with useful reviews by Aldous (1987) and Teicher (1988). The example of wheat yields in Kansas provides a good application of this result, as the wheat yield from an acre of plants is the sum of the yields from the many individual plants that grow on this acre. Given the general uniformity of plant sites within an acre, one would typically regard the individual plant yields exchangeably. Generalizations of this theorem allow that the components of $\mathbf{X}_K$ may derive from any linear combination of their own components.

Your understanding of this mixture form of a central limit theorem will develop in the context of the following discussion of procedures for eliciting the mixing function component of a mixture.

### 7.4.4 Specifying the Mixing Function, $M(\mu, \pi)$

We have shown that the specification of your exchangeable opinion about $\mathbf{X}_K$ as $M\text{-}N_K(\mu, \pi)$ is completed by your specifying a particular mixing distribution function $M(\mu, \pi)$. How is this to be done? There are two major tacks we can follow, each of which is useful in specific practical situations, and both of which are interesting theoretically.

The first tack involves identifying features of $M(\cdot)$ by specifying your prevision and variance for any individual component of $\mathbf{X}_K$, denoted here by $X_i$. Since the array of ordered pairs $(\mu, \pi) \in [\mathbb{R}, \mathbb{R}^+]$ constitute a partition of the domain of $M(\cdot)$, we can use the following results:

**(i)** $P(X_i) = \mathbb{M}_1[\mathbb{M}_1(X_i; \mu, \pi)] = \mathbb{M}_1(\mu)$;

**(ii)** $V(X_i) = \mathbb{M}_1[\mathbb{M}_2^*(X_i; \mu, \pi)] + \mathbb{M}_2^*[\mathbb{M}_1(X_i; \mu, \pi)]$

$$= \mathbb{M}_1(\pi^{-1}) + \mathbb{M}_2^*(\mu) = \mathbb{M}_{-1}(\pi) + \mathbb{M}_2^*(\mu).$$

The first equality of each equation line applies familiar formulas for your prevision and variance of partitioned quantities which we studied in Sections 4.2 and 5.3.2. The second equality in each equation line completes the application of these formulas to the current case under study, the specification of your distribution for $\mathbf{X}$ as $M\text{-}N_K(\mu, \pi)$.

In principle, we might identify in similar manner even more moments of the mixture distribution for $\mu$ and $\pi$ via the specification of even higher-order previsions for $X_i$. For example,

$$P(X^4) = G[\mathbb{M}_1(\mu), \mathbb{M}_2(\mu), \mathbb{M}_3(\mu), \mathbb{M}_4(\mu), \mathbb{M}_{-1}(\pi), \mathbb{M}_{-1.5}(\pi), \mathbb{M}_2(\pi)],$$

for a specific function $G$. But the formulae are complicated, as you can imagine; moreover, your prevision for high powers of the quantity $X_i$ may be difficult to assess.

A more inviting prospect in the same vein is to specify your probabilities for the events that $X_i$ lies in various intervals in the hope that features of the mixing distribution function may be derivable from them. Such a procedure does in fact prove fruitful when some additional constraints are put on the form of the mixing functions under consideration. Useful applied developments of this line of thinking appear in Kadane, Dickey, et al. (1980), Dickey and Chen (1985), and Dickey, Dawid, and Kadane (1986).

A second major tack in specifying your mixing distribution function $M(\mu, \pi)$ is to identify particular observable quantities for which your probability distribution function is approximately identical to your mixing distribution function. Then in specifying your probability distribution for these quantities, you are essentially specifying your mixing distribution function as well. In the context of an assessed $M\text{-}N_K(\mu, \pi)$ distribution, this tack is exemplified in the following argument. Let us state the conclusion as a theorem, prove it, and then discuss it, suggesting a practical problem.

**Theorem 7.11.** If you regard the components of $\mathbf{X}$ exchangeably infinitely extendable and your assessed distribution is $\mathbf{X} \sim M\text{-}N_K(\mu, \pi)$, then your mixing distribution function $M(\mu, \pi)$ must be approximately identical to your assessed probability distribution function for the quantities $(\bar{X}_N, S_N^2)$, where $N$ is a large number. To be precise mathematically, under these conditions

$$\int_0^{\infty}\int_{-\infty}^{\infty} |dF(\bar{x}_N, s_N^2) - dM(\mu, \pi)| \to 0 \qquad \text{as } N \to \infty.$$

*Proof.* For each value of $\mu$ and $\pi$, we can write the component distribution of your mixture probability distribution for $\mathbf{X}_K$ as $(\mathbf{X}; \mu, \pi) \sim N_K(\mu \mathbf{1}_K, \pi^{-1}\mathbf{I}_K)$. Notice that these distributions are in exchangeable Normal form, with variance-covariance matrix $\alpha \mathbf{I}_K + \beta \mathbf{1}_{K,K}$ defined by the special case of $\alpha = \pi^{-1}$ and $\beta = 0$. Thus, for any specific values of $\mu$ and $\pi$, and for the extension of the distribution to any $N > K$ quantities, the implied mixture distribution for $\bar{X}_N$ and $S_{N-1}^2$ treats them independently and insists that $\bar{X}_N$ is practically identical to $\mu$ and that $S_{N-1}^2$ is practically identical to $\pi^{-1}$. For Theorem 7.6 and Corollary7.6.1 identify the $(\mu, \pi)$-specific independent distributions for $\bar{X}_N$ as $N(\mu, \alpha N^{-1} + \beta) = N(\mu, \alpha/N)$, and for $[\pi(N-1)]S_{N-1}^2$ as $\chi_{N-1}^2$. This determines the $(\mu, \pi)$-specific moments of $S_{N-1}^2$ as $\mathbb{M}_1(S_{N-1}^2; \mu, \pi) = \pi^{-1}$ and $\mathbb{M}_2^*(S_{N-1}^2; \mu, \pi) = 2/[\pi^2(N-1)]$. Thus, for large values of $N$, these distributions are virtually degenerate on the specific $(\mu, \pi)$ pair of the mixture component distribution. Formally, we write

$$\lim_{N \to \infty} f(\bar{x}_N, s_{N-1}^2; \mu, \pi) = \delta(\mu, \pi^{-1}),$$

where $\delta$ is Dirac's delta function. Because of the virtual equivalence of $\bar{x}_N$ with $\mu$ and of $s_{N-1}^2$ with $\pi^{-1}$ in the eyes of the $(\mu, \pi)$-specific mixture component, your

$$\begin{aligned} f(\bar{x}_N, s_{N-1}^2) &= \int_0^{\infty}\int_{-\infty}^{\infty} f(\bar{x}_N, s_{N-1}^2; \mu, \pi)\, dM(\mu, \pi) \\ &\approx dM(\mu = \bar{x}_N, \pi = (s_{N-1}^2)^{-1}) \qquad \text{for large } N. \end{aligned}$$

Your probability for the event that $(\bar{X}_N, S_{N-1}^2)$ is within any particular region of $[\mathbb{R}, \mathbb{R}^+]$ approximately equals the amount of the function $M(\mu, \pi^{-1})$ that is amassed in the same region. □

This, too, is a pleasant conclusion. For it allows you to specify your mixture distribution $M(\mu, \pi)$ by reflecting on your probability distribution for the average of many components of $\mathbf{X}$ and for the average squared differences between them and their average. In a problem such as the wheat yields we have been considering, you can distinguish perhaps easily some components of your uncertainty about any $X_i$ by assessing the values of your $P(\bar{X}_N)$, $V(\bar{X}_N)$, and your $P(S^2_{N-1})$ for some large number $N$. The relation among these assertions and your prevision and variance for individual components of $\mathbf{X}$ is described by

$$P(X_i) = P(\bar{X}_N) \qquad \text{and} \qquad V(X_i) \approx P(S^2_{N-1}) + V(\bar{X}_N).$$

This latter result can often be helpful in assessing variances. Your variance for an individual measurement equals approximately your prevision for the average squared deviations of many observations from their average value, plus your variance for the value of their average. The following problem will guide you through some intriguing considerations that make use of these concepts.

## PROBLEM

1. Recall the introductory problems of Sections 7.2 in which you attempted to assess your prevision and variance for the yield of a particular acre of Kansas sown in wheat. Now, supposing that $X_i$ is the yield from an acre of Kansas wheat planted during 1997, try to determine your current variance $V(X_i)$ by assessing your $P[S^2_{N-1}] + V[\bar{X}_N]$, where $N$ is a large number. Which of these two components of your approximate $V(X_i)$ is the larger? What if you were told the exact output from one arbitrary acre of Kansas during 1997, $X_1 = x_1$? What is your $P[S^2_{N-1} | X_1 = x_1]$ and $V[\bar{X}_N | X_1 = x_1]$? Would the relative sizes of your $P$ and $V$ reverse their order? How many quantities $X_i$ would you need to know in order for these two assessments to reverse in their order of magnitude?

Recent computational developments based on Monte Carlo integration procedures described in Gelfand and Smith (1992), Casella and George (1992), Tanner (1993), and Wilks et al. (1996) today allow us the numerical capacity to evaluate inferences based on initial mixing distributions that have any algebraical form whatsoever. The earlier computational tack discussed by Kadane, Dickey et al. (1980) reduces the complexity of computational requirements for approximations at the cost of restricting characteristics of the forms of mixture distributions that can be represented. It allows you to assess several quantiles of your opinions about $\mathbf{X}$ and to compute an approximate posterior density $f(\mathbf{x}_2 | \mathbf{X}_1 = \mathbf{x}_1)$ via specifiable connections among your quantile assessments and your appropriate mixing function. These connections rely on the algebraic results of conjugate Normal distribution theory which will conclude this chapter. Studying the Normal-gamma distribution and its extensions is useful both for numerical approximation to exact inferences in certain situations and for the insight it provides into the general theory of inference under uncertainty. For it provides easily interpretable algebraic details of one very specific coherent strategy of conditioning.

### 7.4.5 The Normal-Gamma Distribution and the Mixture-Normal It Generates, the *t*-Distribution

The convenient natural conjugate form of mixing function over $\mu$ and $\pi$ in the mixture-Normal context is the Normal-gamma distribution, which implies marginally a $t$-distribution for $\mu$. In this section we study the mathematical characteristics of these functions and then discuss their relevance to inference about $\mathbf{X}_2$ from the observation $\mathbf{X}_1 = \mathbf{x}_1$. To understand the algebraic detail, begin by reviewing the properties of the gamma function, $\Gamma(\alpha) \equiv \int_0^\infty x^{\alpha-1}\exp(-x)dx$, which we introduced in Section 5.4.2.

#### 7.4.5.1 The Family of Gamma Distributions: $\Gamma(\alpha, \beta)$ with $\alpha > 0$ and $\beta > 0$

***Definition 7.4.*** Writing $\pi \sim \Gamma(\alpha, \beta)$ means that the distribution function for $\pi$ is the Riemann integral of the density function

$$f(\pi) = [\beta^\alpha/\Gamma(\alpha)]\pi^{\alpha-1}\exp(-\beta\pi) \qquad (\pi > 0)$$

for specified values of $\alpha > 0$ and $\beta > 0$. □

Each point $(\alpha, \beta)$ within the parameter space of this family, $(\mathbb{R}^+, \mathbb{R}^+)$, corresponds to a specific density function of the form displayed in the definition. Analyzing this functional form over various values of $(\alpha, \beta)$ shows the several shapes of functions in the family of gamma distributions.

It is easy to derive the moment generating function for the generic distribution as $M_t(\pi) = \mathbb{M}_1[\exp(t\pi)] = [1 - t/\beta]^{-\alpha}$. [You need to use the fact that, since the gamma density function must integrate to 1, an integral of the form $\int_0^\infty \pi^{\alpha-1}\exp(-\beta\pi)d\pi$ must equal $\Gamma(\alpha)/\beta^\alpha$.] The first two moments of a gamma distribution are $\mathbb{M}_1(\pi) = \alpha/\beta$ and $\mathbb{M}_2^*(\pi) = \alpha/\beta^2$. Moreover, the implied first two moments for $\pi^{-1}$ are

$$\begin{aligned}
\mathbb{M}_1(\pi^{-1}) &= \mathbb{M}_{-1}(\pi) = \beta/(\alpha - 1), \\
\mathbb{M}_2^*(\pi^{-1}) &= \mathbb{M}_2(\pi^{-1}) - [\mathbb{M}_1(\pi^{-1})]^2 = \mathbb{M}_{-2}(\pi) - [\mathbb{M}_{-1}(\pi)]^2 \\
&= \beta^2/[(\alpha - 1)^2(\alpha - 2)].
\end{aligned}$$

We use these details in subsequent analysis.

The family of gamma distributions is a member of the exponential family of distributions, discussed in Section 5.6. It includes as "subfamilies" the families of simple exponential and chi-square distributions, viz., $\alpha = 1$ implies $\pi \sim \exp(\beta)$, and $\alpha = N/2$ and $\beta = 1/2$ imply $\pi \sim \chi_N^2$. Figure 7.4 identifies these subfamilies of $\Gamma(\alpha, \beta)$ within the general family-defining space of $\alpha > 0$ and $\beta > 0$. Figure 7.5 displays the family members of $\chi_N^2$ densities for a few values of $N$.

In general, the density function of any $\Gamma(\alpha, \beta)$ distribution has a single mode at $\pi = (\alpha - 1)/\beta$ presuming $\alpha > 1$, and it has inflection points at $\pi = \beta^{-1}[(\alpha - 1) \pm$

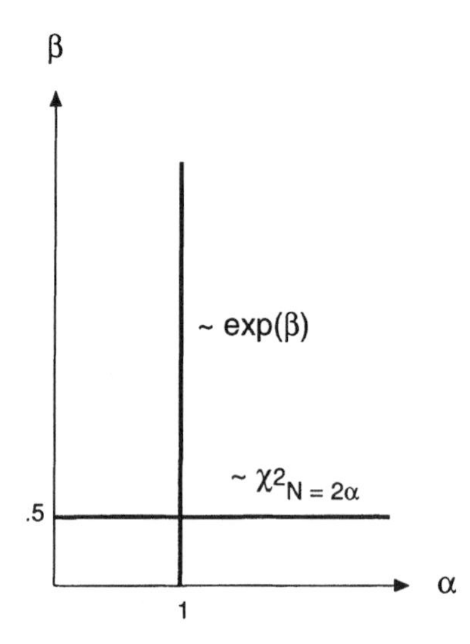


**Figure 7.4.** Two subfamilies of $\Gamma(\alpha, \beta)$ distributions with $(\mathbb{R}^+, \mathbb{R}^+)$.

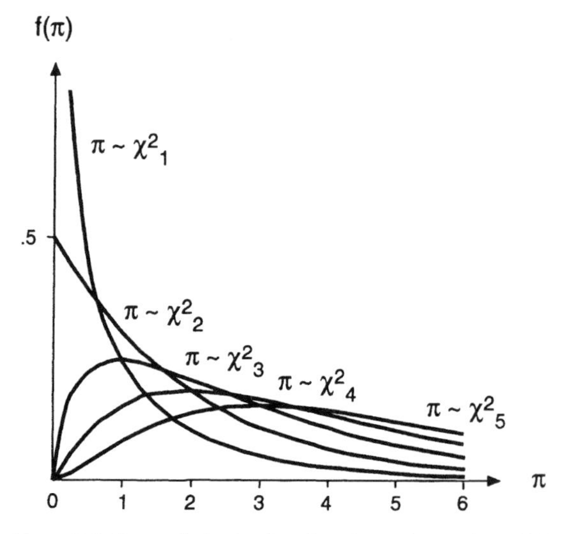


**Figure 7.5.** Some $\chi_N^2$ density functions for various values of $N$.

$(\alpha - 1)^{1/2}]$ presuming $\alpha > 2$. When $\alpha \in (1, 2]$ there is only one inflection point. When $\alpha \leqslant 1$, there are none. These properties are exemplified in the $\chi_N^2$ densities exhibited in Figure 7.5. You can generate graphs of various gamma pdf's in MATLAB, using the command "help gampdf".

#### *7.4.5.2 The Family of t-Distributions: $t(\upsilon, \mu, \tau)$ with $v > 0, \mu \in \mathbb{R}, \tau > 0$*

We examine this family of distributions in two stages; (1) the "standardized" $t(v)$ distribution, and (2) the generic family member.

***Definition 7.5.*** Writing $X \sim t(v)$ means that the distribution function for $X$ is the Riemann integral of the density function

$$f(x) \propto [v + x^2]^{-(v+1)/2} \quad (x \in \mathbb{R}) \qquad \text{for some } v > 0,$$

with a proportionality constant of $\{v^{v/2}\Gamma[(v+1)/2]/\Gamma(1/2)\Gamma(v/2)\}$. We refer to this density function as the *standard $t(v)$ density.* □

(Recall that $\Gamma(1/2) = \sqrt{\Pi}$.)

To begin, it should be apparent that the standard $t(v)$ density is symmetric about $x = 0$, since $x$ is squared in the formula for $f(x)$. The first two moments of the distribution are $\mathbb{M}_1(X) = 0$, provided that $v > 1$, and $\mathbb{M}_2^*(X) = [v/(v-2)]$, provided that $v > 2$. In general, the integral defining the $k$th moment of a $t$-distribution exists only if $v > k$.

The general $t$-distribution family members are specified by a linear transformation of a quantity that is distributed standard $t$, as specified in the following definition.

***Definition 7.6.*** A quantity $Y$ is said to be distributed with the general $t$-distribution, $Y \sim t(v, \mu, \tau)$ if the transformed quantity $Z \equiv \tau^{1/2}(Y - \mu) \sim t(v)$. Thus, the distribution function for $Y$ is the Riemann integral of the density function

$$f(y) \propto [v + \tau(y-\mu)^2]^{-(v+1)/2} \quad (y \in \mathbb{R}) \qquad \text{for some } (v, \mu, \tau) \in (\mathbb{R}^+, \mathbb{R}, \mathbb{R}^+),$$

with a proportionality constant of $\{\tau^{1/2}v^{v/2}\Gamma[(v+1)/2]/\Gamma(1/2)\Gamma(v/2)\}$. □

Two important moments of the general distribution are $\mathbb{M}_1(Y) = \mu$, provided that $v > 1$, and $\mathbb{M}_2^*(Y) = v/[(v-2)\tau]$, provided that $v > 2$. Thus, $\mu$ determines the location of the density function, and $\tau$ determines its scale. Note that $t(v, 0, 1)$ is called the "standard $t$-distribution" despite the fact that its second central moment does not equal 1 but $v/(v-2)$. Usually, a *standardized* distribution is considered to exhibit a first moment of 0 and a second central moment of 1, as the standard Normal distribution, for example. This special feature of the general $t$-distribution will be relevant to the algebraic detail of some subsequent results we shall study.

A final technical result is worth mentioning, since it applies to the use of quadratic scoring functions to assess sequential forecasts in the form of a general $t(v, \mu, \tau)$-distribution. If $Y \sim t(v, \mu, \tau)$ then

$$\int_{-\infty}^{\infty} f_Y^2(y)\, dy = \frac{\tau^{1/2}\Gamma^2(v/2 + 1/2)\Gamma(v + 1/2)}{v^{1/2}\Gamma(1/2)\Gamma^2(v/2)\Gamma(v+1)}.$$

The logic of the proof is simple, since the square of a general $t$-density is proportional to another general $t$-density. The details require your attention to the form of the proportionality constant of the density.

#### 7.4.5.3 *The Family of Normal-Gamma Distributions: $N\Gamma(\mu_0, \tau, \alpha, \beta)$*

Members of the normal-gamma family of distributions are joint distributions for two variables, here denoted as $\mu$ and $\pi$, over the domain $(\mathbb{R}, \mathbb{R}^+)$. The joint density for the two arguments is defined in terms of the product of a conditional and a marginal density. (Remember we now parameterize a normal density by its precision rather than its variance.)

***Definition 7.7.*** Writing $(\mu, \pi) \sim N\Gamma(\mu_0, \tau, \alpha, \beta)$ means that the joint density function for $(\mu, \pi)$ is the product of a conditional density function for $(\mu|\pi) \sim N(\mu_0, \tau\pi)$ and a marginal density for $\pi \sim \Gamma(\alpha, \beta)$. Thus, the joint density function for $(\mu, \pi)$ is

$$f(\mu, \pi) \propto (\tau\pi)^{1/2} \exp[-\tau\pi(\mu - \mu_0)^2/2]\pi^{\alpha-1} \exp(-\beta\pi) \qquad ((\mu, \pi) \in (\mathbb{R}, \mathbb{R}^+)),$$

with a proportionality constant of $(2\Gamma(1/2))^{-1/2}\{\beta^\alpha/\Gamma(\alpha)\}$, for specified values of $\mu_0, \tau > 0$, $\alpha > 0$, and $\beta > 0$. □

Before studying the shape of the joint density function geometrically, let us prove algebraically the important relation between the Normal-gamma distribution and the $t$-distribution, specified by the following theorem.

**Theorem 7.12.** If $(\mu, \pi) \sim N\Gamma(\mu_0, \tau, \alpha, \beta)$, then $\mu \sim t(2\alpha, \mu_0, \tau\alpha/\beta)$, and $\pi|\mu \sim \Gamma(\alpha + 1/2, \beta + \tau(\mu - \mu_0)^2/2)$.

*Proof.* We can determine the density function for $\mu$ by integrating $f(\mu, \pi)$ with respect to $\pi$:

$$f(\mu) = \int_0^\infty f(\mu|\pi) f(\pi) d\pi$$

Ignoring proportionality constants that do not involve $\mu$ or $\pi$, and resolving some simple algebra, this becomes

$$f(\mu) \propto \int_0^\infty \pi^{(\alpha+1/2)-1} \exp\{-\pi[\beta + \tau(\mu - \mu_0)^2/2]\} \, d\pi.$$

As a function of $\pi$, the integrand is proportional to a gamma density function with parameters $\alpha + 1/2$ and $\beta + \tau(\mu - \mu_0)^2/2$. Thus, the integral of this function over $\pi$ must be the *inverse* of the appropriate proportionality constant for a gamma density. Thus, we have

$$f(\mu) \propto \Gamma(\alpha + 1/2)/[\beta + \tau(\mu - \mu_0)^2/2]^{(\alpha+1/2)}.$$

Again ignoring proportionality constants, as a function only of $\mu$ this expression can be reduced algebraically to

$$f(\mu) \propto [2\alpha + (\tau\alpha/\beta)(\mu - \mu_0)^2]^{-(2\alpha+1)/2}.$$

Comparing this density function with the form of the density $t(v, \mu, \tau)$ in Definition 7.6 yields the result that $\mu \sim t(2\alpha, \mu_0, \tau\alpha/\beta)$.

As to the conditional density of $\pi|\mu$, its gamma form is evident merely by considering the joint density formula $f(\mu, \pi)$ as a function of $\pi$ alone for a fixed value of $\mu$ and ignoring all proportionality constants. □

Another way to state this theorem is that a gamma mixture of Normal distributions is a $t$-distribution and is equivalent to a $t$-mixture of gamma distributions.

Graphically, the surface of a Normal-gamma density function describes a Normal density with respect to $\mu$ if you slice the surface with a plane specified by a particular value of $\pi$. Moreover, the definition implies that if you collapse the density surface over values of $\mu$ by integration, the marginal density for $\pi$ also looks a gamma density. Theorem 7.12 then allows that if you slice the surface with a plane specified by a particular value of $\mu$, the intersection exhibits a gamma density whose shape is determined by the specified values of $\alpha$, $\beta$, and $\tau$; and if you collapse the surface over values of $\pi$ by integration, the marginal density for $\mu$ is a $t$-density.

Traditionally, the way to study the shape of a joint density surface would be to study first its ridge lines [contours of points at which the partial derivatives $\partial_\mu f(\mu, \pi) = 0$ and $\partial_\pi f(\mu, \pi) = 0$]. These can be found to be the line $\mu = \mu_0$, and the curve $\pi = (2\alpha - 1)/2[\beta + \tau(\mu - \mu_0)^2/2]$, as long as $\alpha > 1/2$. Then the inflections of the surface surrounding these ridges can be identified from the conditional densities: for each value of $\pi$, the normal conditional density of $\mu|\pi$ shows inflection points at $\mu = \mu_0 \pm (\tau\pi)^{-1/2}$, whereas for each value of $\mu$ the inflection points of the gamma conditional density of occur at $\pi = [\beta + \tau(\mu - \mu_0)^2/2]^{-1}[(\alpha^{-1/2}) \pm (\alpha^{-1/2})^{1/2}]$. The various possible shapes of the joint density can be easily imagined based on these results.

Another useful way to study the joint density is to write a MATLAB program to display the Normal-gamma density in three-dimensions, and to display the contours of constant density. (Use "help mesh" to learn how to produce a graph of a function of two variables. You should produce several graphs, representing various configurations of $\alpha$, $\beta$, and $\tau$, particularly both with $\alpha < 1/2$ and with $\alpha > 1/2$. The commands "subplot" and "contour" are useful.)

We have motivated our interest in the family of Normal-gamma distributions by its accessibility as a mixing distribution over $(\mu, \pi)$ in the mixture characterization of $X|(\mu, \pi)$ as $M\text{-}N(\mu, \pi)$. In this context it is of interest that a Normal-gamma mixture of a Normal distribution is also a $t$-distribution.

**Theorem 7.13.** If a distribution for $X$ is a mixture-Normal, $X \sim M\text{-}N(\mu, \pi)$, and the mixing distribution is $(\mu, \pi) \sim N\Gamma(\mu_0, \tau, \alpha, \beta)$, then the distribution for $X$ is $X \sim t(2\alpha, \mu_0, [\tau\alpha/(1+\tau)\beta])$.

*Proof.* Ignoring proportionality constants, we can derive the density as

$$\begin{aligned} f(x) &= \int_0^\infty \int_{-\infty}^\infty f(x|\mu,\pi) f(\mu,\pi)\, d\mu\, d\pi \\ &\propto \int_0^\infty \int_{-\infty}^\infty \pi^{1/2} \exp\{-\pi(x-\mu)^2/2\} \pi^{(\alpha+1/2)-1} \\ &\quad \times \exp\{-\pi[\beta + \tau(\mu-\mu_0)^2/2]\}\, d\mu\, d\pi \\ &\propto \int_0^\infty \int_{-\infty}^\infty \pi^\alpha \exp\{-\pi[\beta + (x-\mu)^2/2 + \tau(\mu-\mu_0)^2/2]\}\, d\mu\, d\pi. \end{aligned}$$

Now expanding the squares and collecting terms involving $\mu$ yields, equivalently,

$$\begin{aligned} f(x) &\propto \int_0^\infty \pi^\alpha \exp\{-\pi[\beta + \tau(x-\mu_0)^2/2(1+\tau)]\} \\ &\quad \times \int_{-\infty}^\infty \exp\{-\pi(1+\tau)[\mu - (x+\tau\mu_0)/(1+\tau)]^2/2\}\, d\mu\, d\pi \\ &\propto \int_0^\infty \pi^{(\alpha-1/2)} \exp\{-\pi[\beta + \tau(x-\mu_0)^2/2(1+\tau)]\}\, d\pi, \end{aligned}$$

since that Normal integral with respect to $\mu$ is proportional to $[\pi(1+\tau)]^{-1/2}$. Continuing now by integrating the remaining density in gamma form with respect to $\pi$ yields its proportionality constant involving only $x$:

$$\begin{aligned} f(x) &\propto [\beta + \tau(x-\mu_0)^2/2(1+\tau)]^{-(\alpha+1/2)} \\ &\propto \{2\alpha + [\tau\alpha/(1+\tau)\beta](x-\mu_0)^2\}^{-(2\alpha+1)/2}. \end{aligned}$$

This is the general $t$-density identified in the theorem: $t(2\alpha, \mu_0, [\tau\alpha/(1+\tau)\beta])$. Its moments are

$$\begin{aligned} P(X) &= \mathbb{M}_1(X) = \mu_0 && \text{as long as } \alpha > 1/2, \\ V(X) &= \mathbb{M}_2^*(X) = [(1+\tau)\beta/\tau(\alpha-1)] && \text{as long as } \alpha > 1. \end{aligned}$$

□

#### *7.4.5.4 Application to Inference about Quantities That Are Regarded Exchangeably with a Normal-Gamma Mixture-Normal Distribution*

We are now prepared to address the inferential situation of normal exchangeable judgments about **X** that has led us to study the Normal-gamma mixing function.

The analysis proceeds most simply by stating a theorem and its proof, followed by a discussion.

**Theorem 7.14.** Suppose that you regard the components of $\mathbf{X}$ exchangeably and that your assessed distribution is $\mathbf{X} \sim M\text{-}N_K(\mu, \pi)$, with your mixing distribution function $M(\mu, \pi)$ specified as $N\Gamma(\mu_0, \tau, \alpha, \beta)$. Suppose $\mathbf{X}$ is partitioned into $(\mathbf{X}_1, \mathbf{X}_2)$, where $\mathbf{X}_1$ is $N \times 1$ and $\mathbf{X}_2$ is $(K - N) \times 1$. Then coherency requires that your conditional distribution for $\mathbf{X}_2 | (\mathbf{X}_1 = \mathbf{x}_1)$ be also assessed as a mixture Normal distribution, $M\text{-}N_{K-N}(\mu, \pi)$, with the conditional mixing function $M(\mu, \pi | \mathbf{X}_1 = \mathbf{x}_1)$ in the Normal-gamma form

$$N\Gamma\left\{(\tau + N)^{-1}(\tau\mu_0 + N\bar{X}_N), \tau + N, \frac{\alpha + N}{2}, \beta + \left(\frac{N}{2}\right)s_N^2 + \frac{\tau N(\bar{x}_N - \mu_0)^2}{2(N + \tau)}\right\},$$

where $\bar{x}_N$ is the arithmetic average of the $N$ components of $\mathbf{x}_1$, and $s_N^2$ is the average squared difference, $N^{-1}\sum_{i=1}^{N}(x_i - \bar{x})^2$. Thus, the conditional distribution of $X_{N+1} | (\mathbf{X}_N = \mathbf{x}_N)$ is $t(v_N, \mu_N, \tau_N^*)$, where

$$v_N = 2\alpha + N, \qquad \mu_N = (\tau + N)^{-1}(\tau\mu_0 + N\bar{x}_N),$$

$$\tau_N^* = \frac{(\tau + N)(2\alpha + N)}{(\tau + N + 1)[2\beta + Ns_N^2 + \tau N(\bar{x}_N - \mu_0)^2/(N + \tau)]},$$

while the marginal distribution of $X_{N+1}$ is $t(v_0 = 2\alpha,\ \mu_0,\ \tau_0 = \tau\alpha/(1 + \tau)\beta)$.

*Proof.* Theorem 7.7 already establishes that we need only calculate

$$dM(\mu, \pi | \mathbf{X}_1 = \mathbf{x}_1) \propto \pi^{N/2} \exp\{-(N\pi/2)[s_N^2 + (\bar{x}_N - \mu)^2]\}\, dM(\mu, \pi).$$

Since the Normal-gamma mixing function admits the density function

$$dM(\mu, \pi) = m(\mu, \pi)\, d\mu\, d\pi \propto \pi^{\alpha + 1/2 - 1} \exp\{-\pi[\beta + \tau(\mu - \mu_0)^2/2]\}\, d\mu\, d\pi,$$

the multiplication yields $dM(\mu, \pi | \mathbf{X}_1 = \mathbf{x}_1)$ arranged in a form that identifies it as proportional to

$$\pi^{1/2} \exp\{-(N + \tau)\pi(\mu - [(\tau + N)^{-1}(\tau\mu_0 + N\bar{x}_N)])^2/2\}$$
$$\times\, \pi^{(\alpha + N/2) - 1} \exp\{-\pi[\beta + (N/2)s_N^2 + \tau N(\bar{x}_N - \mu_0)^2/2(N + \tau)]\}.$$

This is the form for the Normal-gamma mixing density prescribed in the theorem. □

**Corollary 7.14.1.** Under the conditions of opinions about $\mathbf{X}$ specified in the theorem,

$$P(X_{N+1}|\mathbf{X}_N = \mathbf{x}_N) = (\tau\mu_0 + N\bar{x}_N)/(\tau + N),$$

$$V(X_{N+1}|\mathbf{X}_N = \mathbf{x}_N) = \frac{(\tau + N + 1)[2\beta + Ns_N^2 + \tau N(\bar{x}_N - \mu_0)^2/(N + \tau)]}{(\tau + N)[2\alpha + N - 2]}.$$

These previsions are computed directly as the moments $\mathbb{M}_1$ and $\mathbb{M}_2^*$ of the $t$-density specified in the theorem.

The first moment displayed in the corollary identifies that your posterior (that is, conditional) prevision for $X_{N+1}$ given $(\mathbf{X}_N = \mathbf{x}_N)$ equals a weighted average of $\mu_0$, your prior (marginal) prevision for $X_{N+1}$, and $\bar{x}_N$, the numerical average of the observed conditioning quantities. The weights are proportional to the relative precisions of your opinion regarding $X_{N+1}$, which given a specification of $\pi$ equals $\tau\pi$, and your precision regarding each observation, which given $\pi$ equals $\pi$. Since the ratio of these precisions is $\tau$ to 1 whatever the value of $\pi$, the relative weights on $\mu_0$, and on each conditioning value of $x_i$ are $\tau$ and 1's. [In the language of regression analysis, the regression coefficient of $X_{N+1}$ on each $X_i$ equals $1/(\tau + N)$, and the regression constant equals $\tau\mu_0/(\tau + N)$.] If the number of conditioning quantities is large, the posterior prevision is closer to the average of the observations; if $\tau$ is large relative to the number of conditioning observations, the posterior prevision is closer to the prior prevision.

The second moment displayed in the corollary identifies your posterior variance for $X_{N+1}$ given $(\mathbf{X}_N = \mathbf{x}_N)$ as a linear combination of the prior variance, $(\tau + 1)\beta/\tau\alpha$, the average squared difference of the conditioning quantities from their average, $s_N^2$, and the squared difference of the average conditioning value from the prior prevision, $(\bar{x}_N - \mu_0)^2$, all with positive coefficients. Analyzing these shows that the coefficient on $s_N^2$ tends to 1 as $N$ increases, whereas the coefficients on the other two terms tend toward 0.

To conclude, it is worthwhile to print the sequential computation formulas for the conditional inference regarding $X_{N+1}$ given $\mathbf{X}_N = \mathbf{x}_N$, when the components of $\mathbf{X}$ are observed sequentially. The *sequential* computations of $v_N$, $\mu_N$, and $\tau_N^*$ that identify the conditional distribution for $X_{N+1}|(\mathbf{X}_N = \mathbf{x}_N)$ as $t(v_N, \mu_N, \tau_N^*)$ are specified most simply in this context via sequential formulas for the components of the corresponding $N\Gamma(\mu_N, \tau_N, \alpha_N, \beta_N)$ mixing distribution over $(\mu, \pi)$. These equations are

$$\mu_N = (\tau_{N-1} + 1)^{-1}(\tau_{N-1}\mu_{N-1} + x_N), \quad \tau_N = \tau_{N-1} + 1, \quad \alpha_N = \alpha_{N-1} + 1/2,$$

$$\beta_N = \beta_{N-1} + \tau_{N-1}(x_N - \mu_{N-1})^2/2(\tau_{N-1} + 1).$$

The parameters of the conditional $t(v_N, \mu_N, \tau_N^*)$-distribution for $X_{N+1}|(\mathbf{X}_N = \mathbf{x}_N)$ are calculated from these sequential formulations by $v_N = 2\alpha_N$, $\mu_N = \mu_N$, and

$\tau_N^* = \tau_N\alpha_N/[(\tau_N + 1)\beta_N]$. (Notice in this final equation, the $\tau_N^*$ of the left-hand side is a parameter of the $t$-distribution for $X_{N+1}$, whereas the $\tau_N$ of the right-hand side is a parameter of the Normal-gamma mixing distribution.)

The next subsection defines the multivariate $t$-distribution and shows that the analysis we have just concluded amounts to a characterization of the exchangeable multivariate $t$-distribution for the components of $\mathbf{X}$. The broader applicability of the multivariate $t$ will be made apparent in Section 7.5.

#### 7.4.5.5 *Algebraic Representation of the Joint Distribution for X as Exchangeable Multivariate* $t_K(2\alpha, \mu_0\mathbf{1}_K, (\alpha/\beta)[\mathbf{I}_K - (\tau + K)^{-1}\mathbf{1}_{K,K}])$

We have just concluded a derivation of your sequential conditional distribution for $X_{N+1}|(\mathbf{X}_N = \mathbf{x}_N)$ for $N = 0, 1, 2, \ldots, (K-1)$ under the specification that you regard the entire sequence $X_1, \ldots, X_K$ exchangeably, with the distribution $\mathbf{X} \sim N\Gamma\text{-}N_K(\mu_0, \tau, \alpha, \beta)$. Given the result that for every $N$, your implied "one-step ahead" conditional distributions are in $t$ form, $X_{N+1}|(\mathbf{X}_N = \mathbf{x}_N) \sim t(\nu_N, \mu_N, \tau_N^*)$, it may not surprise you that your *joint* distribution for all $K$ components of $\mathbf{X}$ can be written in a "multivariate **t**" form.

Of course, we could write the joint density function for $\mathbf{X}$ directly as the product of the conditional $t$-densities, $f(x_1, \ldots, x_K) = \prod_N^{K-1} f(X_{N+1}|\mathbf{X}_N = \mathbf{x}_N)$. But algebraically, the product of these conditional $t$-densities looks rather formidable. An equivalent direct derivation of the joint density function results in the very manageable multivariate **t** form. Let us begin by defining a multivariate **t**-distribution in $K$ dimensions, and stating its first two moments.

***Definition 7.8.*** A distribution function for the vector $\mathbf{X} \in \mathbb{R}^K$ that can be expressed as the Riemann integral of the density function

$$f(\mathbf{x}) \propto [\nu + (\mathbf{x} - \boldsymbol{\mu}_K)^{\mathrm{T}}\boldsymbol{\tau}_{K,K}(\mathbf{x} - \boldsymbol{\mu}_K)]^{-(\nu+K)/2} \qquad (\mathbf{x} \in \mathbb{R}^K),$$

with proportionality constnat $C = \{|\tau_{K,K}|^{1/2}\nu^{\nu/2}\Gamma[(\nu + K)/2]/(\Gamma(1/2))^{K/2}\Gamma(\nu/2)\}$, is said to be a *K-dimensional multivariate t-distribution* with family parameter $\nu > 0$, location vector $\boldsymbol{\mu}_K$, and precision matrix $\tau_{K,K}$. We write $\mathbf{X} \sim \mathbf{t}_K(\nu, \boldsymbol{\mu}_K, \tau_{K,K})$ to denote that this distribution applies to $\mathbf{X}$. □

The first two moments for this distribution are $\mathbb{M}_1(\mathbf{X}) = \boldsymbol{\mu}_K (\nu > 1)$ and $\mathbb{M}_2^*(\mathbf{X}) = [\nu/(\nu - 2)]\tau_{K,K}^{-1} (\nu > 2)$. A detailed derivation of these moment results and of the normalizing proportionality constant can be found in Zellner (1971, Appendix B.2.) One oddity of this parameterization of the $\mathbf{t}_K$-distribution is that $\tau_{K,K}$ is called the precision matrix of the distribution, even though the precision matrix for $\mathbf{X}$ is actually proportional to it, specifically $[(\nu - 2)/\nu]\tau_{K,K}$.

The multivariate **t**-distribution is a $K$-dimensional vector generalization of the general univariate $t(\nu, \mu, \tau)$-distribution of Definition 7.6. We also state, without deriving the details, that any subvector of $\mathbf{X}$ is marginally distributed multivariate **t** and the conditional distribution of the remaining subvector is also in multivariate **t** form. We are about to see that assessing the components of a vector $\mathbf{X}$ exchangeably with the mixture distribution $\mathbf{X} \sim N\Gamma\text{-}N_K(\mu_0, \tau, \alpha, \beta)$ is equivalent to assessing a multivariate **t**-distribution for $\mathbf{X}$ with a particular form for its precision matrix:

$$\mathbf{X} \sim \mathbf{t}_K(2\alpha, \mu_0 \mathbf{1}_K, (\alpha/\beta)[\mathbf{I}_K - (\tau + K)^{-1}\mathbf{1}_{K,K}]).$$

Notice in passing that the multivariate $\mathbf{t}_K$-density function is uniform over ellipses of the form $(\mathbf{x} - \boldsymbol{\mu}_K)^{\mathrm{T}} \tau_{K,K}(\mathbf{x}_K - \boldsymbol{\mu}_K) = c$. Thus, the result to be proved exemplifies a generalization of our Theorem 7.8 in Section 7.4.3, that any elliptically symmetric distribution can be written as a mixture of Normal distributions. We develop our result in two steps.

The first stage of the development we state and prove as a lemma.

**Lemma 7.15.** Writing $\mathbf{X} \sim N\Gamma\text{-}N(\mu_0, \tau, \alpha, \beta)$ is equivalent to writing $\mathbf{X} \sim \Gamma\text{-}N_K(\mu_0 \mathbf{1}_K, \pi[\mathbf{I}_K - (\tau + K)^{-1}\mathbf{1}_{K,K}]; \alpha, \beta)$. That is, for each $\pi > 0$, $\mathbf{X}|\pi \sim N_K(\mu_0 \mathbf{1}_K, \pi[\mathbf{I}_K - (\tau + K)^{-1}\mathbf{1}_{K,K}])$, while marginally, $\pi \sim \Gamma(\alpha, \beta)$.

*Proof.* Ignoring proportionality constants, we can write the density function for $\mathbf{X}$ as

$$f(\mathbf{x}) = \int_0^\infty \int_{-\infty}^\infty f(\mathbf{x}; \mu, \pi) f(\mu|\pi) f(\pi)\, d\mu\, d\pi$$

$$\propto \int_0^\infty \int_{-\infty}^\infty \pi^{K/2} \exp\left(-\pi \sum_{i=1}^K (x_i - \mu)^2/2\right) \pi^{1/2} \exp(-\tau\pi(\mu - \mu_0)^2/2) \pi^{\alpha-1} e^{-\beta\pi}\, d\mu\, d\pi$$

$$\propto \int_0^\infty \int_{-\infty}^\infty \pi^{(\alpha + K/2 + 1/2 - 1)} \exp\left(-\pi\left[\beta + \sum_{i=1}^K (x_i - \mu)^2/2 + \tau(\mu - \mu_0)^2/2\right]\right) d\mu\, d\pi.$$

Collecting terms involving $\mu$, algebraic manipulation allows that this expression is proportional to

$$\int_0^\infty \pi^{(\alpha + K/2 - 1)} \exp\left(-(\pi/2)\left[2\beta + \sum_{i=1}^K x_i^2(1 - 1/(\tau + K)) - 2\sum_{j>i=1}^K x_i x_j/(\tau + K)\right.\right.$$
$$\left.\left. - 2\tau\mu_0 \sum_{i=1}^K x_i/(\tau + K)\right]\right)$$
$$\cdot \int_{-\infty}^\infty \pi^{1/2} \exp\left(-\pi[(K + \tau)/2]\left[\mu - \left(\sum_{i=1}^K x_i + \tau\mu_0\right)/(K + \tau)\right]^2\right) d\mu\, d\pi.$$

Performing the inner integration of the Normal density with respect to $\mu$ yields

$$f(\mathbf{x}) \propto \int_0^\infty \pi^{(\alpha+K/2-1)} \cdot \exp\left( -(\pi/2)\left[ 2\beta + \sum_{i=1}^{K} x_i^2(1 - 1/(\tau+K)) - 2\sum_{j>i=1}^{K} x_i x_j/(\tau+K) - 2\tau\mu_0 \sum_{i=1}^{K} x_i/(\tau+K) \right]\right) d\pi$$

since the integral over $\mu$ is a constant that does not depend on $\pi$. Now a matrix simplification of the quadratic function of $\mathbf{x}$ in the exponent yields

$$f(\mathbf{x}) \propto \int_0^\infty \pi^{(\alpha+K/2-1)} \exp(-\pi\{\beta + (1/2)(\mathbf{x} - \mu_0\mathbf{1}_K)^{\mathrm{T}}[\mathbf{I}_K - (\tau+K)^{-1}\mathbf{1}_{K,K}] \times (\mathbf{x} - \mu_0\mathbf{1}_K)\})\, d\pi.$$

This concludes the proof of the lemma, since this integral represents the specified gamma mixture of a multivariate Normal density. □

Our desired theorem now results from performing the final integration, as an immediate consequence of the now well-known properties of the gamma integral.

**Theorem 7.16.** Asserting $\mathbf{X} \sim N\Gamma\text{-}N_K(\mu_0, \tau, \alpha, \beta)$ is equivalent to asserting

$$\mathbf{X} \sim \mathbf{t}_K(2\alpha, \mu_0\mathbf{1}_K, (\alpha/\beta)[\mathbf{I}_K - (\tau+K)^{-1}\mathbf{1}_{K,K}]).$$

*Proof.* Integrating the final line of the proof of the lemma yields

$$f(\mathbf{x}) \propto \{\beta + (1/2)(\mathbf{x} - \mu_0\mathbf{1}_K)^{\mathrm{T}}[\mathbf{I}_K - (\tau+K)^{-1}\mathbf{1}_{K,K}](\mathbf{x} - \mu_0\mathbf{1}_K)\}^{-(2\alpha+K)/2}.$$

Multiplying by the constant $(2\alpha/\beta)^{-(2\alpha+K)/2}$, this can be reexpressed as

$$f(\mathbf{x}) \propto \{2\alpha + (\mathbf{x} - \mu_0\mathbf{1}_K)^{\mathrm{T}}(\alpha/\beta)[\mathbf{I}_K - (\tau+K)^{-1}\mathbf{1}_{K,K}](\mathbf{x} - \mu_0\mathbf{1}_K)\}^{-(2\alpha+K)/2},$$

which is the form of the multivariate **t**-density specified in the theorem.

□

**Corollary 7.16.1.** Under the distributional specification of the theorem, the moments of $\mathbf{X}$ are $\mathbb{M}_1(\mathbf{X}) = \mu_0\mathbf{1}_K\ (\alpha > 1/2)$ and $\mathbb{M}_2^*(\mathbf{X}) = [\beta/(\alpha-1)][\mathbf{I}_K + \tau^{-1}\mathbf{1}_{K,K}]$.

*Proof.* The corollary follows directly from the moments of the multivariate $\mathbf{t}_K$-distribution. You need only show that $[\mathbf{I}_K - (\tau+K)^{-1}\mathbf{1}_{K,K}]^{-1} = [\mathbf{I}_K + \tau^{-1}\mathbf{1}_{K,K}]$, using the general result for the inverse of a matrix in this form which you showed as problem 1c of Section 7.3.1. □

Armed with this result, we can now use Theorem 7.7 to yield the following theorem regarding mixture-Normal exchangeable inference.

**Theorem 7.17.** If you regard the components of the vector $\mathbf{X}$ exchangeably, asserting $\mathbf{X} \sim \mathbf{t}_K(v = 2\alpha, \boldsymbol{\mu}_K = \mu_0 \mathbf{1}_K, \tau_{K,K} = (\alpha/\beta)[\mathbf{I}_K - (\tau + K)^{-1}\mathbf{1}_{K,K}])$, and the vector $\mathbf{X}$ is partitioned into the component vectors $\mathbf{X}_1$ and $\mathbf{X}_2$ with dimensions $N \times 1$, and $(K - N) \times 1$, respectively, then your joint conditional distribution for $\mathbf{X}_2|(\mathbf{X}_1 = \mathbf{x}_1)$ is multivariate:

$$\mathbf{t}_{K-N}(v = 2\alpha + N, \boldsymbol{\mu}_{K-N} = (\tau + N)^{-1}(\tau\mu_0 + N\bar{x}_N)\mathbf{1}_{K-N},$$

$$\tau_{K-N,K-N} = \{(2\alpha + N)/[2\beta + Ns_N^2 + \tau N(\bar{x}_N - \mu_0)^2/(\tau + N)]\}$$

$$\times [\mathbf{I}_{K-N} - (\tau + K)^{-1}\mathbf{1}_{K-N,K-N})].$$

*Proof.* The theorem follows immediately from Theorem 7.7, which states that your conditional distribution for $\mathbf{X}_2|\mathbf{X}_1$ is a mixture of independent normals with the conditional mixing function $M(\mu, \pi|\mathbf{X}_1 = \mathbf{x}_1)$ being

$$N\Gamma[(\tau + N)^{-1}(\tau\mu_0 + N\bar{x}_N), \tau + N, \alpha + N/2, \beta + (N/2)s_N^2 + \tau N(\bar{x}_N - \mu_0)^2/2(\tau + N)].$$

The representation of the conditional multivariate $\mathbf{t}_{K-N}$-distribution specified in the theorem then follows directly from Theorem 7.16. □

**Corollary 7.17.1.** Under the distributional specification of Theorem 7.17, the conditional moments of $\mathbf{X}_2|(\mathbf{X}_1 = \mathbf{x}_1)$ are

$$\mathbb{M}_1(\mathbf{X}_2|\mathbf{X}_1 = \mathbf{x}_1) = (\tau + N)^{-1}(\tau\mu_0 + M\bar{x}_N)\mathbf{1}_{K-N},$$

$$\mathbb{M}_2^*(\mathbf{X}_2|\mathbf{X}_1 = \mathbf{x}_1) = \frac{2\beta + Ns_N^2 + \tau N(\bar{x}_N - \mu_0)^2/(\tau + N)}{2\alpha + N - 2}[\mathbf{I}_{K-N} + (\tau + N)^{-1}\mathbf{1}_{K-N,K-N}].$$

These moment formulas for the distribution of the remaining $K - N$ quantities reduce to the formulas for $P(X_{N+1}|\mathbf{X}_N = \mathbf{x}_N)$ and $V(X_{N+1}|\mathbf{X}_N = \mathbf{x}_N)$ displayed in Corollary 7.14.1 by merely replacing $K$ here with $N + 1$.

## PROBLEMS

1. Perform the algebraic manipulation and the matrix simplification alluded to in the proof of Lemma 7.15.

2. Convince yourself that $[\mathbf{I}_K - (\tau + K)^{-1}\mathbf{1}_{K,K}]^{-1} = [\mathbf{I}_K + \tau^{-1}\mathbf{1}_{K,K}]$. *Hint*: Use the formula you derived in problem 1c of Section 7.3.1.

## 7.5 CONDITIONAL MIXTURE DISTRIBUTIONS FOR VECTORS REGARDED EXCHANGEABLY: THE WISHART AND THE MATRIX-*t*-DISTRIBUTIONS

After the extensive developments of Sections 7.2–7.4, we are verging on the capability to represent forecast distributions for sequentially observed vectors of several measurements, such as the macroeconomic statistics with which we introduced the issues of this chapter. The family of multivariate $\mathbf{t}_K$-distributions, which we have found useful in representing judgments regarding the components of $\mathbf{X}$ exchangeably, also has the capacity to represent distributions that do not entail the assessment of exchangeability. For we have defined the family of multivariate $\mathbf{t}_K(v, \boldsymbol{\mu}_K, \tau_{K,K})$-distributions with specific members for any $\boldsymbol{\mu}_K$ and any positive-definite matrix $\tau_{K,K}$; but the *exchangeable* multivariate $\mathbf{t}_K$-distributions involve the restrictions on $\tau_{K,K}$ and its inverse that they be in the form $(a\mathbf{I}_K + b\mathbf{1}_{K,K})$, and on $\boldsymbol{\mu}_K$ that its components are equal. This is the same form as the restrictions on $\boldsymbol{\Sigma}$ and $\boldsymbol{\mu}$ for the exchangeable multivariate Normal distribution.

The plan of this section is first to develop the general multivariate **t**-distribution as a Normal-Wishart mixture of a multivariate Normal distribution. (The Wishart distribution, to be defined shortly, is a multivariate generalization of the gamma distribution.) Then we address the problem of inference concerning a sequence of vectors that are regarded exchangeably but whose components are not regarded exchangeably. The components of each vector are assessed with a nonexchangeable multivariate $\mathbf{t}_K$-distribution. This problem will introduce the general family of matrix-*t*-distributions over a matrix of column vectors and will conclude this chapter.

***Definition 7.9.*** Let $\mathbf{PDS}_K$ denote the set of all positive-definite symmetric real matrices of dimension $K \times K$. Writing $\boldsymbol{\Pi}_{K,K} \sim W(v, \mathbf{P}_{K,K})$ means that the joint distribution function for the components of the matrix $\boldsymbol{\Pi}_{K,K}$ is the Riemann integral of the density function

$$f(\pi_{K,K}) \propto |\pi_{K,K}|^{(v-K-1)/2} \exp\{-(1/2)\operatorname{tr}(\mathbf{P}_{K,K}^{-1}\pi_{K,K})\} \qquad (\pi_{K,K} \in \mathbf{PDS}_K),$$

with proportionality constant

$$C = |\mathbf{P}_{K,K}|^{-v/2}\{2v^{K/2}(\Gamma(1/2))^{K(K-1)/4} \prod_{j=1}^{K} \Gamma[(v+1-j)/2]\}^{-v}.$$

Restrictions on the parameter values defining Wishart family members are that $v$ may be any positive integer or any real number exceeding $K-1$ and that $\mathbf{P}_{K,K} \in \mathbf{PDS}_K$. The symbolic tr in the density denotes "trace." □

Notice that since $\boldsymbol{\Pi}_{K,K}$ is necessarily symmetric, the number of distinct elements of $\boldsymbol{\Pi}_{K,K}$ is $K(K+1)/2$. Thus, the Wishart distribution is a joint distribution of that many variables. Following are some properties of a Wishart distribution that can be studied in Zellner (1971, Appendix B.3).

PROPERTIES OF THE WISHART DISTRIBUTION. $\mathbf{\Pi}_{K,K} \sim W(v, \mathbf{P}_{K,K})$ implies that

**Moments:** $\mathbb{M}_1(\mathbf{\Pi}_{K,K}) = v\mathbf{P}_{K,K}$, $\mathbb{M}_2^*(\Pi_{ij}) = v(P_{ij}^2 + P_{ii}P_{jj})$, and $\mathbb{M}_{11}^*(\Pi_{ij}, \Pi_{kl}) = v(P_{il}P_{jk} + P_{ik}P_{jl})$.

**Quadratic transformations:** $\mathbf{A}_{M,K}\mathbf{\Pi}_{K,K}\mathbf{A}_{M,K}^{\mathrm{T}} \sim W(v, \mathbf{A}_{M,K}\mathbf{P}_{K,K}\mathbf{A}_{M,K}^{\mathrm{T}})$.

**Conditional and marginal distributions:** If $\mathbf{\Pi}_{K,K}$ and $\mathbf{P}_{K,K}$ are partitioned conformably into component matrices $\mathbf{\Pi}_{11}$, $\mathbf{\Pi}_{12} = \mathbf{\Pi}_{21}^{\mathrm{T}}$, $\mathbf{\Pi}_{22}$ and $\mathbf{P}_{11}$, $\mathbf{P}_{12} = \mathbf{P}_{21}^{\mathrm{T}}$, $\mathbf{P}_{22}$, with respective dimensions $N \times N$, $(K-N) \times N$, and $(K-N) \times (K-N)$ in each lot, then

$$\mathbf{\Pi}_{11} \sim W(v, \mathbf{P}_{11}) \quad \text{and} \quad \mathbf{\Pi}_{22} - \mathbf{\Pi}_{21}\mathbf{\Pi}_{11}^{-1}\mathbf{\Pi}_{12} \sim W(v - N, \mathbf{P}_{22} - \mathbf{P}_{21}\mathbf{P}_{11}^{-1}\mathbf{P}_{12}).$$

Notice that when $K = 1$, the Wishart distribution $W_1(v, P_{11})$ is equivalent to a gamma distribution, parameterized as $\Gamma(v/2, (2P_{11})^{-1})$. Thus, the marginal distributions of the diagonal components of the matrix $\mathbf{\Pi}_{K,K}$ assessed with distribution $W(v, \mathbf{P}_{K,K})$ are assessed as $\Gamma(v/2, (2P_{ii})^{-1})$.

While the main results of this section will be easily stated by using a Wishart mixing distribution over the components of $\mathbf{\Pi}_{K,K}$, applied assessments of a mixing distribution over scale parameters of the multivariate Normal distribution are made more easily in terms of a distribution for the inverse of $\mathbf{\Pi}_{K,K}$, or $\mathbf{\Sigma}_{K,K} \equiv \mathbf{\Pi}_{K,K}^{-1}$. The following equivalence theorem, introducing the "inverted Wishart" distribution, supports the assessment in this elicited form.

**Theorem 7.18.** $\mathbf{\Pi}_{K,K} \sim W(v, \mathbf{P}_{K,K}) \Leftrightarrow \mathbf{\Sigma}_{K,K} \equiv \mathbf{\Pi}_{K,K}^{-1} \sim IW(v, \mathbf{S}_{K,K} \equiv \mathbf{P}_{K,K}^{-1})$, meaning

$$f(\mathbf{\Sigma}_{K,K}) \propto |\mathbf{\Sigma}_{K,K}|^{-(v+K+1)/2} \exp\{-(1/2)\,\mathrm{tr}(\mathbf{\Sigma}_{K,K}^{-1}\mathbf{S}_{K,K})\}.$$

A detailed proof of this theorem using this parameterization of the inverted Wishart appears in Zellner (1971, B.4). A different parameterization of this distribution using $\delta \equiv v - K + 1$ is presented in the beautiful unifying paper of Dawid (1981). But for our short-term purposes here, the standard parameterization in terms of $v$ will suffice. The main useful properties of the inverted Wishart distribution, in our context that it is the elicited form for assessing the mixing distribution over $(\boldsymbol{\mu}, \mathbf{\Pi})$, are its first moments, $\mathbb{M}_1(\mathbf{\Sigma}_{K,K}) = (v-2)^{-1}\mathbf{S}_{K,K}$. If we can elicit this matrix of first moments of $\mathbf{\Sigma}_{K,K}$, once the value of $v$ is determined we would have assessed the values of $v$ and $\mathbf{P}_{K,K} = \mathbf{S}_{K,K}^{-1}$ for the Wishart mixture over $\mathbf{\Pi}$.

The practical meaning of specifying values for $v$ and $\mathbf{P}_{K,K} = \mathbf{S}_{K,K}^{-1}$ in an initial mixing function can be gained from the following result, which relates the Wishart distribution to the sum of outer products of vectors whose distributions are identical and multivariate normal. Let $\mathbf{X}_{K,v} \equiv [\mathbf{X}_1 \quad \mathbf{X}_2 \quad \cdots \quad \mathbf{X}_v]$ denote a matrix of $K$-dimensional column vectors that are assessed exchangeably, each with a mixture multivariate $M\text{-}N_K(\mathbf{0}, \mathbf{\Pi}_{K,K} = \mathbf{\Sigma}_{K,K}^{-1})$ distribution. Then for any specified values of $v$ and $\mathbf{\Sigma}_{K,K}$, the conditional distributions for the sum of their outer products, $\mathbf{X}_{K,v}\mathbf{X}_{K,v}^{\mathrm{T}}$, is Wishart $(v, \mathbf{\Sigma}_{K,K})$. The associated first moments for the average of these outer products are $\mathbb{M}_1(v^{-1}\mathbf{X}_{K,v}\mathbf{X}_{K,v}^{\mathrm{T}} | \mathbf{\Sigma}_{K,K}) = \mathbf{\Sigma}_{K,K}$. Moreover, the variances for the average components all converge to 0 as $v$ increases. Details of these results can be found in Zellner (1971, B.3).

In our application of the Wishart distribution as a mixing distribution over $\boldsymbol{\mu}$ and $\mathbf{\Pi}$, or equivalently $\mathbf{\Sigma}$, these results imply that a mixing distribution over $\mathbf{\Sigma}$ must be virtually identical to one's opinions regarding the average outer product matrix for a large number of vectors that are regarded exchangeably as mixture multivariate Normal, $M\text{-}N_K(\boldsymbol{\mu}_K, \mathbf{\Pi}_{K,K})$. Your prevision for this imagined "long-run average" matrix equals the value of $(v-2)^{-1}\mathbf{S}_{K,K}$. The value of $v$ amounts to a specification of the amount of information you entertain about this average, coded in terms of the number of equivalent observations of vectors $\mathbf{X}$ your information represents. The concept of equivalent observations is discussed in several works of Good (1983).

The central algebraic result of this section is that a Normal-Wishart mixture of a multivariate Normal distribution is a multivariate **t**-distribution.

**Theorem 7.19.** Asserting

$$\begin{aligned}
&\mathbf{X} \sim NW\text{-}N_K(\boldsymbol{\mu}_K, \mathbf{\Pi}_{K,K}; \boldsymbol{\mu}_0, \tau\mathbf{\Pi}_{K,K}, v, \mathbf{P}_{K,K}) \\
\Leftrightarrow &\mathbf{X} \sim W\text{-}N_K(\boldsymbol{\mu}_0, [\tau/(\tau+1)]\mathbf{\Pi}_{K,K}; v, \mathbf{P}_{K,K}) \\
\Leftrightarrow &\mathbf{X} \sim \mathbf{t}_K(v-(K-1), \boldsymbol{\mu}_0, [v\tau/(\tau+1)]\mathbf{P}_{K,K}).
\end{aligned}$$

*Proof.* The proof involves some simple but tedious algebraic reductions, which we identify without completing their details. Derivations rely mainly on the matrix algebra result that trace$(AB)$ = trace$(BA)$ and on a determinant result mentioned later. To begin, we can write the joint density $f(\mathbf{x}, \boldsymbol{\mu}_K, \boldsymbol{\pi}_{K,K})$ as the product $f(\mathbf{x}|\boldsymbol{\mu}_K, \boldsymbol{\pi}_{K,K}) f(\boldsymbol{\mu}_K|\boldsymbol{\pi}_{K,K}) f(\boldsymbol{\pi}_{K,K})$ with the details specified in the first line of the theorem. Collecting terms involving $\boldsymbol{\mu}_K$ into a quadratic form results in the expression (dropping the subscripts on $\boldsymbol{\mu}_K$, $\boldsymbol{\pi}_{K,K}$, and $\mathbf{P}_{K,K}$ for simplicity)

$$\begin{aligned}
f(\mathbf{x}, \boldsymbol{\mu}, \boldsymbol{\pi}) \propto |\boldsymbol{\pi}|^{1/2} &\exp[-(1/2)(\boldsymbol{\mu}-\mathbf{m})^{\mathrm{T}}(1+\tau)\boldsymbol{\pi}(\boldsymbol{\mu}-\mathbf{m})] \\
&\times |\boldsymbol{\pi}|^{(v-K)/2} \exp\{-(1/2)\operatorname{tr}[\mathbf{P}^{-1} + (\tau/(1+\tau))(\mathbf{x}-\boldsymbol{\mu}_0)(\mathbf{x}-\boldsymbol{\mu}_0)^{\mathrm{T}}]\boldsymbol{\pi}\},
\end{aligned}$$

where $\mathbf{m} \equiv (1+\tau)^{-1}(\mathbf{x}+\tau\boldsymbol{\mu}_0)$. Integrating this expression first with respect to $\boldsymbol{\mu}$ yields merely the second multiplicand. For the integral of the first multiplicand (the only portion involving $\boldsymbol{\mu}$) involves neither $\mathbf{x}$ nor $\boldsymbol{\pi}$. Thus,

$$f(\mathbf{x}, \boldsymbol{\pi}) \propto |\boldsymbol{\pi}|^{(v-K)/2} \exp\{-(1/2)\operatorname{tr}[\mathbf{P}^{-1} + (\tau/(1+\tau))(\mathbf{x}-\boldsymbol{\mu}_0)(\mathbf{x}-\boldsymbol{\mu}_0)^{\mathrm{T}}]\boldsymbol{\pi}\},$$

which specifies the Wishart mixture of the multivariate Normal distribution denoted in the second line of the theorem. So finally, integrating with respect to the matrix $\pi$ results in the appropriate proportionality constant associated with the Wishart density:

$$\begin{aligned} f(\mathbf{x}) &\propto |\mathbf{P}^{-1} + (\tau/(1+\tau))(\mathbf{x}-\boldsymbol{\mu}_0)(\mathbf{x}-\boldsymbol{\mu}_0)^{\mathrm{T}}|^{-(\nu+1)/2} \\ &\propto |1 + (\mathbf{x}-\boldsymbol{\mu}_0)^{\mathrm{T}}\mathbf{P}(\tau/(1+\tau))(\mathbf{x}-\boldsymbol{\mu}_0)|^{-(\nu+1)/2} \\ &\propto [\nu + (\mathbf{x}-\boldsymbol{\mu}_0)^{\mathrm{T}}(\nu\tau/(1+\tau))\mathbf{P}_{K,K}(\mathbf{x}-\boldsymbol{\mu}_0)]^{-(\nu-(K-1)+K)/2}. \end{aligned}$$

This second line derives from factoring the determinant $|\mathbf{P}^{-1}|$ out of the first line, using the general matrix result that $|\mathbf{A} + \alpha\mathbf{x}\mathbf{y}^{\mathrm{T}}| = |\mathbf{A}||1 + \alpha\mathbf{y}^{\mathrm{T}}\mathbf{A}^{-1}\mathbf{x}|$ for any invertible $K \times K$ matrix $\mathbf{A}$, scalar $\alpha$, and $K$-dimensional vectors $\mathbf{x}$ and $\mathbf{y}$. (See for example, Dhrymes, 1978, p. 458.) The final line, which identifies the $\mathbf{t}_K$-density for $\mathbf{X}$ specified in the theorem, derives from the fact that the matrix in the determinant of line 2 is a scalar. Factoring into the expression the value of $\nu$ merely affects the proportionality constant. □

We can use this result to establish an understanding of coherent inference about a sequence of vectors of quantities that you regard exchangeably. Suppose, for example, that the state of Kansas is divided into blocks of 900 hectares (about 4 square miles). For each block $i$, components of a three-dimensional vector might represent yields of wheat, of sunflower seed, and of hay from that block. Based on what I've described about Kansas, you might well regard the vectors $\mathbf{X}_{3,1}, \mathbf{X}_{3,2}, \ldots, \mathbf{X}_{3,N}$ exchangeably, although you would probably not regard the components of any vector exchangeably. Furthermore, you may well decide to assess the vectors in terms of a mixture multivariate Normal distribution. This exemplifies the context to which our next theorem applies. It uses the following notation:

$\mathbf{x}_{KN} \equiv (\mathbf{x}_1^{\mathrm{T}}, \mathbf{x}_2^{\mathrm{T}}, \ldots, \mathbf{x}_N^{\mathrm{T}})^{\mathrm{T}}$, a stacked vector of observations,

$\bar{\mathbf{x}} \equiv N^{-1}\sum_{i=1}^{N} \mathbf{x}_i$, the average observation vector,

$\overline{\mathbf{x}\mathbf{x}^{\mathrm{T}}} \equiv N^{-1}\sum_{i=1}^{N} \mathbf{x}_i\mathbf{x}_i^{\mathrm{T}}$, the average outer product observation vector.

**Theorem 7.20.** Suppose you regard the $K$-dimensional vectors $\mathbf{X}_1, \mathbf{X}_2, \ldots, \mathbf{X}_T$ exchangeably. Asserting additionally for them a mixture of conditionally independent $N_K(\boldsymbol{\mu}_K, \boldsymbol{\Pi}_{K,K})$ distributions, with a mixing distribution over $(\boldsymbol{\mu}_K, \boldsymbol{\Pi}_{K,K})$ specified as Normal-Wishart$(\boldsymbol{\mu}_0, \tau, \nu, \mathbf{P}_{K,K})$ implies via coherency that your conditional distribution for any vector $\mathbf{X}_{N+1}|(\mathbf{X}_{KN} = \mathbf{x}_{KN})$ is multivariate $\mathbf{t}_K$ with specific parameters induced by a Normal-Wishart$(\boldsymbol{\mu}_N, \tau_N, \nu_N, \mathbf{P}_{K,K,N})$ mixture:

$$\boldsymbol{\mu}_N = (\tau + N)^{-1}(\tau\boldsymbol{\mu}_0 + N\bar{\mathbf{x}}), \qquad \tau_N = \tau + N, \qquad \nu_N = \nu + N,$$

$$\mathbf{P}_{K,K,N} = \{\mathbf{P}_{K,K}^{-1} + (\tau N/(\tau + N))(\boldsymbol{\mu}_0 - \bar{\mathbf{x}})(\boldsymbol{\mu}_0 - \bar{\mathbf{x}})^{\mathrm{T}} + N(\overline{\mathbf{x}\mathbf{x}^{\mathrm{T}}}) - \bar{\mathbf{x}}\bar{\mathbf{x}}^{\mathrm{T}})\}^{-1}.$$

*Proof*. The exchangeable assessment of the vectors $\mathbf{X}_1, \ldots, \mathbf{X}_T$ requires that any subgroup of vectors $\mathbf{X}_1, \ldots, \mathbf{X}_{N+1}$ must be assessed with the same mixture of conditionally independent multivariate Normal distributions. A routine application of Bayes' theorem then allows that the conditional distribution for $\mathbf{X}_{N+1}|(\mathbf{X}_{KN} = \mathbf{x}_{KN})$ is also mixture-multivariate Normal, with mixing density

$$f(\boldsymbol{\mu}, \boldsymbol{\pi}|\mathbf{X}_{KN} = \mathbf{x}_{KN}) \propto f(\mathbf{x}_{KN}; \boldsymbol{\mu}, \boldsymbol{\pi}) f(\boldsymbol{\mu}, \boldsymbol{\pi}),$$

with

$$f(\mathbf{x}_{KN}; \boldsymbol{\mu}, \boldsymbol{\pi}) \propto |\boldsymbol{\pi}|^{N/2} \exp\left[ -\frac{1}{2} \sum_{i=1}^{N} (\mathbf{x}_i - \boldsymbol{\mu})^{\mathrm{T}} \boldsymbol{\pi} (\mathbf{x}_i - \boldsymbol{\mu}) \right],$$

where each vector $\mathbf{x}_i$ is a $K$-dimensional subcomponent of the stacked vector $\mathbf{x}_{KN}$, and the dimension subscripts on $\boldsymbol{\mu} = \boldsymbol{\mu}_K$ and $\boldsymbol{\pi} = \boldsymbol{\pi}_{K,K}$ have been suppressed. Multiplying this density by the Normal-Wishart mixing density $f(\boldsymbol{\mu}, \boldsymbol{\pi})$, and collecting terms involving $\boldsymbol{\mu}$ into a quadratic form yield

$$\begin{aligned} &f(\boldsymbol{\mu}, \boldsymbol{\pi}|\mathbf{X}_{KN} = \mathbf{x}_{KN}) \\ &\quad \propto |\boldsymbol{\pi}|^{1/2} \exp[-(1/2)(\boldsymbol{\mu} - \mathbf{m})^{\mathrm{T}}(N + \tau)\boldsymbol{\pi}(\boldsymbol{\mu} - \mathbf{m})] \\ &\times |\boldsymbol{\pi}|^{(v+N-K-1)/2} \exp\left\{ -(1/2)\operatorname{tr}\left[ \mathbf{P}^{-1} + \sum_{i=1}^{N} \mathbf{x}_i \mathbf{x}_i^{\mathrm{T}} + \tau \boldsymbol{\mu}_0 \boldsymbol{\mu}_0^{\mathrm{T}} + (N + \tau)\mathbf{m}\mathbf{m}^{\mathrm{T}} \right] \boldsymbol{\pi} \right\}, \end{aligned}$$

where $\mathbf{m} \equiv (N + \tau)^{-1}(N\bar{\mathbf{x}} + \tau\boldsymbol{\mu}_0)$. The equivalence of the matrix $\mathbf{P}_{K,K,N}$ stated in the theorem with $[\mathbf{P}^{-1} + \sum_{i=1}^{N} \mathbf{x}_i \mathbf{x}_i^{\mathrm{T}} + \tau\boldsymbol{\mu}_0 \boldsymbol{\mu}_0^{\mathrm{T}} + (N + \tau)\mathbf{m}\mathbf{m}^{\mathrm{T}}]^{-1}$ is left as an exercise in matrix algebra. □

Still another interesting equivalent representation of $\mathbf{P}_{K,K,N}$ (also left as an exercise) that will be relevant to our concluding comments is

$$\mathbf{P}_{K,K,N} = \{\mathbf{P}_{K,K}^{-1} + (\mathbf{x}_{K,N} - \mathbf{1}_N \otimes \boldsymbol{\mu}_K)[\mathbf{I}_N - (N + \tau)^{-1}\mathbf{1}_{N,N}](\mathbf{x}_{K,N} - \mathbf{1}_N \otimes \boldsymbol{\mu}_K)^{\mathrm{T}}\}^{-1}.$$

The Kronecker product of a matrix $\mathbf{A}$ with $\mathbf{B}$, denoted $A \otimes B \equiv [A_{i,j} B]$, is defined as the matrix generated by multiplying every scalar component of $A$ by the entire matrix $B$.

In terms of the formulas stated in Theorem 7.20, the parameters of the conditional multivariate $\mathbf{t}(v_N, \boldsymbol{\mu}_N, \tau_N)$-distribution for $\mathbf{X}_{N+1}|(\mathbf{X}_{KN} = \mathbf{x}_{KN})$ are $v_N = (v + N - (K - 1))$, $\boldsymbol{\mu}_N = \boldsymbol{\mu}_N$, $\tau_N = [(v + N)(\tau + N)/(\tau + N + 1)]\mathbf{P}_{K,K,N}$. Thus,

$$P(\mathbf{X}_{N+1}|\mathbf{X}_{KN} = \mathbf{x}_{KN}) = (\tau + N)^{-1}(\tau\boldsymbol{\mu}_0 + N\bar{\mathbf{x}}),$$

and

$$V(\mathbf{X}_{N+1}|\mathbf{X}_{KN} = \mathbf{x}_{KN}) = [(v + N - (K - 1))/(v + N - (K + 1))]\mathbf{P}_{K,K,N}^{-1}.$$

This variance formula yields $V(\mathbf{X}_1) = [(v - (K - 1))/(v - (K + 1))]\mathbf{S}_{K,K}$ when

$N = 0$. This suggests another way to identify the appropriate value of $v$ for the initial mixing function in an application. Compare your prevision for the average outer product of many vectors, $\mathbf{S}_{K,K}$, with your variance assertion for any particular vector. The latter must be proportionally larger, and the proportionality factor yields the appropriate value of $v$.

As to the inferential use of the conditioning observations $\mathbf{X}_{KN} = \mathbf{x}_{KN}$, Theorem 7.20 implies that the statistics $\overline{\mathrm{X}}$ and $\overline{\mathbf{XX}^{\mathrm{T}}}$ are regarded as sufficient for inference about the vector $\mathbf{X}_{N+1}$

Some comment on the joint distribution of the entire matrix of vectors, $\mathbf{X}_{K,T}$, will conclude our discussion. Although the conditional distribution for each vector $\mathbf{X}_{N+1}|(\mathbf{X}_{KN} = \mathbf{x}_{KN})$ is multivariate $\mathbf{t}_K$ for each $N = 0, 1, \ldots, T-1$, the joint distribution for the entire stacked column vector $\mathbf{X}_{KT}$ is not multivariate $\mathbf{t}$ but rather "matrix-**t**," to be defined shortly.

Regarding any $K$-dimensional vectors $\mathbf{X}_1, \mathbf{X}_2, \ldots, \mathbf{X}_N$ exchangeably by asserting for them a Normal-Wishart mixture of conditionally independent $N_K(\boldsymbol{\mu}_K, \boldsymbol{\pi}_{K,K})$ distributions is equivalent to assessing the stacked vector $\mathbf{X}_{KN}$ with this specified Normal-Wishart mixture of an $NK$-dimensional multivariate Normal distribution of the form $N_{KN}(\mathbf{1}_N \otimes \boldsymbol{\mu}_K, \mathbf{I}_N \otimes \boldsymbol{\pi}_{K,K})$. However, the Wishart mixing function over the conditional covariance matrix for $\mathbf{X}_{KN}$ does not mix over all its $(KN+1)KN/2$ components but over only $(K+1)K/2$ of them. These compose the parameter matrix $\boldsymbol{\pi}_{K,K}$, the block diagonal component of the covariance matrix $\mathbf{I}_N \otimes \boldsymbol{\pi}_{K,K}$. Understood in this way, the Wishart mixture over the entire covariance matrix is recognized as a "constrained Wishart." This is why the entire vector $\mathbf{X}_{KN}$ is not assessed with a multivariate **t**-distribution. The distribution is more informative than that, because only a few components of the $\boldsymbol{\pi}$ matrix are involved in the mixing.

Algebraically, the joint density $f(\mathbf{x}_{KN})$ can be shown to have the form

$$f(\mathbf{x}_{KN}) \propto |\mathbf{P}_{K,K}^{-1} + (\mathbf{x}_{K,N} - \mathbf{1}_N \otimes \boldsymbol{\mu}_K)[\mathbf{I}_N - (N+\tau)^{-1}\mathbf{1}_{N,N}](\mathbf{x}_{K,N} - \mathbf{1}_N \otimes \boldsymbol{\mu}_K)^{\mathrm{T}}|^{-(v+N)/2}.$$

This is derived by integrating the mixture density $f(\mathbf{x}_{KN}, \boldsymbol{\mu}_K, \boldsymbol{\pi}_{K,K})$ with respect to $\boldsymbol{\mu}$ and $\boldsymbol{\pi}$, using properties of the Wishart distribution. This joint density exemplifies the form of a matrix-**t**-density, according to the following definition.

***Definition 7.10.*** A distribution function for the $NK$ elements of the matrix $\mathbf{X}_{K,N} \in \mathbb{R}^{K,N}$ that can be expressed as the Riemann integral of the density function

$$f(\mathbf{x}_{K,N}) \propto |\mathbf{P}_{K,K}^{-1} + (\mathbf{x}_{K,N} - \boldsymbol{\mu}_{K,N})\,\mathbf{Q}_{N,N}^{-1}(\mathbf{x}_{K,N} - \boldsymbol{\mu}_{K,N})^{\mathrm{T}}|^{-v/2} \qquad (\mathbf{x}_{K,N} \in \mathbb{R}^{K,N}),$$

with proportionality constant

$$C = \frac{|\mathbf{P}_{K,K}^{-1}|^{(\delta+K-1)/2}|\mathbf{Q}_{N,N}^{-1}|^{K/2} \prod_{i=1}^{K} \Gamma[(\delta+N+K-i)/2]}{(\Gamma(1/2))^{KN} \prod_{i=1}^{K} \Gamma[(\delta+N-i)/2]},$$

is said to be a $(K \times N)$-*dimensional matrix*-**t**-*distribution* with family parameter $\delta$, location $\boldsymbol{\mu}_{K,N}$, and precisions $\mathbf{P}_{K,K}$ and $\mathbf{Q}_{N,N}$. Restrictions on the parameter values are $\delta > 0$ and $\mathbf{P}_{K,K}$ and $\mathbf{Q}_{N,N}$ must be symmetric and positive-definite. □

Details of the matrix-**t**-distribution are derived in Dickey (1967) and Zellner (1971). The important reparameterization of the distribution as presented here is motivated in Dawid (1981).

## 7.6 CONCLUDING COMMENTS

We have concluded an extensive journey through the theory of mixture-Normal distributions. The algebraic details we have studied have been developed since the 1950s within the modern tradition of Bayesian statistics. Our references have identified only some of the contributors. The presentation here has been oriented to highlight a completely subjective understanding and characterization of this distribution theory. The mixture theory is appropriate to represent the uncertain knowledge of anyone who regards an array of quantities in a specific identifiable way. It is surely not required generally for everyone's inferences on the basis of some unobservable random generating structure.

Our regular use of natural conjugate mixing functions during this journey has been designed to exhibit at an introductory level both the interconnections among the components of a mixture distribution that most any statistician should know and to produce some practical computational formulas that are appropriate for inference when one's personal mixing function is adequately approximated by one of these forms. Precise account of the extent of approximation error involved can be determined from Dickey (1976), which generalizes the "principle of stable estimation," developed in Edwards, Lindman, and Savage (1963). To some extent, however, these ornate algebraic details have been rendered obsolete by Monte Carlo developments in statistical computing, which we have mentioned, and for a different reason by developments of expert systems of inference based on the fundamental theorem of prevision.

Finally, our focus on exchangeable forms of mixture multivariate Normal distributions has limited the range of details in the applications we have discussed largely to agricultural crop and dairy yields. But in doing so, it has eliminated the allusion to "independent observations" which confuses most presentations of this material, even in typically objectivist Bayesian analysis. However, the vectors of economic measurements that have motivated our study of multivariate distributions to some extent are not typically regarded exchangeably by practical analysts. Considered opinions about them are usually registered in a mixture regression opinion structure, which imbed exchangeable distribution features only into deeper layers of considerations. Our analysis turns to such formulations in the penultimate chapter.

CHAPTER 8

# Sequential Forecasting Based on Linear Conditional Prevision Structures: Theory and Practice of Linear Regression

*When you sow your peas, when you sow your beans, when you sow your potatoes, when you sow your carrots, your turnips, your parsnips, your other root vegetables, do you do so with punctilio?*
*Watt*, SAMUEL BECKETT

## 8.1 INTRODUCTION

In order to understand the purpose and practice of linear regression, we need to focus on several ideas that have been developed in different places in this text. To begin, a regression equation merely identifies an array of conditional prevision assertions. We have been aware since Chapter 3 that it is both meaningful and fruitful to express your uncertain knowledge about a collection of quantities by asserting conditional previsions for some of the quantities given others. These might include assertions such as $P(A|BCD)$, $P(C|ABD)$, and $P(C|AB)$ along with, say, $P(AD)$ and $P(BD)$. Such an array of assertions would determine cohering bounds on your assertions of both $P(B|ACD)$ and $P(B|CD)$, for example, which are computable via the fundamental theorem of prevision. Each of the previsions and conditional previsions mentioned in this paragraph is a distinct, meaningful assertion.

Both historically and in our text, the first specific mention of a "regression equation" has been in the context of the very particular attitudes toward a vector of quantities that motivate the assertion of a multivariate Normal distribution for the vector: $\mathbf{X} \sim N_K(\boldsymbol{\mu}, \boldsymbol{\Sigma})$. In this context, we noted that the conditional prevision vector $P(\mathbf{X}_i|\mathbf{X}_j = \mathbf{x}_j)$ is required to be a linear function of the conditioning observations, and we stressed that both of the equations

$$P(\mathbf{X}_1|\mathbf{X}_2=\mathbf{x}_2)=\boldsymbol{\mu}_1+\boldsymbol{\Sigma}_{12}\boldsymbol{\Sigma}_{22}^{-1}(\mathbf{x}_2-\boldsymbol{\mu}_2),$$
$$P(\mathbf{X}_2|\mathbf{X}_1=\mathbf{x}_1)=\boldsymbol{\mu}_2+\boldsymbol{\Sigma}_{21}\boldsymbol{\Sigma}_{11}^{-1}(\mathbf{x}_1-\boldsymbol{\mu}_1)$$

represent meaningful and cohering assertions. Actually, these equations identify an array of conditional prevision assertions, for they specify the form of asserted conditional previsions applicable to every vector of conditioning observation values, $(\mathbf{x}_1,\mathbf{x}_2)\in\mathscr{R}(\mathbf{X}_1,\mathbf{X}_2)$.

The coefficients of these linear regression equations are numbers, identified in this context by the values of $\boldsymbol{\mu}$ and $\boldsymbol{\Sigma}$ that you specify in asserting your multivariate Normal distribution for $\mathbf{X}$. We have also extended our understanding of mixture-multivariate Normal distribution theory to the case that you do not specify values for $\boldsymbol{\mu}$ and $\boldsymbol{\Sigma}$, but rather use this parametric structure to express your mixture distribution for $\mathbf{X}$. In the conjugate Normal-Wishart mixture context of Section 7.5, the conditional multivariate **t**-distribution for $\mathbf{X}_{N+1}|(\mathbf{X}_{NK}=\mathbf{x}_{NK})$ specifies a regression equation of the form $P(\mathbf{X}_{N+1}|\mathbf{X}_{NK}=\mathbf{x}_{NK})=(\tau+N)^{-1}(\tau\boldsymbol{\mu}_0+N\bar{\mathbf{x}})$. This regression equation, too, is linear in the conditioning observation vectors; and the coefficients are numbers specified by features of your initial mixing distribution, $\boldsymbol{\mu}_0$ and $\tau$. The implied conditional distributions for $\mathbf{X}_{N+1}|(\mathbf{X}_{NK}=\mathbf{x}_{NK})$ involve characteristics of your prior precision matrix, $\mathbf{P}_{K,K}$, as well.

In general, you may specify your conditional previsions of the form $P(X_{K+1}|\mathbf{X}_K=\mathbf{x}_K)$ as any numbers you please, modulo coherency. If such assertions were made for every vector $\mathbf{x}_K\in\mathscr{R}(\mathbf{X}_K)$, the induced function $G$: $\mathscr{R}(\mathbf{X}_K)\to\mathbb{R}$ specified by $G(\mathbf{x}_K)=P(X_{K+1}|\mathbf{X}_K=\mathbf{x}_K)$ is called your regression function for $X_{K+1}$ on $\mathbf{X}_K$. There is no requirement that this function be linear in $\mathbf{x}_K$. However, in the type of cases we elucidate in this chapter, a motivation for a linear regression structure is often found adequately persuasive. Thus derives the relevance of linear regression analysis to a broad array of scientific and commercial applications. Keeping an eye on the role of conditional probability assertions in representing uncertain opinions, we develop in this chapter the computational procedures involved in applied regression studies as they are motivated within the operational subjective construction of statistics.

Mathematically, the general relevance of linear regression structures is identified when appropriate by considering how well the regression function $G(\cdot)$ might be approximated by piecewise linear functions over connected subsets of $\mathscr{R}(\mathbf{X}_K)$. Of course, if conditional previsions were assessed to the extent that the function $G(\cdot)$ were fully specified to begin with, an approximating piecewise linear function could be computed with no problem. But in such a case, the approximation would be irrelevant, since the function $G(\cdot)$ is surely adequate for itself. The value of asserting approximate linear regression structures arises in situations that you can identify subregions of $\mathscr{R}(\mathbf{X}_K)$ over which your conditional previsions $P(X_{K+1}|\mathbf{X}_K=\mathbf{x}_K)$ are approximately linear in $\mathbf{x}_K$, and you can also identify a sequence of vectors $\mathbf{X}_{K+1,1}, \mathbf{X}_{K+1,2}, \ldots, \mathbf{X}_{K+1,N+1}$, for which some form of exchangeability characterizes your judgments. It sounds like we should get ready for some work.

Historically, the concept of a linear regression was introduced within an objectivist understanding of probability by Francis Galton (1886) in his studies of heredity, based on a bivariate Normal distribution of the adult heights of parents and progeny. Galton was the first to construct Figure 7.3 of Section 7.2.2.1, which guided his attitude toward the regression setup, and to understand it with the aid of his mathematician colleague, J. Hamilton Dickson. Galton's biography by Forrest (1974) and the broader historical discussion of Stigler (1986) are intriguing. Both reproduce the original published figure of an elliptical frequency contour with its relevant pair of regression lines. The algebraic detail of mixture-regression theory has developed in an extensive research literature, with notable contributions by Raiffa and Schlaifer (1961), Ando and Kaufman (1965), Geisser (1965), and Hill (1967). Spirited and rightly influential presentations appear in the already classic contributions by Zellner (1971) and Leamer (1978). The present chapter amounts to a concise presentation of the details of mixture-regression analysis, with particular attention to the motivation for their use provided by the operational subjective statistical method, and to the general understanding this method provides. Our computational formulas and applications will be directed appropriately.

## 8.2 MIXTURE LINEAR REGRESSION

Our familiarity with mixture-Normal distribution theory allows us to get to work on details right away. This section will consist of a sequence of definitions and theorems, interspersed by commentary and allusions to applications in economic forecasting. We defer a computational example from a study of currency exchange rates until Section 8.5 after we have studied several types of regression structures.

***Definition 8.1.*** If your opinion regarding the vector of quantities $\mathbf{X}_{K+1}$ with realm $\mathscr{R}(\mathbf{X}_{K+1})$ specifies the conditional previsions $P(X_{K+1}|\mathbf{X}_K = \mathbf{x}_K)$ for every vector $\mathbf{x}_K \in \mathscr{R}(\mathbf{X}_K)$, the induced function $G: \mathscr{R}(\mathbf{X}_K) \to \mathbb{R}$ specified by $G(\mathbf{x}_K) = P(X_{K+1}|\mathbf{X}_K = \mathbf{x}_K)$ is called your *regression function* for $X_{K+1}$ on $\mathbf{X}_K$. If the regression function is linear in $\mathbf{x}_K$, $P(X_{K+1}|\mathbf{X}_K = \mathbf{x}_K) = b_0 + \mathbf{x}_K^{\mathrm{T}}\mathbf{b}_K$, the coefficients $b_0, b_1, \ldots, b_K$ are called the *linear regression coefficients* for $X_{K+1}$ on $\mathbf{X}_K$. □

Every quantity mentioned in Definition 8.1 is represented by a similarly subscripted symbol, $X_i$. This nondistinguishing notation for the object quantity of the regression, $X_{K+1}$, and the regressor quantities $X_1, \ldots, X_K$ is explicitly meant to convey that there is nothing unusual about $X_{K+1}$ relative to the regressors. A quantity is a quantity is a quantity. Having specified your regression function for $X_{K+1}$ on $\mathbf{X}_K$, there is nothing to keep you from concurrently asserting conditional previsions that specify regressions for $X_{K+1}$ on merely a few subcomponents of $\mathbf{X}_K$, nor for any component of $\mathbf{X}_K$ on some remaining subcomponents of $\mathbf{X}_{K+1}$. Of course, such an array of regression assertions would need to be governed by the logic of coherency. Nonetheless, in the computational formulas to be developed, we distinguish the conditioning quantities in a regression specification in a standard way by the denotation $X_1, X_2, \ldots, X_K$, and the object quantity of the conditional previsions that define the regression equation by the denotation $Y$. This distinction will simplify computational formulas that would otherwise need to rely excessively on attention to subscripts; it will also standardize our notation here with commonly used notation of regression analysis, however it is motivated. But the ideas are not standard.

In principle, regression functions may have any algebraical form whatsoever, since they merely represent your opinions about the quantities under consideration. Moreover, Monte Carlo developments in statistical computing today allow the evaluation of mixture integrals of virtually any functional form. Thus, the application of general mixture-regression theory to regressions of any functional form is now within reach. Already, however, the recognized applicability of considerations that specify a mixture *linear* regression is so extensive that the following details are found to be both practically useful and theoretically insightful.

Our theorems and applications concern the simplest way to conceive of a representation of mixture linear regressions. They involve a setup of quantities and a mixture opinion parameterization that has been found useful in a broad array of scientific and commercial contexts. In studying them, we shall apply our understanding of conditional prevision, normal mixture theory, linearity, exchangeability, and inference.

### 8.2.1 Inference Based on Normal-Gamma Regression Mixtures

In the context of opinion representations we are now to study, it will be expedient to modify the notation for quantity vectors and matrices that has satisfied hereto. Let us focus on these notational adjustments directly.

A boldfaced $\mathbf{X}$ will represent an $(N \times K)$-dimensional matrix of quantities of which the first column is composed of 1's. The size of $N$ is meant to be variable, so it will not be expressly denoted until it merits consideration. If a subscript is used, say on $\mathbf{X}_{N+1}$, this would mean to denote an $(N+1)$st *row vector*, $(1 \quad X_{N+1,2} \quad X_{N+1,3} \quad \cdots \quad X_{N+1,K})$, which when appended to $\mathbf{X}$ would generate a matrix $\mathbf{X}$ composed of one more row. In fact, any particular row of matrix $\mathbf{X}$ might be denoted by $\mathbf{X}_i$, with the subscript $i$ representing the row number.

Similarly, $\mathbf{Y}$ will denote an $N \times 1$ vector of quantities, and $Y_i$ any one of its components. Using this notation, our goal in constructing the distributions that follow is to understand the formal characteristics of opinions that motivate a specific computational formula for

$$P[Y_{N+1} | (\mathbf{X}_{N+1} = \mathbf{x}_{N+1})(\mathbf{X} = \mathbf{x})(\mathbf{Y} = \mathbf{y})],$$

and even a complete conditional distribution for $Y_{N+1}$ in this setting. Lowercase vectors and matrices such as $\mathbf{x}_{N+1}$, $\mathbf{x}$, and $\mathbf{y}$ will continue to represent possible values of the associated quantities, situated within the appropriate realm.

In this section we address algebraically the case where the quantities $\mathbf{Y}$ and $\mathbf{X}$ are logically independent: that is, $\mathscr{R}(\mathbf{Y}, \mathbf{X}) = \mathscr{R}(\mathbf{Y}) \otimes \mathscr{R}(\mathbf{X})$. Examples commonly arise in economic forecasting when the values of some economic measurements are asserted as informative regarding the value of some other measurement. The value of gross output in an economy may be regressed on measurements of capital outlay, labor costs, costs of renewable and nonrenewable resources, and indices of relative power and cohesion among labor and among management in contesting the pace of production. The measurement processes among these quantities are logically independent. Although economists would typically not regard these quantities independently, this is another matter entirely. An example in civil engineering may involve a regression of annual maintenance costs per unit of road area on measurements of road composition and location, and measurements of weather conditions. In biology and agronomy, examples often arise in the regression of yield measurements on components of a factorial design. The unifying feature of all such applications relevant just now is the *logical* independence of the prevised measurements, $\mathbf{Y}$, and their informing, conditioning measurements, $\mathbf{X}$. We study the implications of relaxing this condition in our study of inference based on a mixture of autoregressions in the next section.

To foreshadow our current deliberations, we are going to derive the forms of sequential conditional densities $f(y_{N+1} | (\mathbf{X}_{N+1} = \mathbf{x}_{N+1})(\mathbf{X} = \mathbf{x})(\mathbf{Y} = \mathbf{y}))$ that cohere with an assertion of joint distributions in the mixture-density form

$$f(\mathbf{y} | \mathbf{X} = \mathbf{x}) = \int_{\mathbb{R}^+} \int_{\mathbb{R}^K} f(\mathbf{y} | \mathbf{X} = \mathbf{x}; \boldsymbol{\beta}, \pi) f(\boldsymbol{\beta} | \mathbf{X} = \mathbf{x}; \pi) f(\pi | \mathbf{X} = \mathbf{x}) \, d\boldsymbol{\beta} \, d\pi.$$

The meaning of this mixing parameterization will emerge from our discussion.

Suppose you assess your conditional opinions about $\mathbf{Y}$ given $(\mathbf{X} = \mathbf{x})$ with a mixture of conditionally exchangeable translated assertions, representable by $\mathbf{Y} | (\mathbf{X} = \mathbf{x}; \boldsymbol{\beta}, \pi) \sim N(\mathbf{x}\boldsymbol{\beta}, \pi\mathbf{I})$, where $\boldsymbol{\beta}$ has dimension $K \times 1$, $\pi$ is scalar, and $\mathbf{I}$ is the identity matrix of size $N$. (Note that we are continuing with the characterization of the Normal distribution in terms of its precision matrix rather than its variance.) An imbedded feature of this distribution is that it specifies a conditional distribution for the translated vector of quantities $(\mathbf{Y} - \mathbf{x}\boldsymbol{\beta}) | (\mathbf{X} = \mathbf{x}; \boldsymbol{\beta}, \pi)$ as independent normal. As a result, once the mixing with respect to a distribution over $\boldsymbol{\beta}$ and $\pi$ is engaged, the vectors of differences of $\mathbf{Y}$ quantities from their prevision values are recognized to be assessed exchangeably. Furthermore, in the context of the specific mixing function we shall presume, your joint distribution for $\mathbf{Y}$ is multivariate $\mathbf{t}$.

A second feature of this asserted distribution is an implied conditional distribution for the quantity $(\mathbf{x}^T\mathbf{x})^{-1}\mathbf{x}^T\mathbf{Y}$, conditional upon $(\mathbf{X} = \mathbf{x})$ for any specific mixing parameter values, $\boldsymbol{\beta}$ and $\pi$. On the one hand, the coefficient matrix that defines this vector, $(\mathbf{x}^T\mathbf{x})^{-1}\mathbf{x}^T$, merely amounts to a matrix, call it $\mathbf{A} = \mathbf{A}(\mathbf{x})$, of dimension $K \times N$. Thus, when conditioned on $(\mathbf{X} = \mathbf{x})$ for specified values of $\boldsymbol{\beta}$ and $\pi$, our Theorem 7.2 of Section 7.2.3 confirms the conditional distribution of $\mathbf{AY}$ as Normal with first moments $\boldsymbol{\beta}$, and second central moments $\pi^{-1}(\mathbf{x}^{\mathrm{T}}\mathbf{x})^{-1}$, supposing the matrix $\mathbf{x}^{\mathrm{T}}\mathbf{x}$ is invertible:

$$(\mathbf{x}^{\mathrm{T}}\mathbf{x})^{-1}\mathbf{x}^{\mathrm{T}}\mathbf{Y} = \mathbf{AY} \mid (\mathbf{X} = \mathbf{x}; \boldsymbol{\beta}, \pi) \sim N(\mathbf{A}\mathbf{x}\boldsymbol{\beta} = \boldsymbol{\beta}, (\mathbf{A}\pi^{-1}\mathbf{I}\mathbf{A}^{\mathrm{T}})^{-1} = \pi(\mathbf{x}^{\mathrm{T}}\mathbf{x})).$$

The invertibility of $\mathbf{x}^{\mathrm{T}}\mathbf{x}$ amounts to the condition that the columns of $\mathbf{x}$ are linearly independent. Thus, as long as the matrix $\mathbf{X}$ augments in size by appending rows so that $N^{-1}(\mathbf{x}^{\mathrm{T}}\mathbf{x}/N)^{-1}$ converges toward 0 as $N$ increases, your mixing distribution over $\boldsymbol{\beta}$ must be virtually identical to your distribution for this peculiar linear combination of the associated vector $\mathbf{Y}$, when $N$ is large.

The peculiarity of the matrix $\mathbf{A} \equiv (\mathbf{x}^{\mathrm{T}}\mathbf{x})^{-1}\mathbf{x}^{\mathrm{T}}$ derives from the fact that for any vector $\mathbf{Y}$, it generates the unique vector of linear projection coefficients that orthogonally project the vector $\mathbf{Y}$ onto the vector space spanned by the $K$ column vectors of $\mathbf{x}$. You may show for yourself as a simple exercise in linear algebra that if $\mathbf{C}$ is defined as a vector of coefficients satisfying $\mathbf{Y} = \mathbf{xC} + \mathbf{e}$ in such a way that each column of $\mathbf{x}$ is orthogonal to $\mathbf{e}$, that is, $\mathbf{x}^{\mathrm{T}}\mathbf{e} = \mathbf{0}_K$, then $\mathbf{C} = (\mathbf{x}^{\mathrm{T}}\mathbf{x})^{-1}\mathbf{x}^{\mathrm{T}}\mathbf{Y}$. (Simply premultiply both sides of the equation by $\mathbf{x}^{\mathrm{T}}$, and note the consequences.) Thus, with still more meaning, we should reiterate that your mixing distribution over $\boldsymbol{\beta}$ must be the same as your mixing function over the coefficients that project the quantity vector $\mathbf{Y}$ orthogonally onto the space spanned by the columns of its conditioning matrix, $\mathbf{x}$.

Now suppose your joint mixing distribution over $\boldsymbol{\beta}$ and $\pi$ has a Normal-gamma form, irrespective of the conditioning values of $\mathbf{X}$. This is to say that your mixing density $f(\boldsymbol{\beta}, \pi \mid \mathbf{X} = \mathbf{x})$ is representable as the product

$$f(\boldsymbol{\beta} \mid \mathbf{X} = \mathbf{x}; \pi) f(\pi \mid \mathbf{X} = \mathbf{x}),$$

where

$$\boldsymbol{\beta} \mid (\mathbf{X} = \mathbf{x}; \pi) \sim N(\boldsymbol{\beta}_0, \pi\boldsymbol{\varphi}_0), \qquad \text{and } \pi \mid (\mathbf{X} = \mathbf{x}) \sim \Gamma(\alpha_0, \gamma_0).$$

In this context, we can now state the following result.

**Theorem 8.1.** Suppose the $N \times 1$ vector of quantities, $\mathbf{Y}$, and the $N \times K$ matrix of quantities, $\mathbf{X}$, are logically independent, so $\mathcal{R}(\mathbf{Y}, \mathbf{X}) = \mathcal{R}(\mathbf{Y}) \otimes \mathcal{R}(\mathbf{X})$. Asserting joint conditional distributions for $\mathbf{Y} \mid (\mathbf{X} = \mathbf{x})$ by means of the mixture-Normal linear regressions $\mathbf{Y} \mid (\mathbf{X} = \mathbf{x}) \sim M\text{-}N(\mathbf{x}\boldsymbol{\beta}, \pi\mathbf{I})$ with the mixing distributions $\boldsymbol{\beta} \mid (\mathbf{X} = \mathbf{x}; \pi) \sim N(\boldsymbol{\beta}_0, \pi\boldsymbol{\varphi}_0)$ and $\pi \mid (\mathbf{X} = \mathbf{x}) \sim \Gamma(\alpha_0, \gamma_0)$ implies that the asserted joint distributions for the $\mathbf{Y}$ vector are multivariate $\mathbf{t}$ for every $\mathbf{x} \in \mathcal{R}(\mathbf{X})$:

$$\mathbf{Y} \mid (\mathbf{X} = \mathbf{x}) \sim \mathbf{t}_N(v = 2\alpha_0, \boldsymbol{\mu} = \mathbf{x}\boldsymbol{\beta}_0, \tau = (\alpha_0/\gamma_0)[\mathbf{I} - \mathbf{x}(\boldsymbol{\varphi}_0 + \mathbf{x}^{\mathrm{T}}\mathbf{x})^{-1}\mathbf{x}^{\mathrm{T}}]).$$

Moreover, each implied conditional distribution $Y_{N+1} \mid (\mathbf{X}_{N+1} = \mathbf{x}_{N+1})(\mathbf{X} = \mathbf{x})$ $(\mathbf{Y} = \mathbf{y})$ is univariate $t(v = 2\alpha_N, \mu = \mathbf{x}_{N+1}\boldsymbol{\beta}_N,\ \tau = (\alpha_N/\gamma_N)[1 - \mathbf{x}_{N+1}(\boldsymbol{\varphi}_N +$

$\mathbf{x}_{N+1}^{\mathrm{T}}\mathbf{x}_{N+1})^{-1}\mathbf{x}_{N+1}^{\mathrm{T}}]$, where these three parameters are determined by functions of the parameters of the implied posterior Normal-gamma mixing distribution, $M(\boldsymbol{\beta}, \pi|(\mathbf{X}=\mathbf{x})(\mathbf{Y}=\mathbf{y}))$:

$$\boldsymbol{\beta}_N = (\mathbf{x}^{\mathrm{T}}\mathbf{x} + \boldsymbol{\varphi}_0)^{-1}(\mathbf{x}^{\mathrm{T}}\mathbf{y} + \boldsymbol{\varphi}_0\boldsymbol{\beta}_0),$$

$$\boldsymbol{\varphi}_N = (\mathbf{x}^{\mathrm{T}}\mathbf{x} + \boldsymbol{\varphi}_0),$$

$$\alpha_N = (2\alpha_0 + N)/2,$$

$$\gamma_N = \gamma_0 + \left(\frac{1}{2}\right)(\mathbf{y} - \mathbf{x}\boldsymbol{\beta}_0)^{\mathrm{T}}[\mathbf{I} - \mathbf{x}(\boldsymbol{\varphi}_0 + \mathbf{x}^{\mathrm{T}}\mathbf{x})^{-1}\mathbf{x}^{\mathrm{T}}](\mathbf{y} - \mathbf{x}\boldsymbol{\beta}_0).$$

Iterative formulas for computing these parameters sequentially are

$$\boldsymbol{\beta}_{N+1} = (\mathbf{x}_{N+1}^{\mathrm{T}}\mathbf{x}_{N+1} + \boldsymbol{\varphi}_N)^{-1}(\mathbf{x}_{N+1}^{\mathrm{T}}y_{N+1} + \boldsymbol{\varphi}_N\boldsymbol{\beta}_N),$$

$$\boldsymbol{\varphi}_{N+1} = (\mathbf{x}_{N+1}^{\mathrm{T}}\mathbf{x}_{N+1} + \boldsymbol{\varphi}_N),$$

$$\alpha_{N+1} = \alpha_N + 1/2,$$

$$\gamma_{N+1} = \gamma_N + \left(\frac{1}{2}\right)(y_{N+1} - \mathbf{x}_{N+1}\boldsymbol{\beta}_N)^2[1 - \mathbf{x}_{N+1}(\boldsymbol{\varphi}_N + \mathbf{x}_{N+1}^{\mathrm{T}}\mathbf{x}_{N+1})^{-1}\mathbf{x}_{N+1}^{\mathrm{T}}].$$

*Proof*. An uneventful algebraic reduction of the integrand in the representation

$$f(\mathbf{y}|\mathbf{X}=\mathbf{x}) = \int_{\mathbb{R}^+}\int_{\mathbb{R}^K} f(\mathbf{y}|\mathbf{X}=\mathbf{x};\boldsymbol{\beta},\pi)\, f(\boldsymbol{\beta}|\mathbf{X}=\mathbf{x};\pi)\, f(\pi|\mathbf{X}=\mathbf{x})\, d\boldsymbol{\beta}\, d\pi$$

involves collecting terms involving $\boldsymbol{\beta}$ into a quadratic form and integrating the resulting Normal density with respect to $\boldsymbol{\beta}$. One useful matrix result to aid in the derivation is that if $\mathbf{M} \equiv (\boldsymbol{\varphi}_0 + \mathbf{x}^{\mathrm{T}}\mathbf{x})$ then $[\mathbf{I} - \mathbf{x}\mathbf{M}^{-1}\mathbf{x}^{\mathrm{T}}]^{-1} = [\mathbf{I} + \mathbf{x}\boldsymbol{\varphi}_0^{-1}\mathbf{x}^{\mathrm{T}}]$, which can be established easily by direct multiplication. The remaining integral over $\pi$ is recognizable as a gamma integral, which yields the multivariate $\mathbf{t}_N$-density specified in the theorem.

The posterior mixing density over $(\boldsymbol{\beta}, \pi)$ is recognizable from the same algebraic representation of the mixed joint density $f(\mathbf{y}, \boldsymbol{\beta}, \pi|\mathbf{X}=\mathbf{x})$ in the double integrand. Treating the variables $\mathbf{y}$ as constant renders this function proportional to the posterior density $f(\boldsymbol{\beta}, \pi|(\mathbf{X}_{N+1}=\mathbf{x}_{N+1})(\mathbf{X}=\mathbf{x})(\mathbf{Y}=\mathbf{y}))$, which appears in the proposed Normal-gamma form.

The iterative formulas for the conjugate mixing parameters derive from the direct formulas by replacing the subscripts 0 with $N$, replacing the matrix $\mathbf{x}$ with the augmenting row vector $\mathbf{x}_{N+1}$, replacing the vector $\mathbf{y}$ with the augmenting component $y_{N+1}$, and by paying attention to the dimensions of the resulting forms. □

One of the interesting characteristics of this posterior mixing function on $\boldsymbol{\beta}$ is that it is informed by $\mathbf{x}$ and $\mathbf{y}$ through an information transfer function which projects $\mathbf{y}$ into the space spanned by the column vectors of the conditioning $\mathbf{x}$ quantities. Notice that the posterior regression coefficients after $N$ observations, $\boldsymbol{\beta}_N$, can be expressed as a matrix-weighted-average of the initial regression coefficients and the projection of the observed $\mathbf{y}$ vector onto the conditioning $\mathbf{x}$'s:

$$\boldsymbol{\beta}_N = (\mathbf{x}^T\mathbf{x} + \boldsymbol{\varphi}_0)^{-1}[\mathbf{x}^T\mathbf{x}(\mathbf{x}^T\mathbf{x})^{-1}\mathbf{x}^T\mathbf{y} + \boldsymbol{\varphi}_0\boldsymbol{\beta}_0],$$

at least when $\mathbf{x}^T\mathbf{x}$ is invertible. This form of the information transfer makes sense, since we have seen that the mixing distribution over $\boldsymbol{\beta}$ must be identical to the implied distribution for the projection of many $\mathbf{y}$ observations onto the column space of conditioning $\mathbf{x}$'s.

### 8.2.2 Eliciting the Mixing Distributions for Applications

One way of eliciting your prior mixture on $\boldsymbol{\beta}$ within the Normal-gamma mixture specification is to assess your conditional previsions for $\mathbf{Y}$ conditioning on an invertible $K \times K$ array of $\mathbf{x}$ values. Since your asserted $P(\mathbf{Y}|\mathbf{X} = \mathbf{x})$ must equal $\mathbb{M}_1[P(\mathbf{Y}|\mathbf{X} = \mathbf{x}; \boldsymbol{\beta})] = \mathbf{x}\mathbb{M}_1(\boldsymbol{\beta})$, where $\mathbb{M}_1(\cdot)$ is the first moment operation on your mixture over $\boldsymbol{\beta}$, the components of $\mathbb{M}_1(\boldsymbol{\beta}) = \boldsymbol{\beta}_0$ can be retrieved from your assertions as $\mathbf{x}^{-1}P(\mathbf{Y}|\mathbf{X} = \mathbf{x})$.

Similarly, the initial mixture matrix, $\boldsymbol{\varphi}_0$ can be retrieved from variance and covariance assertions regarding $\mathbf{Y}$ values under various conditions. Alternatively, and more simply in practice, the initial variance components can be retrieved from predictive quantile assertions under these conditions. The paper of Kadane, Dickey, et al. (1980) reports on computational details, and an application to assessed costs of road maintenance.

Another procedure for specifying $\boldsymbol{\varphi}_0$ involves the form of a so-called $g$-prior, introduced in Zellner (1986b). This is based on his recognition of the role of the matrix $\mathbf{x}^T\mathbf{x}$ in the sequential updating formula for the matrix $\boldsymbol{\varphi}_N = \mathbf{x}^T\mathbf{x} + \boldsymbol{\varphi}_0$. If you imagine a scenario of a conditioning $\mathbf{X} = \mathbf{x}$ matrix that provides an appropriate context to represent your accumulated knowledge about $\mathbf{Y}$ in terms of the mixing function $f(\boldsymbol{\beta}|\pi)$, you may wish to specify the value of your $\boldsymbol{\varphi}_0$ matrix as proportional to the transformed conditioning matrix $\mathbf{x}^T\mathbf{x}$. The proportionality constant could be determined by your assessing the equivalent number of $\mathbf{X}$, $\mathbf{Y}$ observations that your prior information represents, understood relative to the number of observations for which you are assessing your uncertainty.

To limit the extent of empirical analysis presented here, we defer the presentation of a computational example until we have addressed the structure of autoregressive opinions.

## 8.3 MIXTURE AUTOREGRESSIVE ASSERTIONS

Let the vector $\mathbf{Y}$ represent a temporal sequence of measurements of the same type, for example, a daily sequence of transaction prices for the New Zealand dollar in terms of a U.S. dollar on the international currency exchange at 3 P.M. At the crudest level of economic theorizing, one might think that the sequential values of components of **Y** should well be informed by measured relative values of the commodity price level in the two countries, by the relative levels of interest rates on riskless investments such as Treasury bills in the two countries, and by relative levels of the external trade balances held by production firms located in the two countries. Indeed, a linear regression of some transformed value of exchange rates on transformed values of these economic indicators may well be asserted and assessed inferentially in the style of regression analysis we have just detailed.

One economic theoretical problem with such assertions is that the persons, firms, institutions, national and international banks, and goverments who invest regularly in international currencies as an asset, change their portfolios of currencies and other assets rather frequently based on their changing knowledge about the historical activity these economic measurements summarize. Typically, these portfolio adjustments are made much more frequently than the statistical measurements of these activities are compiled. Thus, to a great extent, by the time the international statistical accounting offices announce measurements of such quantities, on a monthly or quarterly basis, the information in such "news" has long been accommodated by financiers, "up with the play," whose business it is to make hay with the latest information about the social, political, and economic intrigues within the countries whose relative exchange rates are under consideration. They accommodate such information by incorporating it into their own bid and ask prices for the currencies they hold, thus incorporating it into their own trading histories and the recorded history **Y**.

As a result, an informed consideration of conditional previsions and conditional distributions regarding **Y** values may well conclude that the information regarding the next observation, $Y_t$, contained in economic indicators is subsumed in a limited length of the most recent preceding observations of the **Y** quantities themselves, $Y_{t-1}, Y_{t-2}, \ldots, Y_{t-L}$. The same type of attitude would represent considered opinions regarding sequential gross measurements of any systems that progress by ever-varying inflow and outflow. These accumulate new components of unknown composition, while shedding older components at irregular and unknown rates. As a different example, consider the daily measured fiber content at 11 A.M. in wintertime's neverending pot of soup. Such is the context in which the analysis of autoregressive opinion structures are relevant. Let us turn to the formal detail.

***Definition 8.2.*** Let $\mathbf{Y}_t$ denote a $(t + L) \times 1$ column vector of quantities which augments according to $\mathbf{Y}_{t+1} \equiv (\mathbf{Y}_t^{\mathrm{T}}, Y_{t+1})^{\mathrm{T}}$, for $t = 1, 2, \ldots, T$. Your assertions of conditional previsions of the form $P(Y_t | \mathbf{Y}_{t-1} = \mathbf{y}_{t-1}) = G_t(\mathbf{y}_{t-1})$ for $t = 1, 2, \ldots$ are referred to as an *autoregression* of the sequence **Y** components on their own preceding values. If the functions $G_t(\cdot)$ do not vary with $t$, your assertions are said to be *stationary autoregressive*. If your autoregression is linear, of the form $P(Y_t | \mathbf{Y}_{t-1} = \mathbf{y}_{t-1}) = \beta_1(t) y_{t-1} + \beta_2(t) y_{t-2} + \cdots + \beta_L(t) y_{t-L}$ for each $t$, then your opinions about the sequence **Y** are said to be *linear autoregressive of length L*. If the linear autoregression coefficients $\beta_i(t)$ are all constant with respect to $t$, your opinions are *stationary*. □

When your opinions are linear autoregressive of length $L$, we could say in the notational setup of Section 8.2 that you are specifying a regression for $T$ components of $\mathbf{Y}$ on the rows of a $T \times L$ matrix $\mathbf{X}$, in which each row of $\mathbf{X}$ is constrained to equal $L$ appropriately lagged quantities in the sequence $\mathbf{Y}$. In these terms, the vector $\mathbf{Y}$ is not logically independent of the regressors exhibited in $\mathbf{X}$. Rather, the $T \times (L+1)$ matrices in the realm $\mathscr{R}(\mathbf{Y}, \mathbf{X})$ are constrained by definition to be constant on their northwest-southeast diagonals. That is, components $m$ of any matrix $(\mathbf{y}, \mathbf{x}) \in \mathscr{R}(\mathbf{Y}, \mathbf{X})$ satisfy the restriction that $m_{i,j} \equiv m_{i+1,j+1}$ whenever both subscript pairs define matrix components.

In such a setup, however, the detailed forms of opinions we have analyzed in Theorem 8.1 would actually be incoherent. For the joint $T$-dimensional conditional distribution for $\mathbf{Y}|(\mathbf{X} = \mathbf{x})$ would necessarily be degenerate over its first $T-1$ arguments, since the first $T-1$ components of $\mathbf{Y}$ constitute the first column of the matrix $\mathbf{X} = \mathbf{x}$ which conditions the joint distribution. Nonetheless, we can specify a different form of joint opinions about the sequence $\mathbf{Y}$ which supports a sequence of conditional distributions that is computable by the very same formulas we derived there. Let us state the appropriate theorem for the mixture-linear autoregressive setup, and then discuss its content. We shall find that this opinion structure specifies nonlinear and nonstationary autoregressive assertions over the $\mathbf{Y}$ vector, yet it is recognizable as a partially exchangeable opinion structure through the finite number of sufficient statistics it accumulates for inference.

It will be helpful now to expand our notation, using the expression $\mathbf{X}(t)$ to denote the submatrix composed of the first $t$ rows of $\mathbf{X}$, and continuing with the expression $\mathbf{X}_t$ to denote the row $t$ of $\mathbf{X}$. Moreover, let us recognize that the first row of any conditioning $\mathbf{X}$ matrix is $\mathbf{X}_1 = \mathbf{x}_1 \equiv (y_0 \quad y_{-1} \quad \cdots \quad y_{-L+1})$.

**Theorem 8.2.** Suppose the $T \times 1$ vector of quantities, $\mathbf{Y}$, and the $T \times L$ matrix of quantities, $\mathbf{X}$, are logically related by the restriction that components $m_{i,j}$ of any matrix $\mathbf{m} = (\mathbf{y}, \mathbf{x}) \in \mathscr{R}(\mathbf{Y}, \mathbf{X})$ satisfy the restriction that $m_{i,j} \equiv m_{i+1,j+1}$ whenever both subscript pairs define matrix components: $(1 \leqslant i < T)(i < j)(j \leqslant L) = 1$. Asserting your joint distribution for $\mathbf{Y}$ conditional only on $(\mathbf{X}_1 = (y_0 \quad y_{-1} \quad \cdots \quad y_{-L+1}))$ by means of the mixture-Normal linear autoregressions

$$Y_t|(\mathbf{X}(t) = \mathbf{x}(t)) \sim M\text{-}N(\mathbf{x}_t\boldsymbol{\beta}, \pi\mathbf{I}) \qquad \text{for } t = 1, 2, \ldots, T$$

with the mixing distributions $\boldsymbol{\beta}|(\mathbf{X}_1 = \mathbf{x}_1; \pi) \sim N(\boldsymbol{\beta}_0, \pi\boldsymbol{\varphi}_0)$ and $\pi|(\mathbf{X}_1 = \mathbf{x}_1) \sim \Gamma(\alpha_0, \gamma_0)$ implies that your sequence of conditional distributions for $Y_t|(\mathbf{Y}(t-1) = \mathbf{y}(t-1))$ are univariate $t(\nu, \mu, \tau)$, where

$$\nu = 2\alpha_{t-1}, \quad \mu = \mathbf{x}_t\boldsymbol{\beta}_{t-1}, \quad \text{and} \quad \tau = (\alpha_{t-1}/\gamma_{t-1})[1 - \mathbf{x}_t(\boldsymbol{\varphi}_{t-1} + \mathbf{x}_t^{\mathrm{T}}\mathbf{x}_t)^{-1}\mathbf{x}_t^{\mathrm{T}}].$$

The components of these parameter formulas are the parameters of the implied posterior Normal-gamma mixing distribution, $M(\boldsymbol{\beta}, \pi | \mathbf{X}(t) = \mathbf{x}(t))$, computable iteratively by

$$\begin{aligned}
\boldsymbol{\beta}_t &= (\mathbf{x}_t^{\mathrm{T}}\mathbf{x}_t + \boldsymbol{\varphi}_{t-1})^{-1}(\mathbf{x}_t^{\mathrm{T}} y_t + \boldsymbol{\varphi}_{t-1}\boldsymbol{\beta}_{t-1}),\\
\boldsymbol{\varphi}_t &= (\mathbf{x}_t^{\mathrm{T}}\mathbf{x}_t + \boldsymbol{\varphi}_{t-1}),\\
\alpha_t &= \alpha_{t-1} + 1/2,\\
\gamma_t &= \gamma_{t-1} + \left(\frac{1}{2}\right)(y_t - \mathbf{x}_t\boldsymbol{\beta}_{t-1})^2[1 - \mathbf{x}_t(\boldsymbol{\varphi}_{t-1} + \mathbf{x}_t^{\mathrm{T}}\mathbf{x}_t)^{-1}\mathbf{x}_t^{\mathrm{T}}].
\end{aligned}$$

These can also be computed directly from $\boldsymbol{\beta}_0$, $\boldsymbol{\varphi}_0$, $\alpha_0$, $\gamma_0$, $\mathbf{x}(t)$, and $\mathbf{y}(t)$ using the formulas

$$\begin{aligned}
\boldsymbol{\beta}_t &= (\mathbf{x}(t)^{\mathrm{T}}\mathbf{x}(t) + \boldsymbol{\varphi}_0)^{-1}(\mathbf{x}(t)^{\mathrm{T}}\mathbf{y}(t) + \boldsymbol{\varphi}_0\boldsymbol{\beta}_0),\\
\boldsymbol{\varphi}_t &= (\mathbf{x}(t)^{\mathrm{T}}\mathbf{x}(t) + \boldsymbol{\varphi}_0),\\
\alpha_t &= (2\alpha_0 + t)/2,\\
\gamma_t &= \gamma_0 + \left(\frac{1}{2}\right)(\mathbf{y}(t) - \mathbf{x}(t)\boldsymbol{\beta}_0)^{\mathrm{T}}[\mathbf{I} - \mathbf{x}(t)(\boldsymbol{\varphi}_0 + \mathbf{x}(t)^{\mathrm{T}}\mathbf{x}(t))^{-1}\mathbf{x}(t)^{\mathrm{T}}](\mathbf{y}(t) - \mathbf{x}(t)\boldsymbol{\beta}_0).
\end{aligned}$$

*Proof*. For any value of $t$, the joint density $f(\mathbf{y}(t)|\mathbf{X}_1 = \mathbf{x}_1 \equiv (y_0 \;\; y_{-1} \;\; \cdots \;\; y_{-L+1}))$ can be represented in its mixture form as a mixture integral of the product of its conditional parametric densities:

$$\int_{\mathbb{R}^+}\int_{\mathbb{R}^L} \prod_{j=1}^{t} f(y_j|(\mathbf{X}(j) = \mathbf{x}(j)); \boldsymbol{\beta}, \pi) f(\boldsymbol{\beta}|\mathbf{X}_1 = \mathbf{x}_1; \pi) f(\pi|\mathbf{X}_1 = \mathbf{x}_1)\, d\boldsymbol{\beta}\, d\pi.$$

This product integrand can be understood as proportional either to the posterior mixture density for $Y_t$ given the previous values of $\mathbf{Y}(t-1)$, that is, $f(y_t|(\mathbf{X}_t = \mathbf{x}_t); \boldsymbol{\beta}, \pi) f(\boldsymbol{\beta}, \pi|\mathbf{X}(t) = \mathbf{x}(t))$, or to the posterior mixing function $f(\boldsymbol{\beta}, \pi|\mathbf{Y}(t) = \mathbf{y}(t))$ according to Bayes' theorem. Thus, the parameters of the sequential mixing function derive from precisely the same algebraic manipulations we developed in Theorem 8.1. Indeed the formulas stated here are identical to the formulas in that theorem, only with the observation numbers denoted by $t$, as is appropriate to the context of temporal sequential observations presumed here. □

Computationally, the implications of this theorem for determining the sequential forecast distributions for $Y_{t+1}|((\mathbf{Y}(t) = \mathbf{y}(t))(\mathbf{X}_1 = \mathbf{x}_1))$ are identical to the equation system we derived for mixture-Normal regressions in Section 8.2.1 when we presumed $\mathscr{R}(\mathbf{Y}, \mathbf{X}) = \mathscr{R}(\mathbf{Y}) \otimes \mathscr{R}(\mathbf{X})$. In computing a sequential scoring analysis of a Normal-gamma mixture autoregressive opinion, you need only construct the columns of the $\mathbf{X}$ matrix by lagging the vector of measured $\mathbf{Y} = \mathbf{y}$

values to construct its columns, beginning them with the appropriate number of components of $\mathbf{X}_1 = (y_0 \quad y_{-1} \quad \cdots \quad y_{-L+1})$.

Although the sequential conditional distributions are quite simple, the entire joint distribution of $\mathbf{Y}|(\mathbf{X}_1 = \mathbf{x}_1)$ is complicated, and we shall not detain ourselves long with its study here. You can compute easily the components of its first moment vector by recursive equations, beginning with

$$\mathbb{M}_1(Y_1|\mathbf{X}_1 = \mathbf{x}_1) = (y_0 \quad y_{-1} \quad \cdots \quad y_{-L+1})\boldsymbol{\beta}_0,$$

and continuing sequentially for $t = 2, 3, \ldots, T$ with

$$\mathbb{M}_1(Y_t|\mathbf{X}_1 = \mathbf{x}_1) = \mathbb{M}_1(Y_{t-1})\beta_{0,1} + \mathbb{M}_1(Y_{t-2})\beta_{0,2} + \cdots + \mathbb{M}_1(Y_{t-L})\beta_{0,L},$$

replacing $\mathbb{M}_1(Y_{t-j})$ in this expression with $y_{t-j}$ whenever $t - j < 1$. Here the vector $\boldsymbol{\beta}_0$ components are denoted as $\beta_{0,j}$. A similar recursion formula is applicable, to compute the conditional moments $\mathbb{M}_1(Y_{t+K}|(\mathbf{X}_1 = \mathbf{x}_1)(\mathbf{Y}(t) = \mathbf{y}(t)))$ based on the appropriate posterior mixing vector $\boldsymbol{\beta}_t \equiv (\beta_{t,1} \quad \beta_{t,2} \quad \cdots \quad \beta_{t,L})$, for $K = 1, 2, \ldots, T - t$.

However, our knowledge of the relation between sufficient statistics and partial exchangeability (Section 3.12) allows that the joint distribution for $\mathbf{Y}|(\mathbf{X}_1 = \mathbf{x}_1)$ entails a specific form of partial exchangeability, identifiable by the $(L^2 + 3L + 2)/2$ statistics that are sufficient for inference about any $Y_{t+1}$ on the basis of $\mathbf{Y}(t)$ and $\mathbf{x}_1$. These sufficient statistics appear in the direct equations for the posterior mixing parameters in the form of $\mathbf{x}(t)^{\mathrm{T}}\mathbf{x}(t)$, $\mathbf{x}(t)^{\mathrm{T}}\mathbf{y}(t)$, and $\mathbf{y}(t)^{\mathrm{T}}\mathbf{y}(t)$. The first of these involves $L(L+1)/2$ distinct statistics, the second involves $L$ more, and the last makes only one more, irrespective of the size of $t$. You will remember that complete exchangeability of the components of $\mathbf{Y}$ in the mixture-Normal context requires only two sufficient statistics, $\bar{Y}$ and $\sum(Y_i - \bar{Y})^2$. Thus, the mixture-linear autoregressive setup exemplifies a partially exchangeable assessment of the components of the vector $\mathbf{Y}$, since it merely increases the number of sufficient statistics required to a larger finite number. We shall see in the next section how an extension of the mixture-autoregressive assertion structure defies completely the judgment of exchangeability in any form, except by means of an approximation.

A final feature to notice in the mixture-Normal-autoregressive inference structure is that the implied autoregressions are neither linear nor stationary. Although the regression equations specified by $P(Y_t|\mathbf{Y}(t-1) = \mathbf{y}(t-1)) = P(Y_t|\mathbf{X}(t) = \mathbf{x}(t)) = \mathbf{x}_t\boldsymbol{\beta}_t$ appear linear in both $\mathbf{x}_t$ and of $\boldsymbol{\beta}_t$ at first glance, recall that both of these vectors are determined by functions of the conditioning $\mathbf{y}(t-1)$ values, the $\boldsymbol{\beta}_t$ components deriving from rather complicated functions. Thus, the inner product $\mathbf{x}_t\boldsymbol{\beta}_t$ amounts to a fairly complicated and nonlinear function of the conditioning values of $\mathbf{y}$. What are linear and stationary are the parametric regression assertions $P(Y_t|\mathbf{Y}(t-1) = \mathbf{y}(t-1); \boldsymbol{\beta}, \pi)$. But their mixture implies nonstationary assertions of $P(Y_t|\mathbf{Y}(t-1) = \mathbf{y}(t-1))$.

We shall study a numerical example of the mixture-autoregressive inference in Section 8.5, after a final foray into the theory of mixture linear regressions. "Once more unto the breach, dear friends, once more."

## 8.4 MIXTURE AUTOREGRESSIONS OF MOVING AVERAGES

Our familiarity with the workings of mixture distributions allows us now to construct a mixture distribution for $\mathbf{Y}|(\mathbf{X}_1 = \mathbf{x}_1)$ which, surprisingly, does not entail the feature of exchangeability at all, except as an approximation. In this section we construct and examine the implied sequential forecasting distributions, motivating them by the continuing allusion to New Zealand dollar exchange rates for U.S. dollars. Precise details appear in the next section.

We continue with the notation in which the matrix $\mathbf{X}$ is composed of columns whose values are lagged observations of the vector $\mathbf{Y}$, making only minor modifications as required. $\mathbf{X}(t)$ represents a $t \times K$ matrix, and $\mathbf{X}_t$ represents the row $t$ of this matrix. Since the details will become now fairly involved, we shall forgo the format of theorem and proof, treating this section rather as providing the constructive logic of a computational device.

Consider the sequential parametric mixture distribution for $t = 1, \ldots, T$,

$$Y_t|(\mathbf{X}(t) = \mathbf{x}(t); \boldsymbol{\beta}, \pi, \rho) \sim M\text{-}N(\mathbf{x}_t \boldsymbol{\beta} + \rho[Y_{t-1} - \mathbf{x}_{t-1}\boldsymbol{\beta}], \pi),$$

where mixing is with respect to some distribution $M(\boldsymbol{\beta}, \pi, \rho)$ for $\rho \in (-1, 1)$. Each parametric conditional component density of this mixture would convey the attitude that the value of any $Y_t$ should be somewhere around a linear combination of past values of $\mathbf{Y}$, denoted by $\mathbf{x}_t \boldsymbol{\beta}$. But the extent to which it is expected to be close to this projection is adjusted by a fraction, $\rho$, of the difference that $Y_{t-1}$ was from the very same projection function, $\mathbf{x}_{t-1}\boldsymbol{\beta}$. Moreover, this attitude regarding the $\mathbf{Y}$ sequence continues as it progresses through time. Whatever it is about the quantities $\mathbf{Y}$ that makes me think $Y_t$ should be somewhere around $\mathbf{x}_t \boldsymbol{\beta}$ on the basis of all the conditioning observations, $\mathbf{Y}(t-1) = \mathbf{y}(t-1)$, this aspect continues to influence my sense of the temporal accumulation or diminution of $Y$ units recorded in $Y_{t+1}$ by a proportion, $\rho$, of the difference between $Y_t$ and $\mathbf{x}_t \boldsymbol{\beta}$.

In the context of the New Zealand dollar exchange rate with the U.S. dollar, these features would characterize the attitude of someone who felt that information regarding relative economic conditions would pass somewhat slowly and sporadically among traders, requiring a sizable time duration to be accomodated widely in trading activity. We make some brief comments on historical detail when we introduce a numerical application of our regression theory in the next section. For now let us derive the computational implications of the attitudes just outlined for the sequential forecasting distributions they imply.

To begin, the parametric distribution

$$Y_t|(\mathbf{X}(t) = \mathbf{x}(t); \boldsymbol{\beta}, \pi, \rho) \sim N(\mathbf{x}_t \boldsymbol{\beta} + \rho[Y_{t-1} - \mathbf{x}_{t-1}\boldsymbol{\beta}], \pi)$$

embeds within itself the implied distribution

$$(Y_t - \rho Y_{t-1})|(\mathbf{X}(t) = \mathbf{x}(t); \boldsymbol{\beta}, \pi, \rho) \sim N((\mathbf{x}_t - \rho \mathbf{x}_{t-1})\boldsymbol{\beta}, \pi).$$

Written more simply, that is $Y_t^*|(\mathbf{X}(t) = \mathbf{x}(t); \boldsymbol{\beta}, \pi, \rho) \sim N(\mathbf{x}_t^* \boldsymbol{\beta}, \pi)$, where $Y_t^* \equiv Y_t^*(\rho) = Y_t - \rho Y_{t-1}$ and $\mathbf{x}_t^* = \mathbf{x}_t^*(\rho) = \mathbf{x}_t - \rho \mathbf{x}_{t-1}$. Indeed, an entire matrix $\mathbf{X}^*(t) \equiv \mathbf{X}(t) - \rho \mathbf{X}(t-1)$ is computable from $\mathbf{X}(t)$ and $\rho$, modulo another specified "initial row" $\mathbf{X}_0$ which is needed to compute the row $\mathbf{X}_1^*(t)$. In these terms we can now see that the parametric distribution under consideration implies a parametric distribution of $\rho$-transformed quantities as a simple autoregression:

$$Y_t^*|(\mathbf{X}^*(t) = \mathbf{x}^*(t); \boldsymbol{\beta}, \pi, \rho) \sim N(\mathbf{x}_t^* \boldsymbol{\beta}, \pi),$$

which is equivalent to writing

$$Y_t|(\mathbf{X}^*(t) = \mathbf{x}^*(t); \boldsymbol{\beta}; \pi, \rho) \sim N(\rho Y_{t-1} + \mathbf{x}_t^* \boldsymbol{\beta}, \pi)$$

by translation.

From the first of these forms, the joint distribution of the $\mathbf{Y}$ vector derives its name as a mixture *autoregression of moving averages* $Y_t - \rho Y_{t-1}$ on themselves, in the guise of $\mathbf{x}_t - \rho \mathbf{x}_{t-1}$. The second of these forms will be used to specify sequential mixture previsions for successive components of $\mathbf{Y}$.

In order to identify the joint density $f(\mathbf{y}(t)|\mathbf{X}(1) = \mathbf{x}(1))$ and to compute the sequential densities of the form $f(y_t|\mathbf{Y}(t-1) = \mathbf{y}(t-1))$ associated with such attitudes, we need to address the mixture integral

$$\int_{-1}^{+1} \int_{\mathbb{R}^+} \int_{\mathbb{R}^L} \prod_{j=1}^{t} f(y_j|(\mathbf{X}^*(j) = \mathbf{x}^*(j)); \boldsymbol{\beta}, \pi, \rho) f(\boldsymbol{\beta}, \pi, \rho|\mathbf{X}(1) = \mathbf{x}(1))\, d\boldsymbol{\beta}\, d\pi\, d\rho.$$

Now suppose the mixing density function $f(\boldsymbol{\beta}, \pi, \rho|\mathbf{X}_1 = \mathbf{x}_1)$ is specified by the product $f(\boldsymbol{\beta}, \pi|\mathbf{X}_1 = \mathbf{x}_1; \rho) f(\rho|\mathbf{X}_1 = \mathbf{x}_1)$ according to the stipulation that $(\boldsymbol{\beta}, \pi|\mathbf{X}_1 = \mathbf{x}_1; \rho) \sim N\text{-}\Gamma(\boldsymbol{\beta}_0, \boldsymbol{\varphi}_0, \alpha_0, \gamma_0)$ irrespective of $\rho$. Then we can use the algebraical formulas of a simple mixture (with respect to $\boldsymbol{\beta}$ and $\pi$ for constant $\rho$) autoregression of the $\rho$-transformed $\mathbf{Y}^*$ quantities to yield the mixture form of $f(\mathbf{y}(t)|\mathbf{X}(1) = \mathbf{x}(1))$ as

$$\int_{-1}^{+1} \prod_{j=1}^{t} f(y_j|(\mathbf{X}^*(j) = \mathbf{x}^*(j)); \rho) f(\rho|\mathbf{X}(1) = \mathbf{x}(1))\, d\rho.$$

Computationally, what this argument requires is that for each value of $\rho$ under consideration, a transformed vector $\mathbf{Y}^*(\rho)$ and a matrix $\mathbf{X}^*(\rho)$ are constructed and used to compute sequential Normal-gamma parameters $\boldsymbol{\beta}_t(\rho)$, $\boldsymbol{\varphi}_t(\rho)$, $\alpha_t(\rho)$, and $\gamma_t(\rho)$, using the formulas of Theorem 8.2. These then identify the components of the parametric conditional densities for $Y_t|(\mathbf{Y}_{t-1} = \mathbf{y}_{t-1}; \rho)$, based on the parameter $\rho$, as univariate $t$. The only modification to the formulas in the theorem are that occurrences of $\mathbf{x}(t)$ and $\mathbf{y}(t)$ are adjusted to $\mathbf{x}^*(t)$ and $\mathbf{y}^*(t)$, and that the location of the $t$-density for $Y_t$ shifts to $\mu = \rho Y_{t-1} + \mathbf{x}_t^* \boldsymbol{\beta}$.

Both computationally and theoretically, we are left in an interesting pickle. If the mixing distribution over $\pi \in (-1, 1)$ is a continuous distribution, the mixture distribution defies the exhibition of sufficient statistics of fixed dimension. Although finite sufficient statistics are exhibited by the mixed parametric distributions based on $\rho$ through the conditional autoregressions of $\mathbf{Y}^*(\rho)$ on $\mathbf{X}^*(\rho)$, there are still an infinity of $\rho$ values within $(-1, 1)$ to contend with, unless the mixing distribution function $M(\rho | \mathbf{X}(1) = \mathbf{x}(1))$ is itself discrete over a finite grid of $\rho$ values. This is the tack we follow in completing this analysis. A discrete distribution for $\rho$ may be considered as an exact mixing function in its own right or as merely a computational tool for approximating any mixture distribution based on a nearby continuous mixing distribution.

Suppose we specify a digital mixing distribution over, amassing mixture weight only on a finite grid of $D$ units within $(-1, 1)$. In the example of the next section, we shall use a grid of $D = 19$ digits from $-.9$ through $+.9$ in units of $.1$. In such a case the mixture integral over $\rho$ would reduce to the summation

$$f(\mathbf{y}(t) | \mathbf{X}(1) = \mathbf{x}(1)) = \sum_{i=1}^{D} \prod_{j=1}^{t} f(y_j | (\mathbf{X}^*(j) = \mathbf{x}^*(j)); \rho_i) f(\rho_i | \mathbf{X}(1) = \mathbf{x}(1)),$$

with the sequential conditional densities computable from this summation by

$$f(y_t | \mathbf{X}(t) = \mathbf{x}(t)) = f(\mathbf{y}(t) | \mathbf{X}(1) = \mathbf{x}(1)) / f(\mathbf{y}(t-1) | \mathbf{X}(1) = \mathbf{x}(1)),$$

or, equivalently,

$$f(y_t | \mathbf{Y}(t-1) = \mathbf{y}(t-1)) = \sum_{i=1}^{D} f(y_t | (\mathbf{X}^*(t) = \mathbf{x}^*(t)); \rho_i) f(\rho_i | \mathbf{X}^*(t) = \mathbf{x}^*(t)).$$

The equivalence of the last two lines derives from Bayes' theorem applied iteratively to the "prior" mixture $f(\rho_i | \mathbf{X}(t-1) = \mathbf{x}(t-1))$, recognizing the information transfer function from each $Y_{t-1} = y_{t-1}$ to $Y_t$ through the mixture parameter $\rho$ as the $t$-density function value, $f(y_{t-1} | (\mathbf{X}_{t-1}^* = \mathbf{x}_{t-1}^*); \rho_i)$, from the $\rho_i$-contingent autoregression. Computationally, this means that at each iteration of an observation in the time series, you need to compute the $\rho_i$-contingent density function values for all $\rho_i$ values in the grid, multiply them by the corresponding values of the digital mass function $f(\rho_i | \mathbf{X}(t-1) = \mathbf{x}(t-1))$, sum these products, and finally renormalize the products to yield the posterior $f(\rho_i | \mathbf{X}(t) = \mathbf{x}(t))$ as a mass function which sums to 1.

Before presenting a computational example, two comments are in order. First, notice that although the initial mixing function $f(\boldsymbol{\beta}, \pi | \mathbf{X}(1) = \mathbf{x}(1); \rho)$ was specified as Normal-gamma, independent of the value of $\rho$, the posterior mixing function computed via Bayes' theorem is definitely dependent on $\rho$. For the information transfer function from each observation of $Y$ to the next, the $t$-density value $f(y_{t-1}|(\mathbf{X}^*_{t-1}=\mathbf{x}^*_{t-1});\rho_i)$ explicitly depends on $\rho_i$. As information accumulates in the observed times series of observations, the dependence structure among the arguments of the $f(\boldsymbol{\beta},\pi,\rho|\mathbf{Y}(t))$ typically becomes increasingly ornate. We observe this occurring in our example, shortly.

Second, our development of the mixture of moving average autoregressive opinions has been based on the specification of only one period difference in the moving average component of the regressions. It should be readily apparent that this feature can be easily extended to two, three, or more periods of lag, as in $P(Y_t|\mathbf{Y}(t-1);\boldsymbol{\beta},\pi,\boldsymbol{\rho})=\mathbf{x}_t\boldsymbol{\beta}+\sum_{J=1}^{M}(y_{t-1}-\mathbf{x}_{t-j}\boldsymbol{\beta})\rho_j$. Of course the required number of computations increases by a multiplicative factor of $D$ as each additional $\rho_j$ parameter is included into the mixing distribution. In common notation, a mixture-autoregressive distribution is distinguished from a mixture-moving-average-autoregressive distribution by the mnemonic AR($L$) as opposed to ARMA($L$, $M$), where $L$ and $M$ denote the length of the autoregressive and moving average lags, respectively. We are ready for the promised numerical example.

## 8.5 SEQUENTIAL SCORING OF MIXTURE REGRESSION ASSERTIONS REGARDING MONTHLY CURRENCY EXCHANGE RATES

After years of strictly controlling the sale of its currency and even controlling transaction prices by a variety of breadbasket mechanisms, the government of New Zealand abruptly announced the open float of the New Zealand dollar one Friday evening at the beginning of March 1985. A brief history of currency practice and a description of motivation for the float is documented in the *New Zealand Official Statistical Yearbook, 1990.* After an initial period of particular uncertainty regarding the time series of exchange values due to an expected bulk of necessary, awkward, and overdue adjustments, forecasters would have planned to apply current financial theory of exchange rates to express their uncertainties regarding the future course of N.Z.-U.S. exchange rate history. However, financial theory was not exactly univocal in its assessments of these prospects. Differing speculative views among financial economists regarding the efficiency of markets allows the construction of three different types of forecasting distribution regarding the time series, each based on a different form of a mixture regression we have been studying. In this section we discuss the computational results of a sequential scoring analysis of the three sequential forecasting distributions so specified. We outline the three distributions to be examined shortly, once we first specify the transformation of the exchange price data to be studied.

A daily 3 P.M. transaction price series for the New Zealand dollar in terms of the U.S. dollar (compiled by the Reserve Bank of New Zealand) has been abridged to begin with the final controlled price and to continue with the price on the first Wednesday of each month thereafter. The monthly frequency of observations has been chosen for exhibition here to limit the length of the series under consideration for expository purposes and to allow your replication of the statistical computations as an exercise in subjectivist logic and programming. Of course, an analysis of opinions regarding the more frequent daily series would be possible, as well as a less frequent quarterly series that can be linked to concomitant series of economic indicators. For our purposes in presenting a limited computational example here, the monthly time series of the prices themselves will suffice.

The individual prices in the series were transformed to the logarithm of their ratio to the closing controlled transaction price before the dollar was floated. Thus, the very first value in the transformed series is 0, and the series allows positive or negative values thereafter. One reason for this choice of quantity as the transformed price series is to allow comparative analyses of several currency exchange price series together. Although we shall not pursue such analysis just now, the data on exchange rates with the Australian dollar, the Japanese yen, the Deutsche mark, the Swiss franc, and the British pound sterling are available for further comparison and more extensive analysis. This could take the form of self-contained mixture autoregressions for each currency, which mimic the computations presented here; or the analysis could be expanded to a simultaneous vector forecast of all four exchange rates in the context of so-called hierarchical joint distributions, which we mention in the closing section of this chapter, but shall not pursue.

In this section we present the computations of a sequential scoring analysis of three different attitudes regarding the monthly time series of transformed exchange rates, whose components are denoted by $Y_t$. Each of the attitudes involves a mixture distribution based on a different form of the regression theory we have been developing. We denote the distinct mixture distributions by the appellations RW, AR(3), and ARMA(3,1), recognizing that the associated forecast distributions for the sequence of quantities are *mixtures* of these forms. As the first of these, we construct a "random walk" distribution, according to which the conditional distribution for any $Y_t|(\mathbf{Y}_{t-1} = \mathbf{y}_{t-1})$ is a mixture-Normal, $M$-$N(y_{t-1}, \pi)$, with a conditional mixing function over a specified in the gamma form. The second of these is a Normal-gamma mixture AR(3) distribution, precisely of the form studied in Section 8.3; and the third is in the Normal-gamma-digital mixture ARMA(3,1) form whose computations we have detailed in Section 8.4. We need to motivate a specification of the parameters of the initial mixing function for each of these three distributions before presenting the computational results of the forecast scores.

The mixture-*normality* of all three forecasting distributions merits comment. This simplified form of uncertainty has been selected so as to allow direct application of the structures of mixture regressions we have studied. It does have the feature of representing distributions with fatter tails than a normal, a preference of financial statisticians as noted by Timmermann (1995). However, it is admittedly not up to date with the latest work in this area. Recent proposals in statistical financial theory advanced by Montegna and Stanley (1995) would require Monte Carlo integration procedures for their application in the way we are studying here. Thus, we report only on the mixture-Normal, or $t$, forecasting distributions mentioned.

The specification of three periods in the mixture AR and ARMA distributions represents the assertion that the content of a limited history of past prices is particularly informative regarding the value of any following price in the series. Some research lore has suspected that particularly in small markets (the volume of New Zealand currency being small relative to the U.S. dollar), there may substantive information in a price series regarding future movements of the exchange. Moreover, it would be recognized that the float of the currency was motivated by a desire for change in some basic structures of the New Zealand economy. Presumably these changes, too, would move the exchange value in some direction. Assertors of the autoregressions would allow the historical sequence of exchange prices to tell us where it was going. Such speculation is in contrast to the mixture RW distribution, which involves an assertion that virtually all the information contained in the transaction price series is embodied in the most recent price. The only learning about the price series the RW distribution avows concerns the variation in the series, codified by the changing gamma mixture over the precision parameter, $\pi$. In the volatile context of international currency exchange, the three-month lag lengths in the transmission of information asserted by the mixture AR(3) and ARMA(3,1) distributions are rather bold, though uncertain alternatives to the attitude of an RW efficient markets assertion. Of course, the mixture ARMA(3,1) assertion distinguishes itself from the mixture AR(3) by specifying a distributed lag in the time duration over which the information content of the autoregression is transmitted to market transaction price behavior. Financial economists would be interested to observe the sequential scores achieved by these three different distribution assertions.

There is a second relevant and convenient feature of the three-month lag embodied in the information transfer asserted by AR(3) and ARMA(3, 1). During the first three months of open trading of the $NZ, the lumps and kinks of institutional arrangements regarding currency demand that had been built into economic transactions engaged and planned during the pre–float period would have had a chance to work themselves out to some standard of satisfaction. Thus, we begin our sequential scoring analysis of the three attitudes toward the exchange price series starting with the fourth month after the float began, conditioned on the values of the first three-months' closing prices and the end-of-control period trading price. As a result of this choice, the times series $\mathbf{Y}$ vector of quantities is designated to begin with $Y_1$ as the closing price of the fourth month of the free-trading regime. The initial conditioning values of $\mathbf{X}_1 = (y_0, y_{-1}, y_{-2})$ and $Y_{-3} = y_{-3}$, which are required to generate the $\mathbf{X}$ and $\mathbf{X}^*$ matrices for the mixture AR and mixture ARMA distributions, are taken from the third-, second-, and first-month closing prices, and the final closing controlled price, respectively.

### 8.5.1 Initial Mixing Functions

What remains to specify are the initial mixing function designations that begin the analysis. The outlook presumed in the specifications identified below is that all three distribution proponents are fairly uncertain about the course of the new exchange rate history, to the extent that their economic theoretical viewpoints admit such uncertainty. We specify their previsions and their distributions for the exchange price series in such a way that all three of the distributions support the same prevision for the fourth-month's closing price as equal to the conditioning third-month's closing price. But the three different proponents then engage in their learning about the subsequent exchange prices in the series according to their very different parametric mixture distributions. The initial specifications of each mixing function are designed to specify relatively large variances for the exchange price observations, on account of fairly gentle mixing functions that will sequentially become more precise through the information gained as the time series of observations accumulates.

The easiest component of the initial mixing functions to settle on is the first moment vector, $\boldsymbol{\beta}_0$, of the initial mixture over the $\boldsymbol{\beta}$ vector in the parametric AR and ARMA distributions. In both cases these are specified as $\boldsymbol{\beta}_0 = (1, 0, 0)^{\mathrm{T}}$, representing opinions that do allow autoregressive information transfer, but initially locate the autoregressive information precisely and exclusively on the most recent observation. As to the initial covariance matrix for the mixing $f(\boldsymbol{\beta}|\pi)$, we specify this generously as the identity matrix, $\mathbf{I}_3$, to represent very mild concentration of this density relative to the scale of the $\boldsymbol{\beta}$ vector. Since $\mathbb{M}_2^*(\boldsymbol{\beta}|\pi) = (\pi\boldsymbol{\varphi}_0)^{-1}$, this specifies the matrix $\boldsymbol{\varphi}_0$ as $.0002\mathbf{I}_3$. The size of the proportionality factor will be motivated in the context of the specification of $\alpha_0$ and $\gamma_0$. This designation of $f(\boldsymbol{\beta}|\pi)$ specifies a symmetric mixing density over the $\boldsymbol{\beta}$ vectors on spheres about $\boldsymbol{\beta}_0 = (1, 0, 0)^{\mathrm{T}}$. The initial flatness of this density over the central spheres will be ensured by small values of the initial settings of $\alpha_0$ and $\gamma_0$, described later. Such assertions would be appropriate to this setting in which the observation series from the newly floated currency transactions are recognized by the AR and ARMA proponents to be very informative relative to the content of their "prior" knowledge. Relatively speaking, the RW proponent can be recognized to be making rather strong assertions on the basis of economic theoretical conclusions. The extent of RW's uncertainty about the appropriate location of the information transfer through the $\boldsymbol{\beta}$ vector is precisely nil.

A technical feature of these three forecasting distribution structures is that they are nested within one another. The mixture RW distribution has the identical form of the mixture AR(3), specified by a degenerate mixing function on the $\boldsymbol{\beta}$ vector at $(1, 0, 0)^{\mathrm{T}}$. Similarly, the mixture AR(3) is a reduction of the mixture ARMA(3,1) based on a degenerate mixture over the parameter $\rho$. Thus, it is evident from form of these distributions that the variance assertions regarding prices increase in size from the RW to the AR(3) to the ARMA(3,1) mixtures. A scoring of variance assertions as well as prevision assertions will allow us to understand differences in scores that will be assessed for the distributions themselves via logarithmic scores.

Long amounting to an interpretative statistical problem, a quirk of the more standard curve-fitting statistical analysis of such embedded distribution structures is that more ornate structures necessarily specify a smaller residual sum of squared errors to their "fits" than do their degenerate reductions. A pleasant aspect of the sequential scoring of nested distributions which we observe in this example, is that the more ornate forecasting distribution does not necessarily achieve an improvement in its scores relative to its more simplified version.

Initial specification of the gamma mixing function over $\pi$ is designated by $\alpha_0 = 1.2945$ and $\gamma_0 = 5.89(10^{-5})$, the same for each of the three mixture distribution proponents. Recalling the inverse relation between $\sigma^2$ and $\pi$, these values were determined on the basis of presumed assessments of $\mathbb{M}_1(\sigma) = .01$ and $\mathbb{M}_2^*(\sigma) = .0001$, using equations appropriate for the gamma mixing function on $\pi$:

$$\mathbb{M}_1(\sigma^2) = \mathbb{M}_1(\pi^{-1}) = \gamma_0/(\alpha_0 - 1),$$

and

$$\mathbb{M}_1(\sigma) = \mathbb{M}_1(\pi^{-1/2}) = \gamma_0^{1/2}\Gamma(\alpha_0 - 1/2)/\Gamma(\alpha_0).$$

Finally, the implied moment for $\pi$ of $\mathbb{M}_1(\pi^{-1}) = .0002$ is what motivates the initial specification of $\boldsymbol{\varphi}_0$, as $.0002\mathbf{I}_3$. The expected mixing variance for the $\boldsymbol{\beta}$ vector specified as the identity, $\mathbf{I}_3$, must equal $\mathbb{M}_1(\pi\boldsymbol{\varphi}_0)^{-1}$, which is $.0002\mathbf{I}_3$, identifying $\boldsymbol{\varphi}_0$, accordingly.

The implication of this complete mixing function specification for the RW variance assertion for $Y_1$ amounts to $V_{RW}(Y_1) = .0002$.

A final specification of the initial mixing function over $\rho$ is required to fix the mixture ARMA(3,1) distribution. The mild mixing function presumed in the specification is symmetric about 0, amassing weight only on the 19 values of $\rho$ in a discrete grid from $-.9$ through $+.9$ in steps of .1 apart. Mixing function values over this domain for $f(\rho)$ were .035, .04, .045, .05, .05, .55, .06, .065, .065, .07, .065, .065, .06, .055, .05, .05, .045, .04, and .035. Stepping ahead of our report, but for your interest, the corresponding posterior mixing function values $f(\rho \mid \mathbf{Y}_{59} = \mathbf{y}_{59})$ conditioned on the observed 59 monthly exchange price values, stated to three decimals, were .000, .000, .001, .006, .015, .029, .046, .062, .072, .086, .090, .104, .113, .118, .108, .083, .040, .015, .014, which peak at the value of $\rho = .4$.

### 8.5.2 Comparative Scoring Results

In a word, the RW forecasting distribution regularly outperformed both of the autoregressive distributions, in terms of both the quadratic scores of their associated previsions for the exchange rates and in terms of the logarithmic scores of their distributions. To begin to get our bearings on the scale of the differences in their scores, Figures 8.1 and 8.2 present the cumulative scoring results graphically, first for the quadratic scores of the sequential previsions and then for the

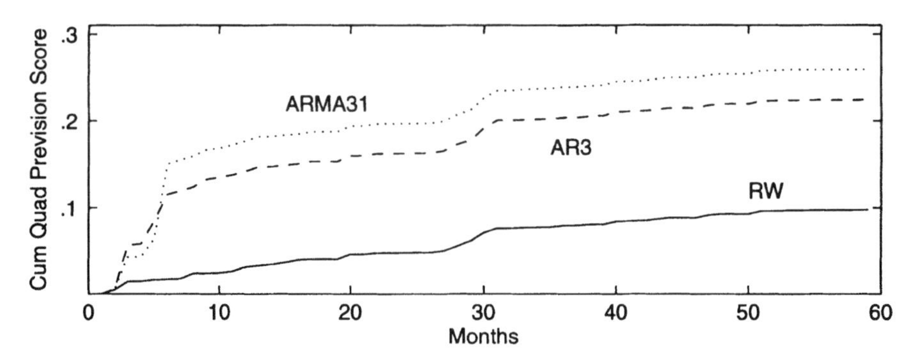


**Figure 8.1.** Cumulating quadratic scores of the sequential previsions for the log exchange rate, computed for the three asserted forecasting distributions.

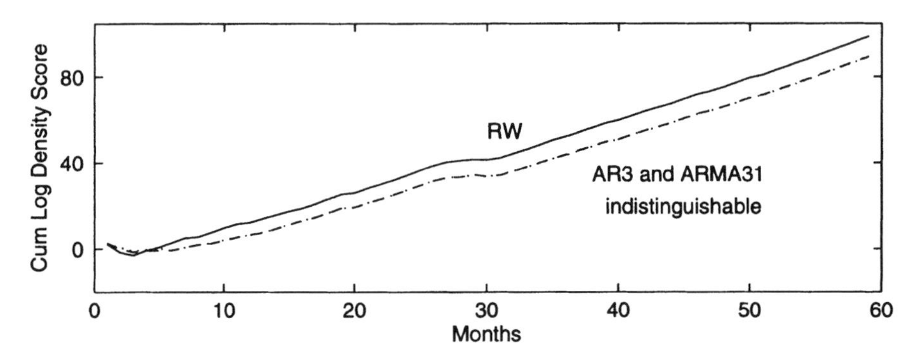


**Figure 8.2.** Cumulating logarithmic scores of the sequential forecast densities for the log exchange rate, computed for the three asserted forecasting distributions. By the sixth month, the cumulating scores for the mixture AR(3) and ARMA(3,1) distributions are indistinguishable to the eye.

logarithmic scores of the distributions. As opposed to our theoretical studies which portrayed the (negative) quadratic score as a gain to be maximized, quadratic scores displayed here are the squared errors themselves, thus representing penalties. The logarithmic density score represents, as always, a desirable gain.

Evaluating the accumulating quadratic scores of the three prevision structures, we ought first notice how poorly the autoregressive mixtures fare in the first six months of scoring. At that point the cumulative scores for RW, AR(3), and ARMA(3, 1), respectively, were .0170, .1154, and .1503. Assertors of poorly scoring autoregressive previsions might well expect such a result. For their mild mixing distributions were meant to convey that they are more uncertain than the RW proponents about the series, and would surely want to learn a bit about the characteristics of the currency market from some observations. Once they had gained some calibrating information regarding the scale of the information transfer, as they see it, they would expect to be faring better. However, if we begin accumulating forecast errors only with the seventh month of the forecast period, the three forecasts still accumulate squared errors over the next 52 months to the extent of .0804, .1086, and .1088, in the same respective order. The squared forecast errors of the autoregressive assertions increase 25% more than those of the random walk forecast.

A second point of interest centers on the relatively poor scores achieved by all three sequential previsions over the period of the months 27 through 32. That this was a period of volatility in the exchange rate is evidenced by the sharply rising blip in the RW score, for this increases directly as the square of the change in the exchange rate. Perhaps most telling for the poor quality of the autoregressive opinions is that the increase in their squared forecast errors is even greater than is the increase in the RW score over this period. For if there is any gain in information from the data series about itself as experience accumulates, it should be most evident and useful in periods when the exchange rate changes a lot, which is explicitly not forecastable on the basis of information in the series according to the RW proponents. Numerically, the squared forecast errors of RW, AR(3), and ARMA(3, 1) increase by .0288, .0383, and .0391 over these six months, suggesting that the information accredited to the series by the autoregressive assertions is bad information!

Figure 8.2 shows that the accumulation of information gains according to the logarithmic score of the three densities occurs more regularly in each case, and that the rate of increase in the RW score above the autoregressive scores is fairly regular at about 10%. Again, the dips in the accumulating scores over the months 27–32, highlight the forecasting difficulties inhering in all three distributions during that particularly volatile period. (Notice that the accumulating log score can decrease, as it does whenever the density function at the observation value is less than 1.)

### 8.5.3 Prevising the Scores of Competing Assertions

Some further understanding of the scoring gains achieved by the RW assertions can be developed by examining a numerical presentation of previsions for ones own score and for "competitors'" scores, as suggested in Section 6.6.4. Comparing the quadratic scores for the RW assertions with the scores for the AR3 assertions, for example, we would do well to examine the pair of column vectors

$$\begin{array}{ll} QS(P_{RW}(Y_t|\mathbf{Y}_{t-1}=\mathbf{y}_{t-1})), & QS(P_{AR}(Y_t|\mathbf{Y}_{t-1}=\mathbf{y}_{t-1})), \\ P_{RW}[QS(P_{RW}(Y_t|\mathbf{Y}_{t-1}=\mathbf{y}_{t-1}))], & P_{AR}[QS(P_{RW}(Y_t|\mathbf{Y}_{t-1}=\mathbf{y}_{t-1}))], \\ P_{RW}[QS(P_{AR}(Y_t|\mathbf{Y}_{t-1}=\mathbf{y}_{t-1}))], & P_{AR}[QS(P_{AR}(Y_t|\mathbf{Y}_{t-1}=\mathbf{y}_{t-1}))]. \end{array}$$

Here QS specifies the quadratic score achieved by the assertion, and the subscripted $P_{RW}$ and $P_{AR}$ symbols designate the prevision associated with the RW and AR(3) assertions respectively.

Without displaying the entire matrix of these vectors over the 59 months, let me comment on its content. Of course everyone's prevision for their own quadratic score must be smaller (when the score is represented as a penalty) than their prevision for anyone else's score. (This is a virtue of the quadratic score being a proper scoring rule for previsions.) Nonetheless, the AR(3) variance (prevised quadratic score) exceeds the RW variance in every month. While not required in general, this is also understandable in the present context that the RW forecast is a partially degenerate member of the family of all mixture AR(3) distributions. What is especially interesting is a comparison of each forecaster's prevision for the score to be achieved by the other forecaster. In the case of this example, it turns out that the AR(3) forecaster always (all 59 months) expects the RW quadratic score to be greater than the RW forecaster expects it to be; in fact, the RW score exceeds the RW forecaster's expectation in only 17 of these months. On the contrary, the RW forecaster expects the AR(3) quadratic score to be greater than AR(3) expects in only six of the months; and in all six of these months the score is greater than AR(3) expects it to be! Of the remaining 53 months in which RW expects the AR(3) quadratic score to be less than AR(3) expects, the score is less than expected by AR(3) during 39 months!

The bottom line of this comparison is that the RW forecaster seems to have a better handle on the size of the AR(3)'s quadratic score than does the proponent of the AR(3) mixture.

The variance assertions, thus, are key to the differences between the mixture RW and the mixture AR distributions for the exchange rate experience. Since variances are themselves previsions, we might evaluate accumulating quadratic scores of the variances. The results are as you might now be expecting. Over the course of the 59 months, the quadratic score of the mixture AR(3) variance accumulates to .0092, whereas the score for the mixture RW variance accumulates only to .0003. Again, as you might imagine from viewing Figure 8.1, the great bulk of the quadratic score for the AR(3) variance is amassed during the first six months, accumulating already to .0088, compared to only .0001 for the RW variance by that point. After that, the rate of score accumulation for the AR(3) variance both subsides and becomes more regular. However, its accumulation of .0004 during the next 53 months doubles that of the RW variance, which is only .0002.

The substantive content of these comparisons may not be too surprising to financial theorists, who find it hard to imagine useful information contained in a past series of exchange prices, on account of the fact that anyone may use such information appropriately to bid the price up or down, if it were there. Nonetheless, the clarity of these results in the relatively small market for New Zealand currency is worth some attention. They tend to discount ideas that small markets are more regularly swayed by speculative ventures. Moreover, they exhibit the swift incorporation of the New Zealand currency into the arena of international exchange, despite the startled uncertainties that may have been generated by the abruptness of the decision to float the currency.

The methodological content of this example, however, is meant to be especially instructive. Comparing differing theoretical attitudes toward any sequence of observable quantities is achieved not by comparing the fits of supposed stochastic generating structures, but by assessing the quality of sequential forecasts specified by the competing understandings as they process information differently in the ways they regard as appropriate. As we have stressed since Chapter 6, the evaluation of the results does not follow some routine checklist, but must be engaged with an eye to understanding the practical importance of the comparisons. The evidence addressed here is not meant to be definitive of a universal methodology for every application. We would all benefit from the sharing of experiences in this type of endeavor.

### 8.5.4 Posterior Mixture Apparatus of Interest

The complete computational details of this example are too extensive to exhibit here, but they can be made available as mentioned in the preface. Two aspects of the posterior mixing distributions are worth exhibiting here, nonetheless, for their exhibition of the intricacies of information transfer within the mixture-autoregressive structures.

First, the location of the posterior mixing function for $\boldsymbol{\beta}$ given $(\mathbf{Y}_{59} = \mathbf{y}_{59})$, which would be denoted according to Theorem 8.2 as $\boldsymbol{\beta}_{59}$, is the column vector $(.9677, -.0046, .0359)^{\mathrm{T}}$. Furthermore, the variance-covariance $(\mathbb{M}_2^*)$ matrix for this posterior mixing function on $\boldsymbol{\beta}$ is

$$\begin{matrix} .0162 & -.0157 & -.0005 \\ -.0157 & .0313 & -.0156 \\ -.0005 & -.0156 & .0164 \end{matrix}$$

which still allows fairly large standard deviations relative to the degeneracy of the corresponding RW mixing function at $\boldsymbol{\beta} = (1, 0, 0)^{\mathrm{T}}$.

Second, we had discussed in the formulation of mixture autoregressions of moving averages how the initial conditional mixing functions $f(\boldsymbol{\beta}|\pi, \rho)$ were specified as independent of $\rho$ with the same location, $(1, 0, 0)^{\mathrm{T}}$. It is of interest to note that this conditional independence is annihilated in the posterior mixtures by data observation, as the locations of $f(\boldsymbol{\beta}|\rho, \mathbf{Y}_{59} = \mathbf{y}_{59})$ for several values of $\rho$ are shown in Table 8.1.

A final comment concerns the limitation of our computation grid for the mixture over $\rho$ to 19 points. Since the resulting computations can be understood as providing a crude approximation to many nearby continuous distributions, computational considerations could appropriately increase the fineness of the grid until further refinement has minimal implications for the computations. In the case of this example, refinement to a 39-point grid has minimal impact.

**Table 8.1. Posterior Locations of Conditional Mixing Functions for $\beta$ under Various Values of $\rho$**

| $\rho$ | −.2 | 0 | .2 | .4 | .6 | .8 |
|---|---|---|---|---|---|---|
| $\beta_{1,59}(\rho)$ | 1.1633 | 0.9677 | 0.7665 | 0.5727 | 0.4137 | 0.2932 |
| $\beta_{2,59}(\rho)$ | −0.2155 | −0.0046 | 0.1390 | 0.2237 | 0.2640 | 0.2529 |
| $\beta_{3,59}(\rho)$ | 0.0513 | 0.0359 | 0.0931 | 0.1991 | 0.3108 | 0.3389 |

## 8.6 HIERARCHICAL FORMS OF MIXTURE REGRESSION ASSERTIONS

The convenience and appropriateness of the mixture-regression format for expressing a wide array of opinion structures regarding an equally wide array of measurement designs assures that we cannot exhaust its study here. Indeed, many complete texts have been published that are devoted exclusively to the analysis of regression statistics. To the extent that you are already familiar with the theory of linear regression as it is motivated by understandings of probability that differ from the subjectivist construction in this text, you will recognize that we would regard many, though surely not all, of these efforts as misdirected. The concluding chapter will expand on this viewpoint.

The focus of the presentation in this chapter has been on the operational subjective logic that motivates the goals and the computations appropriate to mixture-regression analysis. Much more extensive technical details are available in the text of West and Harrison (1989) whose Bayesian outlook is largely compatible with the subjectivist viewpoint followed here. The text of Gelman et al. (1995) contains practical computational examples.

One extension of the regression setup that must be mentioned is the hierarchical form of systems of regressions which has attracted many applications since the original work of Lindley and Smith (1972). It is nicely presented both in the aforementioned text of West and Harrison, and in an actuarial context by Klugman (1992). A historical review of the concept of hierarchical mixtures appears in Good (1983).

The idea behind hierarchically related joint distributions can be expressed easily in the context of the autoregressive information transfer procedures asserted in our study of the $N.Z.-$U.S. currency exchange rate. Two other countries whose trade volume with New Zealand rivals and in fact exceeds that of the United States are Japan and Australia. To be sure, there are obvious differences in the sizes of the financial systems denominated in these national currencies, and there are clear differences in the character of the economic relations each has with New Zealand. Nonetheless, there are also several similarities in the extent to which commercial information is mutually available between them and New Zealand, and in the extent to which trade with New Zealand plays a role in their economies. Transmission of information and of commodities between each of these three and New Zealand would be recognizably different than transmission between France or Chile and New Zealand, to cite two examples.

The implications of recognizing structural similarities in the bilateral relations of the currencies of Japan, Australia, and the United States with New Zealand is that you might be willing to learn something about the structure of information transfer within any one currency exchange series on the basis of evidence within the other two series. Thus, whereas an initial consideration of exchange rate histories to occur in any of these three series might well match the structure of the assertions we have studied in Section 8.5, irrespective of the histories of the other two series, connections among the three series can yet be identified at another level of their joint mixture distribution.

Rather than specifying a mixing distribution over each of the three currency distributions' $\boldsymbol{\beta}$ vectors independently of one another, the logic of exchangeable attitudes can be brought to bear on the three $\boldsymbol{\beta}$ vectors themselves. Conditionally upon a parametric mixture structure for each of the three currencies, the currency series may well be assessed independently of one another. But by linking the joint mixing function over the complete parameterization of opinions regarding all three currency pairs through an exchangeable structure, the time series data from any one currency that sequentially updates the mixture over the $\boldsymbol{\beta}$ vector for that currency concomitantly transforms the mixtures over the other coefficient vectors that are assessed exchangeably with it. Details can be studied in the designated references. A search of the *Current Index to Statistics* for "hierarchical models" will show the many contexts in which these ideas have been applied.

CHAPTER 9

# The Direction of Statistical Research

*Sir*

*A rat, or other small animal, eats of a consecrated wafer.*

*1) Does he ingest the Real Body, or does he not?*

*2) If he does not, what has become of it?*

*3) If he does, what is to be done with him?*

*Yours faithfully,*

*Martin Ignatius MacKenzie*

*(Author of The Chartered Accountant's Saturday Night)*

*Mr Spiro now replied to these questions, that is to say he replied to question one and to question three. He did so at length, quoting from Saint Bonaventura, Peter Lombard, Alexander of Hales, Sanchez, Suarez, Henno, Soto, Diana, Concina and Dens, for he was a man of leisure. ...*

*Personally I would pursue him, said Mr Spiro, if I were sure it was he, with all the rigour of the canon laws. He took his legs off the seat. He put his head out of the window. And pontifical decrees, he cried. A great rush of air drove him back. He was alone, flying through the night.*

*Watt*, SAMUEL BECKETT

## 9.1 INTRODUCTION: SHALL WE NOW TRY TO ARGUE?

We are all uncertain, to some extent about virtually evertything. Scientists do themselves no favor by portraying their uncertainty as something other than it is, as an unobservable stochastic structure that generates history via outcomes of a random process. Our assertions of knowledge about history do not invalidate themselves by the mere fact of the shroud of uncertainty in which we dwell. Statistics is a branch of applied mathematics that identifies the logical restrictions of coherency on the array of our uncertain assertions. In this context, it prescribes the way we can learn from observations, inferentially changing our uncertainty about still other observations yet unmade. Statistics is not a branch of science whose object is a supposed unobservable source of random variation that inheres in nature. Scientists have no method for identifying in-principle-unobservable structures in any field of its interest, any more than do the metaphysicians of religion.

While I have recognized that the content of this text is controversial, I have attempted to present the inferential logic motivated by the subjective foundations of statistics quite frankly, unimpeded by argument. Moreover, I have presented the developments at an introductory, if sophisticated level, without extensive allusion to their deep implications for contemporary programs of statistical research. The student reader has been invited to recognize and admit to your own uncertainty, and to accommodate it in making coherent inferences both on your own and in the scientific community of those who share standards of inquisitiveness, measured investigative behavior, and honesty. I have hoped to develop your understanding of the subjectivist perspective, from which you may judge for yourself the merits of the research programs proposed to you by others in your studies.

Nonetheless, experienced readers of the objectivist statistical persuasion may well have bristled at the unchallenged direction this text has followed, well aware of its implications for your own respected research programs. While not relished, your disconcertion has been intended. May it match the ill-ease felt by subjectvist statisticians who attempt to engage in the professional arena of scientific statistical analysis, still dominated by directed activities of searching for true, unobservable, randomness-generating structures.

When challenged, the proponents of the subjectivist construction cannot solve the myriad well-known methodological problems of statistics that have arisen during the recent half-century of the dominating formalist-objectivist alliance. The reason is that from the subjectivist perspective, many of the proclaimed problems evaporate in their own metaphysical airs. We have no language with which to resolve them, as none is needed. They can be merely dismissed. But this hardly constitutes an alluring basis for productive give and take, and little has been forthcoming.

The popular writings of Thomas Kuhn (1962) provide a way of understanding what is going on here. The operational subjective attitude has not provided the conceptual framework for the development of science as it has occurred during this century. Indeed attempts to report on scientific studies from this perspective have been often suppressed within the professional community by a variety of the protective behaviors Kuhn describes. But the suppression has not been without its attending cancers. Expert statistical modelers have increasingly secluded themselves from and been secluded by "nonstatistical" scientific colleagues who can neither understand nor join in the acclamation of their special powers and access to the mysteries of the unobservable. Science, the empirical activity of attending to nature in all its magnificent forms, is the loser. The reports of empirical studies are increasingly ignored as the output of a black box that can be made to yield whatever you want. It is debatable whether it can. But it is not debatable whether many scientists of many subject areas feel this way.

This chapter will conclude this text first with a commentary on the pseudo-problems that command the attention of many statisticians today, and second with a subjectivist view of meritorious directions for statistical research. Now that randomness is currently ensconced as the domain of statistical attention, it has proliferated in its types, motivating acclaimed distinct "sciences" of variation in econometrics, biometrics, psychometrics, sociometrics, geometrics, geolometrics, and cleometrics to join with statistical physics. My own familiarity with the literature on these "subjects" is the greatest in econometrics, so my commentary on the problems of statistics addresses them as they are construed in that field. However, my practical knowledge of prevalent attitudes in several other fields assures me of a wider relevance of my comments.

To their great credit, economists and econometricians have led the concerns with the foundations of the statistical procedures that have been applied to their subject, though similar debates and concerns have surely been voiced in other fields. From the writings of DeMorgan, Jevons, Edgeworth, Bowley, and Keynes we have mentioned, through the momentous arguments among Tinbergen, Keynes, Friedman, Haavelmo, Koopmans, and Vining nearer this mid-century, well documented by Morgan (1990), discussions of recurring and expanded themes appear in recent works by Hendry (1980, 1993), Leamer (1983, 1985), Swamy, Tinsley, and von zur Muehlen (1985), McAleer, Pagan, and Volker (1985), and Blattenberger (1996) among many others. The subjectivist contributions to the discussion are admittedly relatively recent in making precise suggestions that impinge on practice. Morgan's insightful history of econometric ideas rightfully presents no work of merit that relies on a subjectivist scientific foundation. She notes only (p. 5) that "Edgeworth's peculiar brand of probability did not turn into econometrics." His subjectivist concerns in relating measurements to credibility (Bowley, 1928; Keynes, 1933; Stigler, 1978) have not been the basis for the general understanding and development of statistics during this century. The great body of econometric theory, indeed the very subject matter as commonly conceived, portrays a nonproblem as we see it. Let's talk about it.

## 9.2 WHAT ARE THE PROBLEMS OF STATISTICS?...NOT!

The nonproblems of statistical research stem essentially from the misconstruing of its objective. The most common conclusion of statistical reports is the reminder that "correlation is not causation," and a recommendation of further funding to investigate the elusive evidence on this subject. The subjectivist position, as old as Hume and as modern as Wittgenstein or Russell, that causation is not definable in observable terms, and that science should be weaned of its reliance on the concept, is either dismissed as an amusing oddity or challenged by even more fanciful methaphysical epicycles of probabilistic causality. Philosophical diversions of this sort are described by Suppes (1971, 1984) and Skyrms (1980), whereas their supposed statistical implications are discussed in voluminous publications that can best be searched in the *Current Index to Statistics*. In economic literature the construct is called "Granger causality" after the original work of Granger (1969). If there is a unifying thread to the following assessment of our difficulties, it can be found in the problematic imagination that the goal of statistical analysis is to identify probabilistic causal structures.

Let us begin with a primordial problem, construed as "specification error": either the problem of excluded variables in the model of a process, which induces "omitted variables bias"; or the problem of misspecified functional form, which induces incorrect interpretations of test statistics, and spurious regression relationships. The source of this nonproblem in its many forms is that there *is no* correct stochastic functional form of a system that generates random outcomes. Probabilities, remember, represent the considered uncertain opinions of people who think about observable quantities. Your asserting previsions in whatever form you do, induces via coherency a convex structure on your attitudes toward the many quantities under consideration. The detailed particulars of this structure are representative of you and your opinions, not of the quantities themselves. The quantities are what they are.

A regression structure is merely the codification of a particular group of conditional prevision assertions. It is just as meaningful to assert either, both, or neither of $P(X_1|(X_2=x_2)(X_3=x_3))$ and $P(X_1|(X_2=x_2))$, or $P(X_2|(X_1=x_1))$ for that matter. There is no sense in claiming that these latter assertions involve missing variables. Moreover, to allude to another nonproblem which is often treated quite seriously, it makes no sense to proclaim one of the latter two assertions as the correct regression direction, relegating the other to the status of an incorrect "reverse regression." No one of these assertions is particularly correct or incorrect, well-specified, or spurious. The concern of a statistical consultant is the elicitation and formulation of a data analytic viewpoint and the coherency of the array of regression assertions being made. We can illuminate this recognition in the context of exchangeable judgments, a concept worth introducing into the discussion here on its own merits as well.

The complete characterization of exchangeable and partially exchangeable judgments, via the associated theory of sufficient statistics, has allowed a great advance in our understanding of coherent learning within a community of people who share some feature of symmetry in their judgments. Even when we disagree drastically in our attitudes toward the value of a vector of quantities, $\mathbf{X}$, we know how to arrive coherently at a conjunction in our opinions by observation and experimentation, when we can unearth the commonalities exhibited in our shared symmetries.

As a simple example of the mutual coherence of regressions with supposedly missing regressors, let us consider the regression equation of $X_{N+1}$ on $\mathbf{X}_N$, derived in Secton 7.4.5.4. The context is a sequence of quantities judged exchangeably and assessed with a Normal-gamma mixture distribution. For any size of $N$, the regression equation has the form $P(X_{N+1}|\mathbf{X}_N=\mathbf{x}_N)=(\tau\mu+\sum_{i=1}^N x_i)/(\tau+N)$. Compare the regression coefficient on the quantity $X_1$ in the regression of $X_{N+1}$

on $\mathbf{X}_N$, call it $\beta_1(N) = 1/(\tau + N)$, with the coefficient on the same quantity in the regression of $X_{N+1}$ on $\mathbf{X}_K$, for some value of $K < N$. Call this coefficient $\beta_1(K) = 1/(\tau + K)$. Clearly $\beta_1(N) < \beta_1(K)$. The relevance of this comparison to my comments on omitted variables bias is that, because there is no uniquely correct regression of $X_{N+1}$ on any special variables, neither is $\beta_1(N)$ correct nor is $\beta_1(K)$ correct. Myriad regression structures involving $X_{N+1}$ and $X_1$ can be meaningful simultaneously. Whenever linear, each will specify its own coefficient value on $X_1$. Moaning over "omitted variables bias" when you condition on only some components of $\mathbf{X}_N$ induces no subjectivist tears.

The analysis of this issue yields a similar conclusion when the context is expanded to a sequence of vectors regarded exchangeably, which mimics more directly the situation that bothers the specifiers of "the correct model." Consider our analysis in Section 7.5 concerning the sequence of $K$-dimensional vectors regarded exchangeably. To begin, recognize that precisely the same algebraic formulas would result if only $(K-1)$-dimensional subvectors were assessed. Now consider the predictive distributions for $\mathbf{X}_{N+1}|(\mathbf{X}_{KN} = \mathbf{x}_{KN})$ and for $\mathbf{X}_{N+1}|(\mathbf{X}_{(K-1)N} = \mathbf{x}_{(K-1)N})$. Of course, these two conditional distributions would be different: not only is the first in a multivariate $\mathbf{t}_K$ form, while the second is $\mathbf{t}_{K-1}$, but their precision matrices would be rather different too. Thus, when each of these multivariate **t**-distributions is used to identify the conditional distributions of the first component of $\mathbf{X}_{N+1}$, call it $X_1$ given the remaining $K-1$ components in the first case, and given the remaining $K-2$ in the second case, the coefficient on the shared second component, $X_2 = x_2$, would be different in each regression. This makes good sense. The information content you recognize in the observation of $X_2$ regarding $X_1$ will be different depending on the context of what else you have observed (conditioned on) along with $X_2$. Again there are no tears, but only pleasure induced by this formal result.

As to a supposed specification error in the functional form of a regression, the subjectivist recognizes that you may assert whatever conditional previsions you please, linear or not. Your assertions define the functional form of your regression. It is true that computational limitations have long restricted the tractable forms of assertions to mixture-linear structures on some transformation of measurements. But computational developments of Monte Carlo integration to which we have often alluded have largely eliminated these limitations. Go ahead and say what you want about measurable quantities on the basis of what your scientific theory suggests. The subjective method will allow an evaluation of your opinions by means of appropriate scoring rules. There is no conceivable possibility of your making a specification error in "the functional form," modulo the coherency of all your assertions, as long as you do not confuse yourself. After all, it is your functional form.

Such a simplistic dismissal of a problem for which econometricians have erected such a formidable fortress of tests of functional form merits more extensive comment, but only on that account. Sequential tests of functional form abound, with the goal of ferreting out incorrect models and spurious regressions. As an example of how the simplicity of rejecting the very notion of a correct functional form resolves an objectivist puzzle, let me offer the following reinterpretation of some statistical results published in the exemplary text of Davidson and MacKinnon (1993, pp. 669–672). I shall describe their computations within the objectivist outlook from which they designed them, consider the problem they understand these results to pose, and dismiss it appropriately with a subjectivist interpretation of what their computations portray.

For various sizes of $T = 25$ through 2000, a Monte Carlo generator creates a sequence of 10,000 vectors of pairs, $(\mathbf{x}_T, \mathbf{y}_T)$, according to the equations $x_t = x_{t-1} + u_t$ and $y_t = y_{t-1} + v_t$, where $u_t$ and $v_t$ are i.i.d. $N(0, \sigma^2)$ deviates. (The values of $x_0, y_0$, and $\sigma^2$ used in the generation are not available, but they are irrelevant to our concern.) A linear regression model is then misspecified, in the form $y_t = \alpha + \beta x_t + u_t$, where $u_t$ are presumed i.i.d. $N(0, \sigma^2)$, and the values of $\alpha, \beta$, and $\sigma^2$ are presumed unknown, as objectivists are wont. According to the motivations of this "incorrect model," the least-squares coefficients $\hat{\alpha}$ and $\hat{\beta}$ are computed, along with the residual vector, $\mathbf{e}_T = \mathbf{y}_T - \hat{\alpha}\mathbf{1}_T - \hat{\beta}\mathbf{x}_T$, and an estimate of the error variance, $\hat{\sigma}^2 = (\mathbf{e}_T^T \mathbf{e}_T)/(T-2)$, and the estimated standard error of $\hat{\beta}$ which this implies. For each of the 10,000 such regressions on vectors of various lengths, $T$, it is counted whether a 5% test of the hypothesis that $\beta = 0$ would be rejected in favor of the alternative that $\beta \neq 0$ according to standard procedures. Computationally, that is, the statistic

$$|\hat{\beta}/s(\hat{\beta})| = \frac{|(T\mathbf{x}_T^T\mathbf{y}_T - \mathbf{1}^T\mathbf{x}_T\mathbf{1}^T\mathbf{y}_T)/(T\mathbf{x}_T^T\mathbf{x}_T - (\mathbf{1}^T\mathbf{x}_T)^2)|}{[T\mathbf{e}_T^T\mathbf{e}_T/(T-2)(T\mathbf{x}_T^T\mathbf{x}_T - (\mathbf{1}^T\mathbf{x}_T)^2]^{1/2}}$$

is determined for each of the generated vector pairs, and it is checked whether $|\hat{\beta}/s(\hat{\beta})| > t_{T-2}(.025)$, which would be the appropriate $t$-statistic for this "incorrect test." The results presented are

| $T$ | 25 | 50 | 75 | 100 | 250 | 500 | 750 | 1000 | 2000 |
|---|---|---|---|---|---|---|---|---|---|
| Exceedence proportions | .530 | .662 | .723 | .760 | .847 | .890 | .916 | .928 | .947 |

These results, among others, are presented by Davidson and MacKinnon as surprising. "After all, $x_t$ and $y_t$ are totally independent series, and neither contains a trend. So why do we often—very often indeed for large sample sizes—find evidence of a relationship when we regress $y_t$ on $x_t$?"

Obviously, this question can only make sense in the context of a statistical imagination developed in the preceding 600 pages of the investigators' book, which the subjectivist rejects completely. The simple solution to the problem comes in rejecting as unfounded the assertion that the indiscriminate computation of the statistic $|\hat{\beta}/s(\hat{\beta})|$ and the determination that it exceeds $t_{T-2}(.025)$ provides evidence of "a relationship" between any two series.

Nonetheless, these results do represent an appropriate computation of something. Asserting a complete joint distribution for a vector of pairs, $(\mathbf{X}_T, \mathbf{Y}_T)$, via the assertions $X_t|(\mathbf{X}_{t-1} = \mathbf{x}_{t-1})(\mathbf{Y}_t = \mathbf{y}_t) \sim N(x_{t-1}, \sigma^2)$ and $Y_t|(\mathbf{X}_{t-1} = \mathbf{x}_{t-1})\cdot$

$(\mathbf{Y}_{t-1} = \mathbf{y}_{t-1}) \sim N(y_{t-1}, \sigma^2)$ for $t = 1, 2, \ldots, T$ implies, through coherency, a rather complicated distribution assertion for the functionally related quantity $g(\mathbf{X}_T, \mathbf{Y}_T)$ denoted by $|\hat{\beta}/s(\hat{\beta})|$, but it would be virtually impossible to describe it algebraically. However, the reported Monte Carlo computations identify an approximation for the implied probabilities $P(|\hat{\beta}/s(\hat{\beta})| > t_{T-2}(.025))$ associated with this distribution. These probabilities do not agree with the probabilities of a $t$-distribution. Evidently, it is incoherent to assert independent random walk probabilities for the $(\mathbf{X}_T, \mathbf{Y}_T)$ vector pairs and simultaneously to assert a $t$-distribution for $\hat{\beta}/s(\hat{\beta})$. So what?

Just as we can draw mixed colored balls with replacement from an urn, we can construct sequences of quantities with a computer that virtually everyone, at least everyone with whom we care to speak about these matters, our statistical "community," would assess in the same way. It is a wonderful achievement of computational research that we can use these constructions to determine coherency-implied probabilities for events defined by functions of these sequences. This is a matter quite distinct from scientific argument on any topic regarding sequences of observed quantities defined by measurements. Do you regard the logarithm of the daily \$U.S.–\$N.Z. exchange rate relative to today's as a random walk from now till forever? How about the \$Australia–\$N.Z.? And do you regard this pair of sequences independently? Fair enough, if you do. Moreover, you may regard them this way, even if I do not, and vice versa. Or you and I may agree in our uncertain assessment. But you surely are *not required* to assess them as a random walk. You might well harbor ideas about the future of these three countries that makes you feel quite differently about the course of their future currency exchange rates, perhaps only over a specific time interval. For example, you might express your conditional distribution for the one series given the other via a mixture linear regression!

The goal of statistical analysis cannot be to champion your or my access to the mysteries of stochastic specification, but to provide a data-based procedure to help you or me decide whether we have something to gain by listening more intently to the other talk about nature, in whatever aspect we are studying it. We can achieve this not by estimating the parameters of correctly specified models, but by comparing our sequential forecasts of these series with each other's, and with the series itself. This is a process whereby statistics can be used in the support of science, unadorned by metaphysical delights.

Consider an example of a "spurious relationship" proposed by Hendry (1993) in his characterizing econometric methods that can be distinguished from alchemy. Presenting some statistical results of a regression of the price level on cumulative rainfall in England, he rejects their value as "spurious," by an argument that can only be made more impressive by proclaiming again that the relationship is spurious. I see no problem at all with asserting my regression of the price level on cumulative rainfall with a positive linear coefficient during an era when I believe prices to be rising. I am not ashamed to say that I would use a conditioning rainfall measurement, greater or smaller, to inform myself about how much time has passed, and thus assert a conditional prevision for the paired price level measurement that is accordingly higher or lower. Regressions have nothing to do with "true relationships," but with the uses that can be made of information encoded in various measurements of history.

The problem with the specification searchers is that they rely on fine tuning a statistical method that is designed to estimate accurately an illusion. Consider procedures for estimating the in-principle-unobservable effects of in-principle-unobservable independent variables on an unobservable dependent variable, which can only be measured with unobservable error. Computational procedures are touted as providing "instrumental variables" estimators of such effects. Under the supposition that a dependent variable, $Y_t$, is actually generated by a stochastic structure involving an in-principle-unobservable regressor variable, $X_t$, a search is conducted for an observable instrument variable, $Z_t$, which inheres a specially fortunate stochastic relationship to $X_t$, allowing its use in an estimating formula for the desired coefficient.

If you can believe this type of wild talk, the subjectivist sitting in the stew of life has little to speak about with you. We constrict our scientific analysis to the quantities we can observe, $Y_t$ and $Z_t$, and frankly assert the information content we assess the one to present us about the other. These assertions can be scored on the basis of our experience, to support or to detract from our claims about the information transfer. That the mixture distributions we specify can be more complex than the relatively simple mixture regression structures we have introduced in this text is not at issue. What is at issue is the contribution that statistical theory can make to the assessment of such assertions.

But the list of dismissable contrivences yet proliferates. Another odd categorization of the summary measurements of life that has found favor over the past half-century is the distinction between endogenous and exogenous variables in a stochastic model, based crudely on whether the variable is "determined within the model equations" under consideration or whether it is "determined outside the system." The ostensible purpose of this distinction is to allow the structural identification of parameters in the model from computable statistics. It is embarassingly difficult to know what to say about such talk. Observations are just not determined by models. Full stop.

We could continue, but unfortunately we would too regularly come up with the same short hand, struggling to know what to say.

A final comment in this review concerns the extent to which the subjectivist statistical outlook supports the Bayesian statistical procedures that have reemerged during the past 40 years. While the subjectivist revival has taken place in close relation to this movement, to some extent the Bayesian claims too have a hollow ring. The attitude of many practitioners of Bayesian technique is that, yes indeed, there does exist a true stochastic generating structure that pops out natural history as a random outcome. We are only uncertain about its characterizing parameter values. The recognition, for example, that the second central moment of a posterior mixing function over a parameter structure converges to 0 (as the number of observations increases) is interpreted as an assurance that once the data become extensive enough, we shall surely know the correct value of

$\mu$ or $\sigma^2$, say. It is only a matter of time. As it is now commonly said, without tongue in check, "Bayesian estimators converge to the truth."

Our insistence in this text on the role of mixing functions over parametric structures has shown the very different subjectivist interpretation of some of this shared mathematics, renouncing any claims at all to learning about a nonexistent value of $\mu$ or $\sigma^2$. Whereas an initial mixing function may be rather diffuse, and a posterior mixing function virtually degenerate after many observations, there is nothing that restricts nature to exhibit its next many observations as anything like the many we have just seen. Glorious though our scientific achievements of understanding may be, nature is in no way required to govern itself by restrictions of our making.

A view shared by statistical practitioners of many persuasions is that scientific theories of nature in whatever field should be related more intricately to statistical analysis. Econometric activities must be informed by economic theory, for example. Rather than being led about in our investigations by some required impartial method that resists scientific commitment to a viewpoint, we should openly recognize our scientific viewpoints as informing the statistical procedures we follow in learning from data. When two or more contradictory viewpoints merit favor within the scientific community, they can be scored according to proper scoring rules on the basis of observations of the quantities they previse. Such a procedure can identify whether a specific sequence of observed data is or is not informative enough to distinguish whether any one theory is more insightful than another.

The reconnection of scientific judgment with statistical practice is a program that can be achieved not so much by the development of better statistical tests and estimators but by a reorientation of our statistical understanding toward the subjectivist method. The nonproblems of statistics can be resolved by making a conceptual change to understanding probability as the logic of coherent uncertainty and attending to the analytic consequences of this change in conception. Far from being a completely developed practice, subjectivist procedures are fit subject for creative thinking and new application on many counts. The theory inheres intriguing problems of its own.

## 9.3 WHAT ARE THE PROBLEMS OF STATISTICS?

The operational subjective characterization of uncertain knowledge allows a truly fresh reformulation of the goals and prospects of statistical analysis, fresh at least relative to the statistical outlook that has dominated this century. This is not to say that we need to start all over again. Many promising developments have already been made, stemming largely if not exclusively from the Bayesian statistical movement over the past 40 years. In concluding, let us reflect on some of them.

Undoubtedly a major breakthrough in the scope of feasible computational statistics has recently been achieved with the development of stochastic procedures for computing multidimensional integrals of virtually any form whatsoever. An undiscussed feature of the introduction to statistical inference we have presented via the theory of mixture distributions has been the limited forms of conjugate mixing distribution function we have studied. Natural conjugate mixing functions do allow approximate computations of a range of similar integrals within identifiable error bounds; moreover, as an expository example, they allow a complete understanding of many important features of inference via conditionalization that are relevant quite generally. However, it is fair to say that their development has been motivated largely by their algebraic convenience in deriving exact inferences based on at least some "reference" mixing distribution. The recent development of several procedures of stochastic integration has reduced our reliance upon them.

The so-called Gibbs sampling procedure and others which we have mentioned in references to introductory presentations by Casella and George (1992) and Gelfand and Smith (1992) are quickly developing into a powerful computational tool, with intriguing problems of their own, and great scope for application. An interesting sidelight to their development is their connection with visual rather than algebraical forms of statistical reporting. A more advanced presentation with stimulating discussants appears in the work of Smith and Roberts (1993). A major application with discussion has been presented by Newton and Raftery (1994). The subjectivist outlook developed in our text provides a sometimes different focus for what it is that should be computed, but the technical development of these computing procedures are most fortunate for the expanded applicability of the method.

Most happily, it is more than the family of mixing functions over standard parametric families of distributions that has been enlarged. One of the real problems of public relations in the acceptability of statistical analysis has been a common complaint by reseachers, usually voiced in the form that "my data do not meet the assumptions of any of the statistical models." I understand such concerns to mean that none of the standard parametric families of distributions can represent the considered scientific uncertain opinions about the quantities that have been observed, either by stealth or by design. A major bonus of the new computational procedures is that they can also be applied to information transfer functions (likelihood functions) of any form whatsoever.

Stepping back from the specification of complete distributions for the moment, the generalization of the structure of uncertain assertions provided by the fundamental theorem of prevision, in the form of an asserted partial prevision polytope, has broad scope for application. Rather than having so completely assessed a huge number of important quantities that you assert a complete joint distribution over all their possibilities, you might merely assert an array of interval bounds as your partial prevision. Important developments in the application of mixture regressions deriving from this same spirit of understanding the assessment problem have been made in original works by Leamer (1984), Leamer and Leonard (1983), and others in the same vein, who report sets of posterior distributions cohering with a variety of mixture distributions. Related efforts yielding useful results are catalogued in the *Current Index to Statistics* under the key words of robust Bayesian methods.

The direct application of the fundamental theorem of prevision is currently limited by the huge size of the constraint matrices based on the realm of many quantities. The recent thesis of Myers (1995) makes progress in this area. In this regard we can expect expansion in the feasible domain of application from continuing developments in computing, so work on constructing large-sized applications is already merited. One example pertaining to parole decisions in a justice system is begun in Lad and Coope (1995). Already, applications to large problems can be achieved by reducing our excessive demands on the precision of many of the measurements we make. (I mean here the precision of the measurement, not the precision of your opinions about it.) For example, currently the recording of U.S. gross national product statistics is detailed to the nearest $100 million. With annual measurements totaling in the trillions, that amounts to five significant digits of measurement, more digits than are relevant to anyone's interest in GNP as a measure of product. The use of the fundamental theorem of prevision and the theoretical specification of prevision structures would be facilitated by restricting our attention to measurements in size units that are merited by consequential distinctions in the natural activity under investigation. For many purposes, delimiting the measurement of economic activity by GNP into one of merely three categories would suffice.

The allusion here to gross national product statistics is a sobering reminder of the particular problems and prospects for research in the social sciences. We have discussed in Chapter 2 the concerns that might make one charry of regarding the quantity GNP, operationally defined quantity though it might be, as a useful summary of the productive achievement of an economy. Mathematical wizardry in modeling the sequence of its measurements would do little to resolve the real scientific problems of understanding the world human economy today. Substantive and important statistical research, in which progress is being made, concerns the design of statistical measurements to represent summaries of our experience that are truly relevant to our concerns.

As to partial prevision assertions, there is an interesting theoretical problem that merits attention. Whereas the theory of proper scoring rules is well developed for fully specified previsions and distributions, little success has been achieved in identifying ways to score a partial, interval assertion. With regard to scoring rules per se, the development of experience with the evaluation of comparative scores for conflicting assessments of practical matters is of crucial importance.

There is delightful mathematical wizardry involved in the specification of the operational subjective method of probability and statistics. But such is not the claim to fame in the theory as it is proposed. The central achievement is its highlighting a return of empirical analysis away from technique and toward the fundamental scientific question in any field whatsoever.

What's happening?

# References

Surnames of authors appear in alphabetical order, beginning with the first capitalized letter. Thus, DeGroot and De Morgan are ordered among the D's, whereas de Finetti, von Mises, and van Heijenoort are ordered among the F's, M's, and H's, respectively. The only regularly abbreviated journal titles are *JASA* for the *Journal of the American Statistical Association* and *JRSS* for the *Journal of the Royal Statistical Society*.


Abramovitz, M. and Stegun, I. (eds.) (1964) *Handbook of Mathematical Functions*, Washington, DC: National Bureau of Standards. Also (1965) New York: Dover.

Actes du Congrés International de Philosophie Scientifique (1935) *Actualités Scientifiques et Industrielles*, 8 tms, No. 388–395, Paris: Hermann.

Aitchison, J. (1986) *The Statistical Analysis of Compositional Data*, London: Chapman and Hall.

Aldous, D. J. (1987) Review of "Limit theorems for sums of exchangeable random variables," *JASA*, **82**, 343.

Andersen, E. B. (1970) Sufficiency and exponential families for discrete sample spaces, *JASA*, **65**, 1248–1255.

Ando, A. and Kaufman, G. M. (1965) Bayesian analysis of the independent multinormal process—neither mean nor precision known, *JASA*, **60**, 347–353.

Barnard, G. A. (1958) Thomas Bayes—A biographical note, *Biometrika*, **45**, 295–315. Reprinted in *Studies in the History of Probability and Statistics*, E. S. Pearson and M. G. Kendall (eds.), London: Griffin, 1970.

Bayes, T. (1764) An essay toward solving a problem in the doctrine of chances, *Philosophical Transactions of the Royal Society, 1763*, 370–418. Reproduced in W. E. Deming, *Two Papers by Bayes with Commentaries*, New York: Hafner, 1963.

Beckett, S. (1953) *Watt*, Paris: Olympia. American edition, New York: Grove Press, 1959.

Benacerraf, P. and Putnam, H. (1983) *The Philosophy of Mathematics, Selected Readings*, Cambridge: Cambridge University Press.

Berger, J. O. and Wolpert, R. (1984) *The Likelihood Principle: A Review*, Hayward, CA: Institute of Mathematical Statistics Monograph Series.

Berger, R. L. (1981) A necessary and sufficient condition for reaching consensus using DeGroot's method, *JASA*, **76**, 415–418.

Bernardo, J. M. (1979) Expected information as expected utility, *Annals of Statistics*, **7**, 686–690.

Bernardo, J. M. and Smith, A. F. M. (1994) *Bayes Theory*, Chichester: Wiley.

Birnbaum, A. (1962) On the foundations of statistical inference, with discussion, *JASA*, **57**, 269–306.

Birnbaum, A. (1972) More on concepts of statistical evidence, *JASA*, **67**, 85–861.

Bishop, E. (1967) *Foundations of Constructive Analysis*, New York: McGraw-Hill.

Blackwell, D. (1969) *Basic Statistics*, New York: McGraw-Hill.

Blattenberger, G. (1996) Money demand revisited: an operational subjective approach, *Journal of Applied Econometrics*, **11**, 153–168.

Blattenberger, G. and Lad, F. (1985) Separating the Brier score into calibration and refinement components: a graphical exposition, *The American Statistician*, **39**, 26–32.

Blattenberger, G. and Lad, F. (1986) A subjective Bayesian characterization of aggregate evidence on new views on worker productivity, *Metroeconomica*, **38**, 135–156. Reprinted in *Explorations in Political Economy: Essays in Criticism*, R. K. Kanth and E. K. Hunt (eds.), Savage, MD: Rowman and Littlefield, 1991, 285–313.

Blattenberger, G. and Lad, F. (1988) An application of operational subjective statistical methods to rational expectations, with discussion, *Journal of Business and Economic Statistics*, **6**, 453–477.

Bonferroni, C. E. (1925) Teorie e probabilità, Discorso Inaugurale, annuario 1925–1926, *Regio Istituto Superiore di Scienza Economiche e Commerciali Bari, Bari*: Casa Editrice Cressati.

Boole, G. (1847) *The Mathematical Analysis of Logic*, Cambridge: Macmillan.

Boole, G. (1854a) *An Investigation of the Laws of Thought, on Which Are Founded the Mathematical Theories of Logic and Probability*, Cambridge: Macmillan.

Boole, G. (1854b) A general method in the theory of probabilities, *The Philosophical Magazine*, Series 4, **8**. Reprinted in *Studies in Logic and Probability by George Boole*, R. Rhees (ed.), La Salle, IL: Open Court.

Boole, G. (1862) On the theory of probabilities. *Philosophical Transactions of the Royal Society of London*, **152**. Reprinted in *Studies in Logic and Probability by George Boole*, R. Rhees (ed.), La Salle, IL: Open Court.

Boole, G. (1983) *The Boole–De Morgan Correspondence*, G. C. Smith (ed.), Oxford: Clarendon Press.

Borel, E. (1924) Apropos d'un traité de la probabilité, *Révue Philosophique*, **98**, 321–336. English translation by H. Smokler in Kyburg and Smokler (1964).

Borel, E. (1939) *Traité du Calcul des Probabilités et de Ses Applications: Valeur Pratique et Philosophie des Probabilités*, Paris: Gauthiers–Villars.

Bourbaki, N. (1950) The architecture of mathematics, *American Mathematical Monthly*, **57**, 221–232.

Bowley, A. L. (1928) *F. Y. Edgeworth's Contributions to Mathematical Statistics*, London: Royal Statistical Society.

Bridgman, P. (1927) *The Logic of Modern Physics*, New York: Macmillan. New York: ARNO Press Reprint edition, 1980.

Bridgman, P. (1936) *The Nature of Physical Theory*, Princeton: Princeton University Press. New York: ARNO Press Reprint, 1980.

Brier, G. W. (1950) Verification of forecasts expressed in terms of probability, *Monthly Weather Review*, **78**, 1–3.

Bruno, G. and Gilio, A. (1980) Applicazione del metodo del simplesso al teorema fondamentale per le probabilita nella concezione soggettivistica, *Statistica*, **40(3)**, 337–344.

Buehler, R. (1971) Measuring information and uncertainty, in *Foundations of Statistical Inference*, V. P. Godambe and D. A. Sprott (eds.), Toronto: Holt, Rinehart, and Winston.

Byrne, E. (1968) *Probability and Opinion: A Study in the Medieval Presuppositions of Post medieval Theories of Probability*, The Hague: Martinus Nijhoff.

Carnap, R. (ed.) (1939) *International Encyclopedia of Unified Science*, Vols I–II, Chicago: University of Chicago.

Casella, G. and George, E. (1992) Explaining the Gibbs sampler, *The American Statistician*, **46**, 167–174.

Chow, Y. S. and Teicher, H. (1988) *Probability Theory: Independence, Interchangeability, Martingales*, New York: Springer-Verlag.

Churchman, C. W. and Ratoosh, P. eds. (1959) *Measurement: Theories and Definitions*, New York: Wiley.

Clemens, R. T. and Murphy, A. H. (1986) Objective and subjective precipitation forecasts: improvements via calibration and combination, *Weather and Forecasting*, **1**, 56–65.

Clemens, R. T. and Murphy, A. H. (1987) Calibrating and combining precipitation forecasts, in *Probability and Bayesian Statistics*, R. Viertl (ed.), London: Plenum Press.

Cooke, R. (1991) *Experts in Uncertainty: Opinion and Subjective Probability in Science*, New York: Oxford Press.

Cournot, A. A. (1843) *Exposition de la Théorie des Chances et des Probabilités*, Paris: Librairie Philosophique J. Vrin, 1984.

Couturat, L. (1901) *La Logique de Leibniz, d'apres des Documents Inedits*, Paris: Alcan.

Couturat, L. (1905) *L'Algebre de la Logique*, 2nd ed., 1980, Paris: Blanchard. L. G. Robinson (tr.), *The Algebra of Logic*, London: Open Court, 1914.

Cox, R. T. (1961) *The Algebra of Probable Inference*, Baltimore: Johns Hopkins.

Crépel, P. (1988) Condorcet, la théorie des probabilités et les calculs financiers, *Sciences a l'Epoch de la Révolution Française, Recherches Historiques*, R. Rashed (ed.), Paris: Blanchard.

Crisma, L. (1971) Alcune valutazioni quantitative interessanti la proseguibilità di processi aleatori scambiabili, *Rend. Ist. Mat. Univ. Trieste*, **3**, 96–124.

Crisma, L. (1982) Quantitative analysis of exchangeability in alternative processes, in *Exchangeability in Probability and Statistics*, G. Koch and F. Spizzichino (eds.), Amsterdam: North–Holland, 207–216.

Czuber, E. (1910, 1914) *Wahrscheinlichkeitsrechnung*, 2 volumes, Leipzig: Druck und Verlag von B. G. Teubner.

Daboni, L. (1982) Exchangeability and completely monotone functions, in *Exchangeability in Probability and Statistics*, G. Koch and F. Spizzichino (eds.), Amsterdam: North–Holland, 34–45.

Daboni, L. (1984) On the axiomatic treatment of the utility theory, *Metroeconomica*, **36**, 203–209.

Daboni, L. and Pressacco, F. (1987) Mean-variance, expected utility and run probability in reinsurance decisions: suggestions and comments on the line of de Finetti's seminal work, in *Probability and Bayesian Statistics*, R. Viertl (ed.), New York: Plenum Press, 121–128.

Dale, A. I. (1985) A study of some early investigations of exchangeability, *Historia Mathematica*, **12**, 323–336.

Daly, H. E. and Cobb, J. B. (1989) *For the Common Good: Redirecting the Economy toward Community, the Environment, and a Sustainable Future*, Boston: Beacon Press.

Daston, L. (1988) *Classical Probability in the Enlightenment*, Princeton: Princeton University Press.

Davidson, R. and MacKinnon, J. (1993) *Estimation and Inference in Econometrics*, New York: Oxford University Press.

Dawid, A. P. (1981) Some matrix-variate distribution theory: notational considerations and a Bayesian application, *Biometrika*, **68**, 265–274.

Dawid, A. P. (1982) The well-calibrated Bayesian, *JASA*, **77**, 605–613.

Dawid, A. P. (1985) Calibration-based empirical probability, *Annals of Statistics*, **13**, 1251–1273.

DeGroot, M. (1970) *Optimal Statistical Decisions*, New York: McGraw-Hill.

DeGroot, M. (1974) Reaching a consensus, *JASA*, **69**, 118–121.

DeGroot, M. (1984) Changes in utility as information, *Theory and Decision*, **17**, 287–303. See errata noted in **18** (1985), 319.

DeGroot, M. and Fienberg, S. (1982) Assessing probability assessors: calibration and refinement, in *Statistical Decision Theory and Related Topics III*, Vol. I, S. S. Gupta and J. O. Berger (eds.), New York: Academic Press, 291–314.

DeGroot, M. and Fienberg, S. (1983) The comparison and evaluation of forecasters, *The Statistician*, **32**, 12–22.

Deming, W. E. (1963) *Two Papers by Bayes with Commentaries*, New York: Hafner.

De Morgan, A. (1838) *Essay on Probabilities and their Application to Life Contingencies and Insurance Offices*; reprint, New York: ARNO, 1980.

De Morgan, A. (1847) *Formal Logic or the Calculus of Inference, Necessary and Probable*, A. E. Taylor (ed.), London: Open Court, 1926.

De Morgan, A. (1872) *A Budget of Paradoxes*, 2 volumes, Sonia De Morgan (ed.). Second edition, D. E. Smith (ed.), London: Open Court, 1915.

De Morgan, A. (1966) *On the Syllogism and Other Logical Writings*, Peter Heath (ed.), London: Routledge and Kegan Paul.

Dempster, A. (1967) Upper and lower probabilities induced by a multivalued mapping, *Annals of Mathematical Statistics*, **38**, 325–339.

Descartes, R. (1637) *Discourse on Method*, John Veitch (tr.), La Salle, IL: Open Court, 1962.

Dhrymes, P. (1978) *Introductory Econometrics*, New York: Springer-Verlag.

Diaconis, P. (1977) Finite forms of de Finetti's theorem of exchangeability, *Synthese*, **36**, 271–281.

Diaconis, P. and Freedman, D. (1980a) de Finetti's generalizations of exchangeability, in *Studies in Inductive Logic and Probability*, volume 2, R. Jeffrey (ed.), Berkeley: University of California Press.

Diaconis, P. and Freedman, D. (1980b) de Finetti's theorem for Markov chains, *Annals of Probability*, **8**, 115–130.

Diaconis, P. and Freedman, D. (1984) Partial exchangeability and sufficiency, in *Statistics: Applications and New Directions*, J. K. Ghosh and J. Roy (eds.), Calcutta: Indian Statistical Institute, 205–236.

Diaconis, P. and Freedman, D. (1987) A dozen de Finetti style results in search of a theory, *Annals de l'Institute Henri Poincare*, **23**, 397–423.

Diamond, M. and Stone, M. (1981) Nightingale on Quetelet. I: The passionate statistician. II: The marginalia. III: Essay in memoriam, *JRSSA*, **144**, 66–79, 176–213, 332–351.

Dickey, J. M. (1967) Matricvariate generalizations of the multivariate-$t$ distribution and the inverted multivariate-$t$ distribution, *Annals of Mathematical Statistics*, **38**, 511–518.

Dickey, J. M. (1976) Approximate posterior distributions, *JASA*, **71**, 680–689.

Dickey, J. M. (1983) Multiple hypergeometric functions: probabilistic interpretations and statistical uses, *JASA*, **78**, 628–637.

Dickey, J. M. (1986) A comprehensive generalized mean value based on the Dirichlet distribution, *Quaestiones Mathematicae*, **8**, 343–360.

Dickey, J. M. and Chen, C. H. (1985) Direct subjective-probability modelling using ellipsoidal distributions, in *Bayesian Statistics 2*, J. M. Bernardo, M. H. DeGroot, D. V. Lindley, and A. F. M. Smith (eds.), Amsterdam: Elsevier.

Dickey, J. M., Dawid, P., and Kadane, J. B. (1986) Subjective probability assessment methods for multivariate-$t$ and matrix-$t$ models, in *Bayesian Inference and Decision Techniques*, P. Goel and A. Zellner (eds.), Amsterdam: Elsevier.

Dubins, L. and Savage, L. J. (1965) *Inequalities for Stochastic Processes: How to Gamble if You Must*, New York: McGraw-Hill.

Edwards, A. W. F. (1972) *Likelihood: An Account of the Statistical Concept of Likelihood and Its Application to Scientific Inference*, Cambridge: Cambridge University Press.

Edwards, A. W. F. (1987) *Pascal's Arithmetic Triangle*, New York: Oxford University Press.

Edwards, W., Lindman, H., and Savage, L. J. (1963) Bayesian inference for psychological research, *Psychological Review*, **70**, 193–242.

Eisner, R. (1988) Extended accounts for national income and product, *Journal of Economic Literature*, **23**, 1611–1684.

Epps, T. W. (1993) Characteristic functions and their empirical counterparts: geometrical interpretations and applications to statistical inference, *The American Statistician*, **47**, 33–38.

Estienne, J. E. (1903, 1904) Essai sur l'art de conjecturer, *Revue d'Artillerie*, **61**, 405–409; **62**, 73–117; **64**, 5–39, 65–97.

Eyler, J. M. (1979) *Victorian Social Medicine: The Ideas and Methods of William Farr*, Baltimore: Johns Hopkins University Press.

Fang, J. (1970) *Towards a Philosophy of Modern Mathematics: Bourbaki* (vol. I); *Hilbert* (vol. II), New York: Paideia Press.

Fang, K.-T., Kotz, S., and Ng, K-W. (1990) *Symmetric Multivariate and Related Distributions*, London: Chapman and Hall.

Feller, W. (1958) *An Introduction to Probability Theory and Its Applications*, volume I, New York: Wiley.

de Finetti, B. (1926–1930) *Scritti*, Padova: CEDAM, Editrice Antonio Milani.

de Finetti, B. (1931) Funzioni carattaristica di un fenomeno aleatorio, *Atti della R. Academia Nazionale dei Lincei*, **4**, 86–133.

de Finetti, B. (1936) La logique de la probabilité, Actes du Congrès International de la Philosophie Scientifique, Vol. IV, *Actualités Scientifiques et Industrielles*, #391, 31–39.

de Finetti, B. (1937) La prévision, ses lois logiques, ses sources subjectives, *Annales de L'Institute Henri Poincaré*, **7**, 1–68. H. Kyburg (tr.) Foresight, its logical laws, its subjective sources, in Kyburg and Smokler (1980).

de Finetti, B. (1938a) Probabilists of Cambridge, G. Pelloni (tr.), *The Manchester School* (1985) and *Rivista di Matematica per la Scienza Economiche e Sociali* (1985).

de Finetti, B. (1938b) Sur la condition d'equivalence partielle, *Actualite Scientifique et Industrielles*, #739, Paris: Hermann. English translation in *Studies in Inductive Logic and Probability*, volume II, R. Jeffrey (ed.), P. Benacerraf and R. Jeffrey (trs.), Berkeley: University of California Press.

de Finetti, B. (1939) Compte rendu critique du Colloque de Genève sur la théorie des probabilités, *Actualités Scientifiques et Industrielles*, #766.

de Finetti, B. (1940) Il problem dei pieni, *Giornale Istituto Italiano Attuari*, **11**, 1–88.

de Finetti, B. (1949) On the axiomatization of probability theory, in *Probability, Induction, and Statistics*, B. de Finetti, G. Majone (tr.), New York: Wiley, 1972, 67–113.

de Finetti, B. (1951) Recent suggestions for the reconciliation of theories of probability, *Proceedings of the Second Berkeley Symposium on Mathematical Statistics and Probability*, 217–225.

de Finetti, B. (1952) La notion de 'distribution d'opinions' comme base d'un essai d'interpretation de la statistique, *Publ. Inst. Stat.*, Univ. Paris, I, 2. Translation appears in de Finetti (1993).

de Finetti, B. (1954) Media di decisioni e media di opinioni, *Bulletin of the International Institute of Statistics*, **34**, 144–157.

de Finetti, B. (1961) Commemorazione del prof. C. E. Bonferroni, *Giornale di Matematica Finanziaria*, 1964, 5–24.

de Finetti, B. (1965) Methods for discriminating levels of partial knowledge of a test item, *British Journal of Mathematical and Statistical Psychology*, **18**, 87–123. Reprinted in de Finetti (1972).

de Finetti, B. (1972) *Probability, Induction, and Statistics*, New York: Wiley.

de Finetti, B. (1974a, 1975) *Theory of Probability*, 2 volumes, A. F. M. Smith and A. Machi (trs.), New York: Wiley.

de Finetti, B. (1974b) Utopia: a necessary presupposition for any significant foundation of economics, *Theory and Decision*, **5**, 335–342.

de Finetti, B. (1976) Probability: beware of falsifications!, reprinted in Kyburg and H. Smokler (1980).

de Finetti, B. (1977) Probabilities of Probabilities: a real problem or a misunderstanding?, in *New Developments in the Application of Bayesian methods*, A. Aykaç and C. Brumat (eds.), Amsterdam: North Holland, 1–10.

de Finetti, B. (1993) *Probabilità e Induzione*, original articles and English translations, P. Monari and D. Cocchi (eds.), Biblioteca di Statistica, Bologna: Editrice CLUEB.

Fishburn, P. (1964) Analysis of decisions with incomplete knowledge of probabilities, *Operations Research*, **13(2)**, 217–237.

Fishburn, P. (1985) *Interval Orders and Interval Graphs: A Study of Partially Ordered Sets*, New York: Wiley.

Fisher, R. (1925) *Statistical Methods for Research Workers*, Edinburgh: Oliver and Boyd.

Forrest, D. W. (1974) *Francis Galton: The Life and Work of a Victorian Genius*, New York: Taplinger.

Fountain, J., Macfarlane, A. and McCosker, M. (1995) Framing and incentive effects on risk attitudes when facing uncertain losses, University of Canterbury, Department of Economics research report.

Fountain, J. and McCosker, M. (1995) Are risk attitudes for gains and losses well correlated?, University of Canterbury, Department of Economics research report.

Frank, P. (ed.) (1954) *The Validation of Scientific Theories*, Boston: Beacon.

Franklin, J. (1968) *Matrix Theory*, Englewood Cliffs, NJ: Prentice-Hall.

Fréchet, M. (1937) *Récherche Théoriques Modernes sur le Calcul des Probabilités*: Premier Livre, *Géneralités sur les Probabilités. Variables Aléatoires*; Deuxième Livre, *Méthode des Functions Arbitraire: Théorie des Événement en chaîne dans le Cas d'un Nombre Fini d'États Possibles*, Paris: Gauthiers-Villars.

Fréchet, M. (1940) *Les Probabilités Associées a un Systeme d'Événements Compatibles et Dépendent*. Premier Partie: *Evénements en Nombre Fini Fixe*. Paris: Hermann.

Fréchet, M. (1943) *Les Probabilités Associées a un Systeme d'Événements Compatibles et Dépendent*. Seconde Partie: *Cas Particuliers et Applications*. Paris: Hermann.

Fréchet, M. (1946) Les définitions courantes de la probabilités, *Révue Philosophique*, reprinted in Fréchet (1955).

Fréchet, M. (1947) Rapport sur une enquete internationale relative a l'estimation statistique des parametres, *Proceedings of the International Statistical Conferences*, International Statistical Institute, volume III, 363–422.

Fréchet, M. (1949) Rapport général sur les travaux de la section de calcul des probabilités, *Congrés International de Philosophie des Sciences*, Paris, reprinted in Fréchet (1955).

Fréchet, M. (1955) *Les Mathématiques et le Concret*, Paris: P.U.F.

Frege, G. (1879) *Begriffsschrift, a Formula Language, Modeled upon That of Arithmetic, for Pure Thought*, reprinted in van Heijenoort (1967).

Frege, G. (1884) *The Foundation of Arithmetic, a Logico Mathematical Enquiry into the Concept of Number*, J. L. Austin (tr.), Oxford: Basil Blackwell.

Frege, G. (1893) *The Basic Laws of Arithmetic*, M. Furth (tr. and ed.), Berkeley: University of California.

Frege, G. (1903) *Grundgesetze der Arithmetic*, Jena: H. Pohle.

French, S. (1980) Updating a belief in the light of someone else's opinion, *JRSSA*, **143**, 43–48.

French, S. (1985) Group concensus probability distributions: a survey, *Bayesian Statistics 2*, J. M. Bernardo, et al. (eds.), Amsterdam: North-Holland, 375–390.

Freund, J. E. (1971) *Mathematical Statistics*, Englewood Cliffs, NJ: Prentice-Hall.

Fulks, W. (1979) *Advanced Calculus: An Introduction to Analysis*, 3rd ed., New York: Wiley.

Galton, F. (1886) Regression towards mediocrity in hereditary stature, *Journal of the Anthropological Institute*, **15**, 246–253.

Geisser, S. (1965) Bayesian estimation in multivariate analysis, *Annals of Mathematical Statistics*, **36**, 150–159.

Geisser, S. (1985) On the Prediction of Observables: a Selective Update, in *Bayesian Statistics 2*, J. M. Bernardo, M. H., DeGroot, D. V. Lindley, and A. F. M. Smith (eds.), Amsterdam: North-Holland, 203–229.

Gelfand, J. and Smith, A. F. M. (1992) Bayesian statistics without tears: a sampling—resampling perspective, *The American Statistician*, **46**, 84–88.

Gelman, A., Carlin, J., Stern, H., and Rubin, D. (1995) *Bayesian Data Analysis*, London; Chapman and Hall.

Genest, C., McConway, K., and Schervish, M. (1986) Characterization of externally Bayesian pooling operators, *Annals of Statistics*, **14**, 487–501.

Gigerenzer, G., Swijtink, Z., Porter, T., Daston, L., Beatty, J., and Krüger, L. (1989) *The Empire of Chance: How Probability Changed Science and Everyday Life*, Cambridge: Cambridge University Press.

Gilio, A. and Scozzafava, R. (1988) Le probabilitá condizionate coerenti nei sistemi esperti, *Atti Giornate A. I. R. O*, Pisa, 317–330.

Gilks, W., Richardson, S., and Spiegelhalter, D. (eds.) (1996) *Markov Chain Monte Carlo in Practice*, London: Chapman and Hall.

Gillies, D. (1987) Was Bayes a Bayesian?, *Historia Mathematica*, **14**, 325–346.

Glynn, D. (1987, 1988) Rings of geometries. I, II, *Journal of Combinatorial Theory A*, **44**, 34–48; **49**, 26–66.

Gnedenko, B. V. (1964) *Theory of Probability*, B. D. Seckler (tr.), New York: Chelsea.

Goel, P. and Zellner, A. (eds.) (1986) *Bayesian Inference and Decision Techniques: Essays in Honor of Bruno de Finetti*, Amsterdam: North-Holland.

Goldstein, M. (1981) Revising previsions: a geometric interpretation, *JRSSB*, **43**, 105–130.

Goldstein, M. (1983) The prevision of a prevision, *JASA*, **78**, 817–819.

Goldstein, M. (1985) Temporal Coherence (with discussion), in *Bayesian Statistics 2*, J. M. Bernardo, M. H. DeGroot, D. V. Lindley, and A. F. M. Smith (eds.), Amsterdam: North-Holland, 231–248.

Good, I. J. (1950) *Probability and the Weighing of Evidence*, London: Griffin.

Good, I. J. (1957) *The Appropriate Mathematical Tools for Describing and Measuring Uncertainty*, in *Uncertainty and Business Decisions*, C. F. Carter, G. P. Meredith, and G.L.S. Schackle (eds.) Liverpool: Liverpool University Press, Ch. 3. Reprinted in Good (1983).

Good, I. J. (ed.) (1962) *The Scientist Speculates: An Anthology of Partly-Baked Ideas*, New York: Basic Books.

Good, I. J. (1965) *The Estimation of Probabilities: An Essay on Modern Bayesian Methods*, Cambridge: M. I. T. Press.

Good, I. J. (1971) The probabilistic explication of information, evidence, surprise, causality, explanation, and utility, in *Foundations of Statistical Inference*, V. P. Godambe and D. A. Sprott (eds.), Toronto: Holt, Rinehart, and Winston. Reprinted in Good (1983).

Good, I. J. (1983) *Good Thinking: The Foundations of Probability and Its Applications*, Minneapolis: University of Minnesota Press.

Goodman, I. R., Nguyen, H. T., and Walker, E. A. (1991) *Conditional Inference and Logic for Intelligent Systems*, Amsterdam: North-Holland.

Granger, C. W. J. (1969) Investigating causal relations by econometric models and cross-spectral methods, *Econometrica*, **37**, 424–438.

Haag, J. (1924) Sur un problème général de probabilités et ses diverses applications, *Proceedings of the International Congress of Mathematicians*, Toronto: University of Toronto, 1928, 659–674.

Hacking, I. (1965) *Logic of Statistical Inference*, Cambridge: Cambridge University Press.

Hacking, I. (1975) *The Emergence of Probability: A Philosophical Study of Early Ideas about Probability, Induction, and Statistical Inference*, Cambridge: Cambridge University Press.

Hacking, I. (1990) *The Taming of Chance*, Cambridge: Cambridge University Press.

Hailperin, T. (1965) Best possible inequalities for the probability of a logical function of events, *American Mathematical Monthly*, **72**, 343–359.

Hailperin, T. (1986) *Boole's Logic and Probability: A Critical Exposition from the Standpoint of Contemporary Algebraic Logic and Probability Theory*, 2nd ed., Amsterdam: North-Holland.

Haldane, J. B. S. (1957) The Syadvada system of predication, *Sankhya*, **18**, 195–200.

Hardy, G. H. (1949) *Divergent Series*, Oxford: Clarendon Press.

Hardy, G. H., Littlewood, J. E., and Polya, G. (1934) *Inequalities*, Cambridge: Cambridge University Press.

von Harnack, A. (1885) *History of Dogma*, reprint of first English edition, N. Buchanan (ed.), New York: Dover, 1961.

Heath, D. L. and Sudderth, W. D. (1976) De Finetti's theorem for exchangeable random variables, *The American Statistician*, **30**, 188–189.

van Heijenoort, J. (ed.) (1967) *From Frege to Godel: A Sourcebook of Mathematical Logic, 1879–1931*, Boston: Harvard.

Hempel, C. (1965) *Aspects of Scientific Explanation, and Other Essays*, New York: Free Press.

Hendry, D. (1980) Econometrics: alchemy or science, *Econometrica*, **47**, 397–406.

Hendry, D. (1993) *Econometrics, Alchemy or Science: Essays in Econometric Methodology*, Oxford: Blackwell.

Herman, H. A. (1981) *Improving Cattle by the Millions: NAAB and the Development and Worldwide Application of Artificial Insemination*, Columbia: University of Missouri.

Heyde, C. C. and Seneta, E. (1977) *I. J. Bienayme: Statistical Theory Anticipated*, New York: Springer-Verlag.

Heyting, A. (1956) *Intuitionism: An Introduction*, Amsterdam: North-Holland.

Hilbert, D. and Ackermann, W. (1950) *Mathematical Logic*, L. M. Hammond, G. G. Leckie, and F. Steinhardt (trs.), New York: Chelsea.

Hill, B. M. (1967) Foundations for the theory of least squares, *JRSSB*, **31**, 89–97.

Hill, B. M. (1987) The validity of the likelihood principle, *American Statistician*, **41**, 95–100.

Hill, B. M. (1990) A theory of Bayesian data analysis, in *Bayesian and Likelihood Methods in Statistics and Econometrics*, S. Geisser, J. S. Hodges, S. J. Press, and A. Zellner (eds.), Amsterdam: North-Holland, 49–74.

Hill, B. M. (1994a) On Steinian shrinkage estimators: the finite/infinite problem and formalism in probability and statistics, in *Aspects of Uncertainty: A Tribute to D. V. Lindley*, P. R. Freeman and A. F. M. Smith (eds.), Chichester: Wiley.

Hill, B. M. (1994b) Bayesian forecasting of economic time series, *Econometric Theory*, **10**, 483–513.

Hume, D. (1748) *Enquiry Concerning Human Understanding*, Library of Liberal Arts Edition, New York: Bobbs Merrill, 1955.

Huygens, C. (1657) *Ratiociniis in Aleae Ludo*, see mention in Hacking (1975).

Jeffreys, H. (1939, 1961) *Theory of Probability*, Oxford: Oxford University Press. Third edition, 1961.

Jevons, W. S. (1874) *Principles of Science*, London: Macmillan.

Johnson, N. L. and Kotz, S. (1970) *Continuous Univariate Distributions*, New York: Wiley.

Johnson, N. L. and Kotz, S. (1972) *Continuous Multivariate Distributions*, New York: Wiley.

Johnson, N. L. and Kotz, S. (1977) *Urn Models and Their Applications: An Approach to Modern Discrete Probability Theory*, New York: John Wiley.

Johnson, N. L. and Kotz, S. (1982–89) *Encyclopedia of the Statistical Sciences*, 10 volumes, New York: Wiley.

Johnson, N. L., Kotz, S., and Kemp, A. (1992) *Univariate Discrete Distributions*, New York: Wiley.

Johnson, W. E. (1924) *Logic: Part III, The Logical Foundations of Science*, Cambridge: Cambridge University Press, Appendix I, paragraph 6.2, 183–184.

Jung, K. (1949) Foreword, *The I Ching, or Book of Changes*, R. Wilhelm, and C. Baynes, (trs.), Princeton: Princeton University Press.

Kadane, J. B., Dickey, J. M., Winkler, R. L., Smith, W. S., and Peters, S. C. (1980) Interactive elicitation of opinion for a normal linear model, *JASA*, **75**, 845–854.

Kadane, J. B. and Winkler, R. L. (1988) Separating probability elicitation from utilities, *JASA*, **88**, 354–363.

Kahneman, D., Slovic, P., and Tversky, A. (eds.) (1982) *Judgment Under Uncertainty: Heuristics and Biases*, Cambridge: Cambridge University Press.

Kamlah, A. (1987) The decline of the Laplacian theory of probability: a study of Stumpf, von Kries, and Meinong, in *The Probabilistic Revolution*, vol. I, L. Krüger, L. Daston, and M. Heidelberger (eds.), Cambridge: M. I. T. Press, 91–116.

Karlin, S. and Studden, W. J. (1966) *Tchebycheff Systems: With Applications in Analysis and Statistics*, New York: Wiley.

Keynes, J. M. (1921) *Treatise on Probability*, London: Macmillan.

Keynes, J. M. (1933) *Essays in Biography*, London: Macmillan.

Khinchin, A. I. (1932) Sur les classes d'événements équivalents, *Mathematečeskii Sbornik*, **39**, 40–43.

Khinchin, A. I. (1953) The entropy concept in information theory, in *Mathematical Foundations of Information Theory*, R. A. Silverman and M. D. Friedman (trs.), New York: Dover, 1957.

Klugman, S. A. (1992) *Bayesian Statistics in Actuarial Science*, Boston: Kluwer.

Knobloch, E. (1987) Emile Borel as a probabilist, in *The Probabilistic Revolution*, volume I, L. Krüger, L. Daston, and M. Heidelberger (eds.), Cambridge: M. I. T. Press, 215–236.

Kolmogorov, A. N. (1930) Sur la notion de la moyenne, *Atti della Reale Accademia Nazionale dei Linceii*, **12**, 388–391.

Kolmogorov, A. N. (1933) *The Foundations of Probability*, N. Morrison (tr.), New York: Chelsea.

Kolmogorov, A. N. and Fomin, S. V. (1970) *Introductory Real Analysis*, New York: Dover, 333–340.

Koopman, B. O. (1940) The bases of probability, in *Bulletin of the American Mathematical Society*, **46**. Reprinted in Kyburg and Smokler (1964, 1980).

von Kries, J. (1886) *Die Principien der Wahrscheinlichkeitsrechnung: eine logische untersuchung, 1927*, Tübingen: Verlag von J. C. B. Mohr.

Krüger, L., Daston, L., and Heidelberger, M. (eds.), (1987) *The Probabilistic Revolution*, 2 volumes, Cambridge, Massachusetts: M. I. T. Press.

Kuhn, T. (1962) *The Structure of Scientific Revolutions*, Chicago: University of Chicago.

Kullback, S. (1959) *Information Theory and Statistics*, New York: Wiley.

Kyburg, H. (1974) *Logical Foundations of Statistical Inference*, Boston: Reidell.

Kyburg, H. (1980) Conditionalization, *Journal of Philosophy*, **77**, 98–114.

Kyburg, H. (1983) *Epistemology and Inference*, Minneapolis: University of Minnesota Press.

Kyburg, H. (1984) *Theory and Measurement*, Cambridge: Cambridge University Press.

Kyburg, H. and Smokler, H. (eds.) (1964) *Studies in Subjective Probability*, New York: Wiley; 2nd ed., New York: Krieger, 1980.

Lad, F. (1983a) A subjectivist view of the rational expectations hypothesis: critique and development, *Metroeconomica*, **35**, 29–51.

Lad, F. (1983b) Probability theory: a Marxist discussion, *Science and Society*, **47**, 285–299.

Lad, F. (1984) The calibration question, *British Journal for the Philosophy of Science*, **35**, 213–221.

Lad, F. (1985) Empirical assessment of the efficient markets hypothesis: an operational subjective analysis of the variance bounds approach, *Special Studies Papers #192*, Washington, DC: Federal Reserve Board of Governors.

Lad, F. (1995) Coherent prevision as a linear functional without an underlying measure space: the purely arithmetic structure of conditional quantities, *Mathematical Models for Handling Partial Knowledge in Artificial Intelligence*, G. Coletti, D. Dubois, and R. Scozzafava (eds.), New York: Plenum Press, 101–112.

Lad, F. and Brabyn, M. (1993) Synchronicity of whale strandings with phases of the moon, in *Case Studies in Bayesian Statistics*, C. Gatsonis et al. (eds.), New York: Springer-Verlag, 362–376.

Lad, F. and Coope, I. (1995) Problems and prospects for applying the fundamental theorem of prevision as an expert system: an example of learning about parole decisions, in *Mathematical Models for Handling Partial Knowledge in Artificial Intelligence*, G. Coletti, D. Dubois, and R. Scozzafava (eds.), New York: Plenum Press, 83–100.

Lad, F. and Deely, J. (1994) A subjective utilitarian view of experimental design, in *Aspects of Uncertainty: A Tribute to D. V. Lindley*, A. F. M. Smith and P. R. Freeman (eds.), Chichester: Wiley, 267–281.

Lad, F., Deely, J., and Piesse, A. (1995) Coherency conditions for finite exchangeable inference, *Journal of the Italian Statistical Society*, **4**, 195–213.

Lad, F. and Dickey, J. M. (1990) A general theory of conditional prevision, $P(X \mid Y)$, and the problem of state-dependent preferences, in *Economic Decision-Making: Games, Econometrics and Optimisation, Essays in Honor of Jaques Dreze*, J. J. Gabsewicz, J. F. Richard, and L. A. Wolsey (eds.), Amsterdam: North-Holland, 369–383.

Lad, F., Dickey, J. M., and Rahman, M. A. (1990) The fundamental theorem of prevision, *Statistica*, **50(1)**, 19–39.

Lad, F., Dickey, J. M., and Rahman, M. A. (1992) Numerical application of the fundamental theorem of prevision, *Journal of Statistical Computation and Simulation*, **40**, 131–151.

Lane, D. A. (1980) Fisher, Jeffreys, and the nature of probability, in *R. A. Fisher: An Appreciation*, S. E. Fienberg and D. V. Hinkley (eds.), New York: Springer-Verlag, 148–160.

Lauritzen, S. L. (1988) *Extremal Families and Systems of Sufficient Statistics*, New York: Springer-Verlag.

Leamer, E. (1978) *Specification Searches: Ad Hoc Inference with Non-experimental Data*, New York: Wiley.

Leamer, E. (1983) Let's take the con out of econometrics, *American Economic Review*, **73**, 31–43.

Leamer, E. (1984) Global sensitivity results for generalized least squares estimates, *JASA*, **79**, 867–870.

Leamer, E. (1985) Sensitivity analyses will help, *American Economic Review*, **75**, 308–313.

Leamer, E. (1986) Bid-ask spreads for subjective probabilities, in Goel and Zellner (1986).

Leamer, E. and Leonard, H. (1983) Reporting the fragility of regression estimates, *Review of Economics and Statistics*, **65**, 306–317.

Lee, P. M. (1989) *Bayesian Statistics: An Introduction*, New York: Oxford.

Lehman, E. (1959) *Testing Statistical Hypotheses*, New York: Wiley.

Leibniz, G. W. (1704, 1714) On the logic of probability, in *Leibniz Selections*, P. Weiner (ed.), New York: Scribners, 1951, 82–89.

LePlay, F. (1982) *On Family, Work, and Social Change*, C. B. Silver (tr.), Chicago: University of Chicago.

Levi, I. (1978) Confirmational conditionalization, *Journal of Philosophy*, **75**, 730–737.

Levi, I. (1980) *The Enterprise of Knowledge: An Essay on Knowledge, Credal Probability, and Chance*, Cambridge: M. I. T. Press.

Levi, I. (1981) Direct inference and confirmational conditionalization, *Philosophy of Science*, **48**, 531–552.

Liagre, J. B. J. (1852, 1876) *Calcul des Probabilites et Theorie des Erreurs, avec des Applications aux Sciences d'Observation en Général, et a la Geodesie en Particulier*, Brussels: Alexandre Jamar.

Lichtenstein, S., Fischhoff, B., and Phillips, L. (1982) Calibration of probabilities: The state of the art to 1980, in *Judgment Under Uncertainty: Heuristics and Biases*, D.

Kahneman, P. Slovic and A. Tversky (eds.), Cambridge: Cambridge University Press, 306–334.

Lindley, D. V. (1953) Statistical inference, with discussion, *JRSSB*, **15**, 30–76.

Lindley, D. V. (1965) *Introduction to Probability and Statistics from a Bayesian Viewpoint*, 2 volumes, Cambridge: University Press.

Lindley, D. V. (1982) The improving of probability judgments, *JRSSA*, **145**, 117–126.

Lindley, D. V. (1990) The 1988 Wald lectures: The present position in Bayesian statistics, with discussion, *Statistical Science*, **5**, 44–89.

Lindley, D. V. and Smith, A. F. M. (1972) Bayes estimates for the linear model, with discussion, *JRSSB*, **34**, 1–41.

Lukasievicz, J. (1929) *Elements of Mathematical Logic*, O. Wojtasiewicz (tr.), New York: Pergamon Press, 1963.

Machina, M. and Schmeidler, D. (1992) A more robust definition of subjective probability, *Econometrica*, **60**, 745–780.

Magdoff, H. (1939) The purpose and method of measuring productivity, *JASA*, **34**, 309–318.

Mahalanobis, P. C. (1957) The foundations of statistics, *Sankhya*, **18**, 183–194.

Maistrov, L. E. (1974) *Probability Theory: A Historical Sketch*, S. Kotz (tr.), New York: Academic Press.

Markovitz, H. (1952) Portfolio selection, *Journal of Finance*, **6**, 77–91.

Marshall, A. W. and Olkin, I. (1979) *Inequalities: Theory of Majorization and Its Applications*, New York: Academic Press.

Martin, J. J. (1975) *Bayesian Decision Problems and Markov Chains*, New York: Krieger.

Martz, J. F. and Waller, R. A. (1982) *Bayesian Reliability Analysis*, New York: Wiley.

Matheson, J. and Winkler, R. (1976) Scoring rules for continuous probability distributions, *Management Science*, **22**, 1087–1096.

McAleer, A., Pagan, A., and Volker, P. A. (1985) What will take the con out of econometrics?, *American Economic Review*, **75**, 293–307.

McConway, K. (1978) The combination of experts' opinions in probability assessment: some theoretical considerations, Ph.D. thesis, University College, London.

McConway, K. (1981) Marginalization and linear opinion pools, *JASA*, **76**, 410– 414.

Mill, J. S. (1843) *A System of Logic, Ratiocinative and Inductive, Being a Connected View of the Principles of Evidence and the Methods of Scientific Investigation*, 1893 People's Edition, London: Longmans, Green.

von Mises, R. (1928) *Probability, Statistics and Truth*, H. Geiringer (tr.); 2nd ed., London: Allen and Unwin, 1957.

Montegna, R. N. and Stanley, H. E. (1995) Scaling behaviour in the dynamics of an economic index, *Nature*, **376**, 46–49.

Morgan, M. (1990) *The History of Econometric Ideas*, Cambridge: Cambridge University Press.

Mosteller, F. and Wallace, D. (1964) *Inference and Disputed Authorship: The Federalist*, Reading, MA.: Addison-Wesley.

Moursund, D. and Duris, C. (1967) *Elementary Theory and Applications of Numerical Analysis*, New York: McGraw-Hill.

Muliere, P. and Parmigiani, G. (1993) Utility and means in the 1930's, *Statistical Science*, **8**, 421–432.

Myers, T. (1995) Reasoning with incomplete probabilistic knowledge: the RIP algorithm for de Finetti's fundamental theorem of probability, Ph.D. thesis, M. I. T., Department of Brain and Cognative Sciences.

Nagumo, M. (1930) Über eine klasse von mittelwerte, *Japanese Journal of Mathematics*, **7**, 71–79.

Nalimov, V. V. (1981a) *The Faces of Science*, Philadelphia: ISI Press.

Nalimov, V. V. (1981b) *In the Labyrinths of Language: A Mathematician's Journey*, Philadelphia: ISI Press.

Nalimov, V. V. (1982) *Realms of the Unconscious: The Enchanted Frontier*, Philadelphia: ISI Press.

Nau, R. F. (1992) Indeterminate probabilities on finite sets, *Annals of Statistics*, **20**, 1737–1767.

Needham, J. (1954, 1956, ff.) *Science and Civilization in China*, many volumes, Cambridge: Cambridge University Press.

Neveu, J. (1970) *Bases Mathématique du Calcul des Probabilités*, Paris: Masson.

Newton, M. A. and Raftery, A. E. (1994) Approximate Bayesian inference with the weighted likelihood bootstrap, with discussion, *JRSSB*, **56**, 3–48.

Nilsson, N. J. (1986) Probabilistic logic, *Artificial Intelligence*, **28**, 71–87.

Oakes, D. (1985) Self-calibrating priors do not exist, *JASA*, **80**, 339.

O'Brien, D. P. and Presley, J. R. (eds.) (1981) *Pioneers of Modern Economics in Britain*, Totowa, NJ: Barnes and Noble; and London: Macmillan.

O'Hagan, A. (1988) *Probability: Methods and Measurement*, London: Chapman and Hall.

Pagels, E. (1979) *The Gnostic Gospels*, New York: Random House.

Pearson, K. (1978) *The History of Statistics in 17th and 18th Centuries Against the Changing Background of Intellectual, Scientific and Religious Thought*, E. Pearson (ed.), London: Macmillan.

Pearson, K. (ed.) (1979) *Tables of the Incomplete Beta Function*, 3rd ed., Cambridge: Cambridge University Press.

von Plato, J. (1994) *Creating Modern Probability: Its Mathematics, Physics and Philosophy in Historical Perspective*, Cambridge: Cambridge University Press.

Polanyi, M. (1958) *Personal Knowledge: Toward a Post Critical Philosophy*, London: Routledge and Kegan Paul.

Polanyi, M. (1966) *The Tacit Dimension*, New York: Doubleday.

Polanyi, M. (1974) Scientific thought and social reality, *Psychological Issues*, volume 8, Monograph 32, New York: International University Press.

Polya, G. (1987) *The Polya Picture Album: Encounters of a Mathematician*, G. L. Anderson (ed.), Boston: Birkhauser.

Polya, G. and Szego, G. (1972, 1976) *Problems and Theorems in Analysis*, 2 volumes, D. Aeppli (tr.) of 1925 German first edition, New York: Springer-Verlag.

Popper, K. (1959) *The Logic of Scientific Discovery*, author's translation of German edition of 1934, London: Hutchinson.

Popper, K. (1963) *Conjectures and Refutations: The Growth of Scientific Knowledge*, New York: Basic Books.

Popper, K. (1983) *Realism and the Aims of Science*, London: Hutchinson.

Porter, T. (1986) *The Rise of Statistical Thinking, 1820–1900*, Princeton: Princeton University Press.

Pratt, J. (1962) Must subjective probabilities be realized as relative frequencies?, unpublished seminar paper, Harvard University Graduate School of Business.

Quetelet, A. (1835) *Sur l'homme et le Développement de ses Facultés, ou Essai de Physique Sociale*, Paris: Bachelier.

Quetelet, A. (1869) Physique Sociales, ou Essai sur le Développement des Facultés de l'homme, 2 volumes, Brussels: Muquardt.

Raiffa, H. (1968) *Decision Analysis: Introductory Lectures on Choices Under Uncertainty*, Reading, MA: Addison–Wesley.

Raiffa, H. and Schlaifer, R. (1961) *Applied Statistical Decision Theory*, Cambridge: M. I. T. Press.

Ramsey, F. (1926) *Truth and Probability*, reprinted in Ramsey (1931) and Kyburg and Smokler (1964, 1980).

Ramsey, F. (1931) *The Foundations of Mathematics and Other Logical Essays*, Paterson, NJ: Littlefield-Adams.

Rao, C. R. (1973) *Linear Statistical Inference and Its Applications*, New York: Wiley.

Rappaport, A. (1954) *Operational Philosophy: Integrating Knowledge and Action*, New York: Harper.

Regazzini, E. (1987) Probability theory in Italy between the two wars: a brief historical review, *Metron*, **45**, 5–42.

Reichenbach, H. (1934) *The Theory of Probability*, 2nd English ed. Berkeley: University of California Press, 1971.

Renyi, A. (1961) On measures of entropy and information, *Proceedings of the Fourth Berkeley Symposium on Mathematical Statistics and Probability*, volume I, Berkeley: University of California Press, 498–512.

Renyi, A. (1984) *A Diary on Information Theory*, New York: Wiley.

Resnik, M. (1980) *Frege and the Philosophy of Mathematics*, Ithaca: Cornell University Press.

Sanders, F. (1963) On subjective probability forecasting, *Journal of Applied Meteorology*, **2**, 191–201.

Savage, L. J. (1954) *The Foundations of Statistics*, New York: Wiley; 2nd ed., New York: Dover Books, 1972.

Savage, L. J. (1961) *The Foundations of Statistics*, London: Methuen.

Savage, L. J. (1971) Elicitation of personal probabilities and expectations, *JASA*, **66**, 78–801.

Savage, L. J. (1976) On rereading R. A. Fisher, *Annals of Mathematical Statistics*, **4**, 441–500.

Savage, L. J. (1977) "The Shifting Foundations of Statistics," see Savage (1981).

Savage, L. J. (1981) *The Writings of Leonard Jimmie Savage—A Memorial Selection*, W. A. Ericson (ed.), Washington, DC: American Statistical Association.

Schervish, M. (1985) Comments on "Self-calibrating priors do not exist," *JASA*, **80**, 341–342.

Schneider, I. (1987) Laplace and thereafter: the status of probability calculus in the nineteenth century, in *The Probabilistic Revolution*, volume II, L. Krüger, L. Daston, and M. Heidelberger (eds.), Cambridge: M. I. T. Press, 191–214.

Scott, D. (1964) Measurement structures and linear inequalities, *Journal of Mathematical Psychology*, **1**, 233–247.

Scozzafava, R. (1984) A survey of some common misunderstandings concerning the role and meaning of finitely additive probabilities in statistical inference, *Statistica*, **44**, 21–45.

Scozzafava, R. (1991) A classical analogue of the two-slit model of quantum probability, *Pure Mathematics and Applications Series C*, **2**, 223–235.

Seber, G. (1973) *The Estimation of Animal Abundance*, London: Griffin.

Shafer, G. (1976) *A Mathematical Theory of Evidence*, Princeton: Princeton University.

Shafer, G. (1982) Lindley's paradox, with discussion, *JASA*, **77**, 325–351.

Shannon, C. E. (1948) A mathematical theory of communication, *Bell System Technical Journal*, **27**, 379–423, 623–653; reprint, *The Mathematical Theory of Communication*, C. E. Shannon and W. Weaver (eds.), Urbana: University of Illinois Press, 1949.

Shuford, E., Albert, A., and Massengill, H. (1966) Admissible probability measurement procedures, *Psychometrika*, **31**, 125–145.

Sigwart, C. (1895) *Logic*, 2 volumes, H. Dundy (tr.), New York: Macmillan.

Skyrms, B. (1980) *Causal Necessity*, New Haven: Yale University Press.

Smith, A. F. M. (1981) On random sequences with centred spherical symmetry, *JRSSB*, **43**, 208–209.

Smith, A. F. M. and Roberts, G. O. (1993) Bayesian computation via the Gibbs sampler and related Markov chain Monte Carlo methods, with discussion, *JRSSB*, **55**, 3–23, 53–102.

Smith, C. A. B. (1961) Consistency in statistical inference and decision, *JRSSB*, **23**, 1–25.

Stigler, S. (1978) Francis Ysidro Edgeworth, statistician, *JRSSA*, **141**, 287–322.

Stigler, S. (1982) Thomas Bayes' Bayesian inference, *JRSSA*, **145**, 250–258.

Stigler, S. (1983) Who discovered Bayes' theorem?, *The American Statistician*, **37**, 290–296.

Stigler, S. (1986) *The History of Statistics*, Cambridge, MA: Belknap Press.

Stone, M. (1961) The opinion pool, *Annals of Mathematical Statistics*, **32**, 1339–1342.

Suppes, P. (1971) *A Probabilistic Theory of Causality*, Amsterdam: North-Holland.

Suppes, P. (1974) The measurement of belief, *JRSSB*, **36**, 160–175.

Suppes, P. (1981) *Logique du Probable*, Paris: Flammarion.

Suppes, P. (1984) *Probabilistic Metaphysics*, Oxford: Basil Blackwell.

Swamy, P. A. V. B., Tinsley, P. and von zur Muehlen, P. (1985) The foundations of econometrics: are there any? *Econometric Reviews*, **4**, 1–119.

Tanner, M. (1993) *Tools for Statistical Inference: Methods for the Exploration of Posterior Distributions and Likelihood Functions*, New York: Springer-Verlag.

Tavanec, P. V. (ed.), (1970) *Problems of the Logic of Scientific Knowledge*, T. J. Blakeley (tr.), New York: Humanities Press.

Tavernier, J. B. (1676) *Travels in India*, V. Ball (tr.), New Delhi: Oriental Books, 1977.

Taylor, R. L., Daffer, P. Z., and Patterson, R. F. (1985) *Limit Theorems for Sums of Exchangeable Random Variables*, Totowa, NJ: Rowman and Allanheld.

Teicher, H. (1988) Review of "Limit theorems for sums of exchangeable random variables," *Bulletin of the American Mathematical Society*, **18**, 80–82.

Thapar, R. (1961) *Asoka and the Decline of the Mauryas*, Oxford: Oxford University Press.

Thapar, R. (1966) *A History of India*, volume I, Harmondsworth: Penguin.

Timmermann, A. (1995) Scales and stock markets, *Nature*, **376**, 18–19.

Todhunter, I. (1865) *History of the Theory of Probability*, reprint edition, New York: Chelsea, 1965.

Tukey, J. (1977) *Exploratory Data Analysis*, Reading, MA: Addison-Wesley.

Van Deuren, P. (1934) *Leçons sur le Calcul des Probabilités*, Paris: Gauthiers-Villar.

Varian, H. (1975) A Bayesian approach to real estate assessment, in *Studies in Bayesian Econometrics and Statistics in Honor of Leonard J. Savage*, S. Fienberg and A. Zellner (eds.), Amsterdam: North-Holland, 195–208.

Venn, J. (1866) *The Logic of Chance*, 4th ed., New York: Chelsea, 1962.

Viertl, R. (1987) *Probability and Bayesian Statistics*, R. Viertl (ed.), New York: Plenum Press.

Ville, J. (1939) *Etude Critique de la Notion de Collectif*, Paris: Gauthiers-Villars.

Villegas, C. (1964) On qualitative probability $\sigma$-algebras, *Annals of Mathematical Statistics*, **35**, 1787–1796.

Walley, P. (1991) *Statistical Reasoning with Imprecise Probabilities*, London: Chapman and Hall.

Walley, P. and Fine, T. (1982) Toward a frequentist theory of upper and lower probabilities, *Annals of Statistics*, **10**, 741–761.

Walsh, G. R. (1985) *An Introduction to Linear Programming*, New York: Wiley.

Waring, M. (1988) *Counting for Nothing: What Men Value and What Women Are Worth*, Wellington: Allen and Unwin.

Waring, M. (1989) *If Women Counted: A New Feminist Economics*, London: Macmillan.

West, M. and Harrison, J. (1989) *Bayesian Forecasting and Dynamic Models*, Berlin: Springer-Verlag.

Whitehead, A. N. and Russell, B. (1913) *Principia Mathematica*, Cambridge: Cambridge University Press.

Whitehead, H. (1990) Mark-recapture estimates with emigration and re-immigration, *Biometrics*, **46**, 473–479.

Whittle, P. (1970) *Probability*, Harmondsworth, Middlesex: Penguin.

Whittle, P. (1971) *Optimization under Constraints*, London: Wiley.

Whittle, P. (1982) Non-linear programming, in *Handbook of Applicable Mathematics*, volume IV, ch. 15, W. Ledermann and S. Vajda (eds.), New York: Wiley.

Wilff, H. S. (1994) *Generatingfunctionology*, Boston: Academic Press.

Wood, G. (1992) Binomial mixtures and finite exchangeability, *Annals of Probability*, **20**, 1167–1173.

Wyman, R. (1988) Soil acidity and moisture and the distribution of amphibians in five forests of southcentral New York, *Copeia*, 394–399.

Wyman, R. (1990) What's happening to the amphibians?, *Conservation Biology*, **4**, 350–352.

Wyman, R. and Hawksley-Lescault, D. (1987) Soil acidity affects distribution, behavior, and physiology of the salamander *Plethodon cinereus*, *Ecology*, **68**, 1819–1827.

Zaman, A. (1986) A finite form of de finetti's theorem for stationary Markov chains, *Annals of Probability*, **14**, 1418–1427.

Zellner, A. (1971) *An Introduction to Bayesian Inference in Econometrics*, New York: Wiley.

Zellner, A. (1986a) Bayesian estimation and prediction using asymmetric loss functions, *JASA*, **81**, 446–451.

Zellner, A. (1986b) On assessing prior distributions and Bayesian regression analysis with g-prior distributions, in *Bayesian Inference and Decision Techniques*, P. Goel and A. Zellner (eds.), Amsterdam: Elsevier, 233–243.

# Index

WILEY SERIES IN PROBABILITY AND STATISTICS



*Probability and Statistics*

ANDERSON · An Introduction to Multivariate Statistical Analysis, *Second Edition*
*ANDERSON · The Statistical Analysis of Time Series
ARNOLD, BALAKRISHNAN, and NAGARAJA · A First Course in Order Statistics
BACCELLI, COHEN, OLSDER, and QUADRAT · Synchronization and Linearity: An Algebra for Discrete Event Systems
BARTOSZYNSKI and NIEWIADOMSKA-BUGAJ · Probability and Statistical Inference
BERNARDO and SMITH · Bayesian Statistical Concepts and Theory
BHATTACHARYYA and JOHNSON · Statistical Concepts and Methods
BILLINGSLEY · Convergence of Probability Measures
BILLINGSLEY · Probability and Measure, *Second Edition*
BOROVKOV · Asymptotic Methods in Queuing Theory
BRANDT, FRANKEN, and LISEK · Stationary Stochastic Models
CAINES · Linear Stochastic Systems
CAIROLI and DALANG · Sequential Stochastic Optimization
CHEN · Recursive Estimation and Control for Stochastic Systems
CONSTANTINE · Combinatorial Theory and Statistical Design
COOK and WEISBERG · An Introduction to Regression Graphics
COVER and THOMAS · Elements of Information Theory
CSÖRGŐ and HORVÁTH · Weighted Approximations in Probability Statistics
*DOOB · Stochastic Processes
DUDEWICZ and MISHRA · Modern Mathematical Statistics
ETHIER and KURTZ · Markov Processes: Characterization and Convergence
FELLER · An Introduction to Probability Theory and Its Applications, Volume 1, *Third Edition,* Revised; Volume II, *Second Edition*
FREEMAN and SMITH · Aspects of Uncertainty: A Tribute to D. V. Lindley
FULLER · Introduction to Statistical Time Series, *Second Edition*
FULLER · Measurement Error Models
GIFI · Nonlinear Multivariate Analysis
GUTTORP · Statistical Inference for Branching Processes
HALD · A History of Probability and Statistics and Their Applications before 1750
HALL · Introduction to the Theory of Coverage Processes
HANNAN and DEISTLER · The Statistical Theory of Linear Systems
HEDAYAT and SINHA · Design and Inference in Finite Population Sampling
HOEL · Introduction to Mathematical Statistics, *Fifth Edition*
HUBER · Robust Statistics
IMAN and CONOVER · A Modern Approach to Statistics
JUREK and MASON · Operator-Limit Distributions in Probability Theory
KAUFMAN and ROUSSEEUW · Finding Groups in Data: An Introduction to Cluster Analysis
LAMPERTI · Probability: A Survey of the Mathematical Theory, *Second Edition*
LARSON · Introduction to Probability Theory and Statistical Inference, *Third Edition*
LESSLER and KALSBEEK · Nonsampling Error in Surveys
LINDVALL · Lectures on the Coupling Method
MANTON, WOODBURY, and TOLLEY · Statistical Applications Using Fuzzy Sets

*Now available in a lower priced paperback edition in the Wiley Classics Library.

*Probability and Statistics (Continued)*

MARDIA · The Art of Statistical Science: A Tribute to G. S. Watson
MORGENTHALER and TUKEY · Configural Polysampling: A Route to Practical Robustness
MUIRHEAD · Aspects of Multivariate Statistical Theory
OLIVER and SMITH · Influence Diagrams, Belief Nets and Decision Analysis
*PARZEN · Modern Probability Theory and Its Applications
PRESS · Bayesian Statistics: Principles, Models, and Applications
PUKELSHEIM · Optimal Experimental Design
PURI and SEN · Nonparametric Methods in General Linear Models
PURI, VILAPLANA, and WERTZ · New Perspectives in Theoretical and Applied Statistics
RAO · Asymptotic Theory of Statistical Inference
RAO · Linear Statistical Inference and Its Applications, *Second Edition*
*RAO and SHANBHAG · Choquet-Deny Type Functional Equations with Applications to Stochastic Models
RENCHER · Methods of Multivariate Analysis
ROBERTSON, WRIGHT, and DYKSTRA · Order Restricted Statistical Inference
ROGERS and WILLIAMS · Diffusions, Markov Processes, and Martingales, Volume I: Foundations, *Second Edition;* Volume II: Îto Calculus
ROHATGI · An Introduction to Probability Theory and Mathematical Statistics
ROSS · Stochastic Processes
RUBINSTEIN · Simulation and the Monte Carlo Method
RUBINSTEIN and SHAPIRO · Discrete Event Systems: Sensitivity Analysis and Stochastic Optimization by the Score Function Method
RUZSA and SZEKELY · Algebraic Probability Theory
SCHEFFE · The Analysis of Variance
SEBER · Linear Regression Analysis
SEBER · Multivariate Observations
SEBER and WILD · Nonlinear Regression
SERFLING · Approximation Theorems of Mathematical Statistics
SHORACK and WELLNER · Empirical Processes with Applications to Statistics
SMALL and McLEISH · Hilbert Space Methods in Probability and Statistical Inference
STAPLETON · Linear Statistical Models
STAUDTE and SHEATHER · Robust Estimation and Testing
STOYANOV · Counterexamples in Probability
STYAN · The Collected Papers of T. W. Anderson: 1943–1985
TANAKA · Time Series Analysis: Nonstationary and Noninvertible Distribution Theory
THOMPSON and SEBER · Adaptive Sampling
WELSH · Aspects of Statistical Inference
WHITTAKER · Graphical Models in Applied Multivariate Statistics
YANG · The Construction Theory of Denumerable Markov Processes

*Applied Probability and Statistics*

ABRAHAM and LEDOLTER · Statistical Methods for Forecasting
AGRESTI · Analysis of Ordinal Categorical Data
AGRESTI · Categorical Data Analysis
AGRESTI · An Introduction to Categorical Data Analysis
ANDERSON and LOYNES · The Teaching of Practical Statistics
ANDERSON, AUQUIER, HAUCK, OAKES, VANDAELE, and WEISBERG · Statistical Methods for Comparative Studies
ARMITAGE and DAVID (editors) · Advances in Biometry
*ARTHANARI and DODGE · Mathematical Programming in Statistics
ASMUSSEN · Applied Probability and Queues

*Now available in a lower priced paperback edition in the Wiley Classics Library.

*Applied Probability and Statistics (Continued)*

*BAILEY · The Elements of Stochastic Processes with Applications to the Natural Sciences

BARNETT and LEWIS · Outliers in Statistical Data, *Second Edition*

BARTHOLOMEW, FORBES, and McLEAN · Statistical Techniques for Manpower Planning, *Second Edition*

BATES and WATTS · Nonlinear Regression Analysis and Its Applications

BECHHOFER, SANTNER, and GOLDSMAN · Design and Analysis of Experiments for Statistical Selection, Screening, and Multiple Comparisons

BELSLEY · Conditioning Diagnostics: Collinearity and Weak Data in Regression

BELSLEY, KUH, and WELSCH · Regression Diagnostics: Identifying Influential Data and Sources of Collinearity

BERRY · Bayesian Analysis in Statistics and Econometrics: Essays in Honor of Arnold Zellner

BERRY, CHALONER, and GEWEKE · Bayesian Analysis in Statistics and Econometrics: Essays in Honor of Arnold Zellner

BHAT · Elements of Applied Stochastic Processes, *Second Edition*

BHATTACHARYA and WAYMIRE · Stochastic Processes with Applications

BIEMER, GROVES, LYBERG, MATHIOWETZ, and SUDMAN · Measurement Errors in Surveys

BIRKES and DODGE · Alternative Methods of Regression

BLOOMFIELD · Fourier Analysis of Time Series: An Introduction

BOLLEN · Structural Equations with Latent Variables

BOULEAU · Numerical Methods for Stochastic Processes

BOX · R. A. Fisher, the Life of a Scientist

BOX and DRAPER · Empirical Model-Building and Response Surfaces

BOX and DRAPER · Evolutionary Operation: A Statistical Method for Process Improvement

BOX, HUNTER, and HUNTER · Statistics for Experimenters: An Introduction to Design, Data Analysis, and Model Building

BROWN and HOLLANDER · Statistics: A Biomedical Introduction

BUCKLEW · Large Deviation Techniques in Decision, Simulation, and Estimation

BUNKE and BUNKE · Nonlinear Regression, Functional Relations and Robust Methods: Statistical Methods of Model Building

CHATTERJEE and HADI · Sensitivity Analysis in Linear Regression

CHATTERJEE and PRICE · Regression Analysis by Example, *Second Edition*

CLARKE and DISNEY · Probability and Random Processes: A First Course with Applications, *Second Edition*

COCHRAN · Sampling Techniques, *Third Edition*

*COCHRAN and COX · Experimental Designs, *Second Edition*

CONOVER · Practical Nonparametric Statistics, *Second Edition*

CONOVER and IMAN · Introduction to Modern Business Statistics

CORNELL · Experiments with Mixtures, Designs, Models, and the Analysis of Mixture Data, *Second Edition*

COX · A Handbook of Introductory Statistical Methods

*COX · Planning of Experiments

COX, BINDER, CHINNAPPA, CHRISTIANSON, COLLEDGE, and KOTT · Business Survey Methods

CRESSIE · Statistics for Spatial Data, *Revised Edition*

DANIEL · Applications of Statistics to Industrial Experimentation

DANIEL · Biostatistics: A Foundation for Analysis in the Health Sciences, *Sixth Edition*

DAVID · Order Statistics, *Second Edition*

*DEGROOT, FIENBERG, and KADANE · Statistics and the Law

*DEMING · Sample Design in Business Research

DILLON and GOLDSTEIN · Multivariate Analysis: Methods and Applications

*Now available in a lower priced paperback edition in the Wiley Classics Library.

*Applied Probability and Statistics (Continued)*

DODGE and ROMIG · Sampling Inspection Tables, *Second Edition*
DOWDY and WEARDEN · Statistics for Research, *Second Edition*
DRAPER and SMITH · Applied Regression Analysis, *Second Edition*
DUNN · Basic Statistics: A Primer for the Biomedical Sciences, *Second Edition*
DUNN and CLARK · Applied Statistics: Analysis of Variance and Regression, *Second Edition*
ELANDT-JOHNSON and JOHNSON · Survival Models and Data Analysis
EVANS, PEACOCK, and HASTINGS · Statistical Distributions, *Second Edition*
FISHER and VAN BELLE · Biostatistics: A Methodology for the Health Sciences
FLEISS · The Design and Analysis of Clinical Experiments
FLEISS · Statistical Methods for Rates and Proportions, *Second Edition*
FLEMING and HARRINGTON · Counting Processes and Survival Analysis
FLURY · Common Principal Components and Related Multivariate Models
GALLANT · Nonlinear Statistical Models
GLASSERMAN and YAO · Monotone Structure in Discrete-Event Systems
GREENWOOD and NIKULIN · A Guide to Chi-Squared Testing
GROSS and HARRIS · Fundamentals of Queueing Theory, *Second Edition*
GROVES · Survey Errors and Survey Costs
GROVES, BIEMER, LYBERG, MASSEY, NICHOLLS, and WAKSBERG · Telephone Survey Methodology
HAHN and MEEKER · Statistical Intervals: A Guide for Practitioners
HAND · Discrimination and Classification
*HANSEN, HURWITZ, and MADOW · Sample Survey Methods and Theory, Volume 1: Methods and Applications
*HANSEN, HURWITZ, and MADOW · Sample Survey Methods and Theory, Volume II: Theory
HEIBERGER · Computation for the Analysis of Designed Experiments
HELLER · MACSYMA for Statisticians
HINKELMAN and KEMPTHORNE: · Design and Analysis of Experiments, Volume 1: Introduction to Experimental Design
HOAGLIN, MOSTELLER, and TUKEY · Exploratory Approach to Analysis of Variance
HOAGLIN, MOSTELLER, and TUKEY · Exploring Data Tables, Trends and Shapes
HOAGLIN, MOSTELLER, and TUKEY · Understanding Robust and Exploratory Data Analysis
HOCHBERG and TAMHANE · Multiple Comparison Procedures
HOCKING · Methods and Applications of Linear Models: Regression and the Analysis of Variables
HOEL · Elementary Statistics, *Fifth Edition*
HOGG and KLUGMAN · Loss Distributions
HOLLANDER and WOLFE · Nonparametric Statistical Methods
HOSMER and LEMESHOW · Applied Logistic Regression
HØYLAND and RAUSAND · System Reliability Theory: Models and Statistical Methods
HUBERTY · Applied Discriminant Analysis
IMAN and CONOVER · Modern Business Statistics
JACKSON · A User's Guide to Principle Components
JOHN · Statistical Methods in Engineering and Quality Assurance
JOHNSON · Multivariate Statistical Simulation
JOHNSON and KOTZ · Distributions in Statistics
Continuous Univariate Distributions—2
Continuous Multivariate Distributions
JOHNSON, KOTZ, and BALAKRISHNAN · Continuous Univariate Distributions, Volume 1, *Second Edition*
JOHNSON, KOTZ, and BALAKRISHNAN · Discrete Multivariate Distributions

*Now available in a lower priced paperback edition in the Wiley Classics Library.

*Applied Probability and Statistics (Continued)*

JOHNSON, KOTZ, and KEMP · Univariate Discrete Distributions, *Second Edition*
JUDGE, GRIFFITHS, HILL, LÜTKEPOHL, and LEE · The Theory and Practice of Econometrics, *Second Edition*
JUDGE, HILL, GRIFFITHS, LÜTKEPOHL, and LEE · Introduction to the Theory and Practice of Econometrics, *Second Edition*
JUREČKOVÁ and SEN · Robust Statistical Procedures: Aymptotics and Interrelations
KADANE · Bayesian Methods and Ethics in a Clinical Trial Design
KADANE AND SCHUM · A Probabilistic Analysis of the Sacco and Vanzetti Evidence
KALBFLEISCH and PRENTICE · The Statistical Analysis of Failure Time Data
KASPRZYK, DUNCAN, KALTON, and SINGH · Panel Surveys
KISH · Statistical Design for Research
*KISH · Survey Sampling
LAD · Operational Subjective Statistical Methods: A Mathematical, Philosophical, and Historical Introduction
LANGE, RYAN, BILLARD, BRILLINGER, CONQUEST, and GREENHOUSE · Case Studies in Biometry
LAWLESS · Statistical Models and Methods for Lifetime Data
LEBART, MORINEAU., and WARWICK · Multivariate Descriptive Statistical Analysis: Correspondence Analysis and Related Techniques for Large Matrices
LEE · Statistical Methods for Survival Data Analysis, *Second Edition*
LEPAGE and BILLARD · Exploring the Limits of Bootstrap
LEVY and LEMESHOW · Sampling of Populations: Methods and Applications
LINHART and ZUCCHINI · Model Selection
LITTLE and RUBIN · Statistical Analysis with Missing Data
MAGNUS and NEUDECKER · Matrix Differential Calculus with Applications in Statistics and Econometrics
MAINDONALD · Statistical Computation
MALLOWS · Design, Data, and Analysis by Some Friends of Cuthbert Daniel
MANN, SCHAFER, and SINGPURWALLA · Methods for Statistical Analysis of Reliability and Life Data
MASON, GUNST, and HESS · Statistical Design and Analysis of Experiments with Applications to Engineering and Science
McLACHLAN and KRISHNAN · The EM Algorithm and Extensions
McLACHLAN · Discriminant Analysis and Statistical Pattern Recognition
McNEIL · Epidemiological Research Methods
MILLER · Survival Analysis
MONTGOMERY and MYERS · Response Surface Methodology: Process and Product in Optimization Using Designed Experiments
MONTGOMERY and PECK · Introduction to Linear Regression Analysis, *Second Edition*
NELSON · Accelerated Testing, Statistical Models, Test Plans, and Data Analyses
NELSON · Applied Life Data Analysis
OCHI · Applied Probability and Stochastic Processes in Engineering and Physical Sciences
OKABE, BOOTS, and SUGIHARA · Spatial Tesselations: Concepts and Applications of Voronoi Diagrams
OSBORNE · Finite Algorithms in Optimization and Data Analysis
PANKRATZ · Forecasting with Dynamic Regression Models
PANKRATZ · Forecasting with Univariate Box-Jenkins Models: Concepts and Cases
PORT · Theoretical Probability for Applications
PUTERMAN · Markov Decision Processes: Discrete Stochastic Dynamic Programming
RACHEV · Probability Metrics and the Stability of Stochastic Models
RÉNYI · A Diary on Information Theory
RIPLEY · Spatial Statistics

*Now available in a lower priced paperback edition in the Wiley Classics Library.

*Applied Probability and Statistics (Continued)*

RIPLEY · Stochastic Simulation
ROSS · Introduction to Probability and Statistics for Engineers and Scientists
ROUSSEEUW and LEROY · Robust Regression and Outlier Detection
RUBIN · Multiple Imputation for Nonresponse in Surveys
RYAN · Modern Regression Methods
RYAN · Statistical Methods for Quality Improvement
SCHUSS - Theory and Applications of Stochastic Differential Equations
SCOTT · Multivariate Density Estimation: Theory, Practice, and Visualization
SEARLE · Linear Models
SEARLE · Linear Models for Unbalanced Data
SEARLE · Matrix Algebra Useful for Statistics
SEARLE, CASELLA, and McCULLOCH · Variance Components
SKINNER, HOLT, and SMITH · Analysis of Complex Surveys
STOYAN, KENDALL, and MECKE · Stochastic Geometry and Its Applications, *Second Edition*
STOYAN and STOYAN · Fractals, Random Shapes and Point Fields: Methods of Geometrical Statistics
THOMPSON · Empirical Model Building
THOMPSON · Sampling
TIERNEY · LISP-STAT: An Object-Oriented Environment for Statistical Computing and Dynamic Graphics
TIJMS · Stochastic Modeling and Analysis: A Computational Approach
TITTERINGTON, SMITH, and MAKOV · Statistical Analysis of Finite Mixture Distributions
UPTON and FINGLETON · Spatial Data Analysis by Example, Volume 1: Point Pattern and Quantitative Data
UPTON and FINGLETON · Spatial Data Analysis by Example, Volume II: Categorical and Directional Data
VAN RIJCKEVORSEL and DE LEEUW · Component and Correspondence Analysis
WEISBERG · Applied Linear Regression, *Second Edition*
WESTFALL and YOUNG · Resampling-Based Multiple Testing: Examples and Methods for *p*-Value Adjustment
WHITTLE · Optimization Over Time: Dynamic Programming and Stochastic Control, Volume I and Volume II
WHITTLE · Systems in Stochastic Equilibrium
WONNACOTT and WONNACOTT · Econometrics, *Second Edition*
WONNACOTT and WONNACOTT · Introductory Statistics, *Fifth Edition*
WONNACOTT and WONNACOTT · Introductory Statistics for Business and Economics, *Fourth Edition*
WOODING · Planning Pharmaceutical Clinical Trials: Basic Statistical Principles
WOOLSON · Statistical Methods for the Analysis of Biomedical Data
*ZELLNER · An Introduction to Bayesian Inference in Econometrics

*Tracts on Probability and Statistics*

BILLINGSLEY · Convergence of Probability Measures
KELLY · Reversability and Stochastic Networks
TOUTENBURG · Prior Information in Linear Models